Jak-Stat Signaling: From Basics to Disease

Thomas Decker • Mathias Müller
Editors

Jak-Stat Signaling: From Basics to Disease

Springer

Editors
Thomas Decker
Max F. Perutz Laboratories
Vienna, Austria

Mathias Müller
University of Veterinary Medicine
Vienna, Austria

ISBN 978-3-7091-0890-1 ISBN 978-3-7091-0891-8 (eBook)
DOI 10.1007/978-3-7091-0891-8
Springer Wien Heidelberg New York Dordrecht London

Library of Congress Control Number: 2011945766

Printed on acid-free paper

Springer is part of Springer Science+Business Media (www.springer.com)

Preface

The year 2011 marks the 20th anniversary since the discovery of Jak-Stat signal transduction. During this period we have witnessed the pathway expand from exclusively interferons to other cytokines and beyond. What was once a specialized field has rapidly advanced to become general interest and a must-read in every molecular biology textbook. At the same time a small number of aficionados in a few labs have grown to become a large scientific community meeting at conference series focused on their favorite pathway. The idea to assemble the following book chapters was born at one of these meetings, a FEBS-sponsored event held in February 2010 at Vienna, Austria. We have successfully solicited manuscripts from leaders in the Jak-Stat field, many of whom were participants at the Vienna meeting. We would like to thank them for their work. In addition we gratefully acknowledge input and encouragement by members of the Viennese Jak-Stat special research program and the sponsorship of this program by the Austrian Science Foundation (FWF).

Vienna, January 2012

Thomas Decker
Mathias Müller

Contents

List of Contributors

Ali Abdul-Sater Department of Microbiology and Immunology, Columbia University, New York, NY, USA

Hervé Agaisse Section of Microbial Pathogenesis, Boyer Center for Molecular Medicine, Yale University School of Medicine, New Haven, CT, USA

Lidia Avalle Molecular Biotechnology Center (MBC) and Department of Genetics, Biology and Biochemistry, University of Turin, Turin, Italy

Andrew J. Bannister Gurdon Institute and Department of Pathology, University of Cambridge, Cambridge, UK

Maria M. Caffarel Department of Pathology, University of Cambridge, Cambridge, UK

Mark A. Dawson Department of Haematology, Cambridge Institute for Medical Research, University of Cambridge, Cambridge, UK; Addenbrooke's Hospital, University of Cambridge, Cambridge, UK; Gurdon Institute and Department of Pathology, University of Cambridge, Cambridge, UK

Thomas Decker Department of Microbiology, Immunobiology and Genetics, Max F. Perutz Laboratories, University of Vienna, Vienna, Austria

Mathias Droescher Max Planck Institute of Immunobiology and Epigenetics, Stübeweg 51, 79108 Freiburg, Germany

Ingo Ebersberger Center for Integrative Bioinformatics Vienna, Max F. Perutz Laboratories, University of Vienna, Vienna, Austria

Robert Eferl Ludwig Boltzmann Institute for Cancer Research (LBICR), Vienna, Austria; Medical University of Vienna (MUV), Institute for Cancer Research, Vienna, Austria

Matthias Farlik Department of Microbiology, Immunobiology and Genetics, Max F. Perutz Laboratories, University of Vienna, Vienna, Austria

Katrin Friedbichler Ludwig Boltzmann Institute for Cancer Research (LBI-CR), Vienna, Austria

Kriti Gaur Department of Biomedical Genetics, University of Rochester Medical Center, Rochester, NY, USA

Daniel J. Gough NYU Cancer Institute and Departments of Pathology and Microbiology, NYU School of Medicine, New York, NY, USA

Claude Haan Life Sciences Research Unit, University of Luxembourg, Luxembourg, Luxembourg

Serge Haan Life Sciences Research Unit, University of Luxembourg, Luxembourg, Luxembourg

Kiyoshi Hirahara Molecular Immunology and Inflammation Branch, National Institute of Arthritis, Musculoskeletal and Skin Diseases, National Institutes of Health, Bethesda, MD, USA

Curt M. Horvath Department of Molecular Biosciences, Pancoe Research Pavilion, Northwestern University, Evanston, IL, USA

Lionel B. Ivashkiv Arthritis and Tissue Degeneration Program, Hospital for Special Surgery, New York, NY, USA; Graduate Program in Immunology and Microbial Pathogenesis, Weill Cornell Graduate School of Medical Sciences, New York, NY, USA

Amanda M. Jamieson Department of Microbiology, Immunobiology and Genetics, Max F. Perutz Laboratories, University of Vienna, Vienna, Austria

Yuka Kanno Molecular Immunology and Inflammation Branch, National Institute of Arthritis, Musculoskeletal and Skin Diseases, National Institutes of Health, Bethesda, MD, USA

Robert Kralovics Center for Molecular Medicine, Austrian Academy of Sciences, Vienna, Austria

Arian Laurence Molecular Immunology and Inflammation Branch, National Institute of Arthritis, Musculoskeletal and Skin Diseases, National Institutes of Health, Bethesda, MD, USA

Nicole R. Leitner Institute of Animal Breeding and Genetics, University of Veterinary Medicine Vienna, Vienna, Austria

David E. Levy NYU Cancer Institute and Departments of Pathology and Microbiology, NYU School of Medicine, New York, NY, USA

Haiyan S. Li Department of Immunology, The University of Texas M.D. Anderson Cancer Center, Houston, TX, USA

Willis X. Li Department of Biomedical Genetics, University of Rochester Medical Center, Rochester, NY, USA; Department of Medicine, University of California, San Diego/La Jolla, CA, USA

Richard Moriggl Ludwig Boltzmann Institute for Cancer Research (LBI-CR), Vienna, Austria

Mathias Müller Institute of Animal Breeding and Genetics, University of Veterinary Medicine Vienna, Vienna, Austria; Biomodels Austria, University of Veterinary Medicine Vienna, Vienna, Austria

Harini Nivarthi Ludwig Boltzmann Institute for Cancer Research (LBI-CR), Vienna, Austria

John J. O'Shea Molecular Immunology and Inflammation Branch, National Institute of Arthritis, Musculoskeletal and Skin Diseases, National Institutes of Health, Bethesda, MD, USA

Damla Olcaydu Center for Molecular Medicine, Austrian Academy of Sciences, Vienna, Austria

Tuija Pekkala Institute of Biomedical Technology, University of Tampere and Tampere University Hospital, Tampere, Finland

Valeria Poli Molecular Biotechnology Center (MBC) and Department of Genetics, Biology and Biochemistry, University of Turin, Turin, Italy

Eva Maria Putz University of Veterinary Medicine Vienna, Vienna, Austria

Paulina Rampetsreiter Ludwig Boltzmann Institute for Cancer Research (LBICR), Vienna, Austria

Gabriella Regis Molecular Biotechnology Center (MBC) and Department of Genetics, Biology and Biochemistry, University of Turin, Turin, Italy

Christian Schindler Department of Microbiology and Immunology, Columbia University, New York, NY, USA; Department of Medicine, Columbia University, New York, NY, USA

Jeffrey W. Schmidt Eppley Institute for Research in Cancer and Allied Diseases, University of Nebraska Medical Center, Omaha, NE, USA

Pravin Sehgal Department of Cell Biology and Anatomy, New York Medical College, Valhalla, NY, USA

Veronika Sexl University of Veterinary Medicine Vienna, Vienna, Austria

Priyank Shukla University of Veterinary Medicine, Vienna, Austria

Olli Silvennoinen Institute of Biomedical Technology, University of Tampere and Tampere University Hospital, Tampere, Finland

Birgit Strobl Institute of Animal Breeding and Genetics, University of Veterinary Medicine Vienna, Vienna, Austria

Daniela Ungureanu Institute of Biomedical Technology, University of Tampere and Tampere University Hospital, Tampere, Finland

Alejandro Villarino Molecular Immunology and Inflammation Branch, National Institute of Arthritis, Musculoskeletal and Skin Diseases, National Institutes of Health, Bethesda, MD, USA

Uwe Vinkemeier School of Biomedical Sciences, Nottingham University Medical School, Nottingham, UK

Claus Vogl University of Veterinary Medicine, Vienna, Austria

Kay-Uwe Wagner Eppley Institute for Research in Cancer and Allied Diseases, University of Nebraska Medical Center, Omaha, NE, USA; Department of Pathology and Microbiology, University of Nebraska Medical Center, Omaha, NE, USA

Stephanie S. Watowich Department of Immunology, The University of Texas M.D. Anderson Cancer Center, Houston, TX, USA; The University of Texas Graduate School of Biomedical Sciences, Houston, TX, USA

Christine J. Watson Department of Pathology, University of Cambridge, Cambridge, UK

Xiang-Ping Yang Molecular Immunology and Inflammation Branch, National Institute of Arthritis, Musculoskeletal and Skin Diseases, National Institutes of Health, Bethesda, MD, USA

Eva Zebedin University of Veterinary Medicine Vienna, Vienna, Austria

Feng Zhou Section of Microbial Pathogenesis, Boyer Center for Molecular Medicine, Yale University School of Medicine, New Haven, CT, USA

The Continuing Fascination with Jaks and Stats: An Introduction

Thomas Decker and Mathias Müller

The origins of the discovery of Jak-Stat signal transduction date back to the late 1980s when research groups headed by Jim Darnell, Ian Kerr and George Stark were fascinated by the fact that gene transcription could be induced within minutes after treating cells with type I interferons (IFN-I). The speed with which a signal generated by the plasma membrane–associated IFN-I receptor travelled to nuclear target genes suggested few intermediate steps. The Darnell, Kerr and Stark labs identified bifunctional signal transducers and activators of transcription (Stats) as responsible for IFN-induced transcription by using complementary biochemical and genetic approaches (reviewed in Darnell et al. 1994). Shortly after this seminal discovery, the labs of Sandra Pellegrini, Jim Ihle and Christine Carter-Su independently identified non-receptor protein tyrosine kinases (pTK) in the signaling pathways stimulated by, respectively, the IFN, erythropoietin and growth hormone receptors (Argetsinger et al. 1993; Velazquez et al. 1992; Witthuhn et al. 1993). The same kinases had previously emerged from screens for novel pTKs, conducted in the labs of John Krolewski and Andrew Wilks and named Janus kinases by the latter (Firmbach Kraft et al. 1990; Wilks et al. 1991). With recombinant Jaks and Stats at hand it was possible to reconstitute IFN signaling between receptor and nuclear targets with just two components: receptor associated Jaks that activate Stats by tyrosine phosphorylation. Tyrosine phosphorylated Stats localize to the cell nucleus and bind to promoter DNA of specific target genes (Fig. 1).

Th. Decker (✉)
Department of Microbiology, Immunobiology and Genetics, Max F. Perutz Laboratories, University of Vienna, Dr. Bohr-Gasse 9, 1030 Vienna, Austria
e-mail: thomas.decker@univie.ac.at

M. Müller
Institute of Animal Breeding and Genetics, University of Veterinary Medicine Vienna, Vienna, Austria

Biomodels Austria, University of Veterinary Medicine Vienna, Vienna, Austria
e-mail: mathias.mueller@vetmeduni.ac.at

Th. Decker and M. Müller (eds.), *Jak-Stat Signaling: From Basics to Disease*,
DOI 10.1007/978-3-7091-0891-8_1,

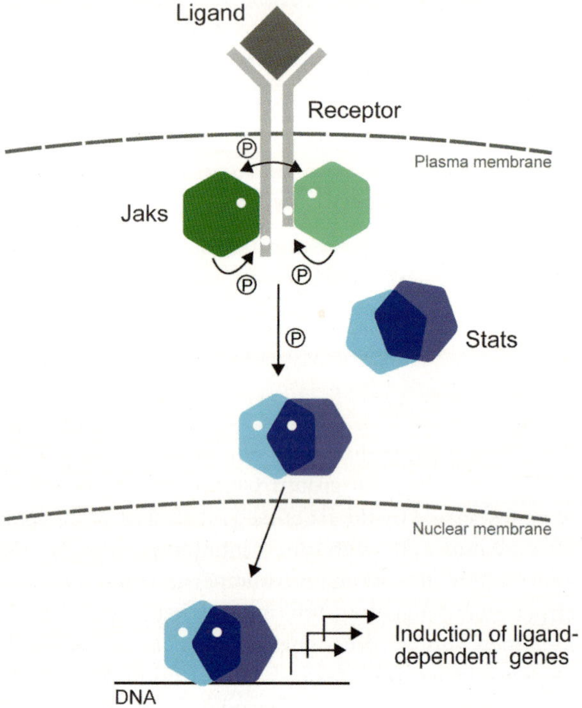

Fig. 1 The essential components of Jak-Stat signal transduction. Cytokine binding alters the conformation of the receptor complex, causing the Jak kinases to phosphorylate and activate each other. Phosphorylation of receptor tyrosines creates docking sites for the Stat SH2 domains. Stats are phosphorylated on a single tyrosine residue where upon they form dimers competent of nuclear translocation and able to associate with DNA binding sites. *White circles* symbolize phosphorylated tyrosines

Today Jak-Stat signal transduction is firmly established as the major route from all class 1 and class 2 cytokine receptors to the cell nucleus. In addition, it contributes to signaling by receptor tyrosine kinases and G-coupled receptors. Therefore it is not surprising that Jaks and Stats play important roles in organisms ranging from slime molds and insects to mammals. Virtually every complex biological process between embryonic development and aging is influenced by Jak-Stat signal transduction. Owing to its overwhelming importance this pathway has rapidly entered text books as a major signaling paradigm. So what is new and justifies yet another book reviewing Jak-Stat signal transduction?

Basics: Jak-Stat research continues to fill knowledge gaps and to produce unexpected findings. Starting with Jaks, whose structure is still not completely solved, the book reviews the state-of-the-art and the intramolecular regulation of Jak activity. Insect Stats were shown to display 'noncanonical' functions independently of tyrosine phosphorylation that regulate chromatin function. In mammalian cells and organisms noncanonical Stat signaling exerts transcriptional control. Moreover, Stat3 molecules lacking phosphotyrosine enter mitochondria and impinge on the respiratory activity of these organelles. This provides one of many links between Stats and cell metabolism that have recently emerged. Mitochondrial Stat3 may contribute to the Warburg effect, the predominant role of glycolysis in the provision of energy to transformed cells.

Early reports addressing mechanistic aspects of Stat activation and dimerization supported the notion that dimers capable of nuclear translocation formed from Stat monomers. The book provides a detailed review of the activation mechanism by dimer reorientation, rather than formation, which accommodates new findings and crystal structures from phosphorylated and unphosphorylated Stats. In addition this chapter presents new ideas of how the subcellular localization of Stats is regulated by postranslational modification.

Basic insight into mechanisms of transcriptional control by Stats has advanced in part because of the much improved general understanding of the molecular machinery regulating the initiation and elongation steps of transcription. Insight into the complex scenario established by the molecular machines mediating nucleosome remodeling, or the activities of histone and RNA polymerase modifying enzymes allows to investigate and understand how Stats interact with these molecules. Detailed analyses of promoter chromatin and associated proteins also provide a better understanding of how different cytokines and signals crosstalk to Stats in the form of gene co-regulation with other transcription factors. Furthermore, the opportunities provided by applying massive parallel sequencing in the context of ChIP-Seq and RNA-Seq technologies open up new prospects of Stat transcription factors embedded into genome-wide landscapes of histone modifications that define distinct functional states of chromatin.

Organismic homeostasis: Articles in this book take a close look at the role of Stats in the generation of hematopoietic cells and, in particular, natural killer (NK) cells, dendritic cells, T and B lymphocytes as regulators and effectors of immunity to infection and cancer. The fascination of Stat biology arises from the fact that the agonistic activity of different family members defines distinct lineages and subpopulations of both DC and T cells. At the same time their antagonistic activity may suppress the development of alternative developmental avenues. Among CD4 + T cells each of the major subsets can be defined by the activity of a different Stat. NK cells provide a striking example how different Stats regulate differentiation, and activation in one cell type.

Stats in disease: Many chapters in this book review new findings that link both physiological Jak-Stat activity to protection from disease and aberrant Jak-Stat signaling to cancer or infectious disease.

Jak2 and Stat5 stand out as regulators of hematopoiesis. Consistent with this both proteins are able to promote leukemic cell transformation if not properly controlled. Several contributions describe molecular mechanisms leading to the leukemic development and highlight the role of mutant Jak kinases as well as the prospect of treating such leukemias with recently identified Jak inhibitors.

Jak-Stat activity at the wrong place or time also favors the development of solid cancers. For example, Jak2, Stat3 and Stat5 promote the establishment or growth of breast cancers each in their own way. Particularly Stat3 is a driving force behind many solid cancers but studies reviewed in a contribution on inflammation-associated colon cancer suggest that cell and organ context determine the net activity of Stat3 as an oncogene or tumor suppressor as well as its antagonistic relationship to Stat1.

IFN signal through the prototypic Jak-Stat pathway to induce an antiviral state. Chapters dealing with the role of Jaks and Stats in infectious disease present the current understanding of the antiviral state as the combined activity of Stat target gene products. They also show how Stats are subject to viral evasion strategies. Since immunological activities of IFN are not limited to the struggle with viral pathogens, recent examples of their impact on bacterial infection are presented to show the pleiotropy of IFN action and the unpredictability of their impact on the course of infection.

This brief description of the book is much less intended to inform comprehensively about its content than it is to convince readers that Jak-Stat research is active, dynamic and timely and that many of the findings described by experts in their field could not have been presented in a similar book a few years ago. We thank our colleagues for their significant time investment in preparing each chapter.

References

Argetsinger LS, Campbell GS, Yang X, Witthuhn BA, Silvennoinen O, Ihle JN, Carter Su C (1993) Identification of JAK2 as a growth hormone receptor-associated tyrosine kinase. Cell 74:237–244

Darnell JE Jr, Kerr IM, Stark GR (1994) Jak-STAT pathways and transcriptional activation in response to IFNs and other extracellular signaling proteins. Science 264:1415–1421

Firmbach-Kraft I, Byers M, Shows T, Dalla-Favera R, Krolewski JJ (1990) tyk2, prototype of a novel class of non-receptor tyrosine kinase genes. Oncogene 5:1329–1336

Velazquez L, Fellous M, Stark GR, Pellegrini S (1992) A protein tyrosine kinase in the interferon alpha/beta signaling pathway. Cell 70:313–322

Wilks AF, Harpur AG, Kurban RR, Ralph SJ, Zurcher G, Ziemiecki A (1991) Two novel protein-tyrosine kinases, each with a second phosphotransferase-related catalytic domain, define a new class of protein kinase. Mol Cell Biol 11:2057–2065

Witthuhn BA, Quelle FW, Silvennoinen O, Yi T, Tang B, Miura O, Ihle JN (1993) JAK2 associates with the erythropoietin receptor and is tyrosine phosphorylated and activated following stimulation with erythropoietin. Cell 74:227–236

Regulation of JAKs: Insights Gleaned from the Functional Protein Domains

Claude Haan, Daniela Ungureanu, Tuija Pekkala, Olli Silvennoinen, and Serge Haan

Abstract

Since their identification in the early 1990s, many studies have investigated the function of Janus kinases as well as their regulation. It took about 15 years until a first crystal structure of a Janus kinase domain was described and by today the structures of all four kinase domains have been explored. In this chapter we discuss the effects of the different JAK domains on the activity, trafficking and localisation of JAKs that were reported in mutagenesis studies in the last 20 years of JAK research. We take into consideration the recently solved crystal structures of the kinase domains as well as other structural information. In addition, we reflect on the lessons that the recently identified activating mutations in patients teach us.

Introduction

The family of Janus kinases (JAK) consists of four mammalian members: JAK1, JAK2, JAK3 and TYK2. JAK1, JAK2 and TYK2 are ubiquitously expressed, but expression of JAK3 is confined mainly to cells of the haematopoietic system (Yeh and Pellegrini 1999; Heinrich et al. 2003; Ihle and Kerr 1995). JAK kinases are involved in a variety of biological processes including haematopoiesis and regulation of the immune system. Cytokine receptors bind different JAKs (Heinrich et al. 2003; O'Sullivan et al. 2007; Pestka et al. 2004; Kovanen and Leonard 2004; Hintzen et al. 2008) and the specificity

C. Haan • S. Haan (✉)
Life Sciences Research Unit, University of Luxembourg, 162A Avenue de la Faïencerie, 1511 Luxembourg, Luxembourg
e-mail: serge.haan@uni.lu

D. Ungureanu • T. Pekkala • O. Silvennoinen
Institute of Biomedical Technology, University of Tampere, Tampere, Finland

Tampere University Hospital, Tampere, Finland

Th. Decker and M. Müller (eds.), *Jak-Stat Signaling: From Basics to Disease*,
DOI 10.1007/978-3-7091-0891-8_2, © Springer-Verlag Wien 2012

of various signalling proteins for phosphotyrosine motifs within this receptor determines the signalling characteristics of the different cytokines.

JAKs are constitutively associated via their FERM domain with the membrane proximal region of the type I and type II hematopoietic cytokine receptors and JAKs are absolutely required for downstream signal transduction. Currently there is no structural information of the cytoplasmic domains of the cytokine receptors and the exact mechanism of binding and activation of the JAKs in the receptor complex are largely theoretical. Ligand binding induces conformational changes in the receptor and allows juxtapositioning and transphosphorylation of the activation loop tyrosines in JAKs resulting in enhancement of catalytic activity. Subsequently, tyrosine residues in the receptors become phosphorylated allowing recruitment of SH2 domain containing signalling proteins such as members of the Signal Transducer and Activator of Transcription (STAT) family transcription factors.

Phosphorylation plays an important role in regulation of JAK activity. As noted, activation of JAKs in response to cytokine stimulation depends on phosphorylation of the activation loop which in all JAKs consists of tandem tyrosine residues. However, JAKs are phosphorylated at multiple sites. JAK2 has been the subject to most thorough phosphor amino acid analysis and approximately 20 tyrosine residues have been identified to be phosphorylated upon cytokine stimulation. Several of these sites have been functionally characterized and in addition to activation loop Y1007/1008, phosphorylation of Y637, Y813, Y868, Y966 and Y972 have been shown to potentiate JAK2 activity, while phosphorylation of Y119, Y221, Y317, Y570 and Y913 regulate JAK2 activity negatively (Argetsinger et al. 2004; Feener et al. 2004; Robertson et al. 2009). Interestingly, in the absence of cytokine stimulation, JAK2 is constitutively phosphorylated on a single residue, S523 which mediates negative regulation of JAK2 activation (Mazurkiewicz-Munoz et al. 2006; Ishida-Takahashi et al. 2006). The precise mechanisms how these phosphorylation events regulate JAK activity is known only for a few residues. Phosphorylation of Y119 in the FERM domain induces dissociation of JAK2 from the Epo receptor, and Y813 binds regulator protein SH2-B and increases JAK2 activity (Funakoshi-Tago et al. 2006; Kurzer et al. 2004).

Due to their critical role in central biological processes such as proliferation, the activity of JAKs needs to be tightly regulated by several mechanisms. Protein tyrosine phosphatases SHP-1 and CD45 are shown to regulate JAK phosphorylation. The family of Suppressor of cytokine signaling (SOCS) proteins plays an important role in negative regulation of JAKs and cytokine signaling. SOCS1-7 and CIS are SH2 domain containing proteins that are transcriptionally induced by cytokine stimulation. SOCS can regulate and control cytokine signaling by different mechanisms (Yoshimura et al. 2007). The hallmark of the family is the C-terminal SOCS-box that possesses Ubiquitin E3 ligase activity, and hyperphosphorylated forms of oncogenic or normal JAKs have been shown to become ubquitinated by SOCS and directed for proteasomal degradation (Kamizono et al. 2001; Ungureanu et al. 2002). SOCS1 and SOCS3 contain also a kinase inhibitory region (KIR) that can inhibit JAK function by serving as a pseudosubstrate (Yasukawa et al. 1999). SOCS proteins can also bind cytokine receptors and compete for SH2 domain binding sites.

Phenotypic analysis of knockout mice of all four JAKs has yielded valuable information for the understanding of their physiological role. These mice show phenotypes that are linked to cytokine signalling deficiencies. JAK1 and JAK2 deficiency is not compatible with life. JAK2 knock-out mice die at day 11 of embryogenesis because of the lack of erythropoiesis (Parganas et al. 1998; Neubauer et al. 1998). JAK1 knock-out mice die perinatally due to motoneuronal defects (Rodig et al. 1998). JAK3 knock-out mice exhibit a SCID (Severe Combined Immuno-Deficiency) phenotype (Nosaka et al. 1995; Park et al. 1995; Thomis et al. 1995). Finally TYK2-deficiency leads to hypersensitivity towards infections due to the absence of pro-inflammatory immune responses (Karaghiosoff et al. 2000; Shimoda et al. 2000).

JAKs are involved in inflammatory and immune disorders in which cytokines play crucial roles (Ghoreschi et al. 2009; Pesu et al. 2008) as well as in cytokine-dependent cancers such as multiple myeloma. JAK3 mutations and deletions lead to severe combined immunodeficiency (SCID) characterised by the absence of circulating T- and NK-cells, normal or increased numbers of nonfunctional B-cells and hypoplasia of lymphoid tissues (Pesu et al. 2008; Macchi et al. 1995). Activating JAK2 fusion proteins (TEL-JAK2, PCM1-JAK2, ETV6-JAK2 and SSBP2-JAK2) evoke lymphoid and myeloid leukemia and MPN-U (Peeters et al. 1997; Lacronique et al. 1997; Reiter et al. 2005; Murati et al. 2005; Bousquet et al. 2005; Adelaide et al. 2006; Griesinger et al. 2005; Poitras et al. 2008; Cirmena et al. 2008). Mutations in the Janus kinase 2 gene were found with high incidence in patients with myeloproliferative neoplasms (MPNs) (JAK2 V617F and a number of point mutations and deletions in exon 12) (James et al. 2005; Kralovics et al. 2005; Levine et al. 2005; Baxter et al. 2005; Zhao et al. 2005), in myeloid leukemia (JAK2-T875N) (Mercher et al. 2006), in acute lymphoblastic leukemia (ALL) (JAK2-L611S) (Kratz et al. 2006), and in acute megakaryoblastic leukemia (AMKL) (JAK2-V617F and JAK2-M535I) (Nishii et al. 2007). These constitutively active JAK2 mutants have been described to activate STAT5 and STAT3, MAP kinases and PI3K/AKT. Activating mutations in JAK1 have also been reported for ALL (Flex et al. 2008; Jeong et al. 2008) and gain of function mutations of JAK3 (A572V, A573V) were found in ALL and AMKL patients (Malinge et al. 2008; Walters et al. 2006). Figure 1 shows a selection of JAK mutations associated with disease (for a more detailed description of more JAK mutations see Pesu et al. 2005; Haan et al. 2010). JAK3 mutations in humans SCID are amino acid changes, a premature stop or frame shift mutations causing altered protein sequence (see also Fig. 1). A point mutation in the pseudokinase domain of TYK2 was reported to impair IL-12 and IFN-mediated signalling and was associated with resistance to collagen-induced arthritis in a murine model (Shaw et al. 2003). Moreover, it has recently been shown that polymorphisms at the TYK2 locus are associated with Systemic Lupus Erythematosus (Sigurdsson et al. 2005).

Sequence similarities between JAK family members initially led to the description of seven JAK homology (JH) domains (Wilks et al. 1991), which only partially match the functional domain structure of JAKs. The JH1 and JH2 domains correspond to the kinase and pseudokinase domain. The JH3 to JH7 regions form a FERM and an SH2 domain (Wilks et al. 1991; Girault et al. 1998).The JAK FERM

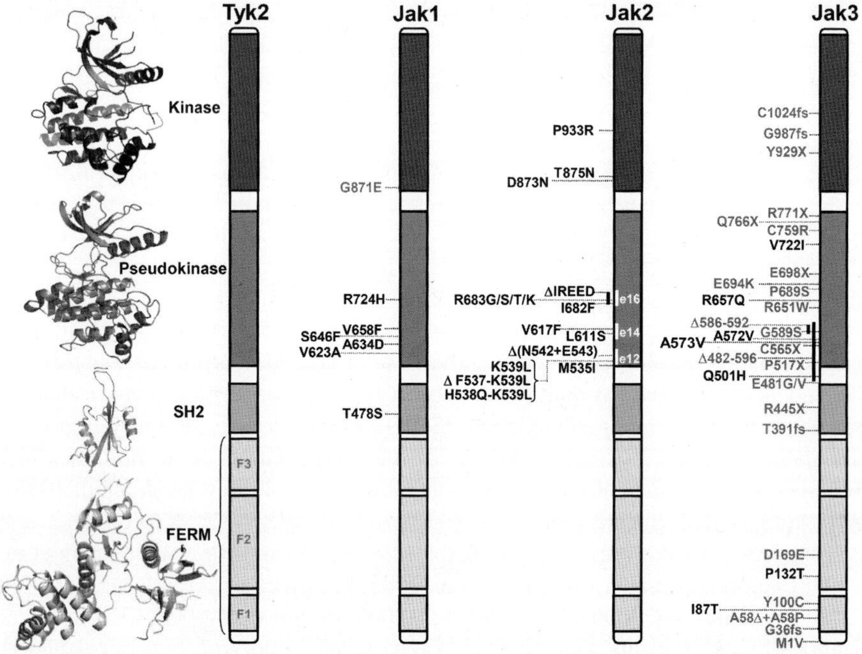

Fig. 1 Domain structure of Janus kinases and of a selection of mutations observed in patients. Model structures of the JAK1-FERM, -SH2, and pseudokinase domain (Haan et al. 2010), the solved crystal structure of the JAK2 kinase domain (PDB entry code: 2B7A) as well as the schematic domain structure for all JAKs are represented. Mutations indicated in black lead to constitutively active JAK proteins (only mutations with validated functions are shown). Mutations in grey represent mutations which lead to a loss of function (in JAK1) or are found in severe combined immunodeficiency (JAK3). Abbreviations used: X = stop codon; fs = frame shift; Δ = deletion; e12/14/16 = exon12/14/16; F1-3 = subdomains of the FERM domain

domain is quite divergent from other FERM domains so that structure prediction is not trivial. The SH2 domain too presents some special features discussed below.

Currently the only structural data available for any of the JAKs are the X-ray structures of the tyrosine kinase domains. While further structural information will be required to obtain complete understanding of the regulation of JAK kinases in physiological and pathogenic signalling, an overall picture of JAK regulation is emerging from various experimental settings. In this review we present biological, biochemical and clinical information about the different functional domains that reveal important information about regulation of JAK kinases.

JAK/Cytokine Receptor Interactions

Crystallographic data on the JAK N-terminal part and of cytokine receptors does not exist, thus the structure/function-relationship between cytokine receptors and Janus kinases still remains elusive as does the exact sequence of events involved in

Janus kinase activation. JAK binding to cytokine receptors is crucial for their function even in the context of constitutively active mutants. The JAK2-V617F mutant is rendered inactive if cytokine receptor binding is abrogated, and concomitantly looses its transforming potential (Lu et al. 2008; Wernig et al. 2008). Activating JAK2 fusion proteins (TEL-JAK2, PCM1-JAK2, ETV6-JAK2 and SSBP2-JAK2) however are constitutively active without cytokine receptor binding, and are activated by oligomerisation of the non-JAK part of the fusion protein, but this is of limited interest for the elucidation of the activation mechanism occurring in full length JAKs.

The FERM Domain

FERM domains are clover-shaped domains comprising three subdomains. The N-terminal subdomain F1 has a ubiquitin-like β-grasp fold. Subdomain F2 has an acyl-CoA-binding-protein-like fold, and subdomain F3 has a PH-domain (pleckstrin homology) fold (Pearson et al. 2000). Structural models of JAK FERM domains (based on structural data of a number of solved FERM domains (Haan et al. 2001, 2008, 2010)) have been used to explore the function of the postulated JAK FERM domain (Girault et al. 1998; Haan et al. 2001, 2008; Hilkens et al. 2001) (reviewed in Haan et al. 2006, 2010). The N-terminal FERM domain in JAKs binds to the membrane-proximal box1/2 region of cytokine receptors (Richter et al. 1998; Zhao et al. 1995; Chen et al. 1997; Cacalano et al. 1999; Kohlhuber et al. 1997). The involvement of rather long sequence stretches within the receptor and JAKs suggests that the interaction is mediated by multiple contacts. A defined JAK orientation on a cytokine receptor ultimately is critical for activation. The receptor-JAK interaction probably induces a restructuring of certain receptor residues into defined interaction interfaces. Such an "induced fit-like" scenario seems necessary to explain the binding of the largely non-structured (according to secondary structure predictions) region of cytokine receptors encompassing the box1 and box2 regions. In receptors such as gp130 or the EpoR this region counts 52 or 61 amino acids and could span a distance of about 19 or 23 nm, respectively. In contrast, the FERM domain of JAKs would at most measure about 6–7 nm across (Fig. 2). Alternatively, a non-structured cytoplasmic tail of a cytokine receptor could adopt a loop structure winding repeatedly through the clefts or along the surface of the FERM domain. Whichever scenario is correct, the involvement of several subdomains (FERM subdomains and SH2 domain) of the JAK and long stretches within the receptor harbours the potential for a very tight and long-lasting interaction. It seems to be a general phenomenon that the mere proximity of JAKs in receptor complexes is not sufficient for their activation and that further conformational changes are required (Constantinescu et al. 2001; Greiser et al. 2002; Watowich et al. 1999; Haan et al. 2002). There is evidence that rigidity of the α-helical transmembrane regions can extend into the intracellular (Constantinescu et al. 2001; Greiser et al. 2002; Zhu and Sizeland 1999) as well as to the extracellular region (Kubatzky et al. 2005). Secondary structure predictions suggest that the

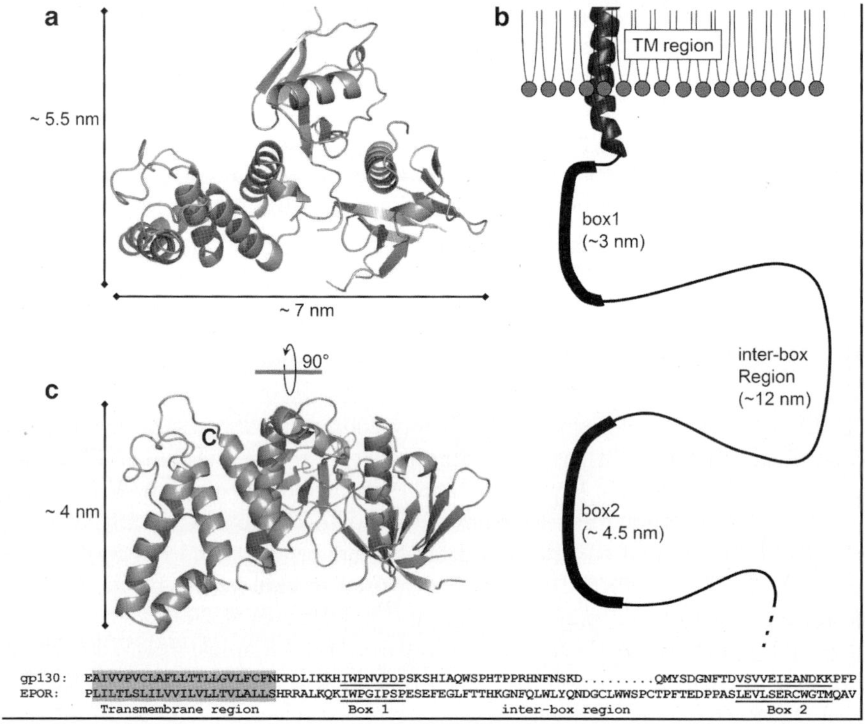

Fig. 2 Model structure of JAK-FERM domain compared to the non structured box region of a cytokine receptor. (**a**) Model structure of the JAK1-FERM domain with indicated dimensions. (**b**) Schematic representation of a non-structured box1–box2 region of the cytokine receptor gp130. The approximate dimensions for a non-structured polypeptide chain are given. (**c**) Alignment of the box1–box2 regions of gp130 and the EpoR

transmembrane α-helix may extend to the beginning of the box1 region (Fig. 2). The proline-rich box1 region might adopt other secondary structures (i.e. polyproline type II helical structure) or the receptor might have a less ordered conformation from there on. Interestingly, mutations or insertions of residues within this putative α-helical region of the cytokine receptor gp130, which did not have an influence on JAK1 binding, were nevertheless crucial for JAK1 phosphorylation and activation (Greiser et al. 2002; Haan et al. 2002). Thus, the role of the membrane proximal region in cytokine receptor signalling is not restricted to mere JAK binding. The W652 mutation in gp130 even behaved dominantly negative, since no signalling occurred when only a single cytoplasmic chain of a gp130 dimer contained the mutation. The corresponding mutation (W258) in the erythropoietin receptor (EpoR) also led to impaired JAK activation and is thought to be part of an α-helically organised region, whose precise orientation is necessary to promote signalling (Constantinescu et al. 2001). Thus, the continuation of the transmembrane helix into the cytoplasm seems to be important for JAK orientation on the cytokine receptor. This of course means that it might be necessary to include receptor sequences into the research effort aiming at structurally solving the JAK N-terminal domains, which again adds a level of complexity to this unsolved problem. In the same line of evidence, not every peptide mimetic mediating EpoR dimerization led to signal transduction (Livnah et al. 1999). Also, signalling through the gp130 homodimer can be elicited using antibodies against the extracellular part of the cytokine receptor. Interestingly, efficient gp130 activation could only be achieved by two distinct agonistic monoclonal antibodies (Müller-Newen et al. 2000; Autissier et al. 1998) again supporting the notion that the sterical information is transduced through the transmembrane region into the cell to ultimately leads to JAK activation (Remy et al. 1999).

ERM proteins (ezrin, radixin, moesin) bind membranes by binding phospholipids with their FERM domains. Interestingly however, the residues which mediate phospholipid binding in the FERM domain of radixin (Hamada et al. 2000) are not conserved in JAKs. Furthermore mutations in Janus kinases which impair receptor binding lead to a cytoplasmic localisation of JAKs. This indicates that the JAKs are recruited to membranes solely by interaction with cytokine receptors.

The SH2 Domain

The FERM domain is followed by a predicted SH2 domain for which secondary structure prediction analysis of the JAK family members reveals the typical pattern of SH2 domains. The conservation of structural (conserved in all) and functional residues (conserved in only some JAKs) within the JAK SH2 domains shows a discrepancy to all other SH2 domains. The essential functional arginine residue at position βB5, conserved to 99.8% in SH2 sequences, is only conserved to 80% in all JAK SH2 sequences. Interestingly no classical SH2 domain function could be shown to date. Neither the IL6 nor the IFN-γ induced signalling capacity of JAK1 was affected by an SH2 domain inactivating point mutation (Radtke et al. 2005). A similar mutation in human JAK2 also did not interfere with IFN-γ signalling (Kohlhuber et al. 1997). JAK SH2 domain sequences show some additional unconventional features. The absence of a well conserved tryptophan which anchors the *N*-terminal tail at the back of the SH2 domain and directs it away from the phosphotyrosine recognition site, indicates that the domain preceding the SH2 domain, namely the FERM domain, could be positioned aside and not behind the SH2 domain. It was postulated that the SH2 domain may act as a spacer and structurally support and stabilise the FERM domain (Radtke et al. 2005). Recently a role for the SH2 domain has been proposed in the context of JAK2-V617F mutant (Gorantla et al. 2010).

Independently of any SH2 specific phosphotyrosine peptide binding function, truncation mutants and SH2 domain swapping mutants showed that the SH2 domain of JAK1 was structurally important for binding to the OSMR and consequently for efficient OSMR surface expression (Radtke et al. 2005). In contrast, for gp130, EpoR and the interferon-α receptor 1 (IFNαR1), the SH2 domain of JAK1,

JAK2 or TYK2, respectively, were not necessary for receptor binding, although the SH2 domain was required for the upregulation of receptor surface expression of EpoR and IFNαR1 (Hilkens et al. 2001; Ragimbeau et al. 2003; Huang et al. 2001).

Trafficking and Localisation of JAK/Cytokine Receptor Complexes

As mentioned above, the structural integrity of the FERM domain (and in some cases the SH2 domain) is crucial for receptor binding and constitutively active oncogenic JAK mutants require receptor interaction to transform cells. Thus, the trafficking and localisation of the JAKs is dependent on their structural features and is intimately linked to the regulation of JAK activity.

The JAK/Receptor Complex Is Comparable to a Receptor Tyrosine Kinase

The data from JAK/cytokine receptor interaction studies, from trafficking studies (Ragimbeau et al. 2003; Huang et al. 2001; Radtke et al. 2002, 2006; He et al. 2005; Royer et al. 2005;) and localisation studies suggest that JAK1 is recruited to membranes by tight association with cytokine receptors. The fact that JAK binding deficient cytokine receptor mutants or JAK mutants impairing receptor binding lead to a cytoplasmic distribution of JAKs shows that JAKs have no significant intrinsic membrane binding potential. A membrane-bound protein, like JAK1, without a transmembrane domain could conceivably also directly bind to the membrane by lipid modifications (e.g. myristoylation, palmitoylation, farnesylation), by lipid binding domains (e.g. FERM-, PH-, FYFE-domains), through membrane penetrating structures, by electrostatic forces, by binding to other membrane-associated proteins, or by a combination of some of these mechanisms. However, this does not seem to be the case for JAKs. As already mentioned above the residues which mediate phospholipid binding in the FERM domain of radixin (Hamada et al. 2000) are not conserved in JAKs. Also, after cytokine stimulation, JAK1 remained localised at the plasma membrane and did not change its localisation (Behrmann et al. 2004). Interestingly, the half-lives of cytokine receptors and JAKs e.g. gp130 and JAK1 are also identical (Siewert et al. 1999) and this again argues in favour of a "common fate" of the two proteins. FRAP experiments showed that the mobilities for overexpressed gp130-YFP and JAK1-YFP were equal. JAK1-YFP diffuses on the plasma membrane with the velocity of a transmembrane protein indicating that there is no rapid exchange of bleached JAKs from a transient cytoplasmic pool. It was also possible to show that immobilisation of gp130-CFP by a pair of cross-linking monoclonal antibodies also led to the immobilisation of JAK1-YFP (Giese et al. 2003). Thus, JAK molecules do not exchange between different receptors at the plasma membrane and the gp130/JAK1 complex at least can be considered as an un-dissociable entity resembling a receptor tyrosine kinase.

The Kinase/Pseudokinase Connection

One surprising finding of the analysis of the kinome consisting of 518 protein kinases was that appr. 10% of them, namely 48 proteins, contained pseudokinase domains (Manning and Cantley 2002). A protein is designated as a pseudokinase if it lacks one or several of the canonical motifs considered to be required for catalysis. In only five of these proteins a pseudokinase domain and an additional functional kinase domain are present in the same protein polypeptide. These are the four Janus kinases and the serine/threonine kinase GCN2. Recent structural data indicate that pseudokinases with significant sequence degeneration adapt a kinase fold that resembles that of their nearest functional relative (Scheeff et al. 2009). The general fold of the pseudokinase domain of JAKs is expected to follow closely a kinase structure.

The Kinase Domain

The kinase activity is mediated by the C-terminal kinase domain. All protein kinases possess a catalytic domain that comprises approximately 300 amino acids. They share the bilobal kinase fold: The N-terminal lobe is composed of five β-strands and a single α-helix. The C-terminal lobe is predominantly α-helical and contains the regulatory activation loop (A-loop). The sequential similarity of the JAK kinase domains is quite high and the solved structures of the JAKs also show little difference in and around the ATP binding pocket. The published crystal structures of all the JAK1, JAK2, JAK3 and TYK2 kinase domains have proven the existence of an additional helix within the C-lobe of the JAK kinase domain which was termed αH-helix for JAK2 and FG-helix in the case of JAK3 (Lucet et al. 2006; Boggon et al. 2005; Williams et al. 2009) and that has been shown to be crucial for kinase activity (Haan et al. 2009). This special feature in JAKs is lining the substrate binding site of the kinase and lies in close proximity to the catalytic cleft of the enzyme. One family member, JAK3 has some special features compared to the other JAKs. It is the only JAK family member in which an alanine residue directly precedes the DFG-motif (in contrast to a glycine residue in the other JAKs). This subtle difference could directly affect the conformation of the A-loop in the way that was already discussed for the inactive insulin receptor (GDFG-motif) and fibroblast growth factor receptor (ADFG-motif) kinase domains (Hubbard et al. 1998). JAK3 is also the only Janus kinase having a cysteine residue at position C909 in close proximity to the ATP binding pocket. Thus JAK3 would be a potential target for ATP-competitive inhibitors with an electrophilic group (so called irreversible inhibitors) (Haan et al. 2010) which would covalently attach to the mentioned JAK3 cysteine. The toxic potential is hard to evaluate but the amount of possible off-kinase-targets potentially reacting with the electrophile is a risk (Rishton 2003).

The Pseudokinase Domain

The sequence of the JH2 pseudokinase domain is conserved to the same extent as the JH1 domain among different JAKs (appr. 30% identity) and during evolution. However, the genomic organization of JH2 differs from JH1 thus suggesting that the domains have evolved individually. The structure of the JH2 domain for any of the JAKs has not yet been published, but the sequence homology to functional kinases suggests that it follows a similar fold. However, differences in some of the conserved sequence motifs considered to be required for catalytic activity are missing or altered in JH2. Specifically, JAKs lack the third Gly in the Glycine-rich (GxGXXG) ATP binding loop, and in the ATP orienting VAIK motif the Alanine is changed to Val, Leu or Ile. In the in DFG cation binding motif the Phenylalanine is changed to Proline. Most dramatic difference, however, is the lack of the catalytic base Aspartic acid in the subdomain VIb. Collectively these alterations have led to the conclusion that JH2 is catalytically inactive and been assigned as pseudokinase domain (Boudeau et al. 2006; Zeqiraj and van Aalten 2010). Recently, however, the pseudokinase status of several proteins, including CASK, haspin, WNK1, VRK3, HER3/ErbB3, and STRADα, has been changed and the studies have shown that ATP-binding and/or catalytic activity can be achieved through non-canonical mechanisms (Mukherjee et al. 2008; Eswaran et al. 2009; Shi et al. 2010; Zeqiraj et al. 2009; Scheeff et al. 2009). Each of these proteins utilizes a distinct mechanism for nucleotide binding and/or catalysis. Interestingly, HER3, which resembles JAKs in lacking the catalytic base aspartate was found to retain low level kinase activity and be able to phosphorylate its intracellular region in vitro (Shi et al. 2010). The crystal structure of HER3 showed that it assumes an atypical conformation for active kinases, particularly in αC helix and activation segment (Shi et al. 2010; Jura et al. 2009). It will be important to evaluate whether the JAK JH2 also possess catalytic activity.

Regulation of Kinase Domain by the Pseudokinase Domain

The domain structure of JAK kinases is conserved from *Drosophila* to mammalians suggesting that the dual kinase domain structure is functionally important. The first insight into the functional role of JH2 domain was obtained from SCID patient, where mutations in JAK3, including those in the JH2 domain, were found to cause abrogation of JAK3 activation and IL-2 mediated signal transduction (Russell et al. 1995; Candotti et al. 1997). The next piece of information related to the function of JH2 domain came from the *Drosophila* system, where a point mutation in the JH2 domain was found to cause hyperactivation of the JAK kinase and hyperproliferation of hemolymph (Luo et al. 1997). Analogous mutation in mammalian JAK2 (E665K) also resulted in hyperactivity though the effect was mild (Luo et al. 1997). Thus, these genetic models provided seemingly controversial conclusion, in JAK3 JH2 domain was required for activity and signalling, while in Drosophila the domain was mediating a negative regulator function. However, biochemical and functional studies have provided additional information about the role of JH2 domain in regulation of JAKs and cytokine signalling (Chen et al. 2000; Yeh et al. 2000). The studies on JAK2 demonstrated that deletion of JH2 domain increased basal activity but abolished the cytokine induced activation of JAK and downstream signalling (Saharinen et al. 2000). The function of JH2 appears to be conserved among JAKs, or at least between JAK2 and JAK3, since chimeric constructs encompassing the JH2 of JAK3 in JAK2 background was able to reconstitute cytokine induced signalling in JAK2 deficient cell line (Saharinen and Silvennoinen 2002). Biochemical and kinetic analysis of the JH2 domain in JAK2 in vitro showed that the JH2 domain did not affect Km but reduced the Vmax of JAK2 catalytic activity thus suggesting a non-competitive mechanism of inhibition (Saharinen et al. 2003). This finding, combined with the cellular interaction between JH1 and JH2 suggested that a physical interaction between JH1 and JH2 is mediating the inhibitory function (Saharinen and Silvennoinen 2002). Furthermore, three inhibitory regions have been identified in JH2 of which the first starts at the loop between β4 and β5 in the N-lobe of JH2. The *Drosophila* Hop mutation as well as the MPN causing V617F mutation reside both in this same region.

The evidence for the requirement of the JH2 domain for JAK activation and functional cytokine signalling is derived from clinical and artificial mutations as well as from functional studies. The underlying mechanism is still unknown but the data from receptor-JAK complex organizations provides insights into this paradigm. The binding of JAK1 and JAK2 to the juxtamembrane regions in gp130 and EpoR, respectively, is necessary but not sufficient for the induction of JAK activation (Constantinescu et al. 2001; Haan et al. 2002). In the case of JAK2 and EpoR, the induction of catalytic activity was suggested to involve an interaction between the active conformation of the α-helical juxtamembrane region and the JH1-JH2 domain. Collectively these data can be summed in a model of JAK regulation in cytokine receptors, where JAKs are maintained inactive through a JH1-JH2 interaction in the absence of cytokine stimulation. Ligand binding to the receptor induces a conformation change in the α-helical hydrophobic juxtamembrane region which relieves the inhibitory JH1-JH2 interaction and allows transphosphorylation of the JAKs and their activation and progression of signal transduction (Fig. 3).

The mechanisms by which JH2 mediates the regulatory functions is currently unknown but a recent study has evaluated the inter-domain interactions in kinase activity and substrate specificity using recombinant JAK2 kinase domains (O. Silvennoinen and I. Touw, personal communication). Using a peptide microarray platform, the JH2 was found to drastically decrease the activity of the JH1 domain by increasing the Km for ATP. JH2 was also found to modulate the peptide preference of JAK2. Interestingly, the V617F mutation partially releases this inhibitory mechanism but did not significantly affect substrate preference or Km for ATP. These results provide the biochemical basis for the interaction between the kinase and the pseudokinase domain of JAK2. In addition, molecular modelling has provided insights into possible mechanism of JH2 function. The model of Lindauer et al based on the crystal structure of the FGF receptor dimer, suggest two interaction interfaces between JH1 and JH2 (Lindauer et al. 2001). It should be noted, that

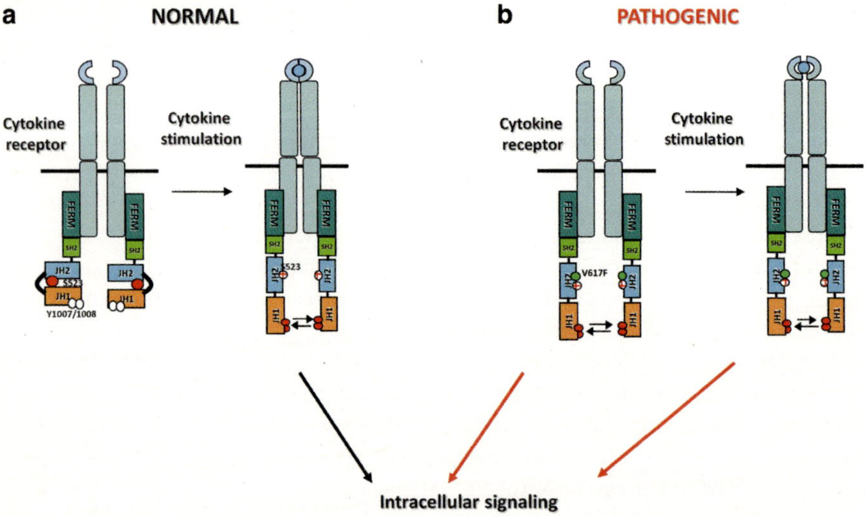

Fig. 3 Schematic representation of the normal and pathological activation of the Janus kinases. (**a**) In the absence of ligand binding, the kinase activity is prevented via the interaction with the pseudokinase domain and involving pS523 (*red dot*) in the JH2 domain. Upon cytokine binding, receptor dimerization leads to kinase activation via transphosphorylation, by releasing the inhibitory JH1-JH2 interaction. (**b**) Cytokine-independent activation of Janus kinases mediated by mutations in the JH2 domain. MPD-causing mutations in the JH2 domain result in displacement of JH1-JH2 inhibitory interaction and altered pSer523 phosphorylation levels. JAKs can transphosphorylate even in the absence of cytokine binding, leading to cytokine-independent signalling and a hyperactive JAK2

there is currently no firm evidence that this model is relevant for JAKs, but nonetheless the model has proven to provide good predictions and explanations to structure/function analysis, particularly related to the V617F mutations. The main interface between JH1 and JH2 is composed of the N-terminal α-helices in both domains. The second interaction site is between the activation loop in JH1 and the loop between β4 and β5 in JH2, starting from V617. This interaction is expected to stabilize the inactive conformation in the activation loop. Recent molecular dynamics simulations largely agree with the original homology model, but provide evidence for additional interfaces consisting of hydrophobic interaction between F595 in JH2 with the activation loop and the interaction between β4 and β5 loop with a loop in JH1 (E1028-S1032) interacting with the activation loop (Lee et al. 2009). The V617F mutation is predicted to inhibit the inhibitory JH1-JH2 interaction by blocking the interaction of F595 and S591 with the activation loop and forcing the activation loop to its active fold. Dusa et al have addressed the function of F595 experimentally and their results also indicate a stacking interaction between F595 and the V617F mutant as a mechanism to activate the kinase (Dusa et al. 2010).

Lessons from Patient Mutations

Mutations Within the Pseudokinase Domain

The majority of the pseudokinase domain mutations affect the N-terminal lobe of the domain and modify residues involved in either the postulated interface with the kinase domain (Levine et al. 2005) or structurally important residues whose mutation can destabilize the N-lobe and thus also affect a possible interface between the pseudokinase and kinase domains. Mutations in the C-lobe of the pseudokinase domain are rare, which could suggest that the structural integrity of this region is essential for JAK function and/or that its surface does not participate in the kinase domain activity regulation. In the case of JAK2 the mutations can be attributed to be part of two structural hotspots (I and II) which are associated with different disease phenotypes. Mutations in hotspot I are associated with MPN while mutations located in hotspot II lead to a different clinical phenotype, namely lymphoblastic leukaemia (discussed in Haan et al. 2010). To date this genotype-phenotype specificity incorporates all activating exon 12, 14 and 16 mutations (Haan et al. 2010; Bercovich et al. 2008). The mutations in the different structural hotspots I and II might influence the recruitment to different signalling complexes including different cytokine receptors and lead to different signalling events. Such genotype-phenotype specificity is not yet obvious for the corresponding mutations in JAK1 and JAK3, where the same structural hotspots are affected by mutations.

A proposed theory, based on a molecular model of the full length JAK2, concerning the effects of the V617F postulates that the residue V617 is part of the binding interface by which the pseudokinase domain contacts the kinase domain and negatively regulates its activity (Levine et al. 2005). Accordingly, mutation of this residue to a larger hydrophobic residue should prevent optimal contact and reduce the affinity of the inhibitory interaction. However, it was shown that a V617Y exchange does not lead to constitutive activity, indicating that the situation may be more complex (Dusa et al. 2008). Although the hypothesis concerning the interface between the pseudokinase and kinase domain makes a lot of sense and explains much of the biological data, it must be noted that the true molecular mechanism could be different and that only a solved structure encompassing at least the pseudokinase and the kinase domains would provide reliable evidence for the mechanism.

Mutations Within the Kinase Domain

Most mutations are confined to a loop-region between the β2 and β3 strands of JAK2 (R867Q, D873N, T875N). Similarly, the other reported mutations (P933R in JAK2 and R879C/H/S in JAK1) affect residues which are exposed on the surface and do not affect the structure of the domain. Considering the kinase-pseudokinase interaction model by Lindauer and colleagues (Lindauer et al. 2001), none of the activating JAK2 mutations can be attributed to the proposed interface between the two domains.

Perspectives

Although there is an overall understanding of the basic functions of the different JAK subdomains it is yet unclear how these domains interact with each other and with the cytokine receptor and which structural changes are imposed on JAKs during activation of the cytokine receptor complex. It still remains mechanistically unclear how the disease-associated mutations in JAKs translate into a gain-of-function phenotype and thus the molecular basis of the mutational hotspots associated with either MPN or leukaemia remains elusive.

Here, we have reviewed data which demonstrate that the FERM domain of JAKs is crucial for receptor association and the SH2-like domain may also be involved in this interaction. Nevertheless, the real situation might still be more complex. The FERM domain has also been described to influence kinase activity. The structural integrity of the pseudokinase domain of TYK2 is essential for high-affinity-binding of cytokines to the IFNAR (Yeh et al. 2000; Gauzzi et al. 1997), pointing to an important role of TYK2 in "organising" the receptor complex. Also, data on JAK3 suggest that the kinase domain may affect receptor binding (Zhou et al. 2001). All this is indicative of a complex interplay of the different JAK subdomains with the cytokine receptor, which very likely reflects different activation states. Interestingly all of these yet unknown intramolecular interactions might be susceptible to interference with allosteric inhibitors.

References

Adelaide J, Perot C, Gelsi-Boyer V, Pautas C, Murati A, Copie-Bergman C, Imbert M, Chaffanet M, Birnbaum D, Mozziconacci MJ (2006) A t(8;9) translocation with PCM1-JAK2 fusion in a patient with T-cell lymphoma. Leukemia 20(3):536–537

Argetsinger LS, Kouadio JL, Steen H, Stensballe A, Jensen ON, Carter-Su C (2004) Autophosphorylation of JAK2 on tyrosines 221 and 570 regulates its activity. Mol Cell Biol 24(11):4955–4967

Autissier P, De Vos J, Liautard J, Tupitsyn N, Jacquet C, Chavdia N, Klein B, Brochier J, Gaillard JP (1998) Dimerization and activation of the common transducing chain (gp130) of the cytokines of the IL-6 family by mAb. Int Immunol 10(12):1881–1889

Baxter EJ, Scott LM, Campbell PJ, East C, Fourouclas N, Swanton S, Vassiliou GS, Bench AJ, Boyd EM, Curtin N, Scott MA, Erber WN, Green AR (2005) Acquired mutation of the tyrosine kinase JAK2 in human myeloproliferative disorders. Lancet 365(9464):1054–1061

Behrmann I, Smyczek T, Heinrich PC, Schmitz-Van de Leur H, Komyod W, Giese B, Muller-Newen G, Haan S, Haan C (2004) Janus kinase (Jak) subcellular localization revisited: the exclusive membrane localization of endogenous Janus kinase 1 by cytokine receptor interaction uncovers the Jak.receptor complex to be equivalent to a receptor tyrosine kinase. J Biol Chem 279(34):35486–35493

Bercovich D, Ganmore I, Scott LM, Wainreb G, Birger Y, Elimelech A, Shochat C, Cazzaniga G, Biondi A, Basso G, Cario G, Schrappe M, Stanulla M, Strehl S, Haas OA, Mann G, Binder V, Borkhardt A, Kempski H, Trka J, Bielorei B, Avigad S, Stark B, Smith O, Dastugue N, Bourquin JP, Tal NB, Green AR, Izraeli S (2008) Mutations of JAK2 in acute lymphoblastic leukaemias associated with Down's syndrome. Lancet 372(9648):1484–1492

Boggon TJ, Li Y, Manley PW, Eck MJ (2005) Crystal structure of the Jak3 kinase domain in complex with a staurosporine analog. Blood 106(3):996–1002

Boudeau J, Miranda-Saavedra D, Barton GJ, Alessi DR (2006) Emerging roles of pseudokinases. Trends Cell Biol 16(9):443–452

Bousquet M, Quelen C, De Mas V, Duchayne E, Roquefeuil B, Delsol G, Laurent G, Dastugue N, Brousset P (2005) The t(8;9)(p22;p24) translocation in atypical chronic myeloid leukaemia yields a new PCM1-JAK2 fusion gene. Oncogene 24(48):7248–7252

Cacalano NA, Migone TS, Bazan F, Hanson EP, Chen M, Candotti F, O'Shea JJ, Johnston JA (1999) Autosomal SCID caused by a point mutation in the N-terminus of Jak3: mapping of the Jak3-receptor interaction domain. EMBO J 18(6):1549–1558

Candotti F, Oakes SA, Johnston JA, Giliani S, Schumacher RF, Mella P, Fiorini M, Ugazio AG, Badolato R, Notarangelo LD, Bozzi F, Macchi P, Strina D, Vezzoni P, Blaese RM, O'Shea JJ, Villa A (1997) Structural and functional basis for JAK3-deficient severe combined immunodeficiency. Blood 90(10):3996–4003

Chen M, Cheng A, Chen YQ, Hymel A, Hanson EP, Kimmel L, Minami Y, Taniguchi T, Changelian PS, O'Shea JJ (1997) The amino terminus of JAK3 is necessary and sufficient for binding to the common gamma chain and confers the ability to transmit interleukin 2-mediated signals. Proc Natl Acad Sci USA 94(13):6910–6915

Chen M, Cheng A, Candotti F, Zhou YJ, Hymel A, Fasth A, Notarangelo LD, O'Shea JJ (2000) Complex effects of naturally occurring mutations in the JAK3 pseudokinase domain: evidence for interactions between the kinase and pseudokinase domains. Mol Cell Biol 20(3):947–956

Cirmena G, Aliano S, Fugazza G, Bruzzone R, Garuti A, Bocciardi R, Bacigalupo A, Ravazzolo R, Ballestrero A, Sessarego M (2008) A BCR-JAK2 fusion gene as the result of a t(9;22)(p24; q11) in a patient with acute myeloid leukemia. Cancer Genet Cytogenet 183(2):105–108

Constantinescu SN, Huang LJ, Nam H, Lodish HF (2001) The erythropoietin receptor cytosolic juxtamembrane domain contains an essential, precisely oriented, hydrophobic motif. Mol Cell 7(2):377–385

Dusa A, Staerk J, Elliott J, Pecquet C, Poirel HA, Johnston JA, Constantinescu SN (2008) Substitution of pseudokinase domain residue Val-617 by large non-polar amino acids causes activation of JAK2. J Biol Chem 283(19):12941–12948

Dusa A, Mouton C, Pecquet C, Herman M, Constantinescu SN (2010) JAK2 V617F constitutive activation requires JH2 residue F595: a pseudokinase domain target for specific inhibitors. PLoS One 5(6):e11157

Eswaran J, Patnaik D, Filippakopoulos P, Wang F, Stein RL, Murray JW, Higgins JM, Knapp S (2009) Structure and functional characterization of the atypical human kinase haspin. Proc Natl Acad Sci USA 106(48):20198–20203

Feener EP, Rosario F, Dunn SL, Stancheva Z, Myers MG Jr (2004) Tyrosine phosphorylation of Jak2 in the JH2 domain inhibits cytokine signaling. Mol Cell Biol 24(11):4968–4978

Flex E, Petrangeli V, Stella L, Chiaretti S, Hornakova T, Knoops L, Ariola C, Fodale V, Clappier E, Paoloni F, Martinelli S, Fragale A, Sanchez M, Tavolaro S, Messina M, Cazzaniga G, Camera A, Pizzolo G, Tornesello A, Vignetti M, Battistini A, Cave H, Gelb BD, Renauld JC, Biondi A, Constantinescu SN, Foa R, Tartaglia M (2008) Somatically acquired JAK1 mutations in adult acute lymphoblastic leukemia. J Exp Med 205(4):751–758

Funakoshi-Tago M, Pelletier S, Matsuda T, Parganas E, Ihle JN (2006) Receptor specific downregulation of cytokine signaling by autophosphorylation in the FERM domain of Jak2. EMBO J 25(20):4763–4772

Gauzzi MC, Barbieri G, Richter MF, Uze G, Ling L, Fellous M, Pellegrini S (1997) The amino-terminal region of Tyk2 sustains the level of interferon alpha receptor 1, a component of the interferon alpha/beta receptor. Proc Natl Acad Sci USA 94(22):11839–11844

Ghoreschi K, Laurence A, O'Shea JJ (2009) Janus kinases in immune cell signaling. Immunol Rev 228(1):273–287

Giese B, Au-Yeung CK, Herrmann A, Diefenbach S, Haan C, Küster A, Wortmann SB, Roderburg C, Heinrich PC, Behrmann I, Müller-Newen G (2003) Long term association of the cytokine receptor gp130 and the Janus kinase Jak1 revealed by FRAP analysis. J Biol Chem 278(40): 39205–39213

Girault JA, Labesse G, Mornon JP, Callebaut I (1998) Janus kinases and focal adhesion kinases play in the 4.1 band: a superfamily of band 4.1 domains important for cell structure and signal transduction. Mol Med 4(12):751–769

Gorantla SP, Dechow TN, Grundler R, Illert AL, Zum Buschenfelde CM, Kremer M, Peschel C, Duyster J (2010) Oncogenic JAK2V617F requires an intact SH2-like domain for constitutive activation and induction of a myeloproliferative disease in mice. Blood 116(22):4600–4611. doi:blood-2009-07-236133

Greiser JS, Stross C, Heinrich PC, Behrmann I, Hermanns HM (2002) Orientational constraints of the gp130 intracellular juxtamembrane domain for signaling. J Biol Chem 277 (30):26959–26965

Griesinger F, Hennig H, Hillmer F, Podleschny M, Steffens R, Pies A, Wormann B, Haase D, Bohlander SK (2005) A BCR-JAK2 fusion gene as the result of a t(9;22)(p24;q11.2) translocation in a patient with a clinically typical chronic myeloid leukemia. Genes Chromosomes Cancer 44(3):329–333. doi:10.1002/gcc.20235

Haan C, Is'harc H, Hermanns HM, Schmitz-Van De Leur H, Kerr IM, Heinrich PC, Grotzinger J, Behrmann I (2001) Mapping of a region within the N terminus of Jak1 involved in cytokine receptor interaction. J Biol Chem 276(40):37451–37458

Haan C, Heinrich PC, Behrmann I (2002) Structural requirements of the interleukin-6 signal transducer gp130 for its interaction with Janus kinase 1: the receptor is crucial for kinase activation. Biochem J 361(Pt 1):105–111

Haan C, Kreis S, Margue C, Behrmann I (2006) Jaks and cytokine receptors-An intimate relationship. Biochem Pharmacol 72(11):1538–1546

Haan S, Margue C, Engrand A, Rolvering C, Schmitz-Van de Leur H, Heinrich PC, Behrmann I, Haan C (2008) Dual role of the Jak1 FERM and kinase domains in cytokine receptor binding and in stimulation-dependent Jak activation. J Immunol 180(2):998–1007

Haan C, Kroy DC, Wuller S, Sommer U, Nocker T, Rolvering C, Behrmann I, Heinrich PC, Haan S (2009) An unusual insertion in Jak2 is crucial for kinase activity and differentially affects cytokine responses. J Immunol 182(5):2969–2977

Haan C, Behrmann I, Haan S (2010) Perspectives for the use of structural information and chemical genetics to develop inhibitors of Janus kinases. J Cell Mol Med 14(3):504–527

Hamada K, Shimizu T, Matsui T, Tsukita S, Hakoshima T (2000) Structural basis of the membrane-targeting and unmasking mechanisms of the radixin FERM domain. EMBO J 19(17):4449–4462

He K, Loesch K, Cowan JW, Li X, Deng L, Wang X, Jiang J, Frank SJ (2005) Janus kinase 2 enhances the stability of the mature growth hormone receptor. Endocrinology 146 (11):4755–4765

Heinrich PC, Behrmann I, Haan S, Hermanns HM, Muller-Newen G, Schaper F (2003) Principles of interleukin (IL)-6-type cytokine signalling and its regulation. Biochem J 374(Pt 1):1–20

Hilkens CM, Is'harc H, Lillemeier BF, Strobl B, Bates PA, Behrmann I, Kerr IM (2001) A region encompassing the FERM domain of Jak1 is necessary for binding to the cytokine receptor gp130. FEBS Lett 505(1):87–91

Hintzen C, Haan C, Tuckermann JP, Heinrich PC, Hermanns HM (2008) Oncostatin M-induced and constitutive activation of the JAK2/STAT5/CIS pathway suppresses CCL1, but not CCL7 and CCL8, chemokine expression. J Immunol 181(10):7341–7349

Huang LJ, Constantinescu SN, Lodish HF (2001) The N-terminal domain of Janus kinase 2 is required for Golgi processing and cell surface expression of erythropoietin receptor. Mol Cell 8 (6):1327–1338

Hubbard SR, Mohammadi M, Schlessinger J (1998) Autoregulatory mechanisms in protein-tyrosine kinases. J Biol Chem 273(20):11987–11990

Ihle JN, Kerr IM (1995) Jaks and Stats in signaling by the cytokine receptor superfamily. Trends Genet 11(2):69–74

Ishida-Takahashi R, Rosario F, Gong Y, Kopp K, Stancheva Z, Chen X, Feener EP, Myers MG Jr (2006) Phosphorylation of Jak2 on Ser(523) inhibits Jak2-dependent leptin receptor signaling. Mol Cell Biol 26(11):4063–4073. doi:26/11/4063

James C, Ugo V, Le Couedic JP, Staerk J, Delhommeau F, Lacout C, Garcon L, Raslova H, Berger R, Bennaceur-Griscelli A, Villeval JL, Constantinescu SN, Casadevall N, Vainchenker W (2005) A unique clonal JAK2 mutation leading to constitutive signalling causes polycythaemia vera. Nature 434(7037):1144–1148

Jeong EG, Kim MS, Nam HK, Min CK, Lee S, Chung YJ, Yoo NJ, Lee SH (2008) Somatic mutations of JAK1 and JAK3 in acute leukemias and solid cancers. Clin Cancer Res 14 (12):3716–3721

Jura N, Shan Y, Cao X, Shaw DE, Kuriyan J (2009) Structural analysis of the catalytically inactive kinase domain of the human EGF receptor 3. Proc Natl Acad Sci USA 106(51):21608–21613. doi:0912101106

Kamizono S, Hanada T, Yasukawa H, Minoguchi S, Kato R, Minoguchi M, Hattori K, Hatakeyama S, Yada M, Morita S, Kitamura T, Kato H, Nakayama K, Yoshimura A (2001) The SOCS box of SOCS-1 accelerates ubiquitin-dependent proteolysis of TEL-JAK2. J Biol Chem 276(16):12530–12538

Karaghiosoff M, Neubauer H, Lassnig C, Kovarik P, Schindler H, Pircher H, McCoy B, Bogdan C, Decker T, Brem G, Pfeffer K, Muller M (2000) Partial impairment of cytokine responses in Tyk2-deficient mice. Immunity 13(4):549–560

Kohlhuber F, Rogers NC, Watling D, Feng J, Gushin D, Briscoe J, Witthuhn BA, Kotenko SV, Pestka S, Stark GR, Ihle JN, Kerr IM (1997) A JAK1/JAK2 chimera can sustain alpha and gamma interferon responses. Mol Cell Biol 17:695–706

Kovanen PE, Leonard WJ (2004) Cytokines and immunodeficiency diseases: critical roles of the gamma(c)-dependent cytokines interleukins 2, 4, 7, 9, 15, and 21, and their signaling pathways. Immunol Rev 202:67–83

Kralovics R, Passamonti F, Buser AS, Teo SS, Tiedt R, Passweg JR, Tichelli A, Cazzola M, Skoda RC (2005) A gain-of-function mutation of JAK2 in myeloproliferative disorders. N Engl J Med 352(17):1779–1790

Kratz CP, Boll S, Kontny U, Schrappe M, Niemeyer CM, Stanulla M (2006) Mutational screen reveals a novel JAK2 mutation, L611S, in a child with acute lymphoblastic leukemia. Leukemia 20(2):381–383

Kubatzky KF, Liu W, Goldgraben K, Simmerling C, Smith SO, Constantinescu SN (2005) Structural requirements of the extracellular to transmembrane domain junction for erythropoietin receptor function. J Biol Chem 280(15):14844–14854

Kurzer JH, Argetsinger LS, Zhou YJ, Kouadio JL, O'Shea JJ, Carter-Su C (2004) Tyrosine 813 is a site of JAK2 autophosphorylation critical for activation of JAK2 by SH2-B beta. Mol Cell Biol 24(10):4557–4570

Lacronique V, Boureux A, Valle VD, Poirel H, Quang CT, Mauchauffe M, Berthou C, Lessard M, Berger R, Ghysdael J, Bernard OA (1997) A TEL-JAK2 fusion protein with constitutive kinase activity in human leukemia. Science 278(5341):1309–1312

Lee TS, Ma W, Zhang X, Giles F, Kantarjian H, Albitar M (2009) Mechanisms of constitutive activation of Janus kinase 2-V617F revealed at the atomic level through molecular dynamics simulations. Cancer 115(8):1692–1700

Levine RL, Wadleigh M, Cools J, Ebert BL, Wernig G, Huntly BJ, Boggon TJ, Wlodarska I, Clark JJ, Moore S, Adelsperger J, Koo S, Lee JC, Gabriel S, Mercher T, D'Andrea A, Frohling S, Dohner K, Marynen P, Vandenberghe P, Mesa RA, Tefferi A, Griffin JD, Eck MJ, Sellers WR, Meyerson M, Golub TR, Lee SJ, Gilliland DG (2005) Activating mutation in the tyrosine kinase JAK2 in polycythemia vera, essential thrombocythemia, and myeloid metaplasia with myelofibrosis. Cancer Cell 7(4):387–397

Lindauer K, Loerting T, Liedl KR, Kroemer RT (2001) Prediction of the structure of human Janus kinase 2 (JAK2) comprising the two carboxy-terminal domains reveals a mechanism for autoregulation. Protein Eng 14(1):27–37

Livnah O, Stura EA, Middleton SA, Johnson DL, Jolliffe LK, Wilson IA (1999) Crystallographic evidence for preformed dimers of erythropoietin receptor before ligand activation. Science 283:987–990

Lu X, Huang LJ, Lodish HF (2008) Dimerization by a cytokine receptor is necessary for constitutive activation of JAK2V617F. J Biol Chem 283(9):5258–5266

Lucet IS, Fantino E, Styles M, Bamert R, Patel O, Broughton SE, Walter M, Burns CJ, Treutlein H, Wilks AF, Rossjohn J (2006) The structural basis of Janus kinase 2 inhibition by a potent and specific pan-Janus kinase inhibitor. Blood 107(1):176–183

Luo H, Rose P, Barber D, Hanratty WP, Lee S, Roberts TM, D'Andrea AD, Dearolf CR (1997) Mutation in the Jak kinase JH2 domain hyperactivates Drosophila and mammalian Jak-Stat pathways. Mol Cell Biol 17(3):1562–1571

Macchi P, Villa A, Giliani S, Sacco MG, Frattini A, Porta F, Ugazio AG, Johnston JA, Candotti F, O'Shea JJ et al (1995) Mutations of Jak-3 gene in patients with autosomal severe combined immune deficiency (SCID). Nature 377(6544):65–68

Malinge S, Ragu C, Della-Valle V, Pisani D, Constantinescu SN, Perez C, Villeval JL, Reinhardt D, Landman-Parker J, Michaux L, Dastugue N, Baruchel A, Vainchenker W, Bourquin JP, Penard-Lacronique V, Bernard OA (2008) Activating mutations in human acute megakaryoblastic leukemia. Blood 112(10):4220–4226

Manning BD, Cantley LC (2002) Hitting the target: emerging technologies in the search for kinase substrates. Sci STKE 2002(162):PE49

Mazurkiewicz-Munoz AM, Argetsinger LS, Kouadio JL, Stensballe A, Jensen ON, Cline JM, Carter-Su C (2006) Phosphorylation of JAK2 at serine 523: a negative regulator of JAK2 that is stimulated by growth hormone and epidermal growth factor. Mol Cell Biol 26(11):4052–4062

Mercher T, Wernig G, Moore SA, Levine RL, Gu TL, Frohling S, Cullen D, Polakiewicz RD, Bernard OA, Boggon TJ, Lee BH, Gilliland DG (2006) JAK2T875N is a novel activating mutation that results in myeloproliferative disease with features of megakaryoblastic leukemia in a murine bone marrow transplantation model. Blood 108(8):2770–2779

Mukherjee K, Sharma M, Urlaub H, Bourenkov GP, Jahn R, Sudhof TC, Wahl MC (2008) CASK functions as a Mg2+−independent neurexin kinase. Cell 133(2):328–339

Müller-Newen G, Küster A, Wijdenes J, Schaper F, Heinrich PC (2000) Studies on the interleukin-6-type cytokine signal transducer gp130 reveal a novel mechanism of receptor activation by monoclonal antibodies. J Biol Chem 275(7):4579–4586

Murati A, Gelsi-Boyer V, Adelaide J, Perot C, Talmant P, Giraudier S, Lode L, Letessier A, Delaval B, Brunel V, Imbert M, Garand R, Xerri L, Birnbaum D, Mozziconacci MJ, Chaffanet M (2005) PCM1-JAK2 fusion in myeloproliferative disorders and acute erythroid leukemia with t(8;9) translocation. Leukemia 19(9):1692–1696

Neubauer H, Cumano A, Muller M, Wu H, Huffstadt U, Pfeffer K (1998) Jak2 deficiency defines an essential developmental checkpoint in definitive hematopoiesis. Cell 93(3):397–409

Nishii K, Nanbu R, Lorenzo VF, Monma F, Kato K, Ryuu H, Katayama N (2007) Expression of the JAK2 V617F mutation is not found in de novo AML and MDS but is detected in MDS-derived leukemia of megakaryoblastic nature. Leukemia 21(6):1337–1338

Nosaka T, van Deursen JM, Tripp RA, Thierfelder WE, Witthuhn BA, McMickle AP, Doherty PC, Grosveld GC, Ihle JN (1995) Defective lymphoid development in mice lacking Jak3. Science 270(5237):800–802

O'Sullivan LA, Liongue C, Lewis RS, Stephenson SE, Ward AC (2007) Cytokine receptor signaling through the Jak-Stat-Socs pathway in disease. Mol Immunol 44(10):2497–2506

Parganas E, Wang D, Stravopodis D, Topham DJ, Marine JC, Teglund S, Vanin EF, Bodner S, Colamonici OR, van Deursen JM, Grosveld G, Ihle JN (1998) Jak2 is essential for signaling through a variety of cytokine receptors. Cell 93(3):385–395

Park SY, Saijo K, Takahashi T, Osawa M, Arase H, Hirayama N, Miyake K, Nakauchi H, Shirasawa T, Saito T (1995) Developmental defects of lymphoid cells in Jak3 kinase-deficient mice. Immunity 3(6):771–782

Pearson MA, Reczek D, Bretscher A, Karplus PA (2000) Structure of the ERM protein moesin reveals the FERM domain fold masked by an extended actin binding tail domain. Cell 101 (3):259–270

Peeters P, Raynaud SD, Cools J, Wlodarska I, Grosgeorge J, Philip P, Monpoux F, Van Rompaey L, Baens M, Van den Berghe H, Marynen P (1997) Fusion of TEL, the ETS-variant gene 6 (ETV6), to the receptor-associated kinase JAK2 as a result of t(9;12) in a lymphoid and t(9;15;12) in a myeloid leukemia. Blood 90(7):2535–2540

Pestka S, Krause CD, Walter MR (2004) Interferons, interferon-like cytokines, and their receptors. Immunol Rev 202:8–32

Pesu M, Candotti F, Husa M, Hofmann SR, Notarangelo LD, O'Shea JJ (2005) Jak3, severe combined immunodeficiency, and a new class of immunosuppressive drugs. Immunol Rev 203:127–142

Pesu M, Laurence A, Kishore N, Zwillich SH, Chan G, O'Shea JJ (2008) Therapeutic targeting of Janus kinases. Immunol Rev 223:132–142

Poitras JL, Dal Cin P, Aster JC, Deangelo DJ, Morton CC (2008) Novel SSBP2-JAK2 fusion gene resulting from a t(5;9)(q14.1;p24.1) in pre-B acute lymphocytic leukemia. Genes Chromosomes Cancer 47(10):884–889

Radtke S, Hermanns HM, Haan C, Schmitz-Van De Leur H, Gascan H, Heinrich PC, Behrmann I (2002) Novel role of Janus kinase 1 in the regulation of oncostatin M receptor surface expression. J Biol Chem 277(13):11297–11305

Radtke S, Haan S, Jorissen A, Hermanns HM, Diefenbach S, Smyczek T, Schmitz-Vandeleur H, Heinrich PC, Behrmann I, Haan C (2005) The Jak1 SH2 domain does not fulfill a classical SH2 function in Jak/STAT signaling but plays a structural role for receptor interaction and up-regulation of receptor surface expression. J Biol Chem 280(27):25760–25768

Radtke S, Jörissen A, Schmitz-Van de Leur H, Heinrich PC, Behrmann I (2006) Three dileucine-like motifs within the interbox1/2 region of the human oncostatin M receptor prevent efficient surface expression in absence of an associated Janus kinase. J Biol Chem 281(7):4024–4034

Ragimbeau J, Dondi E, Alcover A, Eid P, Uze G, Pellegrini S (2003) The tyrosine kinase Tyk2 controls IFNAR1 cell surface expression. EMBO J 22(3):537–547

Reiter A, Walz C, Watmore A, Schoch C, Blau I, Schlegelberger B, Berger U, Telford N, Aruliah S, Yin JA, Vanstraelen D, Barker HF, Taylor PC, O'Driscoll A, Benedetti F, Rudolph C, Kolb HJ, Hochhaus A, Hehlmann R, Chase A, Cross NC (2005) The t(8;9)(p22;p24) is a recurrent abnormality in chronic and acute leukemia that fuses PCM1 to JAK2. Cancer Res 65(7): 2662–2667

Remy I, Wilson IA, Michnick SW (1999) Erythropoietin receptor activation by a ligand-induced conformation change. Science 283(5404):990–993

Richter MF, Dumenil G, Uze G, Fellous M, Pellegrini S (1998) Specific contribution of Tyk2 JH regions to the binding and the expression of the interferon alpha/beta receptor component IFNAR1. J Biol Chem 273(38):24723–24729

Rishton GM (2003) Nonleadlikeness and leadlikeness in biochemical screening. Drug Discov Today 8(2):86–96

Robertson SA, Koleva RI, Argetsinger LS, Carter-Su C, Marto JA, Feener EP, Myers MG Jr (2009) Regulation of Jak2 function by phosphorylation of Tyr317 and Tyr637 during cytokine signaling. Mol Cell Biol 29(12):3367–3378

Rodig SJ, Meraz MA, White JM, Lampe PA, Riley JK, Arthur CD, King KL, Sheehan KC, Yin L, Pennica D, Johnson EM Jr, Schreiber RD (1998) Disruption of the Jak1 gene demonstrates obligatory and nonredundant roles of the Jaks in cytokine-induced biologic responses. Cell 93 (3):373–383

Royer Y, Staerk J, Costuleanu M, Courtoy PJ, Constantinescu SN (2005) Janus kinases affect thrombopoietin receptor cell surface localization and stability. J Biol Chem 280 (29):27251–27261

Russell SM, Tayebi N, Nakajima H, Riedy MC, Roberts JL, Aman MJ, Migone TS, Noguchi M, Markert ML, Buckley RH et al (1995) Mutation of Jak3 in a patient with SCID: essential role of Jak3 in lymphoid development. Science 270(5237):797–800

Saharinen P, Silvennoinen O (2002) The pseudokinase domain is required for suppression of basal activity of Jak2 and Jak3 tyrosine kinases and for cytokine-inducible activation of signal transduction. J Biol Chem 277(49):47954–47963

Saharinen P, Takaluoma K, Silvennoinen O (2000) Regulation of the Jak2 tyrosine kinase by its pseudokinase domain. Mol Cell Biol 20(10):3387–3395

Saharinen P, Vihinen M, Silvennoinen O (2003) Autoinhibition of Jak2 tyrosine kinase is dependent on specific regions in its pseudokinase domain. Mol Biol Cell 14(4):1448–1459

Scheeff ED, Eswaran J, Bunkoczi G, Knapp S, Manning G (2009) Structure of the pseudokinase VRK3 reveals a degraded catalytic site, a highly conserved kinase fold, and a putative regulatory binding site. Structure 17(1):128–138

Shaw MH, Boyartchuk V, Wong S, Karaghiosoff M, Ragimbeau J, Pellegrini S, Muller M, Dietrich WF, Yap GS (2003) A natural mutation in the Tyk2 pseudokinase domain underlies altered susceptibility of B10.Q/J mice to infection and autoimmunity. Proc Natl Acad Sci USA 100(20):11594–11599

Shi F, Telesco SE, Liu Y, Radhakrishnan R, Lemmon MA (2010) ErbB3/HER3 intracellular domain is competent to bind ATP and catalyze autophosphorylation. Proc Natl Acad Sci USA 107(17):7692–7697

Shimoda K, Kato K, Aoki K, Matsuda T, Miyamoto A, Shibamori M, Yamashita M, Numata A, Takase K, Kobayashi S, Shibata S, Asano Y, Gondo H, Sekiguchi K, Nakayama K, Nakayama T, Okamura T, Okamura S, Niho Y (2000) Tyk2 plays a restricted role in IFN alpha signaling, although it is required for IL-12-mediated T cell function. Immunity 13(4):561–571

Siewert E, Müller-Esterl W, Starr R, Heinrich PC, Schaper F (1999) Different protein turnover of interleukin-6-type cytokine signalling components. Eur J Biochem 265:251–257

Sigurdsson S, Nordmark G, Göring HH, Lindroos K, Wiman AC, Sturfelt G, Jönsen A, Rantapää-Dahlqvist S, Möller B, Kere J, Koskenmies S, Widén E, Eloranta ML, Julkunen H, Kristjansdottir H, Steinsson K, Alm G, Rönnblom L, Syvänen AC (2005) Polymorphisms in the tyrosine kinase 2 and interferon regulatory factor 5 genes are associated with systemic lupus erythematosus. Am J Hum Genet 76(3):528–537

Thomis DC, Gurniak CB, Tivol E, Sharpe AH, Berg LJ (1995) Defects in B lymphocyte maturation and T lymphocyte activation in mice lacking Jak3. Science 270(5237):794–797

Ungureanu D, Saharinen P, Junttila I, Hilton DJ, Silvennoinen O (2002) Regulation of Jak2 through the ubiquitin-proteasome pathway involves phosphorylation of Jak2 on Y1007 and interaction with SOCS-1. Mol Cell Biol 22(10):3316–3326

Walters DK, Mercher T, Gu TL, O'Hare T, Tyner JW, Loriaux M, Goss VL, Lee KA, Eide CA, Wong MJ, Stoffregen EP, McGreevey L, Nardone J, Moore SA, Crispino J, Boggon TJ, Heinrich MC, Deininger MW, Polakiewicz RD, Gilliland DG, Druker BJ (2006) Activating alleles of JAK3 in acute megakaryoblastic leukemia. Cancer Cell 10(1):65–75

Watowich SS, Liu KD, Xie X, Lai SY, Mikami A, Longmore GD, Goldsmith MA (1999) Oligomerization and scaffolding functions of the erythropoietin receptor cytoplasmic tail. J Biol Chem 274(9):5415–5421

Wernig G, Gonneville JR, Crowley BJ, Rodrigues MS, Reddy MM, Hudon HE, Walz C, Reiter A, Podar K, Royer Y, Constantinescu SN, Tomasson MH, Griffin JD, Gilliland DG, Sattler M (2008) The Jak2V617F oncogene associated with myeloproliferative diseases requires a functional FERM domain for transformation and for expression of the Myc and Pim proto-oncogenes. Blood 111(7):3751–3759

Wilks AF, Harpur AG, Kurban RR, Ralph SJ, Zurcher G, Ziemiecki A (1991) Two novel protein-tyrosine kinases, each with a second phosphotransferase-related catalytic domain, define a new class of protein kinase. Mol Cell Biol 11(4):2057–2065

Williams NK, Bamert RS, Patel O, Wang C, Walden PM, Wilks AF, Fantino E, Rossjohn J, Lucet IS (2009) Dissecting specificity in the Janus kinases: the structures of JAK-specific inhibitors complexed to the JAK1 and JAK2 protein tyrosine kinase domains. J Mol Biol 387(1):219–232

Yasukawa H, Misawa H, Sakamoto H, Masuhara M, Sasaki A, Wakioka T, Ohtsuka S, Imaizumi T, Matsuda T, Ihle JN, Yoshimura A (1999) The JAK-binding protein JAB inhibits Janus tyrosine kinase activity through binding in the activation loop. EMBO J 18(5):1309–1320

Yeh TC, Pellegrini S (1999) The Janus kinase family of protein tyrosine kinases and their role in signaling. Cell Mol Life Sci 55(12):1523–1534

Yeh TC, Dondi E, Uze G, Pellegrini S (2000) A dual role for the kinase-like domain of the tyrosine kinase Tyk2 in interferon-alpha signaling. Proc Natl Acad Sci USA 97(16):8991–8996

Yoshimura A, Naka T, Kubo M (2007) SOCS proteins, cytokine signalling and immune regulation. Nat Rev Immunol 7(6):454–465

Zeqiraj E, van Aalten DM (2010) Pseudokinases-remnants of evolution or key allosteric regulators? Curr Opin Struct Biol 20(6):772–781

Zeqiraj E, Filippi BM, Deak M, Alessi DR, van Aalten DM (2009) Structure of the LKB1-STRAD-MO25 complex reveals an allosteric mechanism of kinase activation. Science 326 (5960):1707–1711

Zhao Y, Wagner F, Frank SJ, Kraft AS (1995) The amino-terminal portion of the JAK2 protein kinase is necessary for binding and phosphorylation of the granulocyte-macrophage colony-stimulating factor receptor beta c chain. J Biol Chem 270(23):13814–13818

Zhao R, Xing S, Li Z, Fu X, Li Q, Krantz SB, Zhao ZJ (2005) Identification of an acquired JAK2 mutation in polycythemia vera. J Biol Chem 280(24):22788–22792

Zhou YJ, Chen M, Cusack NA, Kimmel LH, Magnuson KS, Boyd JG, Lin W, Roberts JL, Lengi A, Buckley RH, Geahlen RL, Candotti F, Gadina M, Changelian PS, O'Shea JJ (2001) Unexpected effects of FERM domain mutations on catalytic activity of Jak3: structural implication for Janus kinases. Mol Cell 8(5):959–969

Zhu HJ, Sizeland AM (1999) A pivotal role for the transmembrane domain in transforming growth factor-beta receptor activation. J Biol Chem 274(17):11773–11781

Nuclear Functions of the Janus Kinases

Mark A. Dawson and Andrew J. Bannister

Abstract
The Janus kinases are a family of non-receptor tyrosine kinases which are key mediators of cellular signalling pathways. These enzymes regulate the growth, survival, development and differentiation of a number of different tissues including the endocrine, mammary and haematopoietic system. The constitutive activation of these kinases are increasingly being recognised as sentinel events in the initiation and progression of several malignancies. These findings underline the importance of a thorough understanding of the cellular functions of this ancient and essential family of enzymes. The role of the canonical JAK-STAT signalling pathway in conveying extracellular stimuli to the nucleus of cells to alter gene expression has been extensively characterised. However, recent evidence in both drosophila and mammalian cells suggest a previously unrecognised role for the Janus kinases in directly altering gene expression by changing the structure and function of chromatin. In this chapter we review nuclear functions of the Janus kinases with an emphasis on the emerging evidence that these tyrosine kinases have a critical role as chromatin modifying enzymes.

M.A. Dawson (✉)
Department of Haematology, Cambridge Institute for Medical Research, University of Cambridge, Cambridge CB2 0XY, UK

Addenbrooke's Hospital, University of Cambridge, Cambridge, UK

Gurdon Institute and Department of Pathology, University of Cambridge, Tennis Court Road, Cambridge CB2 1QN, UK
e-mail: mafd2@cam.ac.uk

A.J. Bannister
Gurdon Institute and Department of Pathology, University of Cambridge, Tennis Court Road, Cambridge CB2 1QN, UK

Th. Decker and M. Müller (eds.), *Jak-Stat Signaling: From Basics to Disease*,
DOI 10.1007/978-3-7091-0891-8_3, © Springer-Verlag Wien 2012

Introduction

In the late eighties, Andrew Wilks and colleagues in Melbourne, Australia set out to clone the receptors for the newly identified haematopoietic colony stimulating factors (Gough et al. 1984; Nicola and Metcalf 1984). Their first attempt at cloning a critical haematopoietic regulator was successful, however their initial goal was not achieved; instead they had serendipitously cloned two related but as yet unidentified kinases that bore little resemblance to the previously characterised receptor tyrosine kinases. Their disappointment led them to name these two kinases '*J*ust *A*nother *K*inase' one and two respectively (Wilks 1989). Despite their initial reservations the group continued to work on the newly discovered kinases and soon their disappointment gave way to elation as other members of the family were discovered and ascribed important cellular functions including an integral role in cytokine signalling (Firmbach-Kraft et al. 1990; Velazquez et al. 1992). It soon became apparent that the structures of these kinases were unique as they contained two kinase-like domains. It was under this guise that the family adopted their new, more August name, the Janus kinases; named after the Roman God Janus, a deity with two faces who presided over gates and doorways, new beginnings and endings (Wilks et al. 1991).

In mammals the Janus kinase family consists of four members; JAK1, JAK2, JAK3 and TYK2. Of these, JAK1, JAK2 and TYK2 are ubiquitously expressed whereas JAK3 expression is confined to the haematopoietic system (Musso et al. 1995). Two decades of research have now firmly established that one of the primary functions of the Janus kinases is to mediate the intracellular signalling for a multitude of structurally diverse receptors including cytokine receptors, receptor tyrosine kinases and G-protein coupled receptors. Whilst a number of cytokines utilise more than one JAK family member, there are a number of non-reductant functions ascribed to each mammalian JAK kinase. This has now been studied in great detail with knock-out mice having been generated for each of the JAK family members (Karaghiosoff et al. 2000; Neubauer et al. 1998; Nosaka et al. 1995; Parganas et al. 1998; Park et al. 1995; Rodig et al. 1998; Shimoda et al. 2000; Thomis et al. 1995).

The main cytoplasmic signalling pathways initiated by activation of the JAK kinases include the signal transducer and activator of transcription (STAT) family of transcription factors, the Ras/mitogen activated protein kinase (MAPK) signalling pathway and the phosphatidylinositol 3-kinase (PI3K)-AKT pathway. The primary focus for understanding the consequences of JAK activation in development and disease has centred on the STAT family of transcription factors. The STAT family of transcription factors were first identified in the early 1990s when they were demonstrated to play a critical role in mediating transcription after interferon stimulation (Fu et al. 1992; Schindler et al. 1992). Since then seven mammalian STATs have been identified STAT1, STAT2, STAT3, STAT4, STAT5A, STAT5B and STAT6. Moreover, alternative splicing and proteolytic processing results in several biologically active isoforms of each of these family members (Lim and Cao 2006).

Recently, gain of function mutations or translocations involving JAK family members have been described in a myriad of haematological neoplasms (Baxter et al. 2005; Bercovich et al. 2008; James et al. 2005; Kralovics et al. 2005; Lacronique et al. 1997; Levine et al. 2005; Mullighan et al. 2009). The most prevalent of these is the JAK2V617F mutation, which underpins the molecular pathogenesis of the myeloproliferative diseases (Campbell and Green 2006; Levine et al. 2007). Much of the research aimed at understanding the functional outcome of these gain of function mutations in JAK2 have centred on elucidating the consequences of constitutive activation of the downstream cytoplasmic signalling pathways such as STAT, MAPK and PI3K-AKT. However, we have known for over a decade that the Janus kinases are also present within the nucleus of cells.

Nuclear Localisation of the Janus Kinases

In 1996, Peter Lobie and colleagues provided one of the first descriptions of a nuclear localisation for the Janus kinases (Lobie et al. 1996). They demonstrated that JAK1 and JAK2 were present in the nucleus of CHO cells and rat liver hepatocytes. Nuclear localisation in this report was confirmed with immunocytochemistry, immunogold electron microscopy and western blotting of purified nuclear extracts. Within a year these observations were confirmed and extended by Ram and Waxman who demonstrated that nuclear JAK2 in rat liver hepatocytes associates with activated STAT3 (Ram and Waxman 1997). In 2001, Sandra Pellegrini's group demonstrated that a third member of the Janus kinase family, TYK2, was also present within the nucleus (Ragimbeau et al. 2001). Furthermore, they mapped a non-classical nuclear localisation sequence (NLS) to the FERM domain of TYK2. This region of TYK2 is broadly conserved in JAK2 and JAK3, but not in JAK1. Subsequent to these early reports other investigators also described a nuclear localisation for the Janus kinases in various tissues including pancreatic cells (Sorenson and Stout 1995), oocytes (Ito et al. 2004), kidney (Kamakura et al. 2004) and mammary cells (Nilsson et al. 2006). Together, these reports highlight the presence of the Janus kinases within the nucleus but until recently their function within this sub-cellular compartment was largely uncharacterised.

Yukiko Gotoh and colleagues provided some early insights into a nuclear function for JAK2 by demonstrating a direct interaction between nuclear JAK2, STAT3 and the Hes proteins (Kamakura et al. 2004). These findings uncovered a previously unrecognised mechanism of cooperation between Notch signalling and the JAK/STAT pathway and demonstrated a specific function for nuclear JAK2 in this process. Similarly, work from Marie Kannius-Janson and colleagues has demonstrated a novel role for nuclear JAK2 in response to prolactin stimulation in mammary tissues (Nilsson et al. 2006). They demonstrated that prolactin stimulated JAK2 specifically phosphorylates NF1-C2 within the nucleus of cells (Nilsson et al. 2006). NF1-C2 is a transcription factor that regulates the expression of milk genes during pregnancy, and tyrosine phosphorylation by nuclear JAK2 retards its proteosomal degradation (Nilsson et al. 2006). More recently, these authors have demonstrated that breast cancer patients whose tissues demonstrate higher levels of nuclear JAK2 and NF1-C2 have an improved survival (Nilsson et al. 2010). This is in part due to the fact that tyrosine phosphorylated NF1-C2 suppresses epithelial-to-mesenchymal transition (EMT) in breast tissues (Nilsson et al. 2010). Other targets for nuclear JAK2 following prolactin stimulation of breast tissues have started to emerge and include the RUSH transcription factors (Helmer et al. 2010; Hewetson et al. 2002). Together, these data provide a novel insight into a unique function for nuclear JAK2 in both normal and malignant breast tissues. Interestingly, both NF1-C2 and RUSH appear to be STAT independent targets of nuclear JAK2 highlighting the fact that these two elemental mediators of cytokine signalling need not always function together.

A JAK Chromatin Link

A pioneering study performed by Willis Li's laboratory in late 2006 uncovered a fascinating role for JAK signalling by demonstrating that JAK signalling can dramatically alter the structure and function of chromatin. The authors of this study performed a genome wide screen to elucidate moderators of JAK signalling in *D. melanogaster* (Shi et al. 2006). The sole Janus kinase in this organism, called hopscotch, is most closely related to human JAK2 (Arbouzova and Zeidler 2006; Hombria and Brown 2002). Like its human orthologue naturally occurring single missense mutations in *hopscotch* lead to its constitutive activation and culminate in haematopoietic malignancies (Harrison et al. 1995; Luo et al. 1995, 1997). The seminal observation from this work was that chromatin associated proteins were one of the major moderators of JAK signalling and indeed the leukemic phenotype within transgenic flies expressing a constitutively active JAK protein. In particular, they noted that JAK signalling disrupted heterochromatic gene silencing by counteracting the effects of key structural components of heterochromatin including the mislocalisation of heterochromatin protein 1 (HP1).

These important findings demonstrated a novel consequence of signalling through the Janus kinases. However, many questions remained unanswered. These included (1) do the Janus kinases alter chromatin structure directly or indirectly via one of its downstream cytoplasmic signalling pathways and (2) what are the molecular mechanisms by which JAK signalling counteract heterochromatin function?

Nuclear JAK2 in Haematopoietic Cells

The work described above raised the possibility that JAK2, an enzyme frequently mutated and activated in haematological cancers might mediate some of its oncogenic potential within the nucleus of haematopoietic cells. We therefore investigated the sub-cellular localization of both wild type JAK2 and JAK2V617F in a number of different haematopoietic cell lines. Nuclear JAK2 was apparent in all of the haematopoietic cell lines tested and moreover we also demonstrated the presence of JAK2V617F in the nuclei of primary CD34 stem cell antigen positive ($CD34^+$) cells isolated from a patient with JAK2V617F positive post-polycythaemic myelofibrosis (Dawson et al. 2009). Whilst we did not observe a clear difference between the nuclear localisation of wild-type JAK2 and JAK2V617F, subsequent work by Rinaldi et al. suggested that JAK2V61F may preferentially localise to the nucleus within haematopoietic cells (Rinaldi et al. 2010). These authors showed that K562 cells expressing wild type JAK2 contained relatively low levels of nuclear JAK2 whereas K562 cells expressing JAK2V617F contained comparatively higher levels of the kinase within the nucleus. Moreover, this study also suggested that unlike wild-type JAK2, JAK2V617F is more likely to be nuclear in more primitive haematopoietic cells, whereas both forms of JAK2 are largely cytoplasmic in terminally differentiated myeloid cells. The biological implications for these findings are currently unknown.

JAK2 Is a Histone Tyrosine Kinase

The identification of a nuclear pool of JAK2 in haematopoietic cells coupled to the findings in *D. melanogaster* raised the intriguing possibility that JAK2 may directly modify a chromatin substrate. Chromatin is a macromolecular complex of DNA and histone proteins that exists in eukaryotic cells. Generally, two distinct chromatin states have been described; heterochromatin refers to the tightly packaged genomic regions that are by and large transcriptionally silent, whereas euchromatin adopts a more ‘open’ conformation that is supportive of transcriptional activity. The basic functional unit of chromatin is the nucleosome, which consists of a histone octamer containing two each of histones H2A, H2B, H3 and H4, around which 147 base pairs of DNA is wrapped (Kornberg and Lorch 1999; Luger et al. 1997; Woodcock 2006). The repeating nucleosome units are further compacted into a larger 30 nm chromatin fibre, at least in vitro, by incorporating the linker histone H1. In the cell, supercoiling and twisting generates the compact chromatin that constitutes the chromosomes. Importantly, folding and unfolding of the chromatin superstructure is a highly regulated and co-ordinated event that governs key DNA based processes including transcription, repair and replication (Woodcock 2006).

The structure of the nucleosome has shed further light on the organization of the histone octamer (Luger et al. 1997). It provided evidence that each histone possesses a structured globular domain and an unstructured flexible amino-terminus and in some cases an unstructured carboxy-terminus. These histone tails extend from their own nucleosome to contact DNA and histones in adjacent nucleosomes. Efforts to study the coordinated regulation of the nucleosome has demonstrated that all the components of the nucleosome are subject to covalent modifications, which fundamentally alter the structure and function of these basic components of chromatin (Kouzarides 2007). To date there are at least eight different classes of modifications and over 60 distinct modification sites described within the major core histones (Kouzarides 2007). These modifications include lysine acetylation

(Sterner and Berger 2000), lysine and arginine methylation (Bannister et al. 2002), serine and threonine phosphorylation (Nowak and Corces 2004), lysine ubiquitylation (Shilatifard 2006), glutamate poly-ADP ribosylation (Hassa et al. 2006), lysine sumoylation (Nathan et al. 2006), arginine deimination (Cuthbert et al. 2004; Wang et al. 2004) and proline isomerization (Nelson et al. 2006). Until recently, tyrosine phosphorylation of the non-variant histones had not been characterized.

We demonstrated that JAK2 phosphorylates tyrosine 41 in histone H3 (H3Y41ph) both in vitro and in vivo (Dawson et al. 2009). H3Y41 lies at a structurally important region within the nucleosome. It is positioned at the N-terminus of the first helix of H3 (αN1-helix), the region of the nucleosome from which the H3N-terminal tail exits (Luger et al. 1997). At this region the tail of histone H3 makes contact with the nearby DNA. Thus, it seems likely that the region of H3 surrounding Y41, and hence the phosphorylation of this site, will play an important role in nucleosome structure/stability and ultimately perhaps higher order chromatin structure and architecture. Consistent with this suggestion, mutation of Y41 to alanine widens the DNA entry/exit angle of the nucleosome (Ferreira et al. 2007). Moreover, the absolute importance of Y41 has been demonstrated in yeast where a conservative mutation (Y > F) is observed to be lethal (Dai et al. 2008; Nakanishi et al. 2008). Considering these data together it seems likely that phosphorylation of H3Y41 by JAK2 may also be linked to alterations in nucleosomal structure.

In addition to the above structural considerations the region of H3 surrounding Y41 also forms a dynamic binding platform for one or more chromatin associated factors. One such example is Heterochromatin Protein 1 alpha (HP1α) that was recently shown to bind this region of H3 (Dawson et al. 2009; Lavigne et al. 2009). The HP1 family of proteins are relatively small heterochromatin associated proteins. In humans there are three highly similar isoforms, HP1α, HP1β and HP1γ (Fanti and Pimpinelli 2008). Each family member consists of two conserved domains; a chromodomain (CD) and a chromoshadow domain (CSD) separated by a less conserved flexible hinge region. Generally speaking, the three HP1 isoforms have overlapping functions and similar, but not identical, chromosomal locations (Fanti and Pimpinelli 2008; Lomberk et al. 2006). All three isoforms are capable of directly associating with histone H3 di/tri-methylated at K9 (H3K9me2/3) (Bannister et al. 2001; Lachner et al. 2001). This association is mediated via the CD of each protein specifically binding to the methylated lysine 9 within the H3 tail (Bannister et al. 2001; Lachner et al. 2001). Notably, H3K9me2/3 occurs predominantly in heterochromatin and this agrees very well with a high degree of enrichment of HP1 in these regions of the genome. However, HP1 proteins are also present in more euchromatic regions of the genome where they are involved in the regulation of certain genes. Exactly how HP1 proteins are recruited to and maintained at these euchromatic sites is still not fully understood but it almost certainly involves more than just binding to H3K9me2/3. For instance, there is only a partial overlap between HP1 and H3K9me2/3 in these regions indicating multiple recruitment mechanisms (Hediger and Gasser 2006; Li et al. 2002).

As mentioned above, it was recently reported that HP1α directly binds to the region of H3 surrounding Y41 (Dawson et al. 2009; Lavigne et al. 2009). In fact, HP1α uses its CSD to directly bind to this region of H3 (Dawson et al. 2009; Lavigne et al. 2009). Crucially, phosphorylation of H3Y41 by JAK2 disrupts the in vivo association of HP1α with chromatin in haematopoietic cells (HEL cells) and with H3 peptides in vitro (Dawson et al. 2009). Thus, there exists a dynamic and regulatable binding of HP1 to chromatin that may be independent of H3K9 methylation. HP1α is known to reduce mitotic recombination (Cummings et al. 2007), repress gene transcription (Panteleeva et al. 2007) and preserve centromeric architecture leading to the faithful segregation of sister chromatids (Yamagishi et al. 2008). It is therefore tempting to speculate that reversal of these functions via phosphorylation of H3Y41 may help explain, at least in part, the phenotypic consequences of constitutive JAK2 activation in haematological malignancies; increased gene expression, mitotic recombination and genetic instability (Campbell and Green 2006; Levine et al. 2007; Plo et al. 2008) (Fig. 1). In accordance with this suggestion, the phospho-switch regulation of HP1α binding to H3Y41 was found to occur at the *lmo2* gene, an essential gene for normal haematopoietic development and one that has been implicated in leukaemogenesis (McCormack and Rabbitts 2004). Specific small molecule inhibition of endogenous JAK2 resulted in an enrichment of HP1α within the *lmo2* promoter as well as inhibition of transcription. Thus, this represents a clear example of nuclear JAK2 directly regulating gene transcription (of an important oncogene) via phosphorylation of H3Y41 at the chromatin interface. These data have

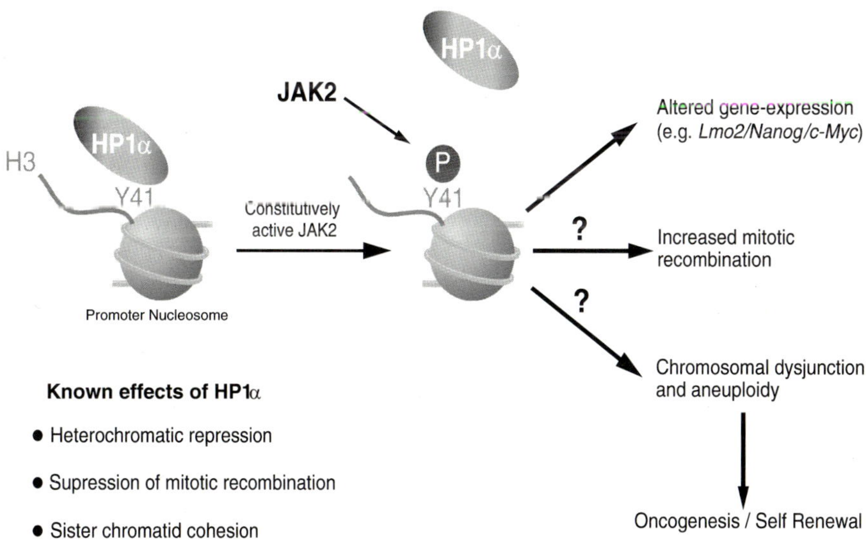

Fig. 1 Nuclear JAK2 phosphorylates histone H3Y41 and prevents the binding of heterochromatin protein 1α (HP1α) to a novel binding site on chromatin. The exclusion of HP1α at the promoters of genes is associated with increased gene expression and may account for other oncogenic phenomena

been independently corroborated by Rinaldi et al. who demonstrated that JAK2V617F increases the expression of *lmo2* in both K562 cells and primary $CD34^+$ haematopoietic stem cells (HSC) (Rinaldi et al. 2010).

The regulation of HP1α binding to the region surrounding H3Y41 by JAK2 may play an important role in several diverse tissues apart from erythroid-leukaemia cells. Evidence in support of this was recently provided by Griffiths et al. who demonstrated that JAK2V617F can completely substitute for leukaemia inhibitory factor (LIF) and Bone Morphogenetic Protein (BMP) signalling in embryonic stem (ES) cells (Griffiths et al. 2011). Murine ES cells are derived from the inner cell mass of the developing mouse blastocyst and they can be maintained in culture indefinitely, while retaining the ability to differentiate into all somatic cell types. Embryonic stem cells are commonly isolated and maintained using a combination of LIF and foetal calf serum (FCS) (Moreau et al. 1988; Smith et al. 1988; Williams et al. 1988). LIF signals via JAK kinases and involves activation of STAT3 (Niwa et al. 1998), which is essential for LIF dependent ES cell self-renewal (Raz et al. 1999). FCS can be replaced by the addition of Bone Morphogenetic Protein (BMP) thus permitting ES cell culture in chemically defined conditions (Ying et al. 2003). More recently it has been demonstrated that two small molecule inhibitors of ERK and GSK3 kinase pathways, referred to as 2i growth conditions, can replace both LIF and BMP (Ying et al. 2008). JAK signalling therefore controls the balance between self-renewal and differentiation of both hematopoietic stem cells (HSC) and ES cells.

Griffiths et al. demonstrated that ES cells expressing the human JAK2V617F mutant kinase remained pluripotent and could self-renew indefinitely in chemically defined conditions without cytokines or small molecule inhibitors (Griffiths et al. 2011). Surprisingly, this phenotype was independent of JAK2 activation of STAT3 or PI3K. Gene-expression analysis showed that ES cells expressing JAK2V617F demonstrated higher levels of the pluripotency regulator *nanog* when compared to wild-type ES cells. *Nanog* was shown to be important for the factor independent growth of JAK2V617F containing ES cells and significantly alteration of JAK2 activity concomitantly altered the expression of *nanog*. Finally, these authors demonstrated that JAK2V617F phosphorylates H3Y41 at nucleosomes surrounding the *nanog* promoter and this phosphorylation dynamically regulates the binding of the HP1α at this site.

Notwithstanding the aforementioned studies, an unresolved question concerning the interplay of JAK2 and HP1α at the chromatin interface is what role, if any does H3K9 methylation play? The importance of understanding the molecular mechanisms governing the binding of HP1 to chromatin was further highlighted by findings from Louis Staudt's laboratory (Rui et al. 2010). These investigators adopted a functional approach to dissecting the molecular pathogenesis of Hodgkin's Lymphoma (HL) and Primary Mediastinal B-Cell Lymphoma (PMBL). Both these diseases frequently demonstrate an amplification of chromosome 9p and these investigators demonstrated using a library of shRNAs that three genes (JAK2, JMJD2C and RANBP6) within this amplicon are primarily responsible for driving the disease. JMJD2C is a histone lysine demethylase that catalyses the removal of methyl groups from lysine 9 on histone H3 (H3K9me2/3)

(Cloos et al. 2006; Loh et al. 2007; Whetstine et al. 2006; Wissmann et al. 2007). As discussed above, H3K9me2/3 serves as a binding site for HP1. The concurrent amplification of JAK2 within these tumours further undermines HP1 binding to chromatin by enhancing H3Y41ph. Indeed, these authors demonstrated that this dual destabilisation of HP1 binding to chromatin results in the increased expression of the potent oncogene *c-myc*, which undoubtedly contributes to malignant transformation in both HL and PMBL (Rui et al. 2010). The results of this study suggest that perturbations of either binding site for HP1 (H3K9me2/3 or the region surrounding H3Y41) may have disastrous consequences and result in malignant transformation. However, what remains unresolved is how HP1 utilises these two distinct binding platforms.

Since HP1 binds to H3K9me via its CD, and to H3Y41 via its CSD, it is possible that both binding events may occur at the same genomic location. If so, a single molecule of HP1 could bind to H3K9 methylation and to H3Y41 on the same H3 tail. On the other hand, one molecule of HP1 could contact two separate H3 tails, perhaps on the same or adjacent nucleosomes; via CD binding to H3K9me on one tail and CSD binding to H3Y41 on the other tail. Alternatively, H3K9me and H3Y41 may be separate binding activities, both in terms of functionality and localization. Whatever the mechanism, it is clear that HP1 is more than just a heterochromatic scaffold protein and that it functions with JAK2 in order to regulate critically important processes such as the transcription of oncogenes and pluripotency factors (Fig. 1).

The identification of *lmo2, c-myc* and *nanog* as JAK2-regulated genes and JAK2's link with HP1α provided a significant insight into the mechanism(s) by which JAK2 may initiate transcriptional programmes. In spite of this, a significant question remains concerning how JAK2 is selectively recruited to specific genomic loci. This may occur via a number of potential mechanisms but perhaps the most likely is via a direct association with a DNA bound transcription factor. In this case, an interaction between JAK2 and a STAT transcription factor seems an attractive hypothesis since JAK2 is known to associate with this family of transcription factors, at least in the cytoplasm. However, both *nanog* and *lmo2* gene do not appear to be regulated by STAT transcription factors leaving unanswered the question concerning JAK2 recruitment to these loci.

Recently, great advances have been made in combining chromatin research with genome wide technologies. It is now possible to couple chromatin immunoprecipitation (the selective immuno-enrichment of specific chromatin fragments via antibody recognition) with massively parallel DNA sequencing of the associated DNA. In this way, one can determine the genomic distribution of any chromatin-associated factor, or indeed histone modification as long as an appropriate antibody is available. By employing this approach we determined the genome-wide distribution of H3Y41ph using chromatin prepared from HEL cells (Dawson et al., unpublished). This analysis generated several interesting and surprising observations. The results indicate that tyrosine phosphorylation behaves differently to other well-characterised histone modifications such as lysine acetylation and methylation. Acetylation and methylation are often associated with active genes, with no apparent preference for a specific pathway. In contrast, H3Y41ph marks a specific set of genes stimulated by a specific signalling pathway. Furthermore, H3Y41ph blankets a set of key haematopoietic genes, a number of which have been implicated in neoplastic transformation. This indicates that H3Y41ph represents a new class of histone modification defining both a signaling pathway and a tissue-specific gene expression programme. The genome-wide H3Y41ph analysis also uncovered a previously unrecognized nuclear component of the JAK/STAT signaling pathway and hinted at the possibility that the STAT family of transcription factors may aid in targeting JAK2 to chromatin.

A key question arising from the aforementioned studies is how is nuclear JAK2 regulated? Certainly, one or more cytokine-dependent pathways are involved since cytokine activation of cells expressing only wild-type JAK2 leads to a global increase in H3Y41ph (Dawson et al. 2009) and also an enrichment of H3Y41ph at the promoters of responsive genes (Dawson et al., unpublished). These data indicate that cytokines lead to the presence of active nuclear JAK2. This response could be achieved by at least two, non-mutually exclusive mechanisms. It may be that cytokine-mediated activation of the JAK/STAT pathway initiates a signal transduction pathway that extends in to the nucleus where it culminates in the activation of a nuclear pool of JAK2. Future work will be needed to investigate this intriguing possibility. Alternatively, activation of cytoplasmic JAK2 via the canonical pathway may in some instances induce its nuclear translocation. Evidence supporting the latter suggestion comes from work analysing the distribution of JAK2V617F in the erythroleukaemic K562 cell line. In this cell line JAK2V617F is predominantly nuclear, but inhibition of the kinase by a small molecule inhibitor (AG490) induces a cytoplasmic localization (Rinaldi et al. 2010). This indicates that JAK2 needs to be in an activated form in order to enter the nucleus, which is indicative of a tightly regulated import mechanism. If so, it is possible that blocking this import mechanism in patients with mutated JAK2 may provide a new opportunity for therapeutic intervention.

Future Directions

The most surprising but informative finding from the genome-wide analyses is the fact that JAK2-STAT5 signalling occurs right at the chromatin template (Dawson et al., unpublished). What remains unclear from the work discussed above is whether there are two distinct pools of cellular JAK2, one within the cytoplasm and one within the nucleus, or whether JAK2 is actively shuttled between these cellular compartments. Furthermore, it is also unclear whether JAK2 differentially phosphorylates STAT5 within the nucleus and the cytoplasm or whether these observations are simply an extension of the well described canonical cytoplasmic pathway. If the latter is proven to be the case, an attractive hypothesis given the striking correlation between STAT5 binding and H3Y41ph is that STAT5 serves as the chaperone and targeting module for JAK2's activity at chromatin. A physical interaction between JAK2 and STAT5 is well established. Several studies using different methodologies such as yeast-two-hybrid screens, co-immunoprecipitation and in vitro binding assays with the recombinant proteins have clearly demonstrated that JAK2 and STAT5 physically interact (Barahmand-Pour et al. 1998; Fujitani et al. 1997). This physical interaction was most elegantly defined by Thomas Decker's laboratory when they introduced mammalian JAK2 and STAT5 into yeast, which lack orthologues to either of these proteins or indeed any components of the canonical JAK-STAT pathway (Barahmand-Pour et al. 1998). Here, they were able to demonstrate that the SH2 domain of STAT5 and the JH1 domain of JAK2 mediate the principal interaction between these proteins. A further ill-defined association between the proteins that was independent of the SH2 and JH1 domains was also noted (Barahmand-Pour et al. 1998).

This physical interaction between JAK2 and STAT5 raises the possibility that STAT5 chaperones JAK2 into the nucleus of cells and targets its kinase activity at chromatin through its high-affinity interaction with its sequence specific elements on DNA. Alternatively, STAT5 may bind and target only the nuclear fraction of JAK2. A third but equally plausible scenario is that JAK2 and STAT5 are independently targeted to their locations by a different factor(s). Clearly, delineating the role of STAT5 in targeting the enzymatic activity of JAK2 at chromatin is an important area for future work. However, dissecting this problem will be a challenging process for many reasons. Firstly, there are four JAKs and seven mammalian STAT family members and several of these have already been demonstrated to have varying degrees of functional redundancy (Benekli et al. 2003; Parganas et al. 1998; Teglund et al. 1998; Yamaoka et al. 2004). Moreover, the subtotal eradication of STAT5 in assessing its contribution to JAK2 targeting is unlikely to be informative as evidenced by the original description of the STAT5A/B double knock out mice. Analysis of these mice surprisingly demonstrated that they possessed a virtually normal haematopoietic system and a practically normal response to cytokines that exclusively signal via JAK2 such as erythropoietin (Teglund et al. 1998). These mice have subsequently been shown to still express low amounts of N-terminally truncated forms of both STAT5A/B (Dolznig et al. 2006; Hoelbl et al. 2006; Moriggl et al. 2005; Yao et al. 2006), which are still capable of activating several target genes, highlighting the difficulties of dissecting the pathway in cells that contain even low levels of functional STAT5. True double knock out STAT5A/B mice have subsequently been generated. However, these mice suffer from severe anaemia and die perinatally which largely precludes the assessment of STAT5A/B in normal haematopoiesis in vivo (Cui et al. 2004). The direct assessment of the function of STAT5A/B in haematopoietic cells will require a clean genetic system such as the conditional knockout of these transcription factors in specified haematopoietic lineages using the Cre/Lox system. A system such as this will be necessary to accurately delineate the role of STAT5, if any, in chaperoning and/or targeting JAK2 to chromatin. It will also allow the phenotypic assessment of the consequence of impaired STAT5 function in communicating the intracellular signals of lineage specific cytokines such as erythropoietin and thrombopoietin that signal exclusively via JAK2.

Negative Regulation of JAK2-STAT5 Within the Nucleus

The findings that JAK2 has a critical role as a chromatin modifying enzyme raises important questions relating to the negative regulation of this essential enzyme within the nucleus. To date much of what we know about the negative regulation of the Janus kinases and indeed the JAK/STAT pathway has focused on events within the cytoplasm of cells. The three main classes of negative regulators of this pathway include phosphatases such as SHP1/2, the SOCS and PIAS family of proteins, each which have varied roles in dampening down the JAK-STAT mediated cytokine response (Wormald and Hilton 2004). Whilst some members of the PIAS family have been well described to have nuclear functions (Sharrocks 2006), until recently the SHP phosphatases and SOCS family were largely thought to be limited to the cytoplasm. However, several recent reports have described an unexpected nuclear localisation for several of these proteins. Both SHP1 and SHP2 have recently been described to have a nuclear localisation (Chughtai et al. 2002; Craggs and Kellie 2001). Interestingly, SHP2 was noted to be co-localised with STAT5 within the nucleus (Chughtai et al. 2002). Similarly, several members of the SOCS family have also been described to be unexpectedly present within the nucleus of different tissues (Koelsche et al. 2009; Lee et al. 2008). These findings raise the intriguing possibility that the negative regulation of the JAK-STAT pathway within the nucleus may be similar to what has previously been described within the cytoplasm of cells.

Related to this issue is the role, if any, of HP1α in regulating the expression of canonical JAK2-STAT5 target genes. It is interesting that recent work by Willis Li's laboratory has demonstrated that STAT92E (the sole STAT member in *D. melanogaster*) physically interacts with and stabilises the binding of HP1 to chromatin (Shi et al. 2008; Yan et al. 2010). More specifically, these authors initially demonstrated that depleting STAT92E unexpectedly induced a marked de-repression of heterochromatin. Conversely, elevated STAT92E levels lead to a stabilisation of heterochromatin. Both of these findings were dependent on HP1 expression as concurrently increasing HP1 was able to counteract the effects of reducing STAT92E. These findings appeared somewhat contradictory to those initially reported by this group where increased JAK signalling had led to heterochromatic de-repression. These seemingly incongruous findings were explained by the fact that unphosphorylated STAT92E physically binds to HP1 and that this binding is abrogated once STAT92E becomes phosphorylated by JAK. The model postulated by these authors suggests that unphosphorylated STAT binds HP1 at chromatin to stabilise its association and complement its function. Then, following phosphorylation by a tyrosine kinase, STAT no longer associates with HP1 and HP1 binding to chromatin is de-stabilised.

These findings raise the interesting possibility that unphosphorylated STAT5 may similarly associate with HP1 and contribute to its chromatin association and function in mammalian cells. Aspects of the genome-wide data discussed above (Dawson et al., unpublished), are complimentary to those reported in *D. melanogaster*; our results suggest that if unphosphorylated STAT5 is indeed

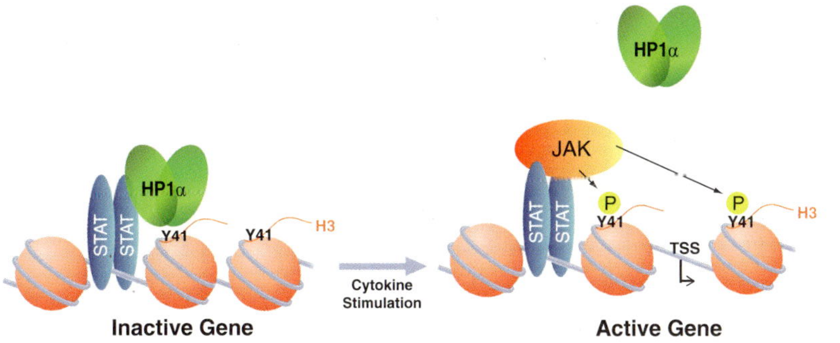

Fig. 2 Hypothetical model of transcriptional regulation at chromatin. Un-phosphorylated STAT proteins bind HP1α and co-operate in transcriptional repression. Following cytokine stimulation, activated JAKs phosphorylate STAT proteins increasing their nuclear import and binding to DNA. Phosphorylated STATs physically associate with activated JAKs and aid in targeting them to chromatin. At chromatin activated JAKs phosphorylate both STAT proteins and histone H3 at tyrosine 41 (H3Y41) resulting in the displacement of HP1α and consequently enhancing transcription

associated with HP1α then phosphorylation of both STAT5 and H3Y41 is required for the displacement of HP1α from chromatin (Fig. 2). This possibility warrants further investigation. It would be interesting to know if any or all of the mammalian STAT family members physically interact with the HP1 family of proteins. In addition, it will be important to determine if any of the HP1 proteins are present and regulated at the canonical JAK2-STAT5 genes.

JAK-STAT Signalling at Chromatin in Development and Oncogenesis

The finding that JAK2 signalling is present at chromatin raises several important questions for the field relating to the wider applicability of this discovery to both normal and malignant processes. A key question is whether these results are peculiar to JAK2 or are they more pervasive. Understanding if all the JAK family members are capable of entering the nucleus of cells to phosphorylate H3Y41 is an essential step in further defining the broader application of this research. Similarly, it is equally important to understand if all the STAT family members communicate and cooperate with their cognate JAK members within the nucleus and how H3Y41ph along with the binding sites of the STAT proteins regulate transcription.

As discussed previously, the JAK-STAT pathway is critical for haematopoiesis, adipogenesis, immune and mammary development (O'Shea et al. 2002). What is currently unknown however is what role, if any, H3Y41ph plays in controlling the transcriptional programmes required for the normal development of these tissues. An attractive system in which to investigate this is the haematopoietic system. Haematopoietic stem cells (HSC) are the best-characterised tissue-specific stem cells and the developmental hierarchy of haematopoiesis has been extensively studied. These, therefore, serve as an excellent blueprint for understanding the molecular mechanisms of self-renewal and differentiation in both a developmental and malignant context. It would be interesting to identify the genes marked by H3Y41ph in HSC and temporally follow the distribution of this modification as these cells subsequently mature to make cell fate choices and differentiate into the various terminally differentiated cells that constitute the haematopoietic system. Linking these findings with functional activity of the various JAK family members and the genomic localisation of the various STAT family members is also of interest. These findings are likely to be particularly informative when addressing the issue of constitutive JAK-STAT signalling and the mechanisms by which this results in oncogenesis. Delineating the key genes marked by both H3Y41ph and STAT may provide better insight in to the crucial transcription targets for this pathway in driving tumorigenesis.

It should be noted that the JAK-STAT pathway is not only important in haematopoietic development and neoplasia. Indeed, recent evidence would suggest that over-activation of this pathway is prevalent in a wide variety of human cancers, including breast, prostate, head and neck, and ovarian cancers, among other solid tumors (Bromberg et al. 1999; Catlett-Falcone et al. 1999; Dhir et al. 2002; Garcia et al. 2001; Grandis et al. 2000; Yu and Jove 2004). These findings have underpinned the investigation of specific JAK2 inhibitors as a novel therapeutic measure to treat these various cancers. Whilst data from human phase 2/3 clinical trial are pending, a number of these JAK2 inhibitors have demonstrated promising results in murine models of both haematological and solid malignancies (Dawson et al. 2010; Geron et al. 2008; Hedvat et al. 2009; Pardanani 2008).

References

Arbouzova NI, Zeidler MP (2006) JAK/STAT signalling in Drosophila: insights into conserved regulatory and cellular functions. Development 133:2605–2616

Bannister AJ, Zegerman P, Partridge JF, Miska EA, Thomas JO, Allshire RC, Kouzarides T (2001) Selective recognition of methylated lysine 9 on histone H3 by the HP1 chromo domain. Nature 410:120–124

Bannister AJ, Schneider R, Kouzarides T (2002) Histone methylation: dynamic or static? Cell 109:801–806

Barahmand-Pour F, Meinke A, Groner B, Decker T (1998) Jak2-Stat5 interactions analyzed in yeast. J Biol Chem 273:12567–12575

Baxter EJ, Scott LM, Campbell PJ, East C, Fourouclas N, Swanton S, Vassiliou GS, Bench AJ, Boyd EM, Curtin N et al (2005) Acquired mutation of the tyrosine kinase JAK2 in human myeloproliferative disorders. Lancet 365:1054–1061

Benekli M, Baer MR, Baumann H, Wetzler M (2003) Signal transducer and activator of transcription proteins in leukemias. Blood 101:2940–2954

Bercovich D, Ganmore I, Scott LM, Wainreb G, Birger Y, Elimelech A, Shochat C, Cazzaniga G, Biondi A, Basso G et al (2008) Mutations of JAK2 in acute lymphoblastic leukaemias associated with Down's syndrome. Lancet 372(9648):1484–1492

Bromberg JF, Wrzeszczynska MH, Devgan G, Zhao Y, Pestell RG, Albanese C, Darnell JE Jr (1999) Stat3 as an oncogene. Cell 98:295–303

Campbell PJ, Green AR (2006) The myeloproliferative disorders. N Engl J Med 355:2452–2466
Catlett-Falcone R, Landowski TH, Oshiro MM, Turkson J, Levitzki A, Savino R, Ciliberto G, Moscinski L, Fernandez-Luna JL, Nunez G et al (1999) Constitutive activation of Stat3 signaling confers resistance to apoptosis in human U266 myeloma cells. Immunity 10:105–115
Chughtai N, Schimchowitsch S, Lebrun JJ, Ali S (2002) Prolactin induces SHP-2 association with Stat5, nuclear translocation, and binding to the beta-casein gene promoter in mammary cells. J Biol Chem 277:31107–31114
Cloos PA, Christensen J, Agger K, Maiolica A, Rappsilber J, Antal T, Hansen KH, Helin K (2006) The putative oncogene GASC1 demethylates tri- and dimethylated lysine 9 on histone H3. Nature 442:307–311
Craggs G, Kellie S (2001) A functional nuclear localization sequence in the C-terminal domain of SHP-1. J Biol Chem 276:23719–23725
Cui Y, Riedlinger G, Miyoshi K, Tang W, Li C, Deng CX, Robinson GW, Hennighausen L (2004) Inactivation of Stat5 in mouse mammary epithelium during pregnancy reveals distinct functions in cell proliferation, survival, and differentiation. Mol Cell Biol 24:8037–8047
Cummings WJ, Yabuki M, Ordinario EC, Bednarski DW, Quay S, Maizels N (2007) Chromatin structure regulates gene conversion. PLoS Biol 5:e246
Cuthbert GL, Daujat S, Snowden AW, Erdjument-Bromage H, Hagiwara T, Yamada M, Schneider R, Gregory PD, Tempst P, Bannister AJ et al (2004) Histone deimination antagonizes arginine methylation. Cell 118:545–553
Dai J, Hyland EM, Yuan DS, Huang H, Bader JS, Boeke JD (2008) Probing nucleosome function: a highly versatile library of synthetic histone H3 and H4 mutants. Cell 134:1066–1078
Dawson MA, Bannister AJ, Gottgens B, Foster SD, Bartke T, Green AR, Kouzarides T (2009) JAK2 phosphorylates histone H3Y41 and excludes HP1alpha from chromatin. Nature 461:819–822
Dawson MA, Curry JE, Barber K, Beer PA, Graham B, Lyons JF, Richardson CJ, Scott MA, Smyth T, Squires MS et al (2010) AT9283, a potent inhibitor of the Aurora kinases and Jak2, has therapeutic potential in myeloproliferative disorders. Br J Haematol 150:46–57
Dhir R, Ni Z, Lou W, DeMiguel F, Grandis JR, Gao AC (2002) Stat3 activation in prostatic carcinomas. Prostate 51:241–246
Dolznig H, Grebien F, Deiner EM, Stangl K, Kolbus A, Habermann B, Kerenyi MA, Kieslinger M, Moriggl R, Beug H et al (2006) Erythroid progenitor renewal versus differentiation: genetic evidence for cell autonomous, essential functions of EpoR, Stat5 and the GR. Oncogene 25:2890–2900
Fanti L, Pimpinelli S (2008) HP1: a functionally multifaceted protein. Curr Opin Genet Dev 18:169–174
Ferreira H, Somers J, Webster R, Flaus A, Owen-Hughes T (2007) Histone tails and the H3 alphaN helix regulate nucleosome mobility and stability. Mol Cell Biol 27:4037–4048
Firmbach-Kraft I, Byers M, Shows T, Dalla-Favera R, Krolewski JJ (1990) tyk2, prototype of a novel class of non-receptor tyrosine kinase genes. Oncogene 5:1329–1336
Fu XY, Schindler C, Improta T, Aebersold R, Darnell JE Jr (1992) The proteins of ISGF-3, the interferon alpha-induced transcriptional activator, define a gene family involved in signal transduction. Proc Natl Acad Sci USA 89:7840–7843
Fujitani Y, Hibi M, Fukada T, Takahashi-Tezuka M, Yoshida H, Yamaguchi T, Sugiyama K, Yamanaka Y, Nakajima K, Hirano T (1997) An alternative pathway for STAT activation that is mediated by the direct interaction between JAK and STAT. Oncogene 14:751–761
Garcia R, Bowman TL, Niu G, Yu H, Minton S, Muro-Cacho CA, Cox CE, Falcone R, Fairclough R, Parsons S et al (2001) Constitutive activation of Stat3 by the Src and JAK tyrosine kinases participates in growth regulation of human breast carcinoma cells. Oncogene 20:2499–2513
Geron I, Abrahamsson AE, Barroga CF, Kavalerchik E, Gotlib J, Hood JD, Durocher J, Mak CC, Noronha G, Soll RM et al (2008) Selective inhibition of JAK2-driven erythroid differentiation of polycythemia vera progenitors. Cancer Cell 13:321–330

Gough NM, Gough J, Metcalf D, Kelso A, Grail D, Nicola NA, Burgess AW, Dunn AR (1984) Molecular cloning of cDNA encoding a murine haematopoietic growth regulator, granulocyte-macrophage colony stimulating factor. Nature 309:763–767

Grandis JR, Drenning SD, Zeng Q, Watkins SC, Melhem MF, Endo S, Johnson DE, Huang L, He Y, Kim JD (2000) Constitutive activation of Stat3 signaling abrogates apoptosis in squamous cell carcinogenesis in vivo. Proc Natl Acad Sci USA 97:4227–4232

Griffiths DS, Li J, Dawson MA, Trotter MW, Cheng YH, Smith AM, Mansfield W, Liu P, Kouzarides T, Nichols J et al (2011) LIF-independent JAK signalling to chromatin in embryonic stem cells uncovered from an adult stem cell disease. Nat Cell Biol 13:13–21

Harrison DA, Binari R, Nahreini TS, Gilman M, Perrimon N (1995) Activation of a Drosophila Janus kinase (JAK) causes hematopoietic neoplasia and developmental defects. EMBO J 14:2857–2865

Hassa PO, Haenni SS, Elser M, Hottiger MO (2006) Nuclear ADP-ribosylation reactions in mammalian cells: where are we today and where are we going? Microbiol Mol Biol Rev 70:789–829

Hediger F, Gasser SM (2006) Heterochromatin protein 1: don't judge the book by its cover! Curr Opin Genet Dev 16:143–150

Hedvat M, Huszar D, Herrmann A, Gozgit JM, Schroeder A, Sheehy A, Buettner R, Proia D, Kowolik CM, Xin H et al (2009) The JAK2 inhibitor AZD1480 potently blocks Stat3 signaling and oncogenesis in solid tumors. Cancer Cell 16:487–497

Helmer RA, Panchoo M, Dertien JS, Bhakta SM, Hewetson A, Chilton BS (2010) Prolactin-induced Jak2 phosphorylation of RUSH: a key element in Jak/RUSH signaling. Mol Cell Endocrinol 325:143–149

Hewetson A, Hendrix EC, Mansharamani M, Lee VH, Chilton BS (2002) Identification of the RUSH consensus-binding site by cyclic amplification and selection of targets: demonstration that RUSH mediates the ability of prolactin to augment progesterone-dependent gene expression. Mol Endocrinol 16:2101–2112

Hoelbl A, Kovacic B, Kerenyi MA, Simma O, Warsch W, Cui Y, Beug H, Hennighausen L, Moriggl R, Sexl V (2006) Clarifying the role of Stat5 in lymphoid development and Abelson-induced transformation. Blood 107:4898–4906

Hombria JC, Brown S (2002) The fertile field of Drosophila Jak/STAT signalling. Curr Biol 12: R569–R575

Ito M, Nakasato M, Suzuki T, Sakai S, Nagata M, Aoki F (2004) Localization of janus kinase 2 to the nuclei of mature oocytes and early cleavage stage mouse embryos. Biol Reprod 71:89–96

James C, Ugo V, Le Couedic JP, Staerk J, Delhommeau F, Lacout C, Garcon L, Raslova H, Berger R, Bennaceur-Griscelli A et al (2005) A unique clonal JAK2 mutation leading to constitutive signalling causes polycythaemia vera. Nature 434:1144–1148

Kamakura S, Oishi K, Yoshimatsu T, Nakafuku M, Masuyama N, Gotoh Y (2004) Hes binding to STAT3 mediates crosstalk between Notch and JAK-STAT signalling. Nat Cell Biol 6:547–554

Karaghiosoff M, Neubauer H, Lassnig C, Kovarik P, Schindler H, Pircher H, McCoy B, Bogdan C, Decker T, Brem G et al (2000) Partial impairment of cytokine responses in Tyk2-deficient mice. Immunity 13:549–560

Koelsche C, Strebovsky J, Baetz A, Dalpke AH (2009) Structural and functional analysis of a nuclear localization signal in SOCS1. Mol Immunol 46:2474–2480

Kornberg RD, Lorch Y (1999) Twenty-five years of the nucleosome, fundamental particle of the eukaryote chromosome. Cell 98:285–294

Kouzarides T (2007) Chromatin modifications and their function. Cell 128:693–705

Kralovics R, Passamonti F, Buser AS, Teo SS, Tiedt R, Passweg JR, Tichelli A, Cazzola M, Skoda RC (2005) A gain-of-function mutation of JAK2 in myeloproliferative disorders. N Engl J Med 352:1779–1790

Lachner M, O'Carroll D, Rea S, Mechtler K, Jenuwein T (2001) Methylation of histone H3 lysine 9 creates a binding site for HP1 proteins. Nature 410:116–120

Lacronique V, Boureux A, Valle VD, Poirel H, Quang CT, Mauchauffe M, Berthou C, Lessard M, Berger R, Ghysdael J et al (1997) A TEL-JAK2 fusion protein with constitutive kinase activity in human leukemia. Science 278:1309–1312

Lavigne M, Eskeland R, Azebi S, Saint-Andre V, Jang SM, Batsche E, Fan HY, Kingston RE, Imhof A, Muchardt C (2009) Interaction of HP1 and Brg1/Brm with the globular domain of histone H3 is required for HP1-mediated repression. PLoS Genet 5:e1000769

Lee KH, Moon KJ, Kim HS, Yoo BC, Park S, Lee H, Kwon S, Lee ES, Yoon S (2008) Increased cytoplasmic levels of CIS, SOCS1, SOCS2, or SOCS3 are required for nuclear translocation. FEBS Lett 582:2319–2324

Levine RL, Wadleigh M, Cools J, Ebert BL, Wernig G, Huntly BJ, Boggon TJ, Wlodarska I, Clark JJ, Moore S et al (2005) Activating mutation in the tyrosine kinase JAK2 in polycythemia vera, essential thrombocythemia, and myeloid metaplasia with myelofibrosis. Cancer Cell 7:387–397

Levine RL, Pardanani A, Tefferi A, Gilliland DG (2007) Role of JAK2 in the pathogenesis and therapy of myeloproliferative disorders. Nat Rev Cancer 7:673–683

Li Y, Kirschmann DA, Wallrath LL (2002) Does heterochromatin protein 1 always follow code? Proc Natl Acad Sci USA 99(Suppl 4):16462–16469

Lim CP, Cao X (2006) Structure, function, and regulation of STAT proteins. Mol Biosyst 2:536–550

Lobie PE, Ronsin B, Silvennoinen O, Haldosen LA, Norstedt G, Morel G (1996) Constitutive nuclear localization of Janus kinases 1 and 2. Endocrinology 137:4037–4045

Loh YH, Zhang W, Chen X, George J, Ng HH (2007) Jmjd1a and Jmjd2c histone H3 Lys 9 demethylases regulate self-renewal in embryonic stem cells. Genes Dev 21:2545–2557

Lomberk G, Wallrath L, Urrutia R (2006) The heterochromatin protein 1 family. Genome Biol 7:228

Luger K, Mader AW, Richmond RK, Sargent DF, Richmond TJ (1997) Crystal structure of the nucleosome core particle at 2.8 A resolution. Nature 389:251–260

Luo H, Hanratty WP, Dearolf CR (1995) An amino acid substitution in the Drosophila hopTum-l Jak kinase causes leukemia-like hematopoietic defects. EMBO J 14:1412–1420

Luo H, Rose P, Barber D, Hanratty WP, Lee S, Roberts TM, D'Andrea AD, Dearolf CR (1997) Mutation in the Jak kinase JH2 domain hyperactivates Drosophila and mammalian Jak-Stat pathways. Mol Cell Biol 17:1562–1571

McCormack MP, Rabbitts TH (2004) Activation of the T-cell oncogene LMO2 after gene therapy for X-linked severe combined immunodeficiency. N Engl J Med 350:913–922

Moreau JF, Donaldson DD, Bennett F, Witek-Giannotti J, Clark SC, Wong GG (1988) Leukaemia inhibitory factor is identical to the myeloid growth factor human interleukin for DA cells. Nature 336:690–692

Moriggl R, Sexl V, Kenner L, Duntsch C, Stangl K, Gingras S, Hoffmeyer A, Bauer A, Piekorz R, Wang D et al (2005) Stat5 tetramer formation is associated with leukemogenesis. Cancer Cell 7:87–99

Mullighan CG, Zhang J, Harvey RC, Collins-Underwood JR, Schulman BA, Phillips LA, Tasian SK, Loh ML, Su X, Liu W et al (2009) JAK mutations in high-risk childhood acute lymphoblastic leukemia. Proc Natl Acad Sci USA 106:9414–9418

Musso T, Johnston JA, Linnekin D, Varesio L, Rowe TK, O'Shea JJ, McVicar DW (1995) Regulation of JAK3 expression in human monocytes: phosphorylation in response to interleukins 2, 4, and 7. J Exp Med 181:1425–1431

Nakanishi S, Sanderson BW, Delventhal KM, Bradford WD, Staehling-Hampton K, Shilatifard A (2008) A comprehensive library of histone mutants identifies nucleosomal residues required for H3K4 methylation. Nat Struct Mol Biol 15(8):881–888

Nathan D, Ingvarsdottir K, Sterner DE, Bylebyl GR, Dokmanovic M, Dorsey JA, Whelan KA, Krsmanovic M, Lane WS, Meluh PB et al (2006) Histone sumoylation is a negative regulator in *Saccharomyces cerevisiae* and shows dynamic interplay with positive-acting histone modifications. Genes Dev 20:966–976

Nelson CJ, Santos-Rosa H, Kouzarides T (2006) Proline isomerization of histone H3 regulates lysine methylation and gene expression. Cell 126:905–916
Neubauer H, Cumano A, Muller M, Wu H, Huffstadt U, Pfeffer K (1998) Jak2 deficiency defines an essential developmental checkpoint in definitive hematopoiesis. Cell 93:397–409
Nicola NA, Metcalf D (1984) Binding of the differentiation-inducer, granulocyte-colony-stimulating factor, to responsive but not unresponsive leukemic cell lines. Proc Natl Acad Sci USA 81:3765–3769
Nilsson J, Bjursell G, Kannius-Janson M (2006) Nuclear Jak2 and transcription factor NF1-C2: a novel mechanism of prolactin signaling in mammary epithelial cells. Mol Cell Biol 26:5663–5674
Nilsson J, Helou K, Kovacs A, Bendahl PO, Bjursell G, Ferno M, Carlsson P, Kannius-Janson M (2010) Nuclear Janus-activated kinase 2/nuclear factor 1-C2 suppresses tumorigenesis and epithelial-to-mesenchymal transition by repressing Forkhead box F1. Cancer Res 70:2020–2029
Niwa H, Burdon T, Chambers I, Smith A (1998) Self-renewal of pluripotent embryonic stem cells is mediated via activation of STAT3. Genes Dev 12:2048–2060
Nosaka T, van Deursen JM, Tripp RA, Thierfelder WE, Witthuhn BA, McMickle AP, Doherty PC, Grosveld GC, Ihle JN (1995) Defective lymphoid development in mice lacking Jak3. Science 270:800–802
Nowak SJ, Corces VG (2004) Phosphorylation of histone H3: a balancing act between chromosome condensation and transcriptional activation. Trends Genet 20:214–220
O'Shea JJ, Gadina M, Schreiber RD (2002) Cytokine signaling in 2002: new surprises in the Jak/Stat pathway. Cell 109(Suppl):S121–S131
Panteleeva I, Boutillier S, See V, Spiller DG, Rouaux C, Almouzni G, Bailly D, Maison C, Lai HC, Loeffler JP et al (2007) HP1alpha guides neuronal fate by timing E2F-targeted genes silencing during terminal differentiation. EMBO J 26:3616–3628
Pardanani A (2008) JAK2 inhibitor therapy in myeloproliferative disorders: rationale, preclinical studies and ongoing clinical trials. Leukemia 22:23–30
Parganas E, Wang D, Stravopodis D, Topham DJ, Marine JC, Teglund S, Vanin EF, Bodner S, Colamonici OR, van Deursen JM et al (1998) Jak2 is essential for signaling through a variety of cytokine receptors. Cell 93:385–395
Park SY, Saijo K, Takahashi T, Osawa M, Arase H, Hirayama N, Miyake K, Nakauchi H, Shirasawa T, Saito T (1995) Developmental defects of lymphoid cells in Jak3 kinase-deficient mice. Immunity 3:771–782
Plo I, Nakatake M, Malivert L, de Villartay JP, Giraudier S, Villeval JL, Wiesmuller L, Vainchenker W (2008) JAK2 stimulates homologous recombination and genetic instability: potential implication in the heterogeneity of myeloproliferative disorders. Blood 112 (4):1402–1412
Ragimbeau J, Dondi E, Vasserot A, Romero P, Uze G, Pellegrini S (2001) The receptor interaction region of Tyk2 contains a motif required for its nuclear localization. J Biol Chem 276:30812–30818
Ram PA, Waxman DJ (1997) Interaction of growth hormone-activated STATs with SH2-containing phosphotyrosine phosphatase SHP-1 and nuclear JAK2 tyrosine kinase. J Biol Chem 272:17694–17702
Raz R, Lee CK, Cannizzaro LA, d'Eustachio P, Levy DE (1999) Essential role of STAT3 for embryonic stem cell pluripotency. Proc Natl Acad Sci USA 96:2846–2851
Rinaldi CR, Rinaldi P, Alagia A, Gemei M, Esposito N, Formiggini F, Martinelli V, Senyuk V, Nucifora G, Pane F (2010) Preferential nuclear accumulation of JAK2V617F in CD34+ but not in granulocytic, megakaryocytic or erythroid cells of patients with Philadelphia-negative myeloproliferative neoplasia. Blood 116(26):6023–6026
Rodig SJ, Meraz MA, White JM, Lampe PA, Riley JK, Arthur CD, King KL, Sheehan KC, Yin L, Pennica D et al (1998) Disruption of the Jak1 gene demonstrates obligatory and nonredundant roles of the Jaks in cytokine-induced biologic responses. Cell 93:373–383

Rui L, Emre NC, Kruhlak MJ, Chung HJ, Steidl C, Slack G, Wright GW, Lenz G, Ngo VN, Shaffer AL et al (2010) Cooperative epigenetic modulation by cancer amplicon genes. Cancer Cell 18:590–605

Schindler C, Fu XY, Improta T, Aebersold R, Darnell JE Jr (1992) Proteins of transcription factor ISGF-3: one gene encodes the 91-and 84-kDa ISGF-3 proteins that are activated by interferon alpha. Proc Natl Acad Sci USA 89:7836–7839

Sharrocks AD (2006) PIAS proteins and transcriptional regulation – more than just SUMO E3 ligases? Genes Dev 20:754–758

Shi S, Calhoun HC, Xia F, Li J, Le L, Li WX (2006) JAK signaling globally counteracts heterochromatic gene silencing. Nat Genet 38:1071–1076

Shi S, Larson K, Guo D, Lim SJ, Dutta P, Yan SJ, Li WX (2008) Drosophila STAT is required for directly maintaining HP1 localization and heterochromatin stability. Nat Cell Biol 10:489–496

Shilatifard A (2006) Chromatin modifications by methylation and ubiquitination: implications in the regulation of gene expression. Annu Rev Biochem 75:243–269

Shimoda K, Kato K, Aoki K, Matsuda T, Miyamoto A, Shibamori M, Yamashita M, Numata A, Takase K, Kobayashi S et al (2000) Tyk2 plays a restricted role in IFN alpha signaling, although it is required for IL-12-mediated T cell function. Immunity 13:561–571

Smith AG, Heath JK, Donaldson DD, Wong GG, Moreau J, Stahl M, Rogers D (1988) Inhibition of pluripotential embryonic stem cell differentiation by purified polypeptides. Nature 336:688–690

Sorenson RL, Stout LE (1995) Prolactin receptors and JAK2 in islets of Langerhans: an immunohistochemical analysis. Endocrinology 136:4092–4098

Sterner DE, Berger SL (2000) Acetylation of histones and transcription-related factors. Microbiol Mol Biol Rev 64:435–459

Teglund S, McKay C, Schuetz E, van Deursen JM, Stravopodis D, Wang D, Brown M, Bodner S, Grosveld G, Ihle JN (1998) Stat5a and Stat5b proteins have essential and nonessential, or redundant, roles in cytokine responses. Cell 93:841–850

Thomis DC, Gurniak CB, Tivol E, Sharpe AH, Berg LJ (1995) Defects in B lymphocyte maturation and T lymphocyte activation in mice lacking Jak3. Science 270:794–797

Velazquez L, Fellous M, Stark GR, Pellegrini S (1992) A protein tyrosine kinase in the interferon alpha/beta signaling pathway. Cell 70:313–322

Wang Y, Wysocka J, Sayegh J, Lee YH, Perlin JR, Leonelli L, Sonbuchner LS, McDonald CH, Cook RG, Dou Y et al (2004) Human PAD4 regulates histone arginine methylation levels via demethylimination. Science 306:279–283

Whetstine JR, Nottke A, Lan F, Huarte M, Smolikov S, Chen Z, Spooner E, Li E, Zhang G, Colaiacovo M et al (2006) Reversal of histone lysine trimethylation by the JMJD2 family of histone demethylases. Cell 125:467–481

Wilks AF (1989) Two putative protein-tyrosine kinases identified by application of the polymerase chain reaction. Proc Natl Acad Sci USA 86:1603–1607

Wilks AF, Harpur AG, Kurban RR, Ralph SJ, Zurcher G, Ziemiecki A (1991) Two novel protein-tyrosine kinases, each with a second phosphotransferase-related catalytic domain, define a new class of protein kinase. Mol Cell Biol 11:2057–2065

Williams RL, Hilton DJ, Pease S, Willson TA, Stewart CL, Gearing DP, Wagner EF, Metcalf D, Nicola NA, Gough NM (1988) Myeloid leukaemia inhibitory factor maintains the developmental potential of embryonic stem cells. Nature 336:684–687

Wissmann M, Yin N, Muller JM, Greschik H, Fodor BD, Jenuwein T, Vogler C, Schneider R, Gunther T, Buettner R et al (2007) Cooperative demethylation by JMJD2C and LSD1 promotes androgen receptor-dependent gene expression. Nat Cell Biol 9:347–353

Woodcock CL (2006) Chromatin architecture. Curr Opin Struct Biol 16:213–220

Wormald S, Hilton DJ (2004) Inhibitors of cytokine signal transduction. J Biol Chem 279:821–824

Yamagishi Y, Sakuno T, Shimura M, Watanabe Y (2008) Heterochromatin links to centromeric protection by recruiting shugoshin. Nature 455:251–255

Yamaoka K, Saharinen P, Pesu M, Holt VE III, Silvennoinen O, O'Shea JJ (2004) The Janus kinases (Jaks). Genome Biol 5:253

Yan SJ, Lim SJ, Shi S, Dutta P, Li WX (2010) Unphosphorylated STAT and heterochromatin protect genome stability. FASEB J 25(1):232–241

Yao Z, Cui Y, Watford WT, Bream JH, Yamaoka K, Hissong BD, Li D, Durum SK, Jiang Q, Bhandoola A et al (2006) Stat5a/b are essential for normal lymphoid development and differentiation. Proc Natl Acad Sci USA 103:1000–1005

Ying QL, Nichols J, Chambers I, Smith A (2003) BMP induction of Id proteins suppresses differentiation and sustains embryonic stem cell self-renewal in collaboration with STAT3. Cell 115:281–292

Ying QL, Wray J, Nichols J, Batlle-Morera L, Doble B, Woodgett J, Cohen P, Smith A (2008) The ground state of embryonic stem cell self-renewal. Nature 453:519–523

Yu H, Jove R (2004) The STATs of cancer – new molecular targets come of age. Nat Rev Cancer 4:97–105

Self-association of STAT Proteins from Monomers to Paracrystals

Mathias Droescher and Uwe Vinkemeier

Abstract

STAT transcription factors assemble dimers of variable solubility and complex conformational dynamics. Unphosphorylated STAT dimers are formed through reciprocal N domain (ND) and core fragment (CF) interactions between protomers, resulting in an antiparallel conformation. Phosphorylated STAT dimers, in contrast, oscillate between this antiparallel and an equally stable parallel conformation that requires reciprocal SH2:pTyr interactions. Moreover, the phosphorylated STAT dimers can polymerize, which occurs both on DNA and off DNA. Polymerization of the parallel phosphodimers is DNA-dependent, and results in cooperative DNA binding, whereas antiparallel phosphodimers can polymerize off DNA, resulting in paracrystals that protect the activated STATs from inactivation. Thus, the central event in cytokine signaling – STAT tyrosine phosphorylation – does not initiate STAT dimerization. Rather, STAT activation regulates the partitioning between different dimer conformations and triggers polymerization of the activated dimers.

Dimerization of STAT Proteins Before and After Activation

The STAT family comprises seven members in mammals, namely STAT1, STAT2, STAT3, STAT4, STAT5a, STAT5b, and STAT6. All have the same structural and functional domain organization. For a detailed discussion of the domain structures we refer the reader to the cited papers describing the crystal structures of STAT

M. Droescher
Max Planck Institute of Immunobiology and Epigenetics, Stübeweg 51, 79108 Freiburg, Germany

U. Vinkemeier (✉)
School of Biomedical Sciences, Nottingham University Medical School,
Nottingham NG7 2UH, UK
e-mail: uwe.vinkemeier@nottingham.ac.uk

Th. Decker and M. Müller (eds.), *Jak-Stat Signaling: From Basics to Disease*,
DOI 10.1007/978-3-7091-0891-8_4, © Springer-Verlag Wien 2012

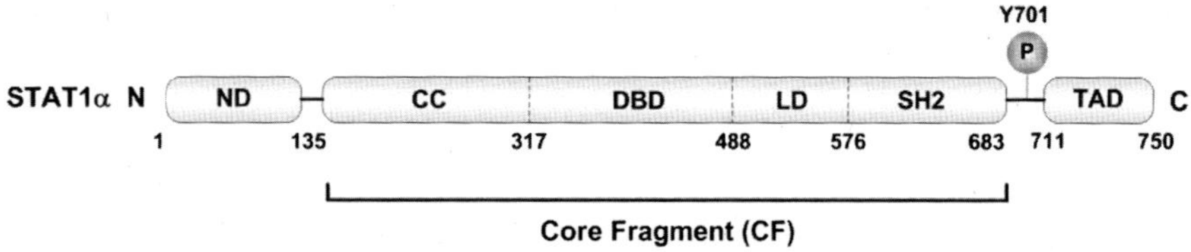

Fig. 1 The functional and structural organisation of STAT proteins is conserved. The domain organization of STAT1α is shown as an example. STAT proteins consist of three structural domains, an aminoterminal N-domain (ND), a core fragment (CF) and a carboxyterminal transactivation domain (TAD). The core fragment encompasses several functional domains: a coiled-coil domain (CC), a DNA binding domain (DBD), a linker domain, and a src homology 2 domain (SH2). The position of the invariant phosphotyrosine residue is highlighted

proteins. In this article we focus on the association of STAT proteins, namely their dimerization and polymerization, which appears to involve the spatial rearrangement of structural domains while leaving the actual domain structures largely unchanged. STAT proteins encompass six well conserved functional domains, which are assembled into three independently folding structural units (Fig. 1). The structural units are the aminoterminal N-domain, a large core fragment comprising four functional domains including a DNA binding domain as well as a src-homology 2 (SH2) domain, and a C-terminal transactivation domain (TAD). The TADs of both STAT1 and STAT2 are intrinsically unstructured and undergo folding transitions upon binding to transcription co-activators (Wojciak et al. 2009). The TADs are least conserved among the STATs, can be absent due to alternative splicing, and are probably dispensable for dimerization and polymerization for most STATs (Lim and Cao 2006; Wenta et al. 2008). N-domain and TAD are connected to the core fragment via short flexible linkers, which contain a conserved tyrosine in position ~700 (residue 701 in STAT1). Phosphorylation of this residue in response to cytokine treatment of cells is critical for subsequent gene transcription, and this event hence is referred to as STAT activation (reviewed in Levy and Darnell 2002).

The two most conserved domains in the STAT family are the SH2 domain and the N-domain, both of which are protein interaction modules required for dimerization and polymerization of STAT proteins. While the SH2 domain is used by many different proteins to mediate tyrosine-phosphorylation-dependent protein interactions (reviewed in Pawson 2004), the all-helical N-domain is unique to the STATs (Vinkemeier et al. 1998). Like the TAD, this domain can interact with multiple proteins including transcription co-regulators, e.g. p300 (Zhang et al. 1996). However, for the purpose of this review, mutual N-domain interactions are most relevant. These interactions are thought to be homotypic in nature. It should be noted, however, that experimental evidence is presently limited to a yeast two-hybrid interaction screen (Ota et al. 2004). Importantly, N-domain interactions involve a single conserved binding interface that appears to participate in all known N-domain-dependent STAT assemblies, be it unphosphorylated dimers, or polymers of activated STATs on and off DNA (Chen et al. 2003; Meyer et al. 2004; Ota et al. 2004).

Initially, it was believed that STATs are monomers prior to their activation by tyrosine phosphorylation. However, accumulating structural and functional evidence indicates that multiple unphosphorylated STATs (U-STATs) are dimers in living cells. The crystal structure of unphosphorylated STAT1 (amino acids 1–683), determined in a complex with a phosphopeptide derived from the α chain of interferon-γ (IFNγ) receptor, indicated two dimer interfaces, one between the N domains, and the other between the core fragments, specifically between the coil–coil and DNA binding domains (Mao et al. 2005). The SH2 domains do not participate in dimerization and are set apart at both ends of the dimer structure; STAT dimers that adopt this conformation are called antiparallel dimers. Crystal structures of U-STAT5a and U-STAT3 core fragments are available, too. The structure of unphosphorylated STAT5a (amino acids 129–712) is analogous to STAT1, as it displays antiparallel dimers maintained through reciprocal core fragment interaction; and an N-domain could be docked onto the core fragment dimer using molecular modeling (Neculai et al. 2005). Small-angle x-ray scattering of unphosphorylated STAT5 core fragment identifies the simultaneous presence of monomers and dimers, and suggests the presence of dimers in the antiparallel assembly (Bernado et al. 2009). Moreover, fluorescence resonance energy transfer and optimized firefly luciferase complementation imaging using living cells indicate dimerization prior to activation (Luker et al. 2004; Neculai et al. 2005). Collectively, these results suggest a similar mode of dimerization for unphosphorylated STAT5 and STAT1. Conversely, the crystal structure of truncated unphosphorylated STAT3 (amino acids 127–688) lacks the core fragment dimer interfaces and appears to be monomeric, a result confirmed by light scattering experiments and native PAGE (Ren et al. 2008; Vogt et al. 2011). Thus, unphosphorylated STAT3, like U-STAT1 (Wenta et al. 2008), is monomeric in the absence of N-domain interactions. Whether full-length U-STAT3 is dimeric at physiological concentrations is less clear, however. Analytical ultracentrifugation and native PAGE using purified proteins suggest that U-STAT1 and U-STAT3 do not differ in terms of dimerization, but the studies could not fully clarify this question (Braunstein et al. 2003). Additionally, dimerization of full-length U-STAT3 in living cells is demonstrated by various analytical techniques (Lackmann et al. 1998; Kretzschmar et al. 2004; Li and Shaw 2004; Schröder et al. 2004; Vogt et al. 2011). Yet, U-STAT1 and U-STAT3 do differ in certain aspects regards dimerization. For one, the N-domain interactions of STAT1 and STAT3 differ by three orders of magnitude (Wenta et al. 2008), and mutations that destabilize the antiparallel conformation of U-STAT1 appear to have little or no effect on the dimerization of U-STAT3 (Ren et al. 2008; Vogt et al. 2011), Hence, whether stable dimerization is common to all STATs, and how this dimerization occurs, is unclear at present, and additional experimentation is required for an answer. In addition to U-STAT homodimers discussed in the preceding, co-precipitation experiments indicate the existence of poorly characterized STAT1/STAT2 and STAT1/STAT3 heterodimers before the cytokine stimulation of cells (Stancato et al. 1996; Haan et al. 2000). Furthermore, using gel filtration chromatography and live cell imaging, the Sehgal lab demonstrated localization of multiple unphosphorylated STATs in high molecular mass complexes and vesicles in the cytoplasm, probably in association with heterologous proteins, but the composition and structural organization of these assemblies remain incompletely understood (Ndubuisi et al. 1999; Guo et al. 2002; Shah et al. 2002; Mukhopadhyay et al. 2008; reviewed in Sehgal 2008).

The dimerization of unphosphorylated STATs was unambiguosly established rather recently. Contrary, that the activation of STATs is associated with their dimerization was demonstrated almost a decade earlier, soon after the discovery of this protein family. Based on sequence comparisons and biochemical experimentation it was possible to deduce early on that dimerization of phosphorylated STAT1 (P-STAT1) entails mutual SH2:pTyr interactions (Shuai et al. 1994; reviewed in Darnell et al. 1994). This mode of dimerization has been confirmed for all STAT proteins (reviewed in Levy and Darnell 2002). The dimer arrangement, which is dependent upon SH2 domain interactions, but where the N-domains are dispensable, is termed parallel dimer conformation. SH2 domain-mediated dimerization is both homo- and heterotypic, and accordingly different combinations of STAT heterodimers can be found in cytokine stimulated cells (reviewed in Platanias 2005). A distinction of parallel dimers is their ability to bind with high affinity to nonameric DNA palindromes termed gamma-activated-sites, GAS (Horvath et al. 1995; Seidel et al. 1995; Ehret et al. 2001; Rivas et al. 2008; reviewed in Decker et al. 1997). This constitutes the basis of the STATs functioning as transcription regulators in cytokine signaling. The crystal structures of DNA-bound truncated activated STAT1 and STAT3 homodimers are essentially identical, strongly enforcing the idea that activated STATs are highly similar structurally (Becker et al. 1998; Chen et al. 1998). This notion extends even to the STAT ortholog of Dictyostelium discoideum, the simplest organism known to employ STAT signaling (Kawata et al. 1997; reviewed in Darnell 1997). The crystal structure of tyrosine phosphorylated D. discoideum STATa homodimers in the DNA-unbound state reveals a domain architecture similar to that of mammalian STATs, as well as dimerization by SH2 domain:pTyr interactions (Soler-Lopez et al. 2004). However, contrary to the DNA-bound mammalian STATs, the STATa dimer adopts a fully extended conformation, implying a significant domain rearrangement would be required for DNA binding. Whether this is indeed the case, or whether mammalian STATs adopt a similarly extended conformation before binding to DNA is not known. It is interesting to note, however, that molecular modeling of paracrystal-forming STAT polymers was only successful when the extended SH2:pTyr arrangement of DNA-unbound STATa was used as a template (Droescher et al. 2011).

Transitions Between STAT Dimer Conformers

With the characterization of different dimer conformations the question of conformational dynamics arose, and hence renewed interest in quantifying dimer association and dissociation. The initial studies considered dimerization via SH2:pTyr interactions, but due to the lack of purified activated STATs these experiments were limited to studying the binding of phosphotyrosine-containing peptide ligands to U-STATs, e.g. using plasmon surface resonance technology. The dissociation constants for "activated" STAT1 obtained in this way, 150 nM (Greenlund et al. 1995), turns out to be remarkably similar to results obtained subsequently using analytical ultracentrifugation and the full length P-STAT1, ~50 nM (Wenta et al. 2008), stressing that the binding energy stabilizing the parallel dimer conformation is provided almost entirely by the phosphotyrosine and its immediate neighboring residues. Aside from SH2:pTyr interactions, additional posttranslational modifications and additional molecular surfaces have been proposed to affect the assembly of activated STAT dimers. It was reported that the acidic tail domain of STAT3α, which is missing in the splice variant STAT3β, may destabilize the dimer resulting in lower DNA-binding activity and more rapid dephosphorylation (Schaefer et al. 1997; Park et al. 2000). A role for the TAD in dimer stability was suggested also for STAT6 (Patel et al. 1998; Sherman et al. 1999). Furthermore, acetylation of activated STAT3 at Lys695 was reported to affect its DNA binding by enhancing dimer formation (Wang et al. 2005; Yuan et al. 2005), an explanation which has been disputed (O'Shea et al. 2005).

The availability of highly pure Tyr701-phosphorylated STAT1 made analytical ultracentrifugation experiments feasible to compare dissociation equilibria before and after activation, which revealed that the thermodynamic stability of U-STAT1 dimers and P-STAT1 dimers are essentially identical (Wenta et al. 2008). These studies indicated that STAT1 constantly oscillates between different dimer conformations, whereby the abundance of conformers is determined by tyrosine phosphorylation. Thus, while U-STAT1 exists almost exclusively in the antiparallel dimer conformation at physiological concentrations, the parallel and antiparallel dimers of P-STAT1 are probably equally abundant. Notably, as could be expected from their crystal structures, the molecular shapes of U-STAT1 and P-STAT1 dimers in solution differ significantly, with the dimers of activated STAT1 adopting a more globular build (Nardozzi et al. 2010). However, P-STAT1 was detected in a single shape only, which differed from U-STAT1 (Nardozzi et al. 2010). Thus, direct experimental proof for the co-existence of P-STAT1 in multiple dimer conformations is still missing. The failure to detect more than one molecular shape for P-STAT1 could result from the fact that the parallel dimer conformation is in fact thermodynamically favored; alternatively, the transition between the two dimer conformations might be kinetically controlled. Current work in our lab is trying to resolve this issue, which is of great physiological relevance, as there is compelling functional and structural evidence that antiparallel dimers of P-STAT1 are indispensable for crucial STAT activities, namely their tyrosine phosphorylation and dephosphorylation, as well as paracrystal assembly. Work in the Darnell lab has convincingly linked the antiparallel dimer conformation to STAT dephosphorylation (Zhong et al. 2005; Mertens et al. 2006). This provided a unifying nonexclusive explanation for earlier reports of experimental N-domain mutants of STAT1 and STAT5 displaying dephosphorylation defects (Shuai et al. 1996; Strehlow and Schindler 1998; Haspel and Darnell 1999; Meissner et al. 2004a; Meyer et al. 2004; Moriggl et al. 2005). Importantly, defective antiparallel dimerization of STAT1 moreover has been identified as the cause of chronic mucocutaneous candidiasis in human patients (Liu et al. 2011; van de Veerdonk et al. 2011). In agreement with its deviating dimerization described above, STAT3 does not require the N-domain for efficient dephosphorylation (Zhang et al. 2006), which might be the case for STAT6, too (Patel et al. 1998). Of note, the t1/2 of the STAT1 dephosphorylation reaction in cells (~15 min) is probably shorter than the half-life of both STAT dimer conformations, which based on kinetic modeling, was estimated to be 20–40 min (Haspel et al. 1996; Wenta et al. 2008). Moreover, in vitro dephosphorylation of STAT1 using the phosphatase TC45 is rather inefficient, although this enzyme very efficiently inactivates STAT1 in living cells (ten Hoeve et al. 2002; Meyer et al. 2003). The details of the transition between parallel and antiparallel dimer conformations are under discussion. Based on our results we favor the idea that the process occurs via dissociation and re-association reactions (Wenta et al. 2008), whereas data of Mertens et al. (2006) indicate that the core domains rotate while the N-domains still keep the dimer together. Irrespective of the molecular details, the current data agree in that the transition between dimer conformations is a crucial and potentially regulated pacemaker determining the decay of cytokine signals. In line with this reasoning, nuclear beta-arrestin1 was proposed as an essential negative regulator of IFN-gamma signaling by acting as a scaffold to facilitate the dephosphorylation of STAT1 by TC45 (Mo et al. 2008). However, our experiments demonstrated that the reported enhancing effect of beta-arrestin1 on STAT1 dephosphorylation is not reproducible (Manuscript under review). Similarly, acetylation of lysine residues in the DNA binding domain was proposed to promote recruitment of TC45 and hence STAT1 dephosphorylation, but these results have been invalidated as well (Antunes et al. 2011). In conclusion, the dimer conformational dynamics are critical for the inactivation of STATs, but the molecular mechanisms remain poorly understood.

The activation of STATs at cytokine receptors has been demonstrated to require antiparallel U-STAT dimers. This was most clearly demonstrated for STAT4 and STAT2, where point mutations or deletion of the N-domain preclude tyrosine phosphorylation in response to cytokine treatment of cell (Qureshi et al. 1996; Murphy et al. 2000; Chang et al. 2003; Ota et al. 2004). The structural requirements of the other STATs, in contrast, appear to be less stringent regards their activation, as they tolerate inactivation of their N-domains and remain responsive to cytokine stimulation, suggesting that they can interact with cytokine receptors as monomers, too (Mikita et al. 1996; Shuai et al. 1996; Strehlow and Schindler 1998; Haspel and Darnell 1999; Moriggl et al. 2005; Zhang et al. 2006). However, the steps leading to STAT activation at the receptors are complex and have not been explored fully

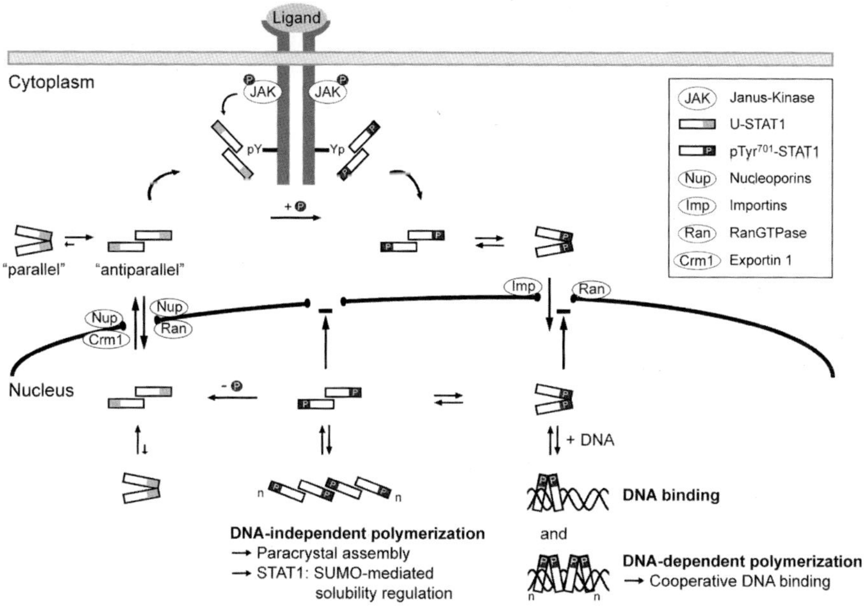

Fig. 2 STAT protein function and conformational dynamics. Unphosphorylated STATs are carrier-independent nucleocytoplasmic shuttling proteins. Translocation occurs via interactions with nuclear pore proteins (Nup); in addition, nuclear export is enhanced by the export factor CRM1 and metabolic energy (Ran). Phosphorylated STATs can enter the nucleus as parallel dimers, a process that requires transport factors (importins) and metabolic energy (Ran). Phosphorylated STATs, in contrast, do not exit the nucleus. STAT activation (+P) can occur through receptor-associated tyrosine kinases (Jak) at the cell membrane. STAT inactivation (−P) occurs predominantly in the nucleus after dissociation from DNA. The cartoon is modelled on data reported for STAT1. Further details are provided in the text

(reviewed in Haan et al. 2006) urging caution as to the interpretation of these results regarding STAT conformational dynamics.

It has also been recognized that the mechanisms of nucleocytoplasmic transport of STAT proteins are determined by their activation state (reviewed in Vinkemeier 2004), but aside from the fact that parallel dimers appear to be required for the nuclear import of activated STAT1, structural information is scarce (Nardozzi et al. 2010). This topic therefore will not be discussed here in further detail. Figure 2 depicts activities of STAT proteins currently linked to their conformational dynamics.

Polymerization of Activated STAT Dimers On and Off DNA

It has long been demonstrated that activated STAT dimers can polymerize on DNA, i.e. form tetramers and higher order polymers. The consequence is strongly reduced dissociation of the STAT dimers from DNA, and hence cooperative DNA binding.

This activity requires phosphodimers in the parallel conformation, which are believed to interact solely through their N-domains when bound to DNA (Vinkemeier et al. 1996; Xu et al. 1996; Li et al. 1998; Vinkemeier et al. 1998; Meyer et al. 2004; reviewed in Leung et al. 1996). In this manner open-ended polymers arise (Chen et al. 1998; reviewed in Rhodes and Schwabe 1998). A number of genes have been identified with multiple STAT binding sites in close proximity in their promoter regions conducive for STAT cooperative DNA binding, i.e. with a center-to-center spacing between GAS elements of ~30 base pairs (Vinkemeier et al. 1996). Examples of genes that require STAT cooperative DNA binding include α2-macroglobulin (Zhang and Darnell 2001) and perforin (Yamamoto et al. 2002). The importance of tetramerization for target gene recognition and gene activation has been particularly well documented for STAT5 (Meyer et al. 1997; John et al. 1999; Soldaini et al. 2000; Hou et al. 2003; Moriggl et al. 2005). Nonetheless, the prevalence of DNA-dependent STAT polymerization in cytokine-mediated gene induction is currently not clear. This refers both to the number of genes where polymerization of phosphodimers is required for gene transcription, as well as to the extent of polymerization at actual gene promoters or other genomic sites. It is of interest in this regard that in mammalian cells a considerable fraction of activated STAT1 is bound to MER41 repeats – primate-specific repetitive gene elements that consist of tandem high-affinity STAT binding sites with optimal spacing for cooperative DNA binding (Schmid and Bucher 2010). In addition, a patient with recurring mycobacterial infections was recently described who expresses a STAT1 protein with a single amino acid exchange in its N-domain (Kristensen et al. 2011), suggesting that DNA-dependent polymerization of STAT1 may be of more widespread significance for interferon-γ signaling.

As detailed in the preceding paragraph, polymerization of STAT phosphodimers in the parallel conformation for some time has been recognized to be of physiological significance. We have recently uncovered that antiparallel STAT phosphodimers can polymerize, too, but this process, albeit normally taking place in the nucleus, occurs entirely independent of DNA (Droescher et al. 2011). In further contrast to the parallel dimers bound cooperatively to DNA, free dimers are interlinked via SH2:pTyr interactions, whereas their N-domains participate in dimerization. These polymers can align laterally, resulting in light-microscopically visible structures that due to their high internal order and apparent self-organized assembly are of paracrystalline nature (Fig. 3). At present a single physiological example of STAT paracrystal formation is known to us, namely the assembly of STAT3 paracrystals during the hepatic acute phase response to systemic bacterial infection (Ray et al. 2005; Droescher et al. 2011). However, cell transfection experiments indicate the disposition of further STATs, i.e. STAT2 and STAT5, to assemble paracrystals in cytokine stimulated cells (Herrington et al. 1999; Frahm et al. 2005). Moreover, mouse oocytes and early embryos harbor large assemblies of activated STAT2, STAT3 and STAT1 (Truchet et al. 2004). Given the highly conserved STAT structure, these results suggest that paracrystals are integral to cytokine signaling.

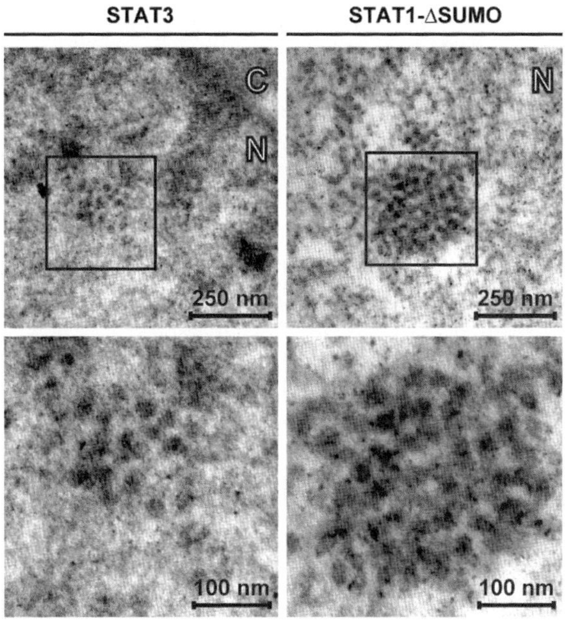

Fig. 3 Activated STATs assemble paracrystalline structures. Overview and close up (*boxed* area) transmission electron micrographs of acute phase liver sections (*left*, STAT3) and IFNγ-stimulated HeLa cells expressing SUMO-free STAT1-E705Q (*right*, STAT1-ΔSUMO). Livers where isolated 2 h after injection of mice with lipopolysaccaride (LPS). Treatment of mice with *Escherichia coli* LPS (serotype 0127:B8, Sigma L3880), liver sectioning, HeLa cell transfections, and electron microscopy were done as described (Droescher et al. 2011). Shown are STAT3 and STAT1 paracrystals, both of which appear to consist of laterally aligned filaments. Nuclear (N) and cytoplasmic (C) compartments are indicated

STAT1 is probably the only STAT family member that does not normally form paracrystals. However, it is perfectly capable to do so, but in order to enable paracrystal assembly its SUMO conjugation, which is unique to STAT1 in the STAT family, needs to be suppressed. Among other experiments, this was demonstrated by reversing the abilities of STAT1 and STAT3 to be SUMO-modified. These experiments serendipitously provided a molecular explanation for earlier experiments performed by Inoue et al. (1997) to determine the domains in STAT3 that contribute to interleukin 6 (IL6)-specific phosphorylation. Using domain swap mutants of STAT3 and STAT1, which is rather unresponsive to IL6, they did not identify the SH2 domains to encode IL6 specificity, contrary to expectations based on results for STAT1 (Heim et al. 1995). Rather, replacement of a single residue in position +4 to the STAT3 activating tyrosine705 with the corresponding residue of STAT1 abolished IL6 activation of STAT3 (Inoue et al. 1997). Unknown to the authors at the time when SUMO had not yet been discovered, the mutation had reconstituted a SUMO target sequence, resulting in the efficient SUMO conjugation of STAT3 (Droescher et al. 2011). Consequently, cytokine-induced-phosphorylation of STAT3 was strongly diminished (Droescher et al. 2011).

In accordance with these results for STAT3, conjugation of the bulky SUMO protein (~100 residues, reviewed in Johnson 2004) was unambiguously demonstrated to preclude the phosphorylation of the proximal tyrosine, which hence constitutes one of two additive mechanisms by which SUMO diminishes STAT activation (Zimnik et al. 2009; Droescher et al. 2011). It is also the basis for another, indirect mechanism, by which SUMO diminishes STAT1 transactivation. Since STAT1 is recruited to the receptors for activation as dimers, SUMO interference that prevents tyrosine phosphorylation increases the abundance of semi-phosphorylated dimers. These, in turn, act as competitive inhibitors that abort the elongation of STAT1 polymers (Droescher et al. 2011). Consequently, paracrystal assembly is prevented, causing increased solubility of the activated STAT1. In this rather indirect way, SUMO becomes a potent regulator of protein solubility. It is critically important, however, that activated STAT1 dimers be relatively stable, that is, have low protomer exchange rates. Otherwise, semi-phosphorylated dimers would readily form regardless of SUMO, as unphosphorylated STAT1 is present in excess even during cytokine stimulation. Moreover, to avoid doubly sumoylated dimers, which are altogether refractory to phosphorylation, the pool of SUMO conjugated molecules ought to be small. Thus, short-lived SUMO conjugation and a seemingly negligible SUMO-modified STAT1 fraction (~2%) must not be used as an argument against a physiological role for SUMO in STAT1 signaling. Rather, these characteristics constitute a prerequisite for the control of STAT1 solubility by SUMO. As many SUMO targets share these features with STAT1, it is conceivable that other SUMO-regulated protein interactions too entail similar competition-based mechanisms.

The reduced solubility of activated STAT is apparent in the formation of paracrystals. Paracrystal formation was identified as a protective mechanism that preserves the phosphorylation of STAT1 not only on tyrosine701, but also on the transcription-enhancing serine727 (Begitt et al. 2011). Moreover, paracrystals buffer the nucleoplasm to a constant level of activated STATs, which profoundly prolongs STAT activity in the nucleus, resulting in increased gene transcription (Begitt et al. 2011; Droescher et al. 2011). Therefore, solubility increase is the second mechanism by which SUMO modification of STAT1 reduces interferon signaling (Begitt et al. 2011). Importantly, SUMO-free STAT1 can integrate into paracrystals of STAT3, and wild-type STAT1 accordingly can dissolve STAT3 paracrystals (manuscript in preparation). This, in turn, can provide a structural basis for the cross-regulation of multiple cytokine signaling pathways by interferon-γ (reviewed in Hu and Ivashkiv 2009).

In the final section of this paper we would like to present experiments that demonstrate the antagonistic relationship between paracrystal incorporation of STAT1 and cooperative DNA binding. The experiments shown in Fig. 4 were performed with autofluorescent SUMO-free STAT1 fused to a canonical nuclear export signal, NES (Lödige et al. 2005). Importantly, fusion of STAT1 to NES overcomes the export block that prevents phosphorylated wild type STAT1 from exiting the nucleus (unpublished observation, reviewed in Vinkemeier 2004).

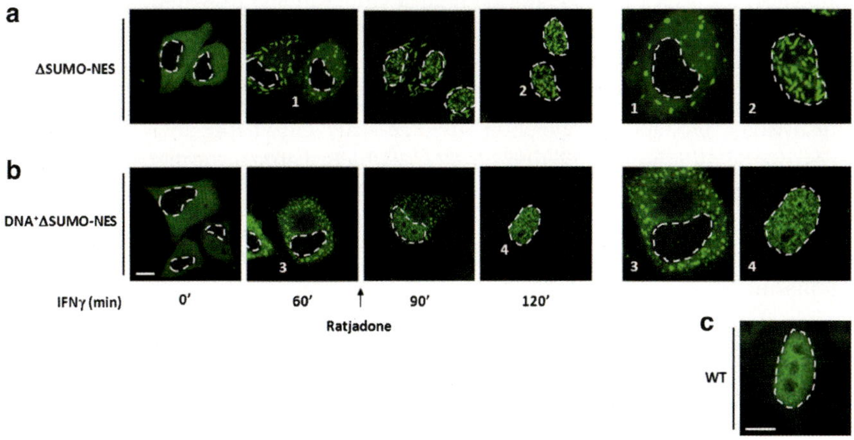

Fig. 4 *DNA binding opposes STAT paracrystal assembly*. Experiments were done using HeLa cells expressing the indicated STAT1 variants fused C-terminally to green fluorescent protein (Droescher et al. 2011). Shown are cells before (panels 0′) and after stimulation with 20 ng/mL human interferon-γ (panels 60′–120′). The NES inhibitor ratjadone (5 ng/mL, Calbiochem) was added after 60 min IFNγ, and was subsequently left on the cells. Panel C depicts the distribution of wild type STAT1 after 60 min stimulation with IFNγ. Enlarged images of the cells labeled 1–4 are shown on the *right*. Nuclei are outlined with *broken lines*; bars, 10 μm

Consequently, both the unphosphorylated and the phosphorylated STAT1 accumulate in the cytoplasm. Accordingly, stimulation of cells with interferon-γ for 60 min did not appreciably alter the nucleocytoplasmic distribution of STAT1 seen in the unstimulated cells (Fig. 4a, panels 0′ and 60′). Nonetheless, the SUMO-free STAT1 variant expectedly assembled paracrystal where P-STAT1 was concentrated, namely in the cytoplasm. We then added ratjadone, which inactivates NES-mediated nuclear export (Meissner et al. 2004b), thus effectively reverting to the STAT1 wild type situation by reinstating the nuclear export block of the phosphorylated protein. Accordingly, STAT1 displayed its characteristic nuclear accumulation within 30–60 min of the addition of ratjadone (Fig. 4a, panels 90′ and 120′). Concomitantly, paracrystals emerged in the nucleus, while extra-nuclear paracrystals gradually dissolved (90′) and eventually disappeared (120′). Of note, directed movement of paracrystals towards the nucleus or their transport across the nuclear envelope was not apparent (unpublished observation). To reveal the opposing effect of DNA binding on paracrystal assembly, we additionally mutated three residues in the DNA binding domain to enhance sequence-unspecific DNA binding. The respective mutant, termed DNAplus, has been described before (Meyer et al. 2003; Meyer and Vinkemeier 2010). As shown in Fig. 4b, enhanced DNA binding had no effect on the cytoplasmic accumulation of SUMO-free STAT1-NES before and after the treatment of cells with interferon (Fig. 4b, panels 0′ and 60′). The abundant paracrystals moreover demonstrated that the DNA binding domain mutations had no adverse effect on STAT1 polymerization in the cytoplasm (Fig. 4b, panel 60′). However, while the addition of ratjadone expectedly diminished the STAT1 concentration in the cytoplasm and triggered nuclear accumulation, reassembly of paracrystals did not occur in the nucleus (Fig. 4b, panels 90′ and 120′). Interestingly, activated DNAplus mutant did not show the homogeneous intranuclear distribution of wild type STAT1 either (see Fig. 4c for a comparison); probably a reflection of large-scale chromatin alterations due to excessive STAT1 polymerization on DNA, which warrants further investigation for clarification.

In conclusion, the increased DNA binding activity of DNAplus mutant was without consequences for paracrystal assembly in the absence of DNA, i.e. in the cytoplasm, but precluded these structures in the nucleus. We inferred that DNA binding and paracrystal incorporation are actually opposing activities, stressing the notion that DNA-dependent and DNA-independent polymerization of activated STATs are mechanistically distinct.

Concluding Remarks

Research pursued in many laboratories has provided a wealth of knowledge about the structure of STAT transcription factors. We have gained a reasonably good understanding of their organization into functional and structural domains, and we have information about the three dimensional structures of all functional and structural units for at least a few STATs, albeit the picture is far from complete. Despite of this, the structure of a full-length STAT protein is still not available, leaving open many questions about the actual inter-domain organization particularly after activation. The recent recognition that STAT proteins undergo complex transitions between different dimer conformations has greatly advanced the understanding of seemingly disparate sets of experimental data concerning the STAT activation/inactivation cycle. Nonetheless, both the molecular details of these events, as well as their biological consequences remain rather ill-defined. Moreover, we are far from knowing whether all STATs are created equal in this regard, but it appears that even apparently subtle structural differences, e.g. affecting the dimer dissociation equilibrium, can have significant functional consequences. Finally, it has also been revealed that dimers are not the final polymerization state of STATs. Indeed, the solubility of STATs is markedly reduced after their tyrosine phosphorylation, which is due to DNA-independent and DNA-dependent polymerization. The latter, also known as cooperative DNA binding, has been known for some time to facilitate cytokine-induced gene transcription. In addition, the discovery of STAT paracrystals and their SUMO-mediated dispersal identifies increased transcription factor solubility as a negative regulatory mechanism in extracellular signaling to the nucleus. Thus, ever more so, STATs find that hanging together can be stimulating.

Acknowledgement The authors would like to thank Dr. N. Wenta for informative discussions and the gifts of purified STAT proteins; and F. Antunes for data collection and figure preparation.

This work was supported by grants from Biotechnology and Biological Sciences Research Council (BB/GO019290/1) and Deutsche Forschungsgemeinschaft (UV218/2-3).

References

Antunes F, Marg A, Vinkemeier U (2011) STAT1 signaling is not regulated by a phosphorylation-acetylation switch. Mol Cell Biol 31:3029–3037

Becker S, Groner B, Müller CW (1998) Three-dimensional structure of the Stat3beta homodimer bound to DNA. Nature 394:145–151

Begitt A, Droescher M, Knobeloch KP, Vinkemeier U (2011) SUMO conjugation of STAT1 protects cells from hyperresponsiveness to IFN-gamma. Blood 118:1002–1007

Bernado P, Perez Y, Blobel J, Fernandez-Recio J, Svergun DI, Pons M (2009) Structural characterization of unphosphorylated STAT5a oligomerization equilibrium in solution by small-angle X-ray scattering. Protein Sci 18:716–726

Braunstein J, Brutsaert S, Olson R, Schindler C (2003) STATs dimerize in the absence of phosphorylation. J Biol Chem 278:34133–34140

Chang HC, Zhang S, Oldham I, Naeger L, Hoey T, Kaplan MH (2003) STAT4 requires the N-terminal domain for efficient phosphorylation. J Biol Chem 278:32471–32477

Chen X, Vinkemeier U, Zhao Y, Jeruzalmi D, Darnell JE Jr, Kuriyan J (1998) Crystal structure of a tyrosine phosphorylated STAT-1 dimer bound to DNA. Cell 93:827–839

Chen X, Bhandari R, Vinkemeier U, Van den Akker F, Darnell JE Jr, Kuriyan J (2003) A reinterpretation of the dimerization interface of the N-terminal domains of STATs. Protein Sci 12:361–365

Darnell JE Jr (1997) Phosphotyrosine signaling and the single cell: metazoan boundary. Proc Natl Acad Sci USA 94:11767–11769

Darnell JE Jr, Kerr IM, Stark GR (1994) Jak-STAT pathways and transcriptional activation in response to IFNs and other extracellular signaling proteins. Science 264:1415–1421

Decker T, Kovarik P, Meinke A (1997) GAS elements: a few nucleotides with a major impact on cytokine-induced gene expression. J Interferon Cytokine Res 117:121–134

Droescher M, Begitt A, Marg A, Zacharias M, Vinkemeier U (2011) Cytokine-induced paracrystals prolong the activity of signal transducers and activators of transcription (STAT) and provide a model for the regulation of protein solubility by small ubiquitin-like modifier (SUMO). J Biol Chem 286:18731–18746

Ehret GB, Reichenbach P, Schindler U, Horvath CM, Fritz S, Nabholz M, Bucher P (2001) DNA binding specificity of different STAT proteins. Comparison of in vitro specificity with natural target sites. J Biol Chem 276:6675–6688

Frahm T, Hauser H, Koster M (2005) IFN-type-I-mediated signaling is regulated by modulation of STAT2 nuclear export. J Cell Sci 119:1092–1104

Greenlund AC, Morales MO, Viviano BL, Yan H, Krolewski J, Schreiber RD (1995) Stat recruitment by tyrosine-phosphorylated cytokine receptors: an ordered reversible affinity-driven process. Immunity 2:677–687

Guo GG, Patel K, Kumar V, Shah M, Fried VA, Etlinger JD, Sehgal PB (2002) Association of the chaperone glucose-regulated protein 58 (GRP58/ER-60/ERp57) with Stat3 in cytosol and plasma membrane complexes. J Interferon Cytokine Res 22:555–563

Haan S, Kortylewski M, Behrmann I, Müller-Esterl W, Heinrich PC, Schaper F (2000) Cytoplasmic STAT proteins associate prior to activation. Biochem J 345(Pt 3):417–421

Haan C, Kreis S, Margue C, Behrmann I (2006) Jaks and cytokine receptors – an intimate relationship. Biochem Pharmacol 72:1538–1546

Haspel RL, Darnell JE Jr (1999) A nuclear protein tyrosine phosphatase is required for the inactivation of Stat1. Proc Natl Acad Sci USA 96:10188–10193

Haspel RL, Salditt-Georgieff M, Darnell JE Jr (1996) The rapid inactivation of nuclear tyrosine phosphorylated Stat1 depends upon a protein tyrosine phosphatase. EMBO J 15:6262–6268

Heim MH, Kerr IM, Stark GR, Darnell JE Jr (1995) Contribution of STAT SH2 groups to specific interferon signaling by the Jak-STAT pathway. Science 267:1347–1349
Herrington J, Rui L, Luo G, Yu-Lee LY, Carter-Su C (1999) A functional DNA binding domain is required for growth hormone-induced nuclear accumulation of Stat5B. J Biol Chem 274: 5138–5145
Horvath CM, Wen Z, Darnell JE Jr (1995) A STAT protein domain that determines DNA sequence recognition suggests a novel DNA-binding domain. Genes Dev 9:984–994
Hou Z, Srivastava S, Mistry MJ, Herbst MP, Bailey JP, Horseman ND (2003) Two tandemly linked interferon-gamma-activated sequence elements in the promoter of glycosylation-dependent cell adhesion molecule 1 gene synergistically respond to prolactin in mouse mammary epithelial cells. Mol Endocrinol 17:1910–1920
Hu X, Ivashkiv LB (2009) Cross-regulation of signaling pathways by interferon-gamma: implications for immune responses and autoimmune diseases. Immunity 31:539–550
Inoue M, Minami M, Matsumoto M, Kishimoto T, Akira S (1997) The amino acid residues immediately carboxyl-terminal to the tyrosine phosphorylation site contribute to interleukin 6-specific activation of signal transducer and activator of transcription 3. J Biol Chem 272:9550–9555
John S, Vinkemeier U, Soldaini E, Darnell JE Jr, Leonard WJ (1999) The significance of tetramerization in promoter recruitment by Stat5. Mol Cell Biol 19:1910–1918
Johnson ES (2004) Protein modification by SUMO. Annu Rev Biochem 73:355–382
Kawata T, Shevchenko A, Fukuzawa M, Jermyn KA, Totty NF, Zhukovskaya NV, Sterling AE, Mann M, Williams JG (1997) SH2 signaling in a lower eukaryote: a STAT protein that regulates stalk cell differentiation in dictyostelium. Cell 89:909–916
Kretzschmar AK, Dinger MC, Henze C, Brocke-Heidrich K, Horn F (2004) Analysis of Stat3 (signal transducer and activator of transcription 3) dimerization by fluorescence resonance energy transfer in living cells. Biochem J 377:289–297
Kristensen IA, Veirum JE, Møller BK, Christiansen M (2011) Novel STAT1 alleles in a patient with impaired resistance to mycobacteria. J Clin Immunol 31:265–271
Lackmann M, Harpur AG, Oates AC, Mann RJ, Gabriel A, Meutermans W, Alewood PF, Kerr IM, Stark GR, Wilks AF (1998) Biomolecular interaction analysis of IFN gamma-induced signaling events in whole-cell lysates: prevalence of latent STAT1 in high-molecular weight complexes. Growth Factors 16:39–51
Leung S, Li X, Stark GR (1996) STATs find that hanging together can be stimulating. Science 273:750–751
Levy DE, Darnell JE Jr (2002) Stats: transcriptional control and biological impact. Nat Rev Mol Cell Biol 3:651–662
Li L, Shaw PE (2004) A STAT3 dimer formed by inter-chain disulphide bridging during oxidative stress. Biochem Biophys Res Commun 322:1005–1011
Li X, Leung S, Burns C, Stark GR (1998) Cooperative binding of Stat1-2 heterodimers and ISGF3 to tandem DNA elements. Biochimie 80:703–710
Lim C, Cao X (2006) Structure, function, and regulation of STAT proteins. Mol Biosyst 2: 536–550
Liu L, Okada S, Kong XF, Kreins AY, Cypowyj S, Abhyankar A, Toubiana J, Itan Y, Audry M, Nitschke P et al (2011) Gain-of-function human STAT1 mutations impair IL-17 immunity and underlie chronic mucocutaneous candidiasis. J Exp Med 208:1635–1648
Lödige I, Marg A, Wiesner B, Malecová B, Oelgeschläger T, Vinkemeier U (2005) Nuclear export determines the cytokine sensitivity of STAT transcription factors. J Biol Chem 280: 43087–43099
Luker KE, Smith MC, Luker GD, Gammon ST, Piwnica-Worms H, Piwnica-Worms D (2004) Kinetics of regulated protein–protein interactions revealed with firefly luciferase complementation imaging in cells and living animals. Proc Natl Acad Sci USA 101:12288–12293

Mao X, Ren Z, Parker GN, Sondermann H, Pastorello MA, Wang W, McMurray JS, Demeler B, Darnell JE Jr, Chen X (2005) Structural bases of unphosphorylated STAT1 association and receptor binding. Mol Cell 17:761–771

Meissner T, Krause E, Lödige I, Vinkemeier U (2004a) Arginine methylation of STAT1: a reassessment. Cell 119:587–589

Meissner T, Krause E, Vinkemeier U (2004b) Ratjadone and leptomycin B block CRM1-dependent nuclear export by identical mechanisms. FEBS Lett 576:27–30

Mertens C, Zhong M, Krishnaraj R, Zou W, Chen X, Darnell JE Jr (2006) Dephosphorylation of phosphotyrosine on STAT1 dimers requires extensive spatial reorientation of the monomers facilitated by the N-terminal domain. Genes Dev 20:3372–3381

Meyer T, Vinkemeier U (2010) Assessing sequence-specific DNA binding and transcriptional activity of STAT1 transcription factor. Methods Mol Biol 647:139–159

Meyer WK, Reichenbach P, Schindler U, Soldaini E, Nabholz M (1997) Interaction of STAT5 dimers on two low affinity binding sites mediates interleukin 2 (IL-2) stimulation of IL-2 receptor alpha gene transcription. J Biol Chem 272:31821–31828

Meyer T, Marg A, Lemke P, Wiesner B, Vinkemeier U (2003) DNA binding controls inactivation and nuclear accumulation of the transcription factor Stat1. Genes Dev 17:1992–2005

Meyer T, Hendry L, Begitt A, John S, Vinkemeier U (2004) A single residue modulates tyrosine dephosphorylation, oligomerization, and nuclear accumulation of Stat transcription factors. J Biol Chem 279:18998–19007

Mikita T, Campbell D, Wu P, Williamson K, Schindler U (1996) Requirements for interleukin-4-induced gene expression and functional characterization of Stat6. Mol Cell Biol 16:5811–5820

Mo W, Zhang L, Yang G, Zhai J, Hu Z, Chen Y, Chen X, Hui L, Huang R, Hu G (2008) Nuclear beta-arrestin1 functions as a scaffold for the dephosphorylation of STAT1 and moderates the antiviral activity of IFN-gamma. Mol Cell 31:695–707

Moriggl R, Sexl V, Kenner L, Duntsch C, Stangl K, Gingras S, Hoffmeyer A, Bauer A, Piekorz R, Wang D, Bunting KD, Wagner EF, Sonneck K, Valent P, Ihle JN, Beug H (2005) Stat5 tetramer formation is associated with leukemogenesis. Cancer Cell 7:87–99

Mukhopadhyay S, Shah M, Xu F, Patel K, Tuder RM, Sehgal PB (2008) Cytoplasmic provenance of STAT3 and PY-STAT3 in the endolysosomal compartments in pulmonary arterial endothelial and smooth muscle cells: implications in pulmonary arterial hypertension. Am J Physiol Lung Cell Mol Physiol 294:L449–L468

Murphy TL, Geissal ED, Farrar JD, Murphy KM (2000) Role of the Stat4 N domain in receptor proximal tyrosine phosphorylation. Mol Cell Biol 20:7121–7131

Nardozzi J, Wenta N, Yasuhara N, Vinkemeier U, Cingolani G (2010) Molecular basis for the recognition of phosphorylated STAT1 by importin alpha5. J Mol Biol 402:83–100

Ndubuisi MI, Guo GG, Fried VA, Etlinger JD, Sehgal PB (1999) Cellular physiology of STAT3: where's the cytoplasmic monomer? J Biol Chem 274:25499–25509

Neculai D, Neculai AM, Verrier S, Straub K, Klumpp K, Pfitzner E, Becker S (2005) Structure of the unphosphorylated STAT5a dimer. J Biol Chem 280:40782–40787

O'Shea JJ, Kanno Y, Chen X, Levy DE (2005) Cell signaling. Stat acetylation – a key facet of cytokine signaling? Science 307:217–218

Ota N, Brett TJ, Murphy TL, Fremont DH, Murphy KM (2004) N-domain-dependent nonphosphorylated STAT4 dimers required for cytokine-driven activation. Nat Immunol 5:208–215

Park OK, Schaefer LK, Wang W, Schaefer TS (2000) Dimer stability as a determinant of differential DNA binding activity of Stat3 isoforms. J Biol Chem 275:32244–32249

Patel BK, Pierce JH, LaRochelle WJ (1998) Regulation of interleukin 4-mediated signaling by naturally occurring dominant negative and attenuated forms of human Stat6. Proc Natl Acad Sci USA 95:172–177

Pawson T (2004) Specificity in signal transduction: from phosphotyrosine-SH2 domain interactions to complex cellular systems. Cell 116:191–203

Platanias LC (2005) Mechanisms of type-I- and type-II-interferon-mediated signalling. Nat Rev Immunol 5:375–386

Qureshi SA, Leung S, Kerr IM, Stark GR, Darnell JE Jr (1996) Function of Stat2 protein in transcriptional activation by alpha interferon. Mol Cell Biol 16:288–293

Ray S, Boldogh I, Brasier AR (2005) STAT3 NH2-terminal acetylation is activated by the hepatic acute-phase response and required for IL-6 induction of angiotensinogen. Gastroenterology 129:1616–1632

Ren Z, Mao X, Mertens C, Krishnaraj R, Qin J, Mandal PK, Romanowski MJ, McMurray JS, Chen X (2008) Crystal structure of unphosphorylated STAT3 core fragment. Biochem Biophys Res Commun 374:1–5

Rhodes D, Schwabe JW (1998) Signal transduction: fast lane to transcriptional activation. Curr Biol 8:R765–R767

Rivas ML, Cobreros L, Zeidler MP, Hombria JC (2008) Plasticity of Drosophila Stat DNA binding shows an evolutionary basis for Stat transcription factor preferences. EMBO Rep 9:1114–1120

Schaefer TS, Sanders LK, Park OK, Nathans D (1997) Functional differences between Stat3alpha and Stat3beta. Mol Cell Biol 17:5307–5316

Schmid CD, Bucher P (2010) MER41 repeat sequences contain inducible STAT1 binding sites. PLoS One 5:e11425

Schröder M, Kroeger KM, Volk HD, Eidne KA, Grütz G (2004) Preassociation of nonactivated STAT3 molecules demonstrated in living cells using bioluminescence resonance energy transfer: a new model of STAT activation? J Leukoc Biol 75:792–797

Sehgal PB (2008) Paradigm shifts in the cell biology of STAT signaling. Semin Cell Dev Biol 19:329–340

Seidel HM, Milocco LH, Lamb P, Darnell JE Jr, Stein RB, Rosen J (1995) Spacing of palindromic half sites as a determinant of selective STAT (signal transducers and activators of transcription) DNA binding and transcriptional activity. Proc Natl Acad Sci USA 92:3041–3045

Shah M, Patel K, Fried VA, Sehgal PB (2002) Interactions of STAT3 with caveolin-1 and heat shock protein 90 in plasma membrane raft and cytosolic complexes. Preservation of cytokine signaling during fever. J Biol Chem 277:45662–45669

Sherman MA, Secor VH, Brown MA (1999) IL-4 preferentially activates a novel STAT6 isoform in mast cells. J Immunol 162:2703–2708

Shuai K, Horvath CM, Huang LH, Qureshi SA, Cowburn D, Darnell JE Jr (1994) Interferon activation of the transcription factor Stat91 involves dimerization through SH2-phosphotyrosyl peptide interactions. Cell 76:821–828

Shuai K, Liao J, Song MM (1996) Enhancement of antiproliferative activity of gamma interferon by the specific inhibition of tyrosine dephosphorylation of Stat1. Mol Cell Biol 16:4932–4941

Soldaini E, John S, Moro S, Bollenbacher J, Schindler U, Leonard WJ (2000) DNA binding site selection of dimeric and tetrameric Stat5 proteins reveals a large repertoire of divergent tetrameric Stat5a binding sites. Mol Cell Biol 20:389–401

Soler-Lopez M, Petosa C, Fukuzawa M, Ravelli R, Williams JG, Müller CW (2004) Structure of an activated Dictyostelium STAT in its DNA-unbound form. Mol Cell 13:791–804

Stancato LF, David M, Carter-Su C, Larner AC, Pratt WB (1996) Preassociation of STAT1 with STAT2 and STAT3 in separate signalling complexes prior to cytokine stimulation. J Biol Chem 271:4134–4137

Strehlow I, Schindler C (1998) Amino-terminal signal transducer and activator of transcription (STAT) domains regulate nuclear translocation and STAT deactivation. J Biol Chem 273: 28049–28056

ten Hoeve J, de Jesus Ibarra-Sanchez M, Fu Y, Zhu W, Tremblay M, David M, Shuai K (2002) Identification of a nuclear Stat1 protein tyrosine phosphatase. Mol Cell Biol 22:5662–5668

Truchet S, Chebrout M, Djediat C, Wietzerbin J, Debey P (2004) Presence of permanently activated signal transducers and activators of transcription in nuclear interchromatin granules of unstimulated mouse oocytes and preimplantation embryos. Biol Reprod 71:1330–1339 van de Veerdonk FL, Plantinga TS, Hoischen A, Smeekens SP, Joosten LA, Gilissen C, Arts P, Rosentul DC, Carmichael AJ, Smits-van der Graaf CA et al (2011) STAT1 mutations in autosomal dominant chronic mucocutaneous candidiasis. N Engl J Med 365:54–61

Vinkemeier U (2004) Getting the message across, STAT! design principles of a molecular signaling circuit. J Cell Biol 167:197–201

Vinkemeier U, Cohen SL, Moarefi I, Chait BT, Kuriyan J, Darnell JE Jr (1996) DNA binding of in vitro activated Stat1 alpha, Stat1 beta and truncated Stat1: interaction between NH2-terminal domains stabilizes binding of two dimers to tandem DNA sites. EMBO J 15:5616–5626

Vinkemeier U, Moarefi I, Darnell JE Jr, Kuriyan J (1998) Structure of the amino-terminal protein interaction domain of STAT-4. Science 279:1048–1052

Vogt M, Domoszlai T, Kleshchanok D, Lehmann S, Schmitt A, Poli V, Richtering W, Muller-Newen G (2011) The role of the N-terminal domain in dimerization and nucleocytoplasmic shuttling of latent STAT3. J Cell Sci 124:900–909

Wang R, Cherukuri P, Luo J (2005) Activation of Stat3 sequence-specific DNA binding and transcription by p300/CREB-binding protein-mediated acetylation. J Biol Chem 280: 11528–11534

Wenta N, Strauss H, Meyer S, Vinkemeier U (2008) Tyrosine phosphorylation regulates the partitioning of STAT1 between different dimer conformations. Proc Natl Acad Sci USA 105:9238–9243

Wojciak JM, Martinez-Yamout MA, Dyson HJ, Wright PE (2009) Structural basis for recruitment of CBP/p300 coactivators by STAT1 and STAT2 transactivation domains. EMBO J 28:948–958

Xu X, Sun YL, Hoey T (1996) Cooperative DNA binding and sequence-selective recognition conferred by the STAT amino-terminal domain. Science 273:794–797

Yamamoto K, Shibata F, Miyasaka N, Miura O (2002) The human perforin gene is a direct target of STAT4 activated by IL-12 in NK cells. Biochem Biophys Res Commun 297:1245–1252

Yuan ZL, Guan YJ, Chatterjee D, Chin YE (2005) Stat3 dimerization regulated by reversible acetylation of a single lysine residue. Science 307:269–273

Zhang X, Darnell JE Jr (2001) Functional importance of Stat3 tetramerization in activation of the alpha 2-macroglobulin gene. J Biol Chem 276:33576–33581

Zhang JJ, Vinkemeier U, Gu W, Chakravarti D, Horvath CM, Darnell JE Jr (1996) Two contact regions between Stat1 and CBP/p300 in interferon gamma signaling. Proc Natl Acad Sci USA 93:15092–15096

Zhang L, Badgwell D, Bevers J, Schlessinger K, Murray P, Levy D, Watowich S (2006) IL-6 signaling via the STAT3/SOCS3 pathway: functional analysis of the conserved STAT3 N-domain. Mol Cell Biochem 288:179–189

Zhong M, Henriksen MA, Takeuchi K, Schaefer O, Liu B, ten Hoeve J, Ren Z, Mao X, Chen X, Shuai K, Darnell JE Jr (2005) Implications of an antiparallel dimeric structure of nonphosphorylated STAT1 for the activation-inactivation cycle. Proc Natl Acad Sci USA 102:3966–3971

Zimnik S, Gaestel M, Niedenthal R (2009) Mutually exclusive STAT1 modifications identified by Ubc9/substrate dimerization-dependent SUMOylation. Nucleic Acids Res 37:e30

How Stats Interact with the Molecular Machinery of Transcriptional Activation

Amanda M. Jamieson, Matthias Farlik, and Thomas Decker

Abstract
The main purpose of Jak-Stat signal transduction is to adjust the fraction of genes expressed in a genome to environmental cues. The focus of this review is to summarize current knowledge and hypotheses concerning the molecular players and mechanisms that allow Stats to regulate gene expression. We pay particular attention not only to modifications of the Stats themselves, but also to interaction partners with relevance to the remodeling or modification of chromatin or to the recruitment of proteins required for transcriptional initiation and elongation.

Introduction

The Jak-Stat paradigm as defined in the 1990s consists of two components only: the receptor associated Janus kinases (Jak 1-3; Tyk2) and the signal transducers and activators of transcription (Stat 1-4, Stat5a, Stat5b, Stat6) that reside in the cytoplasm. Ligand-bound receptors undergo conformational changes to activate Jaks. These, in turn, create phosphotyrosine (pY)-containing receptor docking sites for Stats. In association with the receptor complex Stats are phosphorylated by Jaks on a single tyrosine residue, whereupon they dissociate from the receptor, dimerize via reciprocal SH2 domain-pY interactions and translocate to the cell nucleus. Association with specific nuclear Stat binding sites stimulates gene transcription (Darnell et al. 1994). Although refined, complemented and extended in many ways, the beauty of this simple paradigm has remained unperturbed (Fig. 1a).

A.M. Jamieson • M. Farlik • Th. Decker (✉)
Department of Microbiology, Immunobiology and Genetics, Max F. Perutz Laboratories, University of Vienna, Dr. Bohr-Gasse 9, 1030 Vienna, Austria
e-mail: thomas.decker@univie.ac.at

Th. Decker and M. Müller (eds.), *Jak-Stat Signaling: From Basics to Disease*,
DOI 10.1007/978-3-7091-0891-8_5, © Springer-Verlag Wien 2012

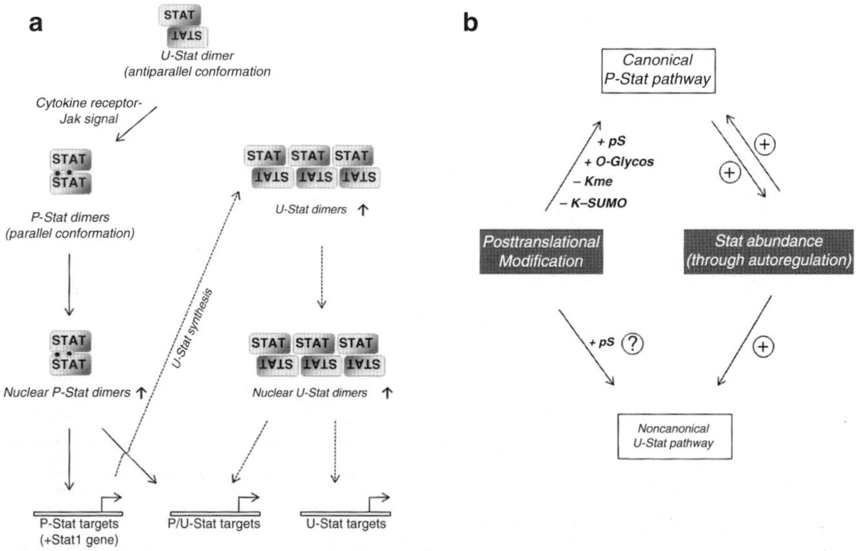

Fig. 1 Gene expression is regulated by tyrosine-phosphorylated P-Stats and by U-Stats that lack tyrosine phosphorylation. *Left*: Signaling through the canonical Jak-Stat pathway produces tyrosine-phosphorylated P-Stat dimers that regulate genes including their own. This produces an increase in U-Stats which increase expression of genes that overlap to various degrees with those induced also by P-Stats. *Right*: Effect of Stat posttranslational modification and abundance on target gene expression. Serine phosphorylation (pS) and O-glycosylation (O-Glycos) increase, whereas Sumoylation (K-SUMO) and lysine methylation (Kme) decrease activity of P-Stats. Serine phosphorylation may be required for U-Stat function. P-Stats increases Stat abundance which initiates U-Stat signaling and provides positive feedback on P-Stat signaling

Stats share a common core structure. To this day they are the only transcription factors using SH2 domains for protein interactions. With exception of Stat2 which borrows the DNA-binding function from associated proteins, all Stats contain homologous DNA-binding domains. Similarly, all Stats contain C-terminal transactivating domains (TAD), but these display no or very little homology to each other. The question of how Stats translate their association with DNA into a transcriptional response has therefore always been studied under the assumption that the answer will not lie in a uniform mechanism.

The basics of Stat activation and their nuclear activity have been addressed in numerous reviews (Darnell 1997; Levy and Darnell 2002; Schindler et al. 2007; Stark 2007) and will be dealt with very briefly here. The main aim of this chapter is to summarize new insight into nuclear responses to Stats and their communication with DNA-associated proteins involved in transcriptional regulation.

Basic Characteristics of Stats and Their Activation

Structural Aspects and DNA Binding Sites

The original concept of latent Stats existing as cytoplasmic monomers prior to activation was challenged not only by biochemical studies but, most notably, by determining the crystal structure of unphosphorylated Stat1 and Stat5 (Mao et al. 2005; Neculai et al. 2005). Together these biochemical and structural studies support the notion that unphosphorylated Stats preexist as dimers. Strikingly, however, unphosphorylated dimers conform to an antiparallel head-to tail alignment whereas the SH2-domain-mediated alignment of tyrosine-phosphorylated Stats is head-to-head or parallel (Mertens et al. 2006; Zhong et al. 2005). The activation-inactivation (or phosphorylation-dephosphorylation) cycle of Stats thus requires the monomers to realign or rotate between parallel and antiparallel conformations. While evidence in agreement with such a mechanism has been provided, its details require further scrutiny.

DNA-bound, tyrosine phosphorylated Stats form highly symmetric dimers with each monomer contacting the SH2 domain on the other (Becker et al. 1998; Chen et al. 1998). The two DNA binding domains display an immunoglobulin fold similar to that of NFκB and p53 transcription factors and do not contact each other. Minor as well as major groove contacts are made on either side of the helix. One consensus DNA sequence forming Stat binding sites is the gamma interferon-activated site [GAS; (Decker et al. 1997)], named after the prototype element in interferon-γ (IFN-γ)-stimulated genes. It is a small $TTCN_xGAA$ palindrome. $X = 3$ sites bind all Stat dimers and $x = 4$ sites select for Stat6 dimers. Variant sequences (e.g. $TTAN_3TAA$ or $TTCN_3TAA$) are also known to be functional as Stat binding sites. For a significant number of target promoters DNA-dependent tetramerization through N-terminus-mediated interaction of adjacent Stat dimers is essential (Moriggl et al. 2005; Vinkemeier et al. 1996; Zhang and Darnell 2001). In this case at least one of the dimers may associate with a highly degenerate variant of the GAS consensus. A second Stat response element is the interferon-stimulated response element [ISRE; (Levy et al. 1988)]. In its consensus version it consists of direct 5′-TTTCNNTTTC-3′ repeats, flanked by AG at the 5′ end and by pyrimidine bases at the 3′ end. Among Stats the ISRE exclusively associates with the ISGF3 complex which is activated in response to type I and type III interferons (IFN) (Darnell 1997; Donnelly and Kotenko 2010; Stark et al. 1998). ISGF3 consists of a Stat1/Stat2 heterodimer in association with IRF9, a member of the interferon regulatory factor (IRF) family. IRF9 contacts the core repeats of the ISRE whereas the Stats make additional contacts to the flanking nucleotides. The employment of IRF9 as the ISGF3 DNA binding subunit reveals a characteristic feature of the ISRE: its core sequence can serve as a binding site for many IRF members and mediate the functional interactions between Stats and IRFs in responses to IFN-γ.

Most members of the Stat family are spliced into different isoforms with a smaller, beta isoform, usually lacking parts of or even the entire TAD. The original notion these might be transcriptionally inactive and form dominant-negative regulators of the full length or alpha isoforms has not held when mice were engineered to reveal the selective effects of Stat isoforms (Maritano et al. 2004; Mo et al. 2008). For both Stat3 and Stat4 the shorter isoforms cause transcription of a specific set of isoform-specific genes. Conversely, another subset of genes can be regulated exclusively by the alpha isoforms. Mice expressing either Stat1α or Stat1β showed that Stat1β does not form transcriptionally inactive dimers as originally suggested. Instead, Stat1β dimers display reduced transcriptional activity, delaying the onset of IFN-γ-induced transcription (Strobl B, Müller M, personal communication). The ways by which alpha and beta isoforms of Stats address different sets of target genes are poorly understood. Likewise it is unclear how the beta isoforms make up for the lack of a bona fide TAD unless the selective recruitment of a "helper" transcription factor, such as JunB in the case of Stat3 (see below), represents a general paradigm.

Regulation by Posttranslational Modification

Apart from tyrosine phosphorylation, activation of Stats is regulated by a number of posttranslational modifications (Fig. 1b). For example, sumoylation of a lysine residue in vicinity of the critical tyrosine has in some Stats the ability to inhibit transcriptional activity (Droescher et al. 2011; Ungureanu et al. 2005). Lysine acetylation was suggested as a regulatory modification controlling Stat tyrosine dephosphorylation and interaction with other transcription factors (Kramer et al. 2009), but recent findings strongly question the validity of the original report (Antunes et al. 2011). Likewise, the role of Stat1 arginine methylation as a mechanism to regulate association with inhibitory protein inhibitors of activated Stats (PIAS) has been challenged (Meissner et al. 2004; Mowen et al. 2001). Phosphorylation of a C-terminal serine impinges on the ability of DNA-bound Stats to contact the transcriptional machinery. Serine727 is located in the TAD of Stat1 and Stat3 [Stats 4, 5 and 6 also contain serine phosphorylation sites at similar positions (Decker and Kovarik 2000)]. In the context of cytokine responses phosphorylation at this residue requires the association with chromatin and the serine kinase itself is similarly associated with DNA (Sadzak et al. 2008; Yang et al. 2010). In fact, recent studies in the lab of Pavel Kovarik suggest that chromatin-associated Stat1 serine kinase activity is provided by CDK8 (personal communication). This S/T kinase and its regulatory subunit cyclin C are found in association with mediator, a multi-subunit complex engaged in orchestrating protein interactions and enzymatic activities in the process of transcriptional initiation (Malik and Roeder 2010). The impact of serine phosphorylation is to enhance cytokine-induced transcription, but this effect varies with different Stat target genes and in different cell types, suggesting its complexity is not yet fully understood. A regulatory mechanism proposed more recently by George Stark and colleagues is based on the methylation of IL-6-activated Stat3 at Lysin 140 by the histone 3K4 methylase Set7/9 (Yang et al. 2010). Methylation of nonhistone proteins is now widely recognized to affect the activation and activity of transcription factors. In keeping with other transcriptional regulators, K140 methylation of Stat3 diminishes its activity through an inhibition of DNA binding. Histone demethylase LSD1 removes the modification of DNA-associated Stat3 dimer. Interestingly, phosphorylation at S727 appears necessary for the methylation of K140 as judged by the lack of methylation of a Stat3S727A mutant. S727 phosphorylation was further suggested to regulate the aforementioned SUMOylation of Stat 1 at Lys703, an activity that may involve the PIAS1 SUMO ligase. Thus, S727 phosphorylation has the potential to enhance transcription through its effect on coactivator binding (see below), but also to inhibit Stat activation and activity, which may explain the gene and cell-specific effects of the S727A mutation of Stats1 and 3 (Shen et al. 2004; Varinou et al. 2003).

A second serine phosphorylation in the carboxy terminus of Stat1 is serine 708, target of the virus/interferon-regulated S/T kinase IKKε. S708 phosphorylation changes the DNA binding site specificity of the ISGF3 complex to increase its target promoter range (Tenoever et al. 2007). The structural basis for this effect is not understood. Speculatively, S708 phosphorylation may determine an alternative alignment of Stat1 and Stat2 within the ISGF3 complex. Knockout of IKKε reduces IFN responsiveness and antiviral immunity of mice and the lack of Stat1 phosphorylation at S708 provides an explanation for this phenotype.

Regulation of Signaling Through Stat Abundance

Apart from regulatory modification the abundance of Stats within a cell determines their transcriptional potential, hence the sensitivity to cytokines employing Jak-Stat signaling for nuclear responses. For Stat1 it is well documented that it regulates its own expression. This autoregulatory loop is established by a tonic signal from type I IFN receptors in absence of their ligands and determines the cellular amount of latent Stat1 (Gough et al. 2010). Evidence for this is provided by the fact that disturbing type I IFN signaling by receptor or Stat2 deficiency causes a strong drop in Stat1 levels (Gough et al. 2010; Park et al. 2000). During responses to IFN or IL-6, Stat1 or Stat3 autoregulation increases their amounts. The increase of Stats is a means by which cytokines alter the hardwiring of their own signals. It is of particular importance for the U-Stat concept discussed below (Fig. 1a).

Lowering the amount of Stats expressed in a cell can be used as a means to adjust the sensitivity to a cytokine. For example, activated CD8 T cells escape the effects of type I IFN produced during a virus infection by lowering amounts of Stat1, thus escaping the antiproliferative effect of the cytokines (Gil et al. 2006). The relative expression levels of one Stat versus another become physiologically relevant where one receptor can employ more than one family member to reprogram gene expression. Resting NK cells express high amounts of Stat4 and owing to its abundance the NK cell type I IFN receptor signals through Stat4 to induce IFN-γ gene expression. During a viral infection the relative amount of Stat1 increases, shifting the type I IFN response from Stat4 to the usual Stat1–Stat2 (ISGF3) activation (Miyagi et al. 2007). Since the IFN-γ gene is not an ISGF3 target, the Stat4-ISGF3 shift thus serves as a means to control IFN-γ production. An inverse situation has been reported for dendritic cells (DC). Upon maturation of immature conventional DC the Stat4:Stat1 ratio increases, causing Stat4 to be preferentially activated by the type I IFN receptor. This leads to a change in gene expression profiles and an increase of IL-12 production by activated DC (Longman et al. 2007).

While Stats are studied predominantly as transcriptional activators numerous gene expression profiles from Stat-deficient cells document the ability of Stats to repress genes they bind to. Prominent examples are the c-myc gene which becomes IFN-γ-inducible in Stat1-deficient cells (Ramana et al. 2001) or the IRF8 gene which is suppressed by Stat5 in response to GM-CSF treatment of bone marrow-derived DC progenitor cells to suppress their plasmacytoid development (Esashi et al. 2008).

Functional and Physical Stat Interactors in the Process of Transcriptional Activation

Primary, Secondary and Tissue-Specific Responses of Stat Target Genes

Transcription factors in the strict sense are only one part of the multi-protein machinery that opens promoter chromatin, forms a transcription initiation complex, and within that complex renders RNA polymerase II (pol II) competent of initiating and elongating the primary transcript. Recent years have identified many players in this process and uncovered their function. Inserting Stats into this complex scenario is a current challenge in the field. In general terms genes regulated by extracellular cues including pathogen or cytokine-induced genes have been classified according to the time flow of their expression as primary if the signals leading to transcriptional induction are hardwired, as secondary if a transcription factor needs to be synthesized and as tertiary if a tissue-specific transcriptional regulator is additionally required to open promoter chromatin. Landmark studies addressing gene regulation by bacterial lipopolysacchride (LPS)/toll-like receptor 4 (TLR4) concluded that a large number of primary response genes contain GC-rich sequences that direct the formation of a transcription initiation complex including a paused pol II prior to the TLR4 signal. In this case the pathogen signal regulates elongation rather than initiation of transcription (Hargreaves et al. 2009; Ramirez-Carrozzi et al. 2009).

Many Stat target genes are primary and respond immediately to a cytokine receptor stimulus. However, a significant number are within the second tier of the transcriptional response to extracellular cues. There is more than one reason for the secondary character of a Stat target gene's response. For example, a cytokine may activate a Stat and at the same time induce the synthesis of a transcription factor required to cooperate with that Stat for gene induction. Prime examples are IFN-γ-induced secondary response genes depending on promoter binding of a Stat1 dimer and, additionally, IRF1. The IRF1 gene itself is a primary response gene to IFN-γ and its product needs to be made for the secondary response gene to be transcribed (Pine et al. 1994). In infected cells Stat target genes belong to the second tier of the transcriptional response because the activating cytokine, usually a type I or type III IFN, needs to be synthesized before it can bind its receptor and cause Stat activation. As an example, signaling by several pattern recognition receptors stimulates type I IFN synthesis, followed by ISGF3 activation. Hence a significant number of secondary pathogen-induced genes are ISGF3 targets (Hertzog et al. 2003). These Stat target genes are secondary according their kinetics of induction during infection, yet the mechanism of induction is primary, because it employs hard-wired signals downstream of the IFN receptor. Finally, Stat target genes can be rendered tissue-specific by cooperation with lineage-restricted transcriptional regulators. In agreement with this induction of numerous Stat1 or ISGF3 target genes by IFN in hematopoietic cells results from the cooperative activity of the Stat1 dimer or the ISGF3 complex with the Ets family protein Pu.1 with either IRF8 alone or with either IRF4 and IRF8, two transcription factors with prominent roles in the maintenance of myeloid and lymphoid cell identity (Kanno et al. 2005). Cooperativity is thought to result from the direct association of some GAS or ISRE elements with Pu.1/IRF complexes and the concomitant enhancement or extension of the IFN response.

Interaction with Non-Stat Transcription Factors

The proteins cooperating in the onset of a transcriptional response can be functionally classified as transcription factors, coactivators, chromatin remodelers, histone-modifying enzymes, variant histones, or as subunits of the mediator or transcription initiation complexes. One or more members of each category has been linked to the induction of Stat target genes (Fig. 2, Table 1).

Cooperativity between Stats and other transcription factors is frequent, but mechanistic insight into the concerted action is relatively scarce. An example of positive cooperativity at the level of DNA binding is presented at the *CIIta* gene encoding the master regulator of MHC II genes. At promoter IV of *CIIta* the E-box-binding factor USF-1 facilitates the interaction of Stat1 with a GAS element and this may be the prelude to the formation of an enhanceosome containing besides Stat1 and USF1 also IRF1 and c-Myc (Morris et al. 2002; Muhlethaler-Mottet et al. 1998; Ni et al. 2005). A more recent report shows that binding of the nuclear receptor PPARγ to promoter chromatin is facilitated by simultaneous association of IL-4-activated Stat6 dimers with a composite DNA element (Szanto et al. 2010). This interaction is in keeping with the common ability of IL-4 and PPARγ ligands to suppress inflammation and to support the anti-inflammatory M2 polarization of macrophages and DC. The ability of Stat6 dimers to suppress IRF1 transcription in response to IFN-γ through inhibition of Stat1 binding (Ohmori and Hamilton 1997), or to reduce E-Selectin expression in response to TNF by inhibition of NFκB

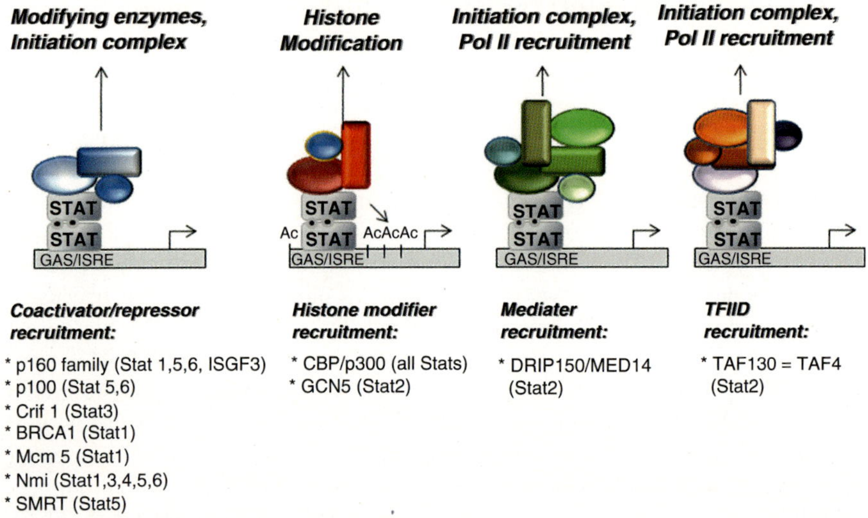

Fig. 2 Stat interaction with protein (complexes) required for transcriptional initiation and elongation. These can be subdivided according to their mechanism of action as nonenzymatic coactivators/corepressors, histone modifiers, or as subunits of the mediator and TFIID complexes. For further explanation see text

association with a site overlapping that of Stat6 (Bennett et al. 1997) is in further agreement with IL4/Stat6′ role in the suppression of inflammation.

Transactivation functions, i.e. enhanced ability to access the transcriptional machinery are a further reason for Stats to interact. A well-documented case is the association of Stat3 with members of the fos/jun transcription factor family. The regulation of hepatic acute phase genes such as the α2Makroglobulin gene requires formation of an enhanceosome containing Oct-1, AP1 (a c-Fos/c-Jun heterodimer), glucocorticoid receptor (GR) and Stat3 (Lerner et al. 2003). Biochemical studies suggest that within this functional unit the coiled-coil domain of Stat3 physically associates with c-Jun (Ginsberg et al. 2007; Schaefer et al. 1995). Therefore the interaction with AP1 provides a platform for enhanced DNA binding of Stat3, but most likely it also increases its transactivation function because in absence of c-Jun Stat3 cannot induce expression of α2Macroglobulin reporter genes even though they contain high affinity Stat3 binding sites (Ginsberg et al. 2007). Similarly, physical interaction of the Stat5 C-terminus with the GR is essential for the expression of Stat5 target genes regulating proper function of hepatocytes (Engblom et al. 2007). This interaction does not require GR to associate with DNA. Although the above examples provide clear evidence for cooperation at the level of transactivation the mechanistic implications for promoter activation and the onset of transcription downstream of their association with target promoters have not been clarified.

Stats can cooperate with other transcription factors in, respectively, the recruitment and phosphorylation of pol II. This was shown by a recent study from our lab

Table 1 Examples of Stat-interacting transcription factors

Protein interactor	Interacting Stats	Molecular mechanism	Consequence	References
IRF1	Stat1	Binding to ISRE, Pol II recruitment	Second tier of transcriptional response to IFN-γ	Ni et al. (2005), Ramsauer et al. (2007)
IRF4	Stat3	Increased Stat binding to low affinity GAS	Tissue specificity of Stat response (lymphoid)	Kwon et al. (2009)
IRF4/IRF8/Pu.1	Stat1, ISGF3	Binding to composite GAS and ISRE sites	Tissue specificity of IFN response (myeloid, lymphoid)	Kanno et al. (2005)
IRF7	Stat1	Binding to ISRE, Pol II recruitment	Second tier of transcriptional response to IFN-γ	Farlik et al. (2012)
USF1	Stat1	Enhanced Stat1 binding	Transcriptional activation of CIIta gene	Morris et al. (2002), Muhlethaler-Mottet et al. (1998), Ni et al. (2005)
NFκB	Stat1	Cooperative initiation complex formation	Increased host response to pathogens	Farlik et al. (2010)
	Stat6	Competitive binding to overlapping sites	Reduced expression of TNF-induced genes in presence of Stat6	Bennett et al. (1997)
PPARγ	Stat6	Binding to composite DNA site	Enhanced M2 polarization of macrophages	Szanto et al. (2010)
c-Jun/JunB	Stat3	Enhanceosome formation, physical association	Enhanced target gene expression (e.g. acute phase genes)	Ginsberg et al. (2007), Lerner et al. (2003), Schaefer et al. (1995)
GR	Stat5	Physical association (DNA-independent)	Enhanced expression of a subset of Stat5 target genes (e.g. in liver)	Engblom et al. (2007), Stocklin et al. (1996)
	ISGF3	Competition for GRIP1 coactivator	Glucocorticoid repression of ISGF3 target genes (cell type-dependent)	Flammer et al. (2010)
SMAD1/SMAD4	Stat3	Cooperative CBP recruitment	Enhanced transcription of LIF/BMP2 target genes	Nakashima et al. (1999)

addressing the regulation of the *Nos2* gene encoding the inducible nitric oxide synthase in macrophages infected with intracellular bacteria (Farlik et al. 2010). In this situation type I IFN is produced and ISGF3 is activated to bind at a distal promoter site, but this complex alone provides only a relatively small increase in Nos2 expression. A full-blown transcriptional response requires the cooperation with NFκB bound to a proximal site. ISGF3 alone can recruit pol II to the Nos2 promoter, but cannot attract the general transcription factor TFIIH with its associated kinase CDK7. The kinase function is needed to phosphorylate the carboxyterminal pol II domain (CTD) and the serine 5 (S5) residue embedded in its hepta-amino acid repeats. Since S5 phosphorylation is an essential prerequisite for promoter clearance and also the subsequent steps leading to elongation competence of pol II, neither ISGF3 nor NFκB activity alone can bring about high levels of Nos2 expression, whereas both signals together are highly efficient. It remains to be clarified whether all ISGF3 target genes, i.e. the classical type I IFN-induced genes (ISGs) require a helper function for TFIIH/CDK7 recruitment, e.g. by a constitutively bound transcription factor, or whether the promoter chromatin of such genes is somehow permissive to the recruitment of an elongation-competent pol II by ISGF3 alone.

Excepting the role of IRF9 in the ISGF3 complex (Horvath et al. 1996), interferon regulatory factors (IRF) are not known to physically interact with activated Stat dimers, but they contribute to the regulation of their target genes. Together IRF1 and IRF7 are essential for the expression of genes representing the secondary response to IFN-γ such as those encoding the antimicrobial guanylate-binding proteins (GBP) 1 and 2 or CIITA, the master regulator of MHC II (Harada et al. 1996; Kamijo et al. 1994; Ni et al. 2005; Ramsauer et al. 2007). In this situation the functional cooperation lies in the recruitment of histone-modifying enzymes by the IFN-γ-activated Stat1 dimer and an unclear role of the IRFs in recruiting pol II (Ramsauer et al. 2007). Interestingly, transcriptional activation of type I IFN-inducible genes by the ISGF3 complex does not require an ancillary function of IRF1 (Matsuyama et al. 1993). This suggests that in the ISGF3 context the Stat2 TAD exerts a pol II-recruiting activity the Stat1 TAD cannot perform in the context of the Stat1 dimer. Hence Stat1 dimers require an IRF to provide this activity. As mentioned above, IRF8 contributes to the IFN-γ inducibility of genes specifically expressed in myeloid cells (Kanno et al. 2005). However, a helper function of IRFs is not restricted to Stat1. The receptor for IL-21, a cytokine regulating B cell differentiation and T cell homeostasis, signals via Stat3. Comprehensive analysis of IL-21 target genes in lymphocytes showed that many of these contain adjacent binding sites for Stat3 and IRF4 (Kwon et al. 2009). Deletion of the *Irf4* gene reduced Stat3 binding to genomic sites and inhibited transcription of IL-21-induced genes. This result suggests that IRFs not only contribute to pol II binding to Stat target genes, but that they may serve to facilitate the association of Stat dimers with target promoter chromatin.

Interaction with General Transcription Factors and Mediator Proteins

The canonical mode of transcription initiation assumes a stepwise assembly of an initiation complex. Activators have the ability to exert their influence on gene transcription by facilitating the binding of one or more components of the initiation complex. The first step in assembly is the binding of the general transcription factor TFIID in close proximity to the transcription start. Most TFIID complexes are composed of the TATA-binding protein (TBP) and up to 14 TBP-associated factors (TAFs, Goodrich and Tjian 2010). In the context of the ISGF3 complex the TAF 130 subunit (new designation: TAF4) potentiates the activity of the Stat2 TAD (Paulson et al. 2002; Fig. 2). In keeping with its function to mount an antiviral response the Stat2 TAD recruits a TBP-free TFIID complex not subject to the TBP-degrading activity of some viruses (Paulson et al. 2002).

Mediator is a multi-subunit complex interacting with the carboxyterminal domain of pol II and contact to a mediator subunit is an additional strategy of transcription factors to facilitate pol II promoter binding (Kim et al. 1994; Malik and Roeder 2010). In vitro purified mediator stimulates transcriptional activity of both Stat1 isoforms (Zakharova et al. 2003), suggesting contact is not via the TAD which is missing in Stat1β. By contrast, the contact of the ISGF3 complex to a multi-subunit mediator complex was shown to be provided by a physical association between the Stat2 TAD and the mediator subunit DRIP150/MED14 (Lau et al. 2003; Fig. 2). Stat2/mediator interaction may influence the rate at which reinitiation occurs at IFN-inducible genes.

Interaction with Coactivators and Histone-Modifying Enzymes

Coactivators fall into two general categories, histone modifiers and nonenzymatic proteins that may serve a bridging or scaffold function for the formation of a transcription initiation complex and an elongation-competent pol II. Within the former category the CBP/P300 histone acetyl transferases (HAT) are widely used, most likely interacting with all Stat TADs (Paulson et al. 1999) and in some instances shown by ChIP to be recruited to promoter chromatin of Stat target genes in a cytokine-regulated fashion (Fig. 2). Studies in Stat1-deficient cells show that both recruitment of CBP and hyperacetylation of histones in response to IFN-γ require Stat1. In vitro, both N- and C terminal portions of Stat1 and Stat3 interact with CBP (Zhang et al. 1996) and the association of CBP with Stat1 does not show a need for S727 phosphorylation (Wojciak et al. 2009). On chromatin-directed in vitro templates (Zakharova et al. 2003) or on cellular chromatin (Ramsauer et al. 2007; Sun et al. 2005; Varinou et al. 2003) association with CBP requires the Stat1 C-terminus and association is enhanced by phosphorylation of S727. P300/CBP can be found in larger coactivator complexes and it is possible that the discrepancy between in vitro and in vivo results is explained by additional proteins that direct such complexes to the Stat1 TAD. In case of Stat5, O-glycosylation of T93, a residue found at a homologous position also in Stats 1, 3 and 6, enhances association with CBP in vitro, consistent with the notion that Stats present more than one critical interface with the coactivator (Gewinner et al. 2004). This assumption is in further agreement with the older observation that the N-myc interactor (Nmi) protein binds to the Stat coiled-coil domains (excepting Stat2) and enhances CBP interaction (Zhu et al. 1999).

The Stat2 TAD is able to associate with P300/CBP (Paulson et al. 1999), but chromatin immunoprecipitation experiments point to the histone acetylase GCN5 as provider of HAT activity and as cellular coactivators for the transcription of type I interferon-stimulated genes (Paulson et al. 2002) (Fig. 2). Like HAT, histone deacetylases (HDAC) can be recruited to target promoters by Stats 1, 2 and 5 and serve as cofactors of transcription (Nusinzon and Horvath 2005; Ramsauer et al. 2007; Rascle et al. 2003). Surprisingly these don't act as HAT antagonists but rather as stimulators of Stat-mediated transcription. The targets of these HDAC are most likely not histones because their recruitment does not cause a concomitant decrease in promoter acetylation. Whether Stats themselves or whether other proteins involved in the activation of transcription are the relevant targets of HDAC activity remains to be clarified. The Zn finger transcription factor PDZF acts as a stimulator of a subset of type I IFN-inducible genes, hence a positive interactor of the ISGF3 complex (Xu et al. 2009). In other contexts the protein is known to associate with repressive complexes containing histone deacetylases. Xu and colleagues suggest that phosphorylation increases association of PDZF with HDAC1 and this may partly explain the stimulatory activity on the expression of interferon-stimulated genes.

A nonenzymatic protein family serving as Stat coactivators is the p160 family of steroid receptor coactivators (York and O'Malley 2010; Fig. 2). Interaction and functional importance of p160 family members SRC-1 and GRIP1/SRC2 were reported for Stats 1, 3, 5 and 6 (Flammer et al. 2010; Giraud et al. 2002; Litterst et al. 2003; Litterst and Pfitzner 2001), although a recent report was not able to confirm relevance for Stat3 target genes (Cvijic et al. 2009). In case of Stat6 the structure of the TAD/SRC1 interface was solved showing interaction of a Stat6 LLXXLL motif with the SRC1 PAS domain (Razeto et al. 2004). A further coactivator without enzymatic activity for both Stat5 and Stat6 is p100, a staphylococcal nuclease-like Tudor domain-containing protein (Paukku et al. 2003; Yang et al. 2002). P100 also interacts with the general transcription factor TFIIE (Leverson et al. 1998). Because TFIIE is considered preassociated with pol II, p100 may serve to bridge the Stat5/6 TADs with the initiation complex. A protein recently shown to enhance Stat3-mediated transcription and biological activity is CR6-interacting factor 1 (Crif1; Kwon et al. 2008a). Crif1, previously known as coactivator for the nuclear orphan receptor Nur77, interacts with the Stat3 coiled-coil domain via its own coiled-coil domain. Absence of Crif 1 reduces the association of Stat3 with DNA, suggesting a mechanism for its function as coactivator.

Two further proteins reportedly enhancing Stat1-mediated transcription are MCM5 and Brca1 (Ouchi et al. 2000; Zhang et al. 1998). Both bind to the S727-phosphorylated C-terminus. Their function in the process of transcriptional activation is unclear. Finally, proteins that bind to Stats and repress their activity have been identified. One example is the nuclear receptor corepressor SMRT, shown to interact with the Stat5 coiled-coil domain, thereby repressing its transcriptional activity (Nakajima et al. 2001). It is unclear whether SMRT activity acts in the context of HDAC complexes or whether it antagonizes the interaction with CBP.

Different transcriptional coactivators are frequently found within multi-subunit complexes that interact with several different transcription factor families.

Assuming such complexes are limiting within the nucleus it is possible that transcription factors need to compete for their coactivators (Horvai et al. 1997). Previous reports are consistent with the idea that competition for CBP-containing activator complexes coordinates Stat1 in inflammatory responses involving NFκB, AP1 or PPARγ activity (Horvai et al. 1997; Kurokawa et al. 1998; Ricote et al. 1998). More recently competition for binding to the steroid receptor coactivator SRC2/GRIP was found to explain the repression of ISGF3 target genes by glucocorticoids (Flammer et al. 2010). The study showed SRC2/GRIP1 to interact with the IRF9 subunit of ISGF3 and to be required for ISGF3-stimulated transcription in a gene and cell type-specific manner. Where levels of SRC2/GRIP1 are limiting, glucocorticoids shift too much of the protein to the glucocorticoid receptor, thus limiting the coactivator for the response to type I IFN.

In addition to competition, Stats and interacting transcription factors can provide a common platform for coactivators as shown for CBP recruitment by Stat3 and the SMAD1/SMAD4 complex. This mechanism is thought to explain synergistic gene induction by LIF and BMP2 (Nakashima et al. 1999).

Interaction with Nucleosome Remodelers

Chromatin remodeling delineates the process of rearranging nucleosomes to alter the availability of DNA for regulatory input (Clapier and Cairns 2009). Numerous studies on the subject of Stat target promoter activation emphasize the critical importance of the mammalian version of the yeast SWI/SNF complex, designated BAF, with two highly related ATPase subunits, BRG1 and BRM (Fig. 3). A related complex, PBAF, shares eight subunits with BAF. However, PBAF has a unique BAF180 subunit whereas BAF250a is exclusively found in BAF (Clapier and Cairns 2009). Human cells lacking BRG1 are available and these have been extensively studied next to cells depleted of subunits BAF47, BAF180 or BAF250a by siRNA-mediated knockdown. The majority of reports addresses the role of BRG1 for Stat1 dimer recruitment and function in the IFN-γ response or for promoter binding and transcriptional activation by the ISGF3 complex, activated by type I IFN. These studies are consistent in showing that for both Stat1 dimer and ISGF3 only a subset of target genes is affected by the lack of BAF function (Huang et al. 2002; Liu et al. 2002; Ni et al. 2005; Pattenden et al. 2002). BAF subunits can be isolated from target promoters by chromatin immunoprecipitation prior to IFN treatment (Liu et al. 2002; Ni et al. 2008) suggesting their binding is mediated by a constitutive factor. Consistent with this finding, BAF-dependent remodeling is a prerequisite to the binding of both Stat1 dimers and ISGF3 to nuclear chromatin (Cui et al. 2004; Ni et al. 2005, 2008). BRG1 association with promoter IV of the *CIIta* gene encoding CIITA, the master regulator of MHC II genes, involves several distal sites and correlates with the increased appearance of active chromatin marks. Recruitment of BRG1 per se does not cause the formation of additional contacts between distant promoter regions (promoter DNA loops), but it is necessary for IFN-γ-dependent long range interactions which may link Stat1 and IRF1 complexes

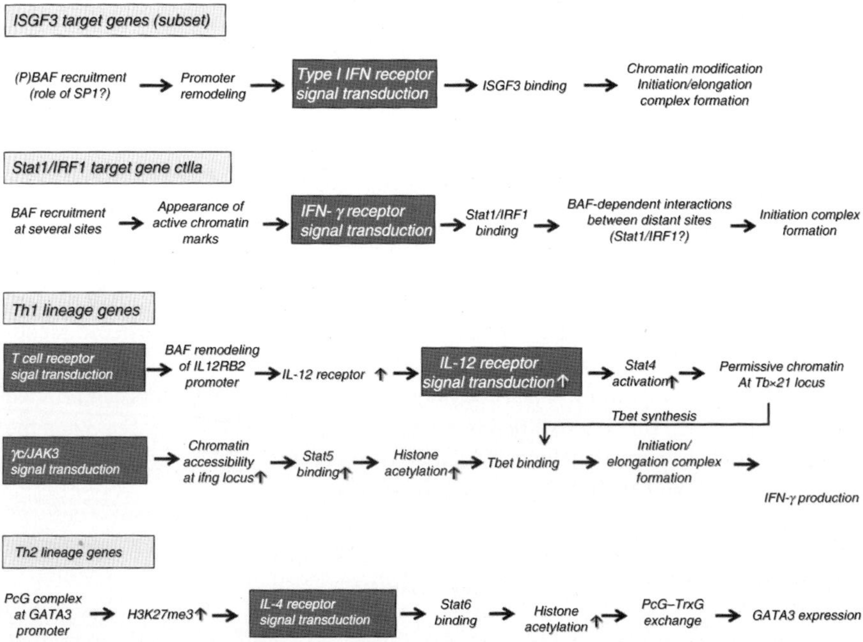

Fig. 3 Sequence of chromatin remodeling, Stat binding and chromatin modification as suggested for interferon-induced genes or for genes specifying helper T cell lineage. For further explanation see text

bound to remote promoter elements and facilitate their interaction with the transcription start site (Ni et al. 2008).

The selectivity of BAF complexes for individual Stat1 dimer or ISGF3 target genes is puzzling. For example, the primary IFN-γ-induced gene *Irf1* does not require BRG1 whereas the secondary response genes *Gbp2* and *CIIta* do. To make matters more complex, induction of the *Irf1* gene by Stat3 in the context of an IL-6 response is BRG1-dependent (Ni and Bremner 2007). Currently there is little insight into the gene and stimulus-dependent BAF requirement. One study demonstrates selectivity of BAF and PBAF complexes for the human type I IFN-induced genes IFIT1 and IFIT3 (Yan et al. 2005), but with both complexes sharing the BRG1 subunit this finding cannot explain the BRG1 independence of some Stat target genes. Transcription factor Sp1 was suggested to recruit BAF selectively to a subset of ISGF3 target genes (Liu et al. 2002), but it is unclear whether BAF recruitment by SP1 or an analogous constitutively associated transcription factor generally explains gene selectivity. Conceivably not all Stat target genes require chromatin remodeling prior to induction by cytokines. Alternatively this function could be exerted by non-BAF remodeling complexes. In this regard a report showing involvement of the NURF chromatin remodeling complex in the regulation of a subset of Stat target genes in Drosophila may be of interest (Kwon et al. 2008b). Similar to gene regulation by Stats 1 and 2, a subset of IL-6-induced genes requires BAF activity as a prerequisite for the association of Stat3 with their promoters (Ni and Bremner 2007).

Stat-mediated chromatin remodeling and modification appears to be of large importance for the development of functional helper T cell (Th) subpopulations (O'Shea et al. 2011; Fig. 3). One example is the locus encoding the IL-12 receptor (IL-12R) β2 chain, which is expressed in the Th1, but not the Th2 lineage. Signaling by the T cell receptor causes BAF-dependent remodeling of the promoter and some IL-12 receptor is made. IL-12 signaling then further increases IL-12Rβ2 chain expression by providing Stat4 activity (Letimier et al. 2007). In addition to the regulation of the IL-12Rβ2 chain, a Jak3-dependent increase of chromatin accessibility at the IFN-γ promoter is thought to involve the recruitment of histone acetylases by Stat5 as a prerequisite for the binding of the Th1 differentiation factor T-bet (Shi et al. 2008a). Permissive chromatin at the T-bet/*Tbx21* gene itself requires Stat4, as judged by the absence of modifications specifying active chromatin in Stat4−/− T cells (Wei et al. 2009).

In analogy to T-bet in Th1 cells, GATA3 is a crucial transcription factor for the development and maintenance of the Th2 fate. An elegant recent study by T. Nakayama and colleagues demonstrates that expression of the *Gata3* gene in the course of Th2 differentiation requires displacement of the PcG (polycomb) complex at the *Gata3* promoter by the TrxG (trithorax)/Menin complex (Onodera et al. 2010; Fig. 3). PcG contains a histone methylase introducing repressive H3K27 marks and maintains a chromatin state that prohibits transcription. By contrast, TrxG/Menin binding leads to the appearance of the activating H3K4me3 mark and serves to establish and maintain a permanently open chromatin state that allows gene expression. Under Th2 conditions, IL-4 activates Stat6 to associate with two sites downstream of the *Gata3* transcription start [binding of Stat6 to the *Gata3* locus was also found by Wei and colleagues (Wei et al. 2009)]. Stat6 promoter binding is necessary for PcG displacement and concomitant binding of TrxG, as both events were inefficient in Stat6−/− T cells. The reported data are further consistent with the notion that the recruitment of CBP/P300 HAT by Stat6 induces H3 hyperacetylation at the relevant promoter region and the subsequent PcG displacement. TrxG binding to the PcG-free promoter abrogates further Stat6 requirement to maintain the Th2 fate. Whether a similar mechanism is responsible for the Stat5/T-bet connection in Th1 cells has not been reported so far.

In addition to moving or depleting nucleosomes, incorporation of variant histones contributes to chromatin remodeling (Clapier and Cairns 2009). Exchange of H3 with the variant histone H3.3 frequently occurs at transcriptionally active genes. A recent report shows that induction of type I IFN-inducible genes by ISGF3 is accompanied by H3.3 incorporation at the coding region (Tamura et al. 2009). Unexpectedly the variant was more prominently found at the distal end of the coding region whereas no exchange occurred at the transcription start. The variant histones persisted beyond the transcriptional response of the genes to a type I IFN stimulus and were replaced by H3 during cell division.

Gene Regulation by Stats in Absence of Their Tyrosine Phosphorylation: The U-Stat Concept

The idea that tyrosine phosphorylation-independent noncanonical U-Stat pathways might be biologically relevant stems from experiments with genetically reconstituted Stat1-deficient cells. As expected, a Stat1Y701F mutant did not restore responses to interferons in such cells, but, unexpectedly, it restored the cells' ability to express apoptosis-related genes and to respond to an apoptotic stimulus (Kumar et al. 1997). Restoration of this response required the S727 phosphorylation site in the Stat1 TAD, suggesting that at least some tyrosine phosphorylation-independent U-Stat signaling requires phosphorylation of S727. Additional impetus for the U-Stat idea came from findings in Drosophila showing that unphosphorylated Stat92E (the only Stat homologue in this organism) interacted with heterochromatin protein 1 (HP1) to maintain transcriptional silencing of parts of the genome and to avoid position effect variegation (Shi et al. 2006, 2008b).

In mammals Stat autoregulation plays an important role in U-Stat signaling (Fig. 1). Both Stat1 and Stat3 dimers when activated by, respectively, interferons or IL-6 upregulate their own expression (Yang et al. 2005, 2007). Therefore, high Stat levels are present in the late phase of these cytokine responses and persist beyond the stage at which Stats are phosphorylated on tyrosine. George Stark and colleagues demonstrated that high U-Stat amounts correlate with the expression of late phase interferon or IL-6-induced genes. These genes are only partially redundant with targets of the Stat3 dimer in the IL-6 response, but largely redundant with IFN-induced genes in case of U-Stat1 (Cheon and Stark 2009; Yang et al. 2005). Thus U-Stat3 partially alters gene expression in the late phase of the IL-6 response, but the function of U-Stat1 (also within a U-Stat1/U-Stat2 complex) appears to maintain expression of a subset of IFN-induced genes beyond the phase of Jak signaling (Cheon and Stark 2009). The mechanisms by which U-Stats translocate to the cell nucleus and activate gene expression are only partially understood. U-Stat3 can interact with the NFκB subunit p65 and use this as a vehicle to the cell nucleus. In fact, many, but not all of the late stage IL-6 targets of U-Stat3 contain NFκB sites (Yang et al. 2007). Some of these encode oncogenes and are highly upregulated in phospho-Stat3-positive human tumors (Yang et al. 2005). Reportedly, U-Stat1 interacts with IRF1 and the complex binds to promoters with overlapping binding sites for Stat1 dimers and IRF1 (Chatterjee-Kishore et al. 2000). However, Stark and colleagues have not found evidence for such sequences in all U-Stat1-regulated promoters, leaving the U-Stat1 potential for gene regulation a challenge for future studies.

A strikingly different mode of employing U-Stats for signaling was reported for helper T cell precursor cells (Maldonado et al. 2009). When such cells commit to the Th1 lineage, T cell receptor signaling causes Stat1 to bind first to the immunological synapse at the cell membrane and, thereafter, translocate to the nucleus. In this process Stat1 is not phosphorylated on tyrosine, but constitutively phosphorylated at S727. Cells deficient for Stat1 or expressing a Stat1S727A mutant show defects in Th1 differentiation. The data suggest a hitherto undefined nuclear role of serine-phosphorylated U-Stat1 in the process of Th1 differentiation.

Genome-Wide Analyses of Stat Association with Nuclear Chromatin

During recent years the field of transcriptional regulation has entered an era during which the scope of identifying regulatory mechanisms has shifted away from individual genes to large groups of genes or all genes of a genome. Whole genome expression profiles obtained from microarrays or, increasingly, from deep sequencing of total cellular RNA (RNA-Seq; Ozsolak and Milos 2010) are combined with transcription factor binding profiles. Originally such profiles were obtained by hybridizing DNA isolated by chromatin immunoprecipitation with anti-transcription factor antibodies (ChIP) to oligonucleotide tiling arrays spanning portions of the genome (ChIP-chip). A few years ago this technology was largely supplanted by ChIP-Seq, i.e. massive parallel sequencing of ChIP-derived genomic DNA (Macquarrie et al. 2011; Valouev et al. 2008). The strength of this approach lies in the possibility to combine expression profiles with transcription factor binding profiles, pol II binding profiles, variant histones and the distribution of activating or repressive chromatin marks. Both ChIP-chip (Hartman et al. 2005; Wormald et al. 2006) and ChIP-Seq (Durant et al. 2010; Elo et al. 2010; Kwon et al. 2009; O'Shea et al. 2011; Robertson et al. 2008; Wei et al. 2010) have been applied to Stats. The biological implications of these studies are reviewed elsewhere, but some implications for transcriptional regulation are briefly summarized in the following paragraphs.

All studies analyzing either chromosome- or genome-wide Stat target genes concur in finding or extrapolating several thousand genomic binding sites for the Stat under study and, after combining these with expression data, in confirming previously known target genes as well as in identifying hitherto unrecognized Stat-regulated genes. The notion from numerous microarrays that a large proportion of genes are negatively regulated by Stats is supported by ChIP chip and ChIP-Seq (Elo et al. 2010; Hartman et al. 2005; Wei et al. 2010). Within the resolution of the technology, most Stat binding appears to involve GAS or ISRE elements, although these are not always perfect consensus sequences (Elo et al. 2010; Hartman et al. 2005; Kwon et al. 2009; Wormald et al. 2006). As predicted from studies prior to the -omics era, deviation from the consensus sequence can indicate cooperative binding with other transcription factors, as convincingly shown for the coregulation of IL-21-induced genes by Stat3 and IRF4 (Kwon et al. 2009). The majority of binding was assigned to regulated genes, with some variability both between Stats and studies of the same Stat e.g. (Hartman et al. 2005; Wormald et al. 2006). There are no rules about the location of binding sites relative to the transcribed region. In fact, many sites are within, far upstream or downstream of the transcribed region, and a fairly small proportion is proximal to the transcription start (Durant et al. 2010; Elo et al. 2010; Hartman et al. 2005; Kwon et al. 2009; Robertson et al. 2008; Wei et al. 2010). However, this view was challenged for Stat1 after combining the data set on Stat1 binding to chromosome 21 (Hartman et al. 2005) with an analysis of pol II binding (Wormald et al. 2006). The authors concluded that most Stat1 binding after IFN-γ treatment occurs in vicinity to a pol II molecule. Depending on Stat and study a variable proportion of Stat binding occurs in intergenic no man's land. Finding the significance of these associations is a future challenge. Likewise the significant amount of Stat binding in absence of a cytokine stimulus requires explanation. These could be U-Stats or dimers formed in response to tonic tyrosine kinase signaling.

Both active (H3Ac, H3K4me3, H3K36me3) or repressive (H3K27me3) chromatin modifications at or in the vicinity of target genes can, but must not require Stat binding (Durant et al. 2010; Wei et al. 2010). Strikingly, Stats can promote active or repressive marks and, as described above for Stat6, they can help remove or prevent repressive marks. In this context, Th differentiation is a well studied arena for the antagonistic activities of Stat4, promoting Th1 development and Stat6, a Th2 lineage transcriptional activator (O'Shea et al. 2011; Wei et al. 2010). Stat4 promotes active marks on genes required for the Th1 lineage and induces repressive marks at a subset of Stat6 target genes specifying the Th2 lineage. The exact opposite is true for Stat6. The molecular basis for the gene-specific antagonistic effects of Stats as transcriptional activators or repressors is not understood but, judged by this and other studies, likely to be transported via target gene chromatin.

An analysis of IFN-induced Stat binding to chromosome 22 produced the expected result that in response to type I IFN most binding of Stat1 occurred together with Stat2 (i.e. in the ISGF3 context), but some binding was Stat2-independent, presumably representing binding by the Stat1 dimer which is formed in response to both type I IFN and IFN-γ. Unexpectedly, some of the Stat2-independent Stat1 binding in IFN-α treated cells occurred at sites not occupied by a Stat1 dimer after IFN-γ treatment (Hartman et al. 2005), suggesting that either a differential Stat1 modification or differential employment of ancillary activators/ coactivators by type I IFN and IFN-γ directs the Stat1 dimer to distinct DNA sites. In correspondence with numerous reports, the vast majority of IFN-induced genes are pol II-free in the uninduced state, confirming the notion that IFN-activated Stats and their partners actively recruit the enzyme to the transcriptional start site (Wormald et al. 2006).

Concluding Remarks

Twenty years of Stat research have accumulated a wealth of knowledge about Stats, their structure, modification, target genes and mechanisms of action. At the same time these 20 years have seen an enormous pace at which new technologies have become available for studies of transcriptional regulation. At the current stage we can take genome-wide snapshots of Stats acting at their target promoters and overlay these with similar images describing chromatin dynamics. Applying this technology is likely to help us understand the complexity of biological effects produced by Stat signaling in different cell types. It has already produced significant challenges for future studies including frequent Stat binding to nuclear chromatin in unstimulated cells, binding to intergenic regions without an obvious functional context, or the association of Stat1 with a subset of chromatin barriers that demarcate the border between active and repressed chromatin (Cuddapah et al. 2009). In addition to this genomic technology-driven advance the findings summarized in our article demonstrate that new discoveries concerning Stat structure, modification and regulation are still being made with more conventional approaches. The excitement of Stat research will continue for years to come.

Acknowledgements Our work is supported by the Austrian Science fund (FWF SFB-28) and GEN-AU III ("Inflammobiota").

References

Antunes F, Marg A, Vinkemeier U (2011) STAT1 signaling is not regulated by a phosphorylation-acetylation switch. Mol Cell Biol 31:3029–3037

Becker S, Groner B, Muller CW (1998) Three-dimensional structure of the Stat3beta homodimer bound to DNA. Nature 394:145–151

Bennett BL, Cruz R, Lacson RG, Manning AM (1997) Interleukin-4 suppression of tumor necrosis factor alpha-stimulated E- selectin gene transcription is mediated by STAT6 antagonism of NF-kappaB. J Biol Chem 272:10212–10219

Chatterjee-Kishore M, Wright KL, Ting JP, Stark GR (2000) How Stat1 mediates constitutive gene expression: a complex of unphosphorylated Stat1 and IRF1 supports transcription of the LMP2 gene. EMBO J 19:4111–4122

Chen X, Vinkemeier U, Zhao Y, Jeruzalmi D, Darnell JE Jr, Kuriyan J (1998) Crystal structure of a tyrosine phosphorylated STAT-1 dimer bound to DNA. Cell 93:827–839

Cheon H, Stark GR (2009) Unphosphorylated STAT1 prolongs the expression of interferon-induced immune regulatory genes. Proc Natl Acad Sci USA 106:9373–9378

Clapier CR, Cairns BR (2009) The biology of chromatin remodeling complexes. Annu Rev Biochem 78:273–304

Cuddapah S, Jothi R, Schones DE, Roh TY, Cui K, Zhao K (2009) Global analysis of the insulator binding protein CTCF in chromatin barrier regions reveals demarcation of active and repressive domains. Genome Res 19:24–32

Cui K, Tailor P, Liu H, Chen X, Ozato K, Zhao K (2004) The chromatin-remodeling BAF complex mediates cellular antiviral activities by promoter priming. Mol Cell Biol 24:4476–4486

Cvijic H, Bauer K, Loffler D, Pfeifer G, Blumert C, Kretzschmar AK, Henze C, Brocke-Heidrich K, Horn F (2009) Co-activator SRC-1 is dispensable for transcriptional control by STAT3. Biochem J 420:123–132

Darnell JE Jr (1997) STATs and gene regulation. Science 277:1630–1635

Darnell JE Jr, Kerr IM, Stark GR (1994) Jak-STAT pathways and transcriptional activation in response to IFNs and other extracellular signaling proteins. Science 264:1415–1421

Decker T, Kovarik P (2000) Serine phosphorylation of STATs. Oncogene 19:2628–2637

Decker T, Kovarik P, Meinke A (1997) Gas elements: a few nucleotides with a major impact on cytokine-induced gene expression. J Interferon Cytokine Res 17:121–134

Donnelly RP, Kotenko SV (2010) Interferon-lambda: a new addition to an old family. J Interferon Cytokine Res 30:555–564

Droescher M, Begitt A, Marg A, Zacharias M, Vinkemeier U (2011) Cytokine-induced paracrystals prolong the activity of STAT transcription factors and provide a model for the regulation of protein-solubility by SUMO. J Biol Chem 286:18731–18746

Durant L, Watford WT, Ramos HL, Laurence A, Vahedi G, Wei L, Takahashi H, Sun HW, Kanno Y, Powrie F, O'Shea JJ (2010) Diverse targets of the transcription factor STAT3 contribute to T cell pathogenicity and homeostasis. Immunity 32:605–615

Elo LL, Jarvenpaa H, Tuomela S, Raghav S, Ahlfors H, Laurila K, Gupta B, Lund RJ, Tahvanainen J, Hawkins RD, Oresic M, Lahdesmaki H, Rasool O, Rao KV, Aittokallio T, Lahesmaa R (2010) Genome-wide profiling of interleukin-4 and STAT6 transcription factor regulation of human Th2 cell programming. Immunity 32:852–862

Engblom D, Kornfeld JW, Schwake L, Tronche F, Reimann A, Beug H, Hennighausen L, Moriggl R, Schutz G (2007) Direct glucocorticoid receptor-Stat5 interaction in hepatocytes controls body size and maturation-related gene expression. Genes Dev 21:1157–1162

Esashi E, Wang YH, Perng O, Qin XF, Liu YJ, Watowich SS (2008) The signal transducer STAT5 inhibits plasmacytoid dendritic cell development by suppressing transcription factor IRF8. Immunity 28:509–520

Farlik M, Reutterer B, Schindler C, Greten F, Vogl C, Muller M, Decker T (2010) Nonconventional initiation complex assembly by STAT and NF-kappaB transcription factors regulates nitric oxide synthase expression. Immunity 33:25–34

Farlik M, Rapp B, Marie I, Levy DE, Jamieson AM, Decker T (2012) Contribution of a TANK-Binding Kinase 1-Interferon (IFN) Regulatory Factor 7 Pathway to IFN-γ-Induced Gene Expression. Mol Cell Biol 32:1032–1043

Flammer JR, Dobrovolna J, Kennedy MA, Chinenov Y, Glass CK, Ivashkiv LB, Rogatsky I (2010) The type I interferon signaling pathway is a target for glucocorticoid inhibition. Mol Cell Biol 30:4564–4574

Gewinner C, Hart G, Zachara N, Cole R, Beisenherz-Huss C, Groner B (2004) The coactivator of transcription CREB-binding protein interacts preferentially with the glycosylated form of Stat5. J Biol Chem 279:3563–3572

Gil MP, Salomon R, Louten J, Biron CA (2006) Modulation of STAT1 protein levels: a mechanism shaping CD8 T-cell responses in vivo. Blood 107:987–993

Ginsberg M, Czeko E, Muller P, Ren Z, Chen X, Darnell JE Jr (2007) Amino acid residues required for physical and cooperative transcriptional interaction of STAT3 and AP-1 proteins c-Jun and c-Fos. Mol Cell Biol 27:6300–6308

Giraud S, Bienvenu F, Avril S, Gascan H, Heery DM, Coqueret O (2002) Functional interaction of STAT3 transcription factor with the coactivator NcoA/SRC1a. J Biol Chem 277:8004–8011

Goodrich JA, Tjian R (2010) Unexpected roles for core promoter recognition factors in cell-type-specific transcription and gene regulation. Nat Rev Genet 11:549–558

Gough DJ, Messina NL, Hii L, Gould JA, Sabapathy K, Robertson AP, Trapani JA, Levy DE, Hertzog PJ, Clarke CJ, Johnstone RW (2010) Functional crosstalk between type I and II interferon through the regulated expression of STAT1. PLoS Biol 8:e1000361

Harada H, Matsumoto M, Sato M, Kashiwazaki Y, Kimura T, Kitagawa M, Yokochi T, Tan RS, Takasugi T, Kadokawa Y, Schindler C, Schreiber RD, Noguchi S, Taniguchi T (1996) Regulation of IFN-alpha/beta genes: evidence for a dual function of the transcription factor complex ISGF3 in the production and action of IFN-alpha/beta. Genes Cells 1:995–1005

Hargreaves DC, Horng T, Medzhitov R (2009) Control of inducible gene expression by signal-dependent transcriptional elongation. Cell 138:129–145

Hartman SE, Bertone P, Nath AK, Royce TE, Gerstein M, Weissman S, Snyder M (2005) Global changes in STAT target selection and transcription regulation upon interferon treatments. Genes Dev 19:2953–2968

Hertzog PJ, O'Neill LA, Hamilton JA (2003) The interferon in TLR signaling: more than just antiviral. Trends Immunol 24:534–539

Horvai AE, Xu L, Korzus E, Brard G, Kalafus D, Mullen TM, Rose DW, Rosenfeld MG, Glass CK (1997) Nuclear integration of jak/stat and ras/ap 1 signaling by cbp and p300. Proc Natl Acad Sci USA 94:1074–1079

Horvath CM, Stark GR, Kerr IM, Darnell JE Jr (1996) Interactions between STAT and non-STAT proteins in the interferon-stimulated gene factor 3 transcription complex. Mol Cell Biol 16:6957–6964

Huang M, Qian F, Hu Y, Ang C, Li Z, Wen Z (2002) Chromatin remodelling factor BRG1 selectively activates a subset of interferon-alpha-inducible genes. Nat Cell Biol 4:774–781

Kamijo R, Harada H, Matsuyama T, Bosland M, Gerecitano J, Shapiro D, Le J, Koh SI, Kimura T, Green SJ et al (1994) Requirement for transcription factor IRF-1 in NO synthase induction in macrophages. Science 263:1612–1615

Kanno Y, Levi BZ, Tamura T, Ozato K (2005) Immune cell-specific amplification of interferon signaling by the IRF-4/8-PU.1 complex. J Interferon Cytokine Res 25:770–779

Kim YJ, Bjorklund S, Li Y, Sayre MH, Kornberg RD (1994) A multiprotein mediator of transcriptional activation and its interaction with the C-terminal repeat domain of RNA polymerase II. Cell 77:599–608

Kramer OH, Knauer SK, Greiner G, Jandt E, Reichardt S, Guhrs KH, Stauber RH, Bohmer FD, Heinzel T (2009) A phosphorylation-acetylation switch regulates STAT1 signaling. Genes Dev 23:223–235

Kumar A, Commane M, Flickinger TW, Horvath CM, Stark GR (1997) Defective TNF-alpha-induced apoptosis in STAT1-null cells due to low constitutive levels of caspases. Science 278:1630–1632

Kurokawa R, Kalafus D, Ogliastro MH, Kioussi C, Xu L, Torchia J, Rosenfeld MG, Glass CK (1998) Differential use of CREB binding protein-coactivator complexes. Science 279:700–703

Kwon MC, Koo BK, Moon JS, Kim YY, Park KC, Kim NS, Kwon MY, Kong MP, Yoon KJ, Im SK, Ghim J, Han YM, Jang SK, Shong M, Kong YY (2008a) Crif1 is a novel transcriptional coactivator of STAT3. EMBO J 27:642–653

Kwon SY, Xiao H, Glover BP, Tjian R, Wu C, Badenhorst P (2008b) The nucleosome remodeling factor (NURF) regulates genes involved in Drosophila innate immunity. Dev Biol 316:538–547

Kwon H, Thierry-Mieg D, Thierry-Mieg J, Kim HP, Oh J, Tunyaplin C, Carotta S, Donovan CE, Goldman ML, Tailor P, Ozato K, Levy DE, Nutt SL, Calame K, Leonard WJ (2009) Analysis of interleukin-21-induced Prdm1 gene regulation reveals functional cooperation of STAT3 and IRF4 transcription factors. Immunity 31:941–952

Lau JF, Nusinzon I, Burakov D, Freedman LP, Horvath CM (2003) Role of metazoan mediator proteins in interferon-responsive transcription. Mol Cell Biol 23:620–628

Lerner L, Henriksen MA, Zhang X, Darnell JE Jr (2003) STAT3-dependent enhanceosome assembly and disassembly: synergy with GR for full transcriptional increase of the alpha 2-macroglobulin gene. Genes Dev 17:2564–2577

Letimier FA, Passini N, Gasparian S, Bianchi E, Rogge L (2007) Chromatin remodeling by the SWI/SNF-like BAF complex and STAT4 activation synergistically induce IL-12Rbeta2 expression during human Th1 cell differentiation. EMBO J 26:1292–1302

Leverson JD, Koskinen PJ, Orrico FC, Rainio EM, Jalkanen KJ, Dash AB, Eisenman RN, Ness SA (1998) Pim-1 kinase and p100 cooperate to enhance c-Myb activity. Mol Cell 2:417–425

Levy DE, Darnell JE Jr (2002) Stats: transcriptional control and biological impact. Nat Rev Mol Cell Biol 3:651–662

Levy DE, Kessler DS, Pine R, Reich N, Darnell JE Jr (1988) Interferon-induced nuclear factors that bind a shared promoter element correlate with positive and negative transcriptional control. Genes Dev 2:383–393

Litterst CM, Pfitzner E (2001) Transcriptional activation by STAT6 requires the direct interaction with NCoA-1. J Biol Chem 276:45713–45721

Litterst CM, Kliem S, Marilley D, Pfitzner E (2003) NCoA-1/SRC-1 is an essential coactivator of STAT5 that binds to the FDL motif in the alpha-helical region of the STAT5 transactivation domain. J Biol Chem 278:45340–45351

Liu H, Kang H, Liu R, Chen X, Zhao K (2002) Maximal induction of a subset of interferon target genes requires the chromatin-remodeling activity of the BAF complex. Mol Cell Biol 22:6471–6479

Longman RS, Braun D, Pellegrini S, Rice CM, Darnell RB, Albert ML (2007) Dendritic-cell maturation alters intracellular signaling networks, enabling differential effects of IFN-alpha/beta on antigen cross-presentation. Blood 109:1113–1122

Macquarrie KL, Fong AP, Morse RH, Tapscott SJ (2011) Genome-wide transcription factor binding: beyond direct target regulation. Trends Genet 27:141–148

Maldonado RA, Soriano MA, Perdomo LC, Sigrist K, Irvine DJ, Decker T, Glimcher LH (2009) Control of T helper cell differentiation through cytokine receptor inclusion in the immunological synapse. J Exp Med 206:877–892

Malik S, Roeder RG (2010) The metazoan mediator co-activator complex as an integrative hub for transcriptional regulation. Nat Rev Genet 11:761–772

Mao X, Ren Z, Parker GN, Sondermann H, Pastorello MA, Wang W, McMurray JS, Demeler B, Darnell JE Jr, Chen X (2005) Structural bases of unphosphorylated STAT1 association and receptor binding. Mol Cell 17:761–771

Maritano D, Sugrue ML, Tininini S, Dewilde S, Strobl B, Fu X, Murray-Tait V, Chiarle R, Poli V (2004) The STAT3 isoforms alpha and beta have unique and specific functions. Nat Immunol 5:401–409

Matsuyama T, Kimura T, Kitagawa M, Pfeffer K, Kawakami T, Watanabe N, Kundig TM, Amakawa R, Kishihara K, Wakeham A, Potter J, Furlonger LC, Narendran A, Suzuki H, Ohashi PS, Paige CJ, Taniguchi T, Mak TW (1993) Targeted disruption of IRF-1 or IRF-2 results in abnormal type I IFN gene induction and aberrant lymphocyte development. Cell 75:83–97

Meissner T, Krause E, Lodige I, Vinkemeier U (2004) Arginine methylation of STAT1: a reassessment. Cell 119:587–589, discussion 589–590

Mertens C, Zhong M, Krishnaraj R, Zou W, Chen X, Darnell JE Jr (2006) Dephosphorylation of phosphotyrosine on STAT1 dimers requires extensive spatial reorientation of the monomers facilitated by the N-terminal domain. Genes Dev 20:3372–3381

Miyagi T, Gil MP, Wang X, Louten J, Chu WM, Biron CA (2007) High basal STAT4 balanced by STAT1 induction to control type 1 interferon effects in natural killer cells. J Exp Med 204:2383–2396

Mo C, Chearwae W, O'Malley JT, Adams SM, Kanakasabai S, Walline CC, Stritesky GL, Good SR, Perumal NB, Kaplan MH, Bright JJ (2008) Stat4 isoforms differentially regulate inflammation and demyelination in experimental allergic encephalomyelitis. J Immunol 181:5681–5690

Moriggl R, Sexl V, Kenner L, Duntsch C, Stangl K, Gingras S, Hoffmeyer A, Bauer A, Piekorz R, Wang DM, Bunting KD, Wagner EF, Sonneck K, Valent P, Ihle JN, Beug H (2005) Stat5 tetramer formation is associated with leukemogenesis. Cancer Cell 7:87–99

Morris AC, Beresford GW, Mooney MR, Boss JM (2002) Kinetics of a gamma interferon response: expression and assembly of CIITA promoter IV and inhibition by methylation. Mol Cell Biol 22:4781–4791

Mowen KA, Tang J, Zhu W, Schurter BT, Shuai K, Herschman HR, David M (2001) Arginine methylation of STAT1 modulates IFNalpha/beta-induced transcription. Cell 104:731–741

Muhlethaler-Mottet A, Di Berardino W, Otten LA, Mach B (1998) Activation of the MHC class II transactivator CIITA by interferon-gamma requires cooperative interaction between Stat1 and USF-1. Immunity 8:157–166

Nakajima H, Brindle PK, Handa M, Ihle JN (2001) Functional interaction of STAT5 and nuclear receptor co-repressor SMRT: implications in negative regulation of STAT5-dependent transcription. EMBO J 20:6836–6844

Nakashima K, Yanagisawa M, Arakawa H, Kimura N, Hisatsune T, Kawabata M, Miyazono K, Taga T (1999) Synergistic signaling in fetal brain by STAT3-smad1 complex bridged by p300. Science 284:479–482

Neculai D, Neculai AM, Verrier S, Straub K, Klumpp K, Pfitzner E, Becker S (2005) Structure of the unphosphorylated STAT5a dimer. J Biol Chem 280:40782–40787

Ni Z, Bremner R (2007) Brahma-related gene 1-dependent STAT3 recruitment at IL-6-inducible genes. J Immunol 178:345–351

Ni Z, Karaskov E, Yu T, Callaghan SM, Der S, Park DS, Xu Z, Pattenden SG, Bremner R (2005) Apical role for BRG1 in cytokine-induced promoter assembly. Proc Natl Acad Sci USA 102:14611–14616

Ni Z, Abou El Hassan M, Xu Z, Yu T, Bremner R (2008) The chromatin-remodeling enzyme BRG1 coordinates CIITA induction through many interdependent distal enhancers. Nat Immunol 9:785–793

Nusinzon I, Horvath CM (2005) Unexpected roles for deacetylation in interferon- and cytokine-induced transcription. J Interferon Cytokine Res 25:745–748

O'Shea JJ, Lahesmaa R, Vahedi G, Laurence A, Kanno Y (2011) Genomic views of STAT function in CD4(+) T helper cell differentiation. Nat Rev Immunol 11:239–250

Ohmori Y, Hamilton TA (1997) IL-4-induced STAT6 suppresses IFN-gamma-stimulated STAT1-dependent transcription in mouse macrophages. J Immunol 159:5474–5482

Onodera A, Yamashita M, Endo Y, Kuwahara M, Tofukuji S, Hosokawa H, Kanai A, Suzuki Y, Nakayama T (2010) STAT6-mediated displacement of polycomb by trithorax complex establishes long-term maintenance of GATA3 expression in T helper type 2 cells. J Exp Med 207:2493–2506

Ouchi T, Lee SW, Ouchi M, Aaronson SA, Horvath CM (2000) Collaboration of signal transducer and activator of transcription 1 (STAT1) and BRCA1 in differential regulation of IFN-gamma target genes. Proc Natl Acad Sci USA 97:5208–5213

Ozsolak F, Milos PM (2010) RNA sequencing: advances, challenges and opportunities. Nat Rev Genet 12:87–98

Park C, Li S, Cha E, Schindler C (2000) Immune response in Stat2 knockout mice. Immunity 13:795–804

Pattenden SG, Klose R, Karaskov E, Bremner R (2002) Interferon-gamma-induced chromatin remodeling at the CIITA locus is BRG1 dependent. EMBO J 21:1978–1986

Paukku K, Yang J, Silvennoinen O (2003) Tudor and nuclease-like domains containing protein p100 function as coactivators for signal transducer and activator of transcription 5. Mol Endocrinol 17:1805–1814

Paulson M, Pisharody S, Pan L, Guadagno S, Mui AL, Levy DE (1999) Stat protein transactivation domains recruit p300/CBP through widely divergent sequences. J Biol Chem 274:25343–25349

Paulson M, Press C, Smith E, Tanese N, Levy DE (2002) IFN-stimulated transcription through a TBP-free acetyltransferase complex escapes viral shutoff. Nat Cell Biol 4:140–147

Pine R, Canova A, Schindler C (1994) Tyrosine phosphorylated p91 binds to a single element in the ISGF2/IRF-1 promoter to mediate induction by IFN alpha and IFN gamma, and is likely to autoregulate the p91 gene. EMBO J 13:158–167

Ramana CV, Gil MP, Han Y, Ransohoff RM, Schreiber RD, Stark GR (2001) Stat1-independent regulation of gene expression in response to IFN-gamma. Proc Natl Acad Sci USA 98:6674–6679

Ramirez-Carrozzi VR, Braas D, Bhatt DM, Cheng CS, Hong C, Doty KR, Black JC, Hoffmann A, Carey M, Smale ST (2009) A unifying model for the selective regulation of inducible transcription by CpG islands and nucleosome remodeling. Cell 138:114–128

Ramsauer K, Farlik M, Zupkovitz G, Seiser C, Kroger A, Hauser H, Decker T (2007) Distinct modes of action applied by transcription factors STAT1 and IRF1 to initiate transcription of the IFN-gamma-inducible gbp2 gene. Proc Natl Acad Sci USA 104:2849–2854

Rascle A, Johnston JA, Amati B (2003) Deacetylase activity is required for recruitment of the basal transcription machinery and transactivation by STAT5. Mol Cell Biol 23:4162–4173

Razeto A, Ramakrishnan V, Litterst CM, Giller K, Griesinger C, Carlomagno T, Lakomek N, Heimburg T, Lodrini M, Pfitzner E, Becker S (2004) Structure of the NCoA-1/SRC-1 PAS-B domain bound to the LXXLL motif of the STAT6 transactivation domain. J Mol Biol 336:319–329

Ricote M, Li AC, Willson TM, Kelly CJ, Glass CK (1998) The peroxisome proliferator-activated receptor-gamma is a negative regulator of macrophage activation. Nature 391:79–82

Robertson AG, Bilenky M, Tam A, Zhao Y, Zeng T, Thiessen N, Cezard T, Fejes AP, Wederell ED, Cullum R, Euskirchen G, Krzywinski M, Birol I, Snyder M, Hoodless PA, Hirst M, Marra MA, Jones SJ (2008) Genome-wide relationship between histone H3 lysine 4 mono- and tri-methylation and transcription factor binding. Genome Res 18:1906–1917

Sadzak I, Schiff M, Gattermeier I, Glinitzer R, Sauer I, Saalmuller A, Yang E, Schaljo B, Kovarik P (2008) Recruitment of Stat1 to chromatin is required for interferon-induced serine phosphorylation of Stat1 transactivation domain. Proc Natl Acad Sci USA 105:8944–8949

Schaefer TS, Sanders LK, Nathans D (1995) Cooperative transcriptional activity of Jun and Stat3 beta, a short form of Stat3. Proc Natl Acad Sci USA 92:9097–9101

Schindler C, Levy DE, Decker T (2007) JAK-STAT signaling: from interferons to cytokines. J Biol Chem 282:20059–20063

Shen Y, Schlessinger K, Zhu X, Meffre E, Quimby F, Levy DE, Darnell JE Jr (2004) Essential role of STAT3 in postnatal survival and growth revealed by mice lacking STAT3 serine 727 phosphorylation. Mol Cell Biol 24:407–419

Shi S, Calhoun HC, Xia F, Li J, Le L, Li WX (2006) JAK signaling globally counteracts heterochromatic gene silencing. Nat Genet 38:1071–1076

Shi M, Lin TH, Appell KC, Berg LJ (2008a) Janus-kinase-3-dependent signals induce chromatin remodeling at the Ifng locus during T helper 1 cell differentiation. Immunity 28:763–773

Shi S, Larson K, Guo D, Lim SJ, Dutta P, Yan SJ, Li WX (2008b) Drosophila STAT is required for directly maintaining HP1 localization and heterochromatin stability. Nat Cell Biol 10:489–496

Stark GR (2007) How cells respond to interferons revisited: from early history to current complexity. Cytokine Growth Factor Rev 18:419–423

Stark GR, Kerr IM, Williams BR, Silverman RH, Schreiber RD (1998) How cells respond to interferons. Annu Rev Biochem 67:227–264

Stocklin E, Wissler M, Gouilleux F, Groner B (1996) Functional interactions between Stat5 and the glucocorticoid receptor. Nature 383:726–728, ISSN: 0028-0836

Sun W, Xu W, Snyder M, He W, Ho H, Ivashkiv LB, Zhang JJ (2005) The conserved Leu-724 residue is required for both serine phosphorylation and co-activator recruitment for Stat1-mediated transcription activation in response to interferon-gamma. J Biol Chem 280:41844–41851

Szanto A, Balint BL, Nagy ZS, Barta E, Dezso B, Pap A, Szeles L, Poliska S, Oros M, Evans RM, Barak Y, Schwabe J, Nagy L (2010) STAT6 transcription factor is a facilitator of the nuclear receptor PPARgamma-regulated gene expression in macrophages and dendritic cells. Immunity 33:699–712

Tamura T, Smith M, Kanno T, Dasenbrock H, Nishiyama A, Ozato K (2009) Inducible deposition of the histone variant H3.3 in interferon-stimulated genes. J Biol Chem 284:12217–12225

Tenoever BR, Ng SL, Chua MA, McWhirter SM, Garcia-Sastre A, Maniatis T (2007) Multiple functions of the IKK-related kinase IKKepsilon in interferon-mediated antiviral immunity. Science 315:1274–1278

Ungureanu D, Vanhatupa S, Gronholm J, Palvimo JJ, Silvennoinen O (2005) SUMO-1 conjugation selectively modulates STAT1-mediated gene responses. Blood 106:224–226

Valouev A, Johnson DS, Sundquist A, Medina C, Anton E, Batzoglou S, Myers RM, Sidow A (2008) Genome-wide analysis of transcription factor binding sites based on ChIP-Seq data. Nat Methods 5:829–834

Varinou L, Ramsauer K, Karaghiosoff M, Kolbe T, Pfeffer K, Muller M, Decker T (2003) Phosphorylation of the Stat1 transactivation domain is required for full-fledged IFN-gamma-dependent innate immunity. Immunity 19:793–802

Vinkemeier U, Cohen SL, Moarefi I, Chait BT, Kuriyan J, Darnell JE Jr (1996) DNA binding of in vitro activated Stat1 alpha, Stat1 beta and truncated Stat1: interaction between NH2-terminal domains stabilizes binding of two dimers to tandem DNA sites. EMBO J 15:5616–5626

Wei G, Wei L, Zhu J, Zang C, Hu-Li J, Yao Z, Cui K, Kanno Y, Roh TY, Watford WT, Schones DE, Peng W, Sun HW, Paul WE, O'Shea JJ, Zhao K (2009) Global mapping of H3K4me3 and H3K27me3 reveals specificity and plasticity in lineage fate determination of differentiating CD4+ T cells. Immunity 30(1):155–167

Wei L, Vahedi G, Sun HW, Watford WT, Takatori H, Ramos HL, Takahashi H, Liang J, Gutierrez-Cruz G, Zang C, Peng W, O'Shea JJ, Kanno Y (2010) Discrete roles of STAT4 and STAT6 transcription factors in tuning epigenetic modifications and transcription during T helper cell differentiation. Immunity 32:840–851

Wojciak JM, Martinez-Yamout MA, Dyson HJ, Wright PE (2009) Structural basis for recruitment of CBP/p300 coactivators by STAT1 and STAT2 transactivation domains. EMBO J 28: 948–958

Wormald S, Hilton DJ, Smyth GK, Speed TP (2006) Proximal genomic localization of STAT1 binding and regulated transcriptional activity. BMC Genomics 7:254

Xu D, Holko M, Sadler AJ, Scott B, Higashiyama S, Berkofsky-Fessler W, McConnell MJ, Pandolfi PP, Licht JD, Williams BR (2009) Promyelocytic leukemia zinc finger protein regulates interferon-mediated innate immunity. Immunity 30:802–816

Yan Z, Cui K, Murray DM, Ling C, Xue Y, Gerstein A, Parsons R, Zhao K, Wang W (2005) PBAF chromatin-remodeling complex requires a novel specificity subunit, BAF200, to regulate expression of selective interferon-responsive genes. Genes Dev 19:1662–1667

Yang J, Aittomaki S, Pesu M, Carter K, Saarinen J, Kalkkinen N, Kieff E, Silvennoinen O (2002) Identification of p100 as a coactivator for STAT6 that bridges STAT6 with RNA polymerase II. EMBO J 21:4950–4958

Yang J, Chatterjee-Kishore M, Staugaitis SM, Nguyen H, Schlessinger K, Levy DE, Stark GR (2005) Novel roles of unphosphorylated STAT3 in oncogenesis and transcriptional regulation. Cancer Res 65:939–947

Yang J, Liao X, Agarwal MK, Barnes L, Auron PE, Stark GR (2007) Unphosphorylated STAT3 accumulates in response to IL-6 and activates transcription by binding to NFkappaB. Genes Dev 21:1396–1408

Yang J, Huang J, Dasgupta M, Sears N, Miyagi M, Wang B, Chance MR, Chen X, Du Y, Wang Y, An L, Wang Q, Lu T, Zhang X, Wang Z, Stark GR (2010) Reversible methylation of promoter bound STAT3 by histone-modifying enzymes. Proc Natl Acad Sci USA 107:21499–21504

York B, O'Malley BW (2010) Steroid receptor coactivator (SRC) family: masters of systems biology. J Biol Chem 285:38743–38750

Zakharova N, Lymar ES, Yang E, Zhang JJ, Roeder RG, Darnell JE Jr (2003) Distinct transcriptional activation functions of STAT1alpha and beta on DNA and chromatin templates. J Biol Chem 278:43067–43073

Zhang X, Darnell JE Jr (2001) Functional importance of Stat3 tetramerization in activation of the alpha 2-macroglobulin gene. J Biol Chem 276:33576–33581

Zhang JJ, Vinkemeier U, Gu W, Chakravarti D, Horvath CM, Darnell JE Jr (1996) Two contact regions between Stat1 and CBP/p300 in interferon gamma signaling. Proc Natl Acad Sci USA 93:15092–15096

Zhang JJ, Zhao Y, Chait BT, Lathem WW, Ritzi M, Knippers R, Darnell JE Jr (1998) Ser727-dependent recruitment of MCM5 by Stat1alpha in IFN-gamma-induced transcriptional activation. EMBO J 17:6963–6971

Zhong M, Henriksen MA, Takeuchi K, Schaefer O, Liu B, ten Hoeve J, Ren Z, Mao X, Chen X, Shuai K, Darnell JE Jr (2005) Implications of an antiparallel dimeric structure of nonphosphorylated STAT1 for the activation-inactivation cycle. Proc Natl Acad Sci USA 102:3966–3971

Zhu M-h, John S, Berg M, Leonard WJ (1999) Functional association of Nmi with Stat5 and Stat1 in IL-2 and IFN-γ-mediated signaling. Cell 96:121–130

Nongenomic Functions of STAT3

Daniel J. Gough, Pravin Sehgal, and David E. Levy

Abstract

STAT3 was discovered as a transcription factor activated during inflammatory cytokine signaling, largely through the action of cytokines that utilize receptors that engage the gp130 signal transducer. While the paradigm of receptor-dependent protein tyrosine phosphorylation of STAT3 has explained many of the biological activities ascribed to this protein, additional functions have been discovered in recent years. These functions involve actions of STAT3 outside the nucleus and do not rely on its ability to interact with chromatin and trigger gene expression. Nongenomic functions of STAT3 include action in the cell cytoplasm and in mitochondria, where it impacts metabolic activities involved in a diverse set of cell functions, including malignant transformation.

Signal Transducer and Activators of Transcription (STAT) proteins were originally identified as latent cytoplasmic proteins that were acutely activated by tyrosine phosphorylation following cytokine stimulation. Tyrosine phosphorylation of a single conserved residue near the carboxyl terminus of a STAT protein (Y705 in STAT3) was found to be critical for a structural alteration that facilitated the nuclear translocation of STAT protein dimers (Levy and Darnell 2002). In the nucleus, these activated transcription factors exhibited sequence-specific DNA binding, coactivator protein recruitment, and stimulation of gene expression. The first members of the multiple gene family encoding STAT proteins, which numbers 7 conserved genes in mammals, were STAT1 and STAT2 that mediate responses to Type I and Type II IFN. The discovery of STAT1 and STAT2 was quickly followed

D.J. Gough • D.E. Levy (✉)
NYU Cancer Institute and Departments of Pathology and Microbiology, NYU School of Medicine, 550 First Avenue, New York, NY 10016, USA
e-mail: david.levy@nyumc.org

P. Sehgal
Department of Cell Biology and Anatomy, New York Medical College, Valhalla, NY, USA

Th. Decker and M. Müller (eds.), *Jak-Stat Signaling: From Basics to Disease*,
DOI 10.1007/978-3-7091-0891-8_6, © Springer-Verlag Wien 2012

by identification of STAT3, activated most prominently by IL6-family cytokines, and by STATs 4–6, activated by IL12, IL2, prolactin and growth hormone, and IL4. All of these proteins were found to function through the paradigm described for IFN-dependent gene activation, involving ctyokine receptor-associated protein tyrosine kinase-dependent site-specific tyrosine phosphorylation, dimer formation, nuclear translocation, and transcription machinery assembly on chromatin (Schindler et al. 2007).

This view of canonical functions for tyrosine-phosphorylated STAT3 can now be expanded to include STAT3 functions that are independent of tyrosine phosphorylation and some that are independent of intrinsic nuclear localization. Unphosphorylated STAT3 can be chaperoned into the nucleus by NF-κB where it can alter the transcription of a subset of genes (Yang et al. 2005, 2007). The first indications that STAT3 may be biologically relevant outside of the nucleus came from the discovery that STAT3 (as well as other STAT proteins) associate with a variety of cytoplasmic structures, including focal adhesions, microtubules, and the mitotic spindle, as well as with membrane elements, such as plasma membrane rafts and endo-lysosomes in addition to mitochondria (Guo et al. 2002; Sehgal 2008; Shah et al. 2006). Some of the studies of cytoplasmic STAT3 have documented the presence of phosphotyrosine, and these studies have been interpreted as representing way stations for activated STAT3 en route to the nucleus or perhaps serving non-canonical functions in the cytoplasm (Inghirami et al. 2005; Sehgal 2008).

Early evidence for the association of STAT3 with cytoplasmic structures documented the protein in sedimentable membrane fractions from cytoplasmic extracts from both untreated and cytokine-stimulated cells (Guo et al. 2002). These fractions were also enriched for markers of endosomal, mitochondrial, or endoplasmic reticulum components. It was estimated that in some cells up to two-thirds of cytoplasmic STAT3 was associated with an endosomal-like compartment (Dang 2010). This evidence of STAT3 within cytoplasmic structures prompted a reexamination of the assumption that STAT3 function exclusively involved phosphorylation of freely diffusible cytoplasmic protein and a search for the functional relevance of membrane associated STAT3. Fundamental findings in this area reported that STAT3 was functionally associated with mitochondria in mouse cardiomyocytes, pro-B cells and in Ras oncogene transformed cells (Gough et al. 2009; Wegrzyn et al. 2009). STAT3 was found in the mitochondrial fractions of all cell types tested and recent evidence suggests that the mitochondrial pool of STAT3 is required for diverse biological systems including cardiac activity and the response to ischemia/reperfusion, neurological activity, and cellular transformation by Ras oncogenes.

STAT3 protects cardiac muscle following ischemia/reperfusion (I/R) as a result of acute myocardial infarction (Haghikia et al. 2011). The mitochondria are central to I/R induced injury due to the release of reactive oxygen species (ROS) and by opening of the mitochondrial transition pore (mTP) (Perrelli et al. 2011). Both are processes that have been found to be influenced by the activity of the mitochondrial pool of STAT3. Mice lacking STAT3 in cardiomyocytes had reduced activity of complex I and II of the electron transport chain (ETC) (Wegrzyn et al. 2009). Moreover, it was found that regions of STAT3 previously documented as essential for its transcriptional function, such as the site for tyrosine phosphorylation and residues implicated in DNA binding or SH2 domain function, were superfluous for this mitochondrial role; however, phosphorylation on a single carboxyl-terminal serine residue (S727) was required for the mitochondrial activity of STAT3. Complex II is a major site of ROS production and it is conceivable that in the absence of appropriate complex II activity, increased ROS may be produced and/or released contributing to I/R injury. Indeed, in vivo I/R experiments on a mouse strain over-expressing mutant STAT3 that was constitutively phosphorylated on S727 and targeted to mitochondria showed a lack of ROS production when compared to wild type animals (Szczepanek et al. 2011). The mTP is a pore spanning the mitochondrial inner membrane. Its opening leads to a loss of mitochondrial membrane potential and if sustained leads to mitochondrial swelling and apoptotic cell death. I/R induces mTP opening; however, mitochondrial STAT3 delays or inhibits mTP opening thus protecting cells from apoptosis in response to I/R (Boengler et al. 2010).

Both STAT3 and the regulation of ROS concentration are also critical for the appropriate functioning of the nervous system. Nerve growth factor (NGF) induces neurite outgrowth, ROS production, and phosphorylation of STAT3 on S727. Interestingly, following NGF stimulation the pool of pS727 STAT3 does not relocate to the nucleus and was instead observed in mitochondria. Moreover, mutation of a mitochondrially restricted STAT3 S727 to alanine resulted in a lack of neurite outgrowth and ROS production in response to NGF (Zhou and Too 2011). In contrast, STAT3-deficient astrocytes produce more superoxide and other ROS species concomitant with a decrease in glutathione concentration and mitochondrial membrane potential (Sarafian et al. 2010), although it has not been determined if these effects require cytoplasmic or nuclear STAT3. It is important to note that a nuclear function of STAT3 capable of indirectly altering mitochondrial activity has been documented by Demaria et al. (2010), an activity that can also contribute to malignant transformation.

As described above, it is becoming increasingly clear that STAT proteins in general and STAT3 in particular function in additional pathways beyond the innate immune cytokine responses in which they were originally discovered. In particular, there is a growing realization that STAT3 plays a critical role in a number of human cancers (Bowman et al. 2000). Many of these cancer-related functions rely on the canonical activity of tyrosine-phosphorylated STAT3 and its ability to be recruited to chromatin and stimulate transcription of genes critical for the malignant state, particularly genes important for cell proliferation and cell survival. Some of these functions depend on cytokine signaling, because some cancer cells secrete a STAT3-activating cytokine and respond to it in an autocrine or paracrine manner. Prominent among these autocrine factors is IL6 and its close relatives. However, STAT3 can also be tyrosine phosphorylated in a cytokine-independent manner, both by growth factor receptor tyrosine kinases and by protein tyrosine kinase oncoproteins, for example, Src and ALK (Inghirami et al. 2005; Schlessinger and

Levy 2005). Both of these patterns of STAT3 activation by cancer cells (cytokine-dependent and oncoprotein-dependent) produce tumor cell-autonomous and non-autonomous changes in STAT3-dependent gene expression that promote tumor formation and maintenance. Typical amongst the STAT3 target genes that contribute to malignancy are cell-autonomous functions that promote cell proliferation and survival and cell-non-autonomous functions mediated by tumor infiltrating leukocytes or surrounding stromal cells that contribute to inflammation and immune suppression (Yu et al. 2009).

However, another contribution that STAT3 makes toward malignancy is in support of the altered cell metabolism that is a hallmark of cellular transformation. It is clear that STAT3 is capable of supporting the altered metabolic requirements in tumor cells through both canonical nuclear (Demaria et al. 2010) and non-canonical mitochondrial roles (Gough et al. 2009). The non-genomic role for STAT3 emanating from mitochondria in Ras-transformed cells was found following the surprising discovery that loss of STAT3 compromised Ras-dependent malignancy. STAT3-null mouse embryo fibroblasts (MEFs) displayed impaired anchorage-independent growth following expression of oncogenic Ras, in spite of the absence of tyrosine phosphorylated STAT3 in wild type cells transformed by H-, N-, or K-Ras. Indeed, STAT3-null cells reconstituted with mutant STAT3 lacking tyrosine 705 (Y705) recovered their ability to grow in soft agar or as solid tumors in nude mice. Similar molecular genetic studies found no requirement for the ability of STAT3 to interact with phosphotyrosine substrates through its SH2 domain, accumulate in the nucleus, or bind to DNA in support of Ras-dependent tumorigenesis (Gough et al. 2009).

Since this Ras-transformation function of STAT3 appeared to emanate from the cytoplasm rather than the nucleus, we investigated the cell biology of this process. Cell fractionation studies revealed that STAT3 accumulated in mitochondrial fractions in Ras-transformed cells, and molecular genetic studies demonstrated that phosphorylation of S727 but not of Y705 was critical. Moreover, phospho-S727 but not phospho-Y705 was increased in Ras-transformed cells, and phospho-S727 was detected in mitochondria. In fact, even following cytokine stimulation of cells, when phospho-Y705 is abundant in both the cytoplasmic and nuclear compartments, no phospho-Y705 was detectable in mitochondrial fractions.

With these data in hand, we investigated the requirement for mitochondrial STAT3 during Ras transformation. Using chimeric STAT3 expression constructs that accumulate exclusively in mitochondria to reconstitute STAT3 knockout mouse cells or STAT3 knockdown human cancer cells, we found that mitochondrial phopho-S727 STAT3 was sufficient to complement the absence of total STAT3 for Ras-dependent anchorage-independent cell and tumor growth. Mitochondrially localized STAT3 was sufficient to sustain the altered glycolytic and oxidative phosphorylation activities of Ras-transformed cells, including increased levels of ATP, higher lactate dehydrogenase activity leading to increased production of lactate, and increased flux through the ETC, particularly with respect to the enzymatic activities of complexes II and V. All of these STAT3 functions depended on phospho-S727, a modification that depended on activated MAP kinase signaling from the Ras oncogene.

These data document a previously unforeseen role for STAT3 as a mitochondrially-localized protein acting in a non-genomic manner in a function that is critical for metabolic aspects of Ras-dependent malignancy. It has been long recognized that many tumors display altered glucose metabolism relative to normal tissue, manifested as a propensity to avidly consume glucose and ferment it to lactic acid, even under aerobic conditions. This property of aerobic glycolysis, known as the Warburg effect (Warburg 1956), provides the mechanism for the clinical localization of human tumors in vivo by positron emission tomography of radiolabeled 2-deoxyglucose. However, its biochemical basis and functional importance, while receiving increasing attention in recent years, have remained elusive.

A number of aspects of the Warburg effect appear contradictory, but are reminiscent of the apparent role of mitochondrial STAT3 in transformed cells. One might think the energy requirements of proliferating tumor cells would be better served by oxidative phosphorylation (OXPHOS), generating a theoretical yield of 36 moles of ATP per mole of glucose consumed, as opposed to two by glycolysis, but bioenergetics don't appear to drive the Warburg effect. The Warburg effect is also distinct from the hypoxic response that is another common property of tumors due to insufficient angiogenesis (Dang 2010; Hsu and Sabatini 2008; Vander Heiden et al. 2009). Enhanced glycolysis is observed in fully oxygenated tumors, such as blood-borne and lung tumors that are exposed to higher oxygen tension than normal tissues that efficiently undergo oxidative phosphorylation. Thus, while tumors deprived of adequate blood supply exhibit enhanced glycolytic activity due to a hypoxic response, it is incorrect to assume that enhanced tumor glycolysis characteristic of the Warburg effect is necessarily an adaptation to hypoxia, and it presumably benefits tumor growth beyond bioenergetics. In fact, enhanced glycolytic activity of tumors often precedes the increase in size that outstrips their blood supply. We have observed altered metabolism in tumor cells dependent on STAT3 under both normoxic and hypoxic conditions.

Understanding the Warburg effect and the role of mitochondrial STAT3 may require a broader view of tumor metabolism. The mitochondrion is more than an energy generator, since the TCA cycle and OXPHOS also provide precursors for anabolic processes required for cell proliferation and increased biomass, such as amino acids, nucleotides, and lipids. There is also evidence that proliferation of non-transformed cells is accompanied by a metabolic shift that bears similarity to tumor cell metabolism, including enhanced glycolysis and increased lactate production in the presence of adequate oxygen and continued OXPHOS activity. Thus, the Warburg effect may be a direct consequence of, and perhaps required for, the enhanced proliferation of tumor cells. While the basis for the metabolic shift of proliferating and transformed cells remains ill defined, there are a number of points where growth factor signaling pathways and the actions of oncoproteins and tumor suppressors impact on the process. For instance, PI3K/AKT signaling in response to growth factor stimulation enhances glucose transporter expression and glycolytic enzyme activities, while p53 functions to balance glycolysis and respiration and its loss in tumors may contribute to a glycolytic shift (Levine et al. 2006). Lactate dehydrogenase (LDH), which catalyzes the reduction of pyruvate to lactate to regenerate oxidized NAD+ and is therefore essential for ongoing fermentation, is required for proliferating cells and reduction in its activity impairs tumor cell growth (Fantin et al. 2006). Another major control point in glycolysis, pyruvate kinase (PK), which is required for production of pyruvate and ATP, is also regulated by growth factor signaling and in cancer. The activity of the isoform of PK that predominates in tumor cells (PKM2) is regulated by tyrosine phosphorylation (Christofk et al. 2008). Paradoxically, tumor and growth factor stimulated cells appear to express reduced activity forms of PKM2 rather than enhanced activity forms, in spite of increased glycolysis and lactate production. This observation suggests that either impaired PKM2 activity, which should result in reduced abundance of pyruvate, actually favors lactate fermentation at the expense of respiration, or that the pyruvate for lactate production is derived from an alternative source. One such source could be glutamine, which can also be converted to pyruvate and whose metabolism is often enhanced in cancer cells (Gao et al. 2009).

Considering the Warburg effect in this broader context suggests that it might be the manifestation of a number of metabolic processes contributing to enhanced tumor cell proliferation and survival by supporting metabolic output that exceeds energy needs in order to provide an adequate supply of precursors for anabolic processes. In this sense, aerobic glycolysis is not a fallback metabolic state due to impaired respiration or defective mitochondria but is rather an adaptive response to proliferation. The metabolic shift characteristic of tumors cells is an important aspect of tumor progression, represents a possible pressure point of vulnerability that could be exploited therapeutically, and demands a deeper understanding at the biochemical and mechanistic level. Similarly, the mechanistic basis for the similarity between functions of mitochondrial STAT3 and the Warburg effect represents an important area for future research.

Concluding Remarks

Many unanswered questions and possible contradictory findings regarding mitochondrial STAT3 accumulation and function remain to be answered. Most prominent amongst these is how STAT3 translocates to mitochondria and how it influences mitochondrial functions, such as activity of ETC complexes I, II and V and the mTP. It has been found that cells lacking heat shock protein H11 kinase/ Hsp22 (Hsp22), a potential component of organelle import, have a diminished reservoir of mitochondrial STAT3. However, Hsp22 can be only part of the STAT3 mitochondrial import process, because Hsp22 is a cytosolic chaperone and STAT3 appears to accumulate in the mitochondrial matrix or inner mitochondrial membrane (Boengler et al. 2010), beyond the reach of Hsp22-mediated import processes. For STAT3 to reach the matrix or inner membrane it must cross two membranes, requiring the help of an army of pore proteins and chaperones, none of which has been yet identified. While it is clear that STAT3 alters the activity of the ETC and can impede mTP opening, it probably does not do so by direct binding to these complexes. Numerically STAT3 may exist in mitochondria of some cell types at a stoichiometric deficit to the proteins of these complexes (Phillips et al. 2010), suggesting that it may function catalytically rather than structurally. A possibly related enigma has been an apparent inability to detect ectopically expressed, fluorescently-tagged recombinant STAT3 fusion proteins in association with mitochondria by live cell imaging (Cimica et al. 2011). Another possible explanation for mitochondrial STAT3 function has been suggested, that STAT3 alters the abundance of mitochondrial proteins by acting as a transcription factor for the mitochondrial genome. This suggestion is based on evidence that STAT3 can bind to D-loop structures of mitochondrial DNA in cell free experiments (Vassilev et al. 2002). However, even if mitochondrial STAT3 does bind to mitochondrial DNA, it does not appear to be required for the transcription of any mitochondrially-encoded genes or for the maintenance or replication of the mitochondrial genome (Gough et al. 2009; Wegrzyn et al. 2009) and therefore is unlikely to be functioning in this organelle as a transcription or replication factor. Thus, while we are beginning to understand the magnitude of the biological significance of mitochondrial STAT3, we have a long way to go toward understanding the very fundamental questions of how STAT3 enters mitochondria and how it orchestrates tumor-promoting effects on ETC and mTP activities.

References

Boengler K, Hilfiker-Kleiner D, Heusch G, Schulz R (2010) Inhibition of permeability transition pore opening by mitochondrial STAT3 and its role in myocardial ischemia/reperfusion. Basic Res Cardiol 105:771–785

Bowman T, Garcia R, Turkson J, Jove R (2000) STATs in oncogenesis. Oncogene 19:2474–2488

Christofk HR, Vander Heiden MG, Wu N, Asara JM, Cantley LC (2008) Pyruvate kinase M2 is a phosphotyrosine-binding protein. Nature 452:181–186

Cimica V, Chen HC, Iyer JK, Reich NC (2011) Dynamics of the STAT3 transcription factor: nuclear import dependent on Ran and importin-beta1. PLoS One 6:e20188

Dang CV (2010) Rethinking the Warburg effect with Myc micromanaging glutamine metabolism. Cancer Res 70:859–862

Demaria M, Giorgi C, Lebiedzinska M, Esposito G, D'Angeli L, Bartoli A, Gough DJ, Turkson J, Levy DE, Watson CJ, Wieckowski MR, Provero P, Pinton P, Poli V (2010) A STAT3-mediated metabolic switch is involved in tumour transformation and STAT3 addiction. Aging (Albany NY) 2:823–842

Fantin VR, St-Pierre J, Leder P (2006) Attenuation of LDH-A expression uncovers a link between glycolysis, mitochondrial physiology, and tumor maintenance. Cancer Cell 9:425–434

Gao P, Tchernyshyov I, Chang TC, Lee YS, Kita K, Ochi T, Zeller KI, De Marzo AM, Van Eyk JE, Mendell JT, Dang CV (2009) c-Myc suppression of miR-23a/b enhances mitochondrial glutaminase expression and glutamine metabolism. Nature 458:762–765

Gough DJ, Corlett A, Schlessinger K, Wegrzyn J, Larner AC, Levy DE (2009) Mitochondrial STAT3 supports Ras-dependent oncogenic transformation. Science 324:1713–1716

Guo GG, Patel K, Kumar V, Shah M, Fried VA, Etlinger JD, Sehgal PB (2002) Association of the chaperone glucose-regulated protein 58 (GRP58/ER-60/ERp57) with Stat3 in cytosol and plasma membrane complexes. J Interferon Cytokine Res 22:555–563

Haghikia A, Stapel B, Hoch M, Hilfiker-Kleiner D (2011) STAT3 and cardiac remodeling. Heart Fail Rev 16:35–47

Hsu PP, Sabatini DM (2008) Cancer cell metabolism: Warburg and beyond. Cell 134:703–707

Inghirami G, Chiarle R, Simmons WJ, Piva R, Schlessinger K, Levy DE (2005) New and old functions of STAT3: a pivotal target for individualized treatment of cancer. Cell Cycle 4:1131–1133

Levine AJ, Hu W, Feng Z (2006) The P53 pathway: what questions remain to be explored? Cell Death Differ 13:1027–1036

Levy DE, Darnell JE Jr (2002) STATs: transcriptional control and biological impact. Nat Rev Mol Cell Biol 3:651–662

Perrelli MG, Pagliaro P, Penna C (2011) Ischemia/reperfusion injury and cardioprotective mechanisms: role of mitochondria and reactive oxygen species. World J Cardiol 3:186–200

Phillips D, Reilley MJ, Aponte AM, Wang G, Boja E, Gucek M, Balaban RS (2010) Stoichiometry of STAT3 and mitochondrial proteins: implications for the regulation of oxidative phosphorylation by protein–protein interactions. J Biol Chem 285:23532–23536

Sarafian TA, Montes C, Imura T, Qi J, Coppola G, Geschwind DH, Sofroniew MV (2010) Disruption of astrocyte STAT3 signaling decreases mitochondrial function and increases oxidative stress in vitro. PLoS One 5:e9532

Schindler C, Levy DE, Decker T (2007) JAK-STAT signaling: from interferons to cytokines. J Biol Chem 282:20059–20063

Schlessinger K, Levy DE (2005) Malignant transformation but not normal cell growth depends on STAT3. Cancer Res 65:5828–5834

Sehgal PB (2008) Paradigm shifts in the cell biology of STAT signaling. Semin Cell Dev Biol 19:329–340

Shah M, Patel K, Mukhopadhyay S, Xu F, Guo G, Sehgal PB (2006) Membrane-associated STAT3 and PY-STAT3 in the cytoplasm. J Biol Chem 281:7302–7308

Szczepanek K, Chen Q, Derecka M, Salloum FN, Zhang Q, Szelag M, Cichy J, Kukreja RC, Dulak J, Lesnefsky EJ, Larner AC (2011) Mitochondrial-targeted Signal transducer and activator of transcription 3 (STAT3) protects against ischemia-induced changes in the electron transport chain and the generation of reactive oxygen species. J Biol Chem 286:29610–29620

Vander Heiden MG, Cantley LC, Thompson CB (2009) Understanding the Warburg effect: the metabolic requirements of cell proliferation. Science 324:1029–1033

Vassilev AO, Lorenz DR, Tibbles HE, Uckun FM (2002) Role of the leukemia-associated transcription factor STAT3 in platelet physiology. Leuk Lymphoma 43:1461–1467

Warburg O (1956) On the origin of cancer cells. Science 123:309–314

Wegrzyn J, Potla R, Chwae YJ, Sepuri NB, Zhang Q, Koeck T, Derecka M, Szczepanek K, Szelag M, Gornicka A, Moh A, Moghaddas S, Chen Q, Bobbili S, Cichy J, Dulak J, Baker DP, Wolfman A, Stuehr D, Hassan MO, Fu XY, Avadhani N, Drake JI, Fawcett P, Lesnefsky EJ, Larner AC (2009) Function of mitochondrial Stat3 in cellular respiration. Science 323: 793–797

Yang J, Chatterjee-Kishore M, Staugaitis SM, Nguyen H, Schlessinger K, Levy DE, Stark GR (2005) Novel roles of unphosphorylated STAT3 in oncogenesis and transcriptional regulation. Cancer Res 65:939–947

Yang J, Liao X, Agarwal MK, Barnes L, Auron PE, Stark GR (2007) Unphosphorylated STAT3 accumulates in response to IL-6 and activates transcription by binding to NFkappaB. Genes Dev 21:1396–1408

Yu H, Pardoll D, Jove R (2009) STATs in cancer inflammation and immunity: a leading role for STAT3. Nat Rev Cancer 9:798–809

Zhou L, Too HP (2011) Mitochondrial localized STAT3 is involved in NGF induced neurite outgrowth. PLoS One 6:e21680

Evolution of Jak and Stat Proteins

Claus Vogl, Priyank Shukla, and Ingo Ebersberger

Abstract

In this article, we examine the evolution of Jak and Stat proteins, and of the Jak-Stat pathway. We first introduce the protein families involved and the signaling pathway in general. We then oppose the simply structured pathway in *Drosophila melanogaster* to the more complex situation in mammals. Furthermore, we compare the Jak-Stat system between mammals and teleost fishes. Finally, we move to the less well investigated roles of Stats in the worm *Caenorhabditis elegans* and the slime mold *Dictyostelium discoideum*. We also survey the distribution of Jaks and Stats among metazoans and other eukaryotes. Orthologs of Stats are widely distributed among metazoans and are also found in choanoflagellates and in slime molds. In contrast, Jaks seem to be confined to the bilaterians and are apparently absent in molluscs, round- and flatworms. This indicates that the Jak-Stat pathway evolved at the base of the bilaterians, but has been lost in some invertebrate groups.

Introduction

The general function of the Jak-Stat pathway is the transduction of a signal induced by the binding of an extracellular ligand at the cell surface to the nucleus, and the subsequent activation of transcription. In Fig. 1, we outline the canonical Jak-Stat signaling pathway. Three functional classes of proteins are involved in this

---

C. Vogl (✉) • P. Shukla
University of Veterinary Medicine, Veterinärplatz 1, 1210 Vienna, Austria
e-mail: claus.vogl@vetmeduni.ac.at

I. Ebersberger (✉)
Center for Integrative Bioinformatics Vienna, Max F. Perutz Laboratories, University of Vienna, Vienna, Austria

Th. Decker and M. Müller (eds.), *Jak-Stat Signaling: From Basics to Disease*,
DOI 10.1007/978-3-7091-0891-8_7,

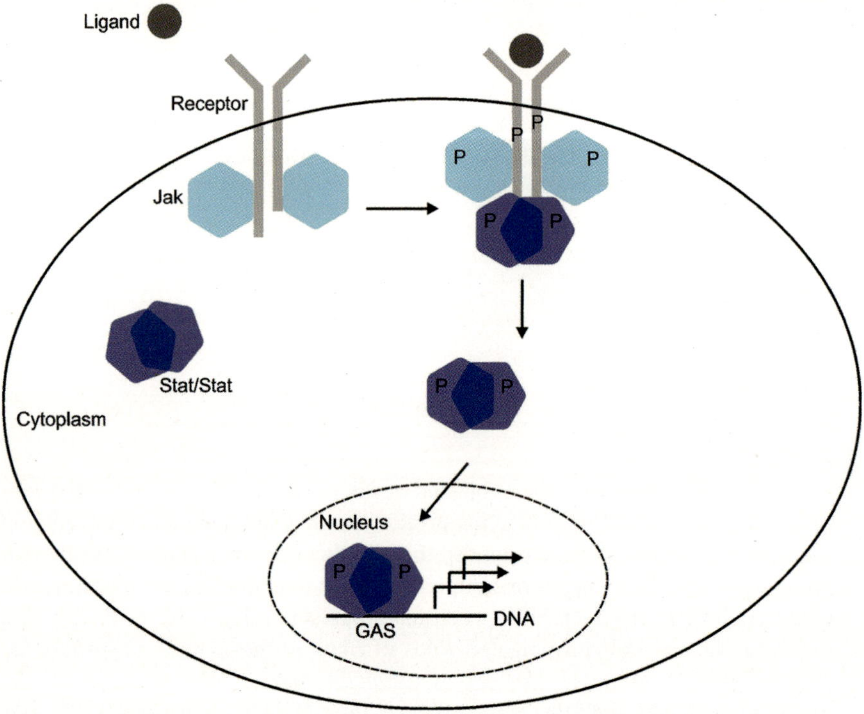

Fig. 1 The canonical model of the Jak-Stat pathway. Binding of an extracellular ligand (*dark grey*) to the trans-membrane receptor (*light grey*) and its associated Jak kinases (*light blue*) activates the receptor complex through phosphorylation (P), which in turn leads to the phosphorylation of cytoplasmic Stats (*dark blue*). This causes a change in the configuration of STATS, which then translocate to the nucleus, and activate transcription by binding to specific transcription factor binding sites (e.g., the palindromic GAS family transcription factor binding sites)

pathway: (1) a trans-membrane receptor complex, (2) the receptor-associated Janus kinase (Jak) proteins, and (3) the signal transducer and activator of transcription (Stat) proteins. While Jak-Stat signal transduction may involve different ligands and receptors among different organisms or also within an organism, the Jak and Stat components of the pathway and their interaction are generally conserved in structure and function. We will, therefore, proceed by explaining the general structure and function of Jaks and Stats and of a canonical Jak-Stat pathway. We will comment on the receptors and ligands, when referring to the situation in specific organisms.

The Structure and Function of Jak

A Jak consists of four domains (Fig. 2a). At the N-terminus, an *erythrocyte protein 4.1 ezrin/radixin/moiesin* domain (*FERM* or *ERM* domain) (Chishti et al. 1998) is

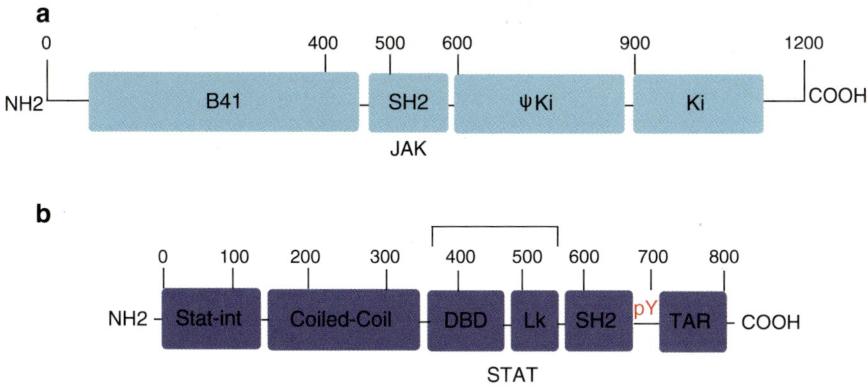

Fig. 2 The structure of (**a**) Jak and (**b**) Stat proteins. From N-terminus to C-terminus: (**a**) B41: *erythrocyte protein 4.1 ezrin/radixin/moiesin* domain; SH2: *Src homology 2 (SH2)*; ΨKi: *pseudo-kinase* domain; Ki: *kinase* domain. (**b**) Stat-int: *Stat protein interaction*; Coiled-coil: *Stat-protein all alpha* domain, also called *coiled-coil* domain; DBD: *Stat protein DNA-binding* domain; Lk: *linker* domain; (the bracket uniting these two domains indicates that they are united as *DNA-binding* domain in the NCBI CDD); SH2: *SH2* domain; pY: a tyrosine residue; TAR: transcriptional activation region

present. Via this domain the Jak interacts with the membrane bound receptor chains. The NCBI conserved domain database (CDD; http://www.ncbi.nlm.nih.gov/cdd) refers to this domain as *band 4.1 homolog* domain (B41, smart00295). The FERM domain is followed by the *Src homology 2 (SH2)* domain (SH2 superfamily: cl00138; smart00252; pfam00017), which binds to phosphorylated tyrosines. Towards the carboxy-terminus, first a *pseudo-kinase* domain (cd05037) and then a *kinase* domain (pfam07714) are found. In the NCBI CDD, both the *kinase* and *pseudo-kinase* domains are annotated as *PKc-like* domains (cl09925).

The Structure and Function of Stat

In the NCBI CDD many Stats are annotated to contain four domains. However, Schindler et al. (2007) annotate it with six domains (Fig. 2b). The *Stat protein interaction* domain (*Stat-int*, pfam02865) is located at the N-terminus and directs dimerization of Stats that are not tyrosine phosphorylated (cf. Fig. 1). *Stat-int* is followed by the *Stat-protein all alpha* domain (pfam01017), which provides a large hydrophilic surface and interacts with other proteins. Note that Schindler et al. (2007) refer to this domain as *coiled-coil* domain due to its structural properties. The third domain is the *Stat protein DNA-binding* domain (pfam02864). Schindler et al. (2007) differentiated two sub-domains, the N-terminal *DNA-binding* domain and a *linker* domain. The *linker* domain connects the DNA-binding domain to the next domain, the *SH2* domain (cl00138). As it is the case for the Jaks, the *SH2* domain binds to phosphorylated tyrosine residues. A functionally relevant tyrosine residue is located immediately adjacent to the *SH2* domain. Phosphorylation of this tyrosine marks the Stat's transition to the active state. The carboxy-terminal region is quite variable. Schindler et al. (2007) annotate this region as transcriptional activation "domain", because it associates with a broad variety of transcriptional regulators. The function of this domain to activate transcription is evolutionarily conserved, however neither is its sequence nor its structure. We will, thus, refer to it as Transcriptional Activation Region (TAR).

The Jak-Stat Pathway

In the inactive form, Stats shuttle into and out of the nucleus, but are found primarily in the cytosol. Already in the inactive form, Stat dimers are formed through the N-terminal Stat protein interaction domain (Vinkemeier et al. 1998). Upon binding of a ligand, the trans-membrane receptor undergoes a conformational change and associated Jaks come into proximity. The Jak kinases then auto- and cross-phosporylate each other as well as the receptor chains. Upon recruitment of cytosolic Stat dimers to the phosphorylated receptor via the Stat *SH2* domains, the Jak kinases activate the Stats via phosphorylation on tyrosine residues. In the activated conformation, the two Stat monomers make contact through the phosphorylated tyrosines and the *SH2* domains. The activated Stat dimer dissociates from the Jak-receptor complex and relocates to the nucleus. Here it binds to a family of palindromic sequence motifs (canonically: TTCNNNGAA), the so-called GAS family of transcription factor binding sites (TFBS). If GAS family TFBSs are located next to each other, cooperative binding may be promoted by the *Stat protein interaction* domain (Vinkemeier et al. 1998).

The Jak-Stat Pathway in Model Organisms

The Jak-Stat Pathway in *Drosophila*

The Jak-Stat pathway in *Drosophila* conforms to the canonical situation described in the previous paragraphs. It involves a receptor, a Jak, and a Stat, each encoded by a single gene (Arbouzova and Zeidler 2006). The Jak in *Drosophila* is named Hopscotch. Hopscotch is generally abbreviated as Hop, which must not be confused with the Hsp70/Hsp90 organizing protein. Its SH2 domain is aberrant and not annotated as such in the NCBI CCD. Still it is recognizable in secondary structure analysis (Gao et al. 2004). The *Drosophila* Jak, Hopscotch, is associated with the trans-membrane receptor Domeless (Dome). Dome has Fibronectin type 3 (cl00065) domains characteristic of cytokine receptors, a transmembrane region, and a coiled-coil domain (Arbouzova and Zeidler 2006). Three ligands of Dome have been characterized, all from the Unpaired gene family: Outstretched (Os, also annotated as Upd1), Unpaired-2 (Upd2), and Unpaired-3 (Upd3). The *Drosophila* Stat92e has the canonical domain structure described previously, and activated Stat dimers bind to the canonical palindromic GAS family TFBS and initiate transcription.

In *Drosophila*, Jak-Stat signaling is involved in a number of elementary cellular processes, such as immune response, stem cell signaling, and germ cell development. Its role has been mainly investigated by mutating individual genes of the pathway (Arbouzova and Zeidler 2006). Loss-of-function mutations of the genes encoding the three ligands each have only a modest phenotype. Due to their similarity, knockouts of either Udp2 or Udp3 can be partially compensated by the other genes. Only a loss-of-function mutation of Os has a pronounced effect, and the affected individuals show strong segmentation and posterior spiracle phenotypes. For the genes downstream to the ligands in the signaling pathway no redundancy exists. Thus, their mutations result in more severe phenotypes. For example, gain-of-function mutations of Jak induce over-proliferation and premature differentiation of larval blood cells leading to formation of melanotic tumors. Moreover, they cause unspecific lamellocyte differentiation and proliferation, an effect that is also part of normal immune response to wounding and infection. Loss-of-function mutations of the Jak lead to small imaginal discs, and thus to small eye phenotypes. Partial loss-of-function mutations of both Jak and Stat result in a reduced immune response. In the fly, immune response to septic injury is mediated by Udp3 expressed in hematocytes circulating at the site of injury. Udp3 activates Stat in the insect fat body resulting in an expression of anti-microbial peptides. This response can be blocked by inactivating the receptor Dome.

The Jak-Stat Pathway in Vertebrates

Genomic Arrangement, Structure, and Function of Mammalian Jaks and Stats

In mammals, both *Jak* and *Stat* genes are present in multiple copies (paralogs) that arose by gene duplications in the vertebrate lineage. Instead of just three ligands and one receptor, as in the fly, many different cytokines, chemokines and hormones bind to a diverse set of receptors. The interaction of the different Jaks and Stats with these receptors leads to a complex system of partially overlapping and partially antagonistic interactions that lead to the activation of distinct sets of genes. This signaling pathway is therefore considerably more complex than in *Drosophila*. We will follow Schindler et al. (2007) in our brief description of the mammalian Jak and Stat gene family members.

The *Jak* gene family of mammals comprises four members, *Tyk2*, *Jak1*, *Jak2*, and *Jak3*. *Tyk2* and *Jak3* are clustered on the same chromosome; the other two are scattered. The encoded proteins are similar in domain architecture to the *Drosophila* Jak (cf. Fig. 2a). The first three genes are ubiquitously expressed; expression of Jak3 is restricted to leukocytes. Tyk2, Jak1 and Jak2 associate with various cytokine and hormone receptors while Jak3 displays the most discrete function by binding solely to the common gamma chain (γ-c) receptor (Pesu et al. 2008). Null mutants of Tyk2 and Jak3 are viable but suffer from immunodeficiencies, whereas Jak1 knockout mice die perinatally due to neuronal defects and Jak2 deficiency is lethal at an embryonic stage due to lack of definitive erythropoiesis.

The mammalian Stat gene family comprises seven members, *Stat1, Stat2, Stat3, Stat4, Stat5a, Stat5b,* and *Stat6,* which are clustered into three groups in the genome. The encoded proteins are in general similar in domain architecture to the *Drosophila* Stat (cf. Fig. 2b). Only the carboxy-terminus harboring the transcriptional activation region differs; a number of additional domains have been described for the mammalian Stats. Stat1 has a *Stat1 TAZ2 binding* domain (pfam12162; Woijziak 2009). This domain binds to the *TAZ2* domain of the CREB-binding protein CRB. In this context it becomes a transcriptional activator. Stat2 has a *Stat2C terminal* domain (pfam12188; Banninger and Reich 2004), which aids nuclear export. The mouse Stat2 contains additionally a *POU domain, class 2, associating factor 1* domain (pfam09310; Chasman et al. 1999) at the C terminus. This domain is involved in activation of transcription and associates with either OCT1 or OCT2. Furthermore, the Stat6 of the mouse has a *topoisomerase II-associated protein PAT1* domain (pfam09770) (Wang et al. 1996). In the NCBI CDD, this domain is annotated as necessary for accurate chromosome transmission during cell division, but no such function has been described for murine Stat6. Thus the role of this domain in the function of Stat6 of mice remains unclear.

Stat1 and *Stat4* genes are found immediately adjacent to each other on the same chromosome. The Stat1 protein is observed in two isoforms: the longer Stat1α, and the shorter Stat1β lacking the carboxy-terminus of Stat1α. Stat1 forms heterodimers with Stat2. Upon stimulation through type I interferon, Stat1, Stat2, and interferon regulatory factor (Irf) 9 form the ISGF3 (interferon stimulated gene factor 3) trimer (Schindler et al. 1992). ISGF3 then binds to a non-palindromic TFBS called ISRE (interferon stimulated response element) through Irf9 (canonically: AGTTTNNNTTTCC). Stat1 also forms homodimers after IFN-γ stimulation, which bind to the canonical GAS family TFBS (Shuai et al. 1992). Activation of downstream target genes of Stat1 generally promotes inflammation and antagonizes proliferation. Mutations in this gene therefore generally lead to increased susceptibility to viral and bacterial infections. Stat4, the genomic neighbor of Stat1, also has a full-length and a shorter isoform. It homodimerises and is important in the response to IL-12 in innate and adaptive immunity (Kisseleva et al. 2002).

The next cluster of Stat genes is that of *Stat3*, *Stat5a*, and *Stat5b* (Miyoshi et al. 2001). Stat3 transduces signals from the entire IL-6 and IL-10 families, as well as from granulocyte(G)-CSF, leptin, IL-21, IL-22, IL23, and IL-27 (Kisseleva et al. 2002; Levy and Darnell 2002). Like *Stat1* and *Stat4, Stat3* exists in two isoforms that differ in their length. It is important during development, as Stat3-/- mouse embryos die around embryonic day 7 (Kisseleva et al. 2002; Levy and Darnell 2002). Tissue-specific knockouts suggest an important anti-inflammatory role of Stat3. Furthermore, Stat3 promotes cell proliferation, tumorigenesis, and cancers (Kisseleva et al. 2002; Levy and Darnell 2002). The anti-inflammatory and proliferative action of Stat3 counteracts the effects of Stat1, which also gets activated by IL-6. Stat5a and Stat5b associate with cytokine and single chain receptors, including erythropoietin, growth hormone and prolactin. Stat5 null mice are lethal mainly through the lack of the single chain receptor responses. *Stat5a* and *Stat5b* genes have been duplicated relatively recently (see below). Their corresponding proteins regulate hematopoiseis, metabolism and growth in partially overlapping fashion (Yao et al. 2006).

Stat2 and *Stat6* genes make up the third group of Stats. They are located on the same chromosome separated by only few genes. Stat2 does not bind to DNA by itself, but only with Stat1 and Irf9 as part of the ISGF3 complex (Decker et al. 2005). Stat6 transduces signals of the γ-c for both IL-4 and IL-13, whose respective receptors share common chains. The *Stat6* homodimer seems to bind to a slightly different palindromic GAS family TFBS with an additional central nucleotide (Kisseleva et al. 2002; Levy and Darnell 2002; O'Shea et al. 2002).

Genomic Arrangement of Jak and Stat Genes in Teleosts

Orthologs to six of the seven mammalian Stats and to all four Jaks are present in teleost fishes. This indicates that all but one gene duplication giving rise to the seven Stats and four Jaks in mammals occurred before the split of the teleost lineage about 450 Mio years ago. Only the duplications of *Stat5* giving rise to *Stat5a* and *Stat5b* giving mammals, and to *Stat5.1* and *Stat5.2* in teleosts, occurred twice independently (Gorissen et al. 2011). Note that the pufferfish *Takifugu rubripes* apparently has lost one of the two copies of *Stat5* (Jaillon et al. 2004). In the zebrafish genome five Jaks are found: all four Jaks also present in mammals plus an additional copy of *Jak2*.

While the seven mammalian Stats are arranged in three clusters and *Tyk2* and *Jak3* are also clustered, no such clustering is seen for the teleosts (Gorissen et al. 2011). This is most likely due to the whole genome duplication that occurred in the common ancestor of the teleosts (Meyer A 2005; Wittbrodt and Schartl 1998), which was followed by a reciprocal loss of Stat paralogs in the duplicated gene clusters. This scenario is exemplified by the genomic localizations of the *Stat1* and *Stat4* genes. In humans and mice the two genes are located immediately adjacent on the same chromosome. Likewise, only a single *Stat1* gene and a single *Stat4* gene is present in zebrafish, despite the whole genome duplication that occurred on the teleost lineage. However, the two genes are now located on different chromosomes (chr 9 and 22, respectively). Notably, in both locations, the paralogs of the genes that originally flanked the *Stat1-Stat4* gene cluster are preserved in the original order, i.e., in shared synteny. Thus a copy of *Stat1* must have been lost from the region on zebrafish chromosome 22 and a copy of *Stat4* from zebrafish chromosome 9 (Gorissen et al. 2011).

Evolutionary Rates of Jak and Stat Paralogs in Vertebrates

Among vertebrate Stat genes, *Stat1*, *Stat2*, *Stat4*, and *Stat6* display an elevated rate of evolution compared to *Stat3*, *Stat5a*, and *Stat5b*; among vertebrate Jak genes, *Tyk2* and *Jak3* evolve faster than *Jak1* and *Jak2* (Gorissen et al. 2011). Genes involved in immunity are often the target of pathogens. If a pathogen incapacitates Stat1, and thus avoids host immunity, functional Stat1 mutants that escape this inactivation and restore immune response may be positively selected. This in turn induces a positive selection of mutants on the pathogen side that can cope with the modified Stat1. The circle starts then over again. The resulting host-parasite arms race (Van Valen 1973) serves as explanation for the frequently observed accelerated rate of evolution of genes involved in immunity (Sackton et al. 2007).

Jaks and Stats in Tunicates

The tunicate *Ciona intestinalis*, a close relation to the vertebrates, has one *Jak* and two functional *Stats* (Hino et al. 2003). This suggests that the gene family encoding the Jaks in vertebrates has expanded after the tunicate-vertebrate split. This would be in line with similar gene duplication events at the base of the vertebrates (Hughes and Friedman 2003; Hughes and Friedman 2004; Meyer A. 2005; Sharman and Holland 1996; Sidow 1996). Sequence similarities indicate that the *Stat* gene family also has diversified after the vertebrate-tunicate split (Hino et al. 2003).

Stats in the Worm *Caenorhabditis elegans*

The lineages of vertebrates and flies split early in bilaterian evolution. This suggests that Jak-Stat mediated signal transduction was originally present in most or all bilaterian lineages (see section "The Phylogenetic Profile of Jaks and Stats" below). In particular, this holds for the worm *Caenorhabditis elegans*, which has been recently placed together with the fly into the monophyletic clade of molting animals (*Ecdysozoa;* e.g. (Dunn et al. 2008)). However, in the genome of *C. elegans* no Jaks are found, while two Stats have been reported: *Sta-1* and *Sta-2* (also known as: F58E6.1). Sta-1 is quite diverged from both the insect and vertebrate forms. It lacks the N-terminal *Stat protein interaction* domain. Therefore, it is presumably not able to form dimers in its unphosphorylated, i.e., inactive state. All other domains and functional regions are conserved. This includes a short carboxy-terminal region that seems to act in transcriptional activation, as it could drive expression of a luciferase reporter gene (Wang and Levy 2006). Expression of Sta-1 is non-uniform and especially high in head and pharynx, and it shows patterns of nuclear exclusion or nuclear accumulation in subsets of neurons (Wang and Levy 2006). As with Stats in organisms with a functional Jak-Stat signaling pathway, unphosphorylated Sta-1 is found primarily in the cytoplasm and relocates to the nucleus upon phosphorylation (Wang and Levy 2006). DNA binding does not seem necessary for the relocation, as mutants without a functional DNA binding domain are also found in the nucleus upon activation (Wang and Levy 2006). Using EMSA assays, it was found that tyrosine-phosphorylated Sta-1 binds to a GAS family TFBS (Wang and Levy 2006). Since GAS family TFBSs are palindromic, we presume that Sta-1 forms homodimers. In summary, it is to date unclear how activation of Sta-1 is accomplished without a Jak, and what the exact function of Sta-1 is. In this context it is interesting to note that *Sta-1* knockouts are viable. While Sta-1 is only lacking the N-terminal *Stat protein interaction* domain, Sta-2 is also lacking a *Stat-protein all alpha* domain. In the Interpro database (http://www.ebi.ac.uk/interpro/), it is annotated as containing an EF-hand-like domain, which is in turn annotated as part of the *STAT transcription factor, DNA-binding domain*. Sta-2 was recently shown to be activated by the p38 MAPK pathway and to be involved in innate immunity (Dierking et al. 2011).

Stats in the Slime Mold Dictyostelium

The slime mold *Dictyostelium discoideum* alternates between a unicellular and a multicellular mode of life. Three proteins are annotated as Stats: dstA, dstB, and dstC. In contrast to nearly all analyzed metazoan Stats, none of the three proteins contains a transcriptional activation region. Rather, they all act as repressors of transcription. dstA functions in signaling during multicellular stages of the life cycle in response to cyclic AMP signaling (Kawata et al. 1997). Activation of dstA is dependent on a serpentine (or G protein-coupled) receptor, yet is unaffected by knocking out the single known gene encoding a heterotrimeric G protein β subunit (Araki et al. 1998). As with metazoan Stats, activation through phosphorylation leads to a relocation into the nucleus facilitating DNA binding. Activated dstA binds to the activator and the repressor domain in the promoter of ecmA, which contains direct and inverted repeats of TTGA. Surprisingly to us, it also binds to the ISRE TFBS, apparently without an Irf-like protein (Araki et al. 1998). Via this binding dstA suppresses any further transcription of ecmA (Kawata et al. 1997). The dstB has an aberrant *SH2* domain, yet seems to be able to form dimers and accumulates in the nucleus upon activation. Notably, both functions are unaffected when the predicted site of tyrosine phosphorylation is substituted by phenylalanine (Zhukovskaya et al. 2004). Thus, its mode of activation seems different from that of other Stats. dstC is a key regulator of the transcriptional response to hyperosmotic shock (Na et al. 2007). The phosphatase ptpC interacts directly with dstC and negatively regulates activation. This stands in contrast to the situation in metazoa, where kinases (Jaks) activate Stats. - In summary, Stats in Dictyostelium are similar enough to metazoan Stats to infer their common evolutionary origin. However they are quite deviant in a number of functionally relevant key properties. They lack a transcriptional activating region, they have different DNA-binding properties, and they suppress rather than activate gene expression. Essentially, they differ in the regulation of their effector genes from their metazoan homologs.

The Phylogenetic Profile of Jaks and Stats

The Jak-Stat pathway is a key player in metazoan signal transduction and has been thoroughly studied in a number of model organisms such as *Drosophila* and mouse. Still we know surprisingly little about its evolution. This is mainly due to two reasons: First, the few studied organisms already show a remarkable variation in the complexity of this signaling cascade raising the question what the proto-Jak-Stat pathway may have looked like. Second, the number of species, which underwent an in-depth analysis of the Jak-Stat pathway, is low and the chosen organisms represent only a small part of the eukaryotic tree of life. This leaves open when during evolution the Jak-Stat pathway emerged, and to what extent this signaling cascade varies across species groups. The problem is best exemplified by the nematode *Caenorhabditis elegans*. If a Jak-Stat pathway is present at all, it is highly reduced in this worm (see above). In fact so far no Jak could be identified in its genome. According to recent evidence *C. elegans* shares a common ancestry with *Drosophila* forming the monophyletic group of molting animals (Ecdysozoa) to the exclusion of the deuterostomes (e.g. humans). Jaks are present both in humans and in *Drosophila*. Hence this protein is older than the split of Ecdysozoa and deuterostomes and must have been lost within the Ecdysozoa on the lineage leading to *C. elegans*. However, when exactly this loss has occurred remains unclear. Therefore we have little idea whether the reduced or absent Jak-Stat pathway is specific to the genus Caenorhabditis, or whether it is a feature that is shared by all nematodes or even by a larger systematic group.

To shed light on the evolutionary history of the Jak-Stat pathway we determined the phylogenetic profile of its two key players, Jak and Stat. In brief, we used HaMSTR (Ebersberger et al. 2009) to search for orthologs to the *Drosophila* Jak and *Drosophila* Stat in protein coding data from 1,476 eukaryotic species distributed over the entire eukaryotic tree (Database: dbDMP at http://www.deep-phylogeny.org). The resulting presence-absence patterns generate a good impression about the evolutionary history of both proteins, and thus of the Jak-Stat pathway. In the following, we will briefly discuss our findings. We will follow Dunn et al. (2008) in the systematic classification of species groups and their evolutionary relationships.

The Evolutionary History of Jak

The phylogenetic profile of Jak reveals this protein as a metazoan invention. No Jak could be identified in a species outside the metazoa (Fig. 3). This agrees with previous findings that e.g. fungi do not have proteins with an *SH2* domain (Hunter and Plowman 1997). Within the animals, Jak appears confined to the Bilateria. Neither corals or their cnidarian allies, nor sponges or comb jellies (Ctenophora) – all of which split prior to the diversification of the Bilateria – show signs of a Jak. Among the Bilateria, Jaks are present in both the deuterostomes (vertebrates, uro- and hemi-chordates, echinoderms) and the protostomes (Ecdysozoa and Lophotrochozoa). Consequently, we can date the latest possible emergence of Jak during metazoan evolution back to the last common ancestor of the Bilateria. However, while Jaks are prevalent within the deuterostomes, their presence in the protostomes appears limited to the ecdysozoa and here specifically to the arthropods (Crustacea and Hexapoda). We found Jaks in the data from many insects and also in the crustacean *Daphnia pulex*. However, we could find Jaks neither in nematodes or flat worms (Plathyhelminthes), which jointly represent the

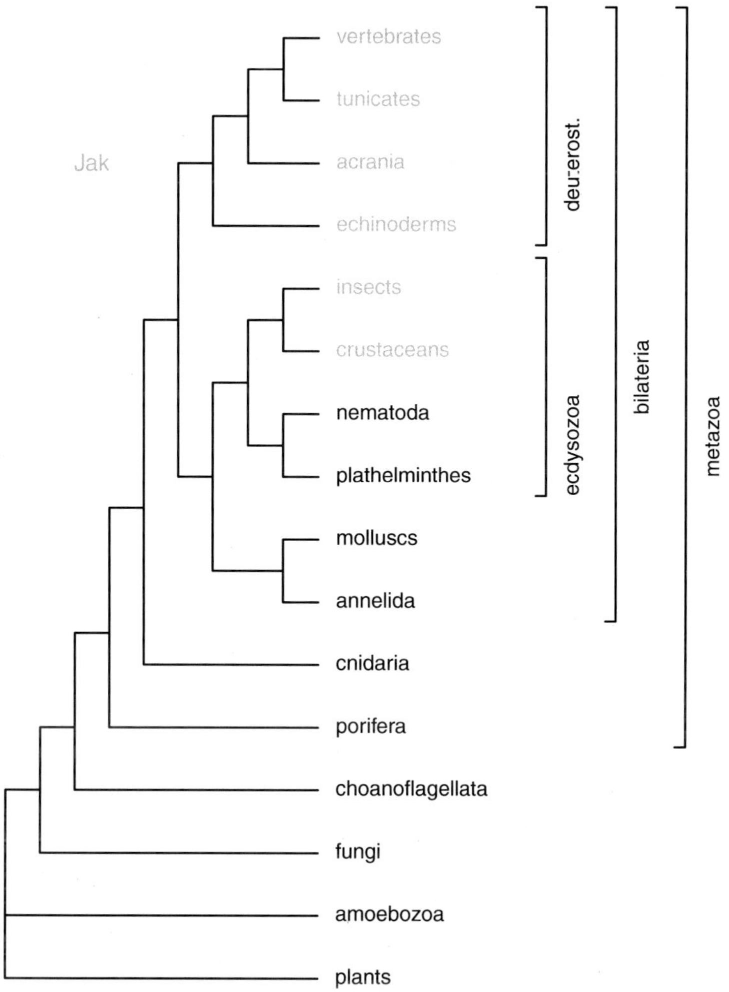

Fig. 3 The phylogenetic profile of Jak. The phylogenetic profile of Jak was established by searching for orthologs to the *Drosophila* Jak in 260 completely sequenced eukaryotic species and additionally in EST data from further 1,216 eukaryotes. Systematic groups where a Jak ortholog could be identified are colored in *light blue*

ecdysozoan sister taxon of the arthropods, nor in our representatives of the Lophotrochozoa (annelids and mollusks). Therefore we conclude that Jaks must have been lost at least twice independently during invertebrate evolution (Fig. 3). One loss must have occurred on the lophotrochozoan lineage after its split from the last common ancestor shared with the Ecdysozoa. A second loss must then have occurred within the Ecdysozoa on the lineage leading to the nematodes and Plathyhelminthes. Thus, the missing Jak in *C. elegans* seems to reflect a situation that most likely attributes to all Ecdysozoa except the arthropods.

The Evolutionary History of Stat

The phylogenetic profile of the Stat stands in strong contrast to that of its counterpart, Jak. Stat is found throughout the metazoan tree and it is also present in the earliest branching animals (Fig. 4). Apparently, loss of this protein is, in the long run, incompatible with the metazoan lifestyle. To assess the evolutionary age of Stat we followed the eukaryotic tree further back in time and determined for each split whether or not a Stat was likely to be present in the corresponding ancestral species. We detected Stats in two close relatives to the animals, the

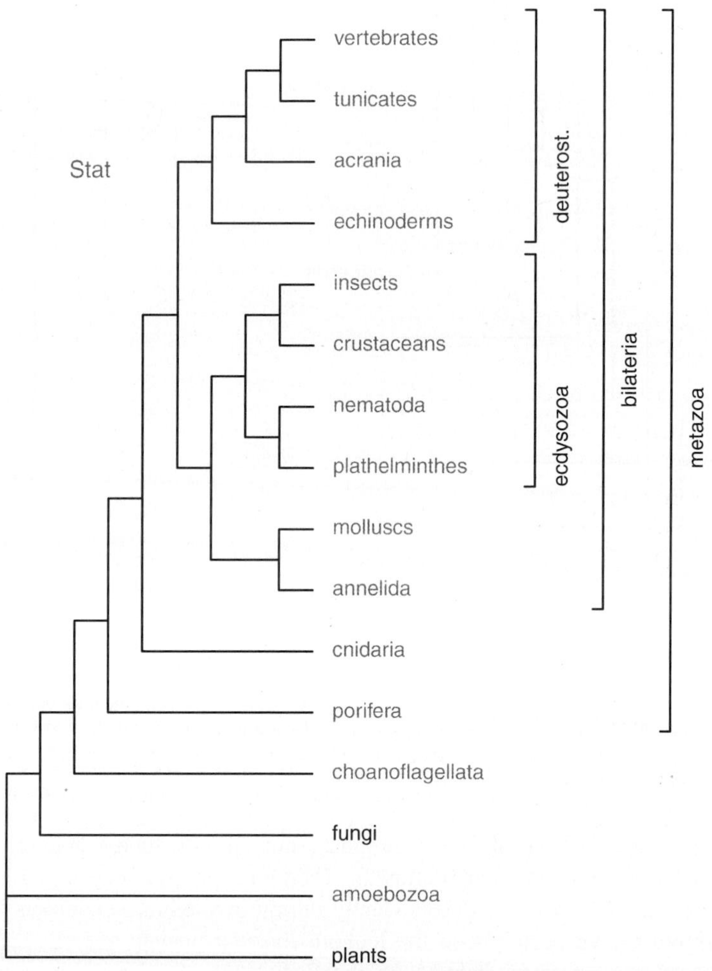

Fig. 4 The phylogenetic profile of Stat. The phylogenetic profile of Stat was established by searching for orthologs to the *Drosophila* Stat in 260 completely sequenced eukaryotic species, and additionally in EST data from further 1,216 eukaryotes. Systematic groups where a Jak ortholog could be identified are colored in *dark blue*

choanoflagellate *Monosiga brevicollis* and *Capsaspora owczarzaki*. This indicates that its evolutionary age predates the emergence of animals. Interestingly, no Stat could be detected in the fungi. Our data includes more than 90 fungal species from different parts of the fungal phylogeny for which a completely sequenced genome is available. Thus it can be safely considered as representative for the fungi in general. We therefore conclude that Stat is missing in the entire fungal kingdom. It follows that Stat was either invented after the split of the fungi or it was lost in the common ancestor of the contemporary fungi. The data from two slime molds, *Dictyostelium discoideum* and *Dictyostelium purpureum*, and from an amoeba *Acanthamoeba castellanii* unequivocally argue for the loss hypothesis. These three species are grouped in the Amoebozoa, the earliest branching lineage within the unikonts (a systematic group unifying the Metazoa, Fungi and Amoebozoa) and all three species contain a Stat. Thus, Stat has already existed in the primordial founder species of the unikonts. With the method at hand we could find no Stat ortholog in a species outside the unikonts. Thus, any further trace of Stat evolution remains in the dark.

The Evolutionary History of the Jak-Stat Signaling Cascade

Jak and Stat differ substantially in their phylogenetic distribution. Stat is an evolutionarily old protein that is present in all unikonts except the fungi. Jak on the other hand appears as a more recent invention. Its confinement to the deuterostomes and the arthropods suggests that the Jak-Stat signaling cascade is not substantially older than the Bilateria. However, not all Bilateria apparently make use of the standard Jak-Stat signaling. Within the protostomes, most groups have lost their Jaks in the course of evolution. The retention of a classical Jak-Stat signaling in the arthropods seems, therefore, rather an exception than a rule. In contrast, the function of Stat in *C. elegans* (see above) may be more representative for the protostomes than hitherto anticipated.

This raises the question about the functional role of Stats in the time prior to the diversification of the Bilateria. This protein has been present throughout opisthokont evolution and has only been lost in the fungi. However, functional studies of this protein are confined to the few species listed above. In particular studies on non-bilaterian animals, e.g. cnidarians, that could help to infer the likely function of Stat in the metazoan ancestor are lacking. The only non-bilaterian species in which Stat has been functionally studied belongs to the genus *Dictyostelium*. As we have noted above, however, *Dictyostelium* Stats are quite different from metazoan Stats. Based on their analysis, it is hard to make any predictions to the situation at the base of the metazoans. In particular, regulation of phosphorylation and de-phosphorylation of the Stat protein is unknown. One aspect however stands out: multicellularity and Stat signaling need not be associated, as the unicellular choanoflagellates also possess a Stat.

Concluding Remarks

Stat is an evolutionarily ancient protein that is present in all opisthokonts except the fungi. Jak, in contrast. seems confined to the deuterostomes and the arthropods. In both taxonomic groups, a canonical Jak-Stat signaling cascade has been described. This dates the emergence of this signal transduction pathway at least back to the base of the bilateria. Within the protostomes, most groups have lost their Jak. Nevertheless, whether a Jak is present or not, Stats mainly act as transcriptional activators and are often involved in immunity in all bilateria. Early during vertebrate evolution both Jaks and Stats diverged into families of paralogous genes. These form a network of partly overlapping partly antagonistic functions in Jak-Stat signaling cascades that not only are involved in immunity but also in the sensing of hormones.

Acknowledgments We would like to thank Mathias Müller, Thomas Decker, and Birgit Strobl for critically reading the manuscript and Nicole Leitner and again Birgit Strobl for help during various stages of the manuscript. The research of CV is supported by and PS is funded by the FWF SFB-F28. IE acknowledges funding by the Wiener Wissenschafts-, Forschungs- und Technologie Fonds (WWTF), and from the DFG priority program SPP 1174 Deep Metazoan Phylogeny [Grant HA 1628/9].

References

Araki T et al (1998) Developmentally and spatially regulated activation of a Dictyostelium STAT protein by a serpentine receptor. EMBO J 17(14):4018–4028

Arbouzova NI, Zeidler MP (2006) JAK/STAT signalling in *Drosophila*: insights into conserved regulatory and cellular functions. Development 133:2605–2616

Banninger G, Reich NC (2004) STAT2 nuclear trafficking. J Biol Chem 279:39199–39206

Chasman D et al (1999) Crystal structure of an OCA-B peptide bound to an Oct-1 POU domain/ octamer DNA complex: specific recognition of a protein-DNA interface. Genes Dev 13 (20):2650–2657

Chishti AH, Kim AC, Marfatia SM, Lutchman M, Hanspal M, Jindal H, Liu SC, Low PS, Rouleau GA, Mohandas N, Chasis JA, Conboy JG, Gascard P, Takakuwa Y, Huang SC, Benz EJ Jr, Bretscher A, Fehon RG, Gusella JF, Ramesh V, Solomon F, Marchesi VT, Tsukita S, Hoover KB et al (1998) The FERM domain: a unique module involved in the linkage of cytoplasmic proteins to the membrane. Trends Biochem Sci 23:281–282

Decker T, Muller M, Stockinger S (2005) The yin and yang of type I interferon activity in bacterial infection. Nat Rev Immunol 5:675–687

Dierking K, Polanowska J, Omi S, Engelmann I, Gut M, Lembo F, Ewbank JJ, Pujol N (2011) Unusual regulation of a STAT protein by an SLC6 family transporter in *C. elegans* epidermal innate immunity. Cell Host Microbe 9:425–435

Dunn CW, Hejnol A, Matus DQ, Pang K, Browne WE, Smith SA, Seaver E, Rouse GW, Obst M, Edgecombe GD, Sorensen MV, Haddock SH, Schmidt-Rhaesa A, Okusu A, Kristensen RM, Wheeler WC, Martindale MQ, Giribet G (2008) Broad phylogenomic sampling improves resolution of the animal tree of life. Nature 452:745–749

Ebersberger I, Strauss S, von Haeseler A (2009) HaMStR: profile hidden markov model based search for orthologs in ESTs. BMC Evol Biol 9:157

Gao Q, Hua J, Kimura R, Headd JJ, Fu XY, Chin YE (2004) Identification of the linker-SH2 domain of STAT as the origin of the SH2 domain using two-dimensional structural alignment. Mol Cell Proteomics 3:704–714

Gorissen M, de Vrieze E, Flik G, Huising MO (2011) STAT genes display differential evolutionary rates that correlate with their roles in the endocrine and immune system. J Endocrinol 209(2):175–184

Hino K, Satou Y, Yagi K, Satoh N (2003) A genomewide survey of developmentally relevant genes in Ciona intestinalis. VI. Genes for Wnt, TGFbeta, Hedgehog and JAK/STAT signaling pathways. Dev Genes Evol 213:264–272

Hughes AL, Friedman R (2003) 2R or not 2R: testing hypotheses of genome duplication in early vertebrates. J Struct Funct Genomics 3:85–93

Hughes AL, Friedman R (2004) Pattern of divergence of amino acid sequences encoded by paralogous genes in human and pufferfish. Mol Phylogenet Evol 32:337–343

Hunter T, Plowman GD (1997) The protein kinases of budding yeast: six score and more. Trends Biochem Sci 22:18–22

Jaillon O, Aury JM, Brunet F, Petit JL, Stange-Thomann N, Mauceli E, Bouneau L, Fischer C, Ozouf-Costaz C, Bernot A, Nicaud S, Jaffe D, Fisher S, Lutfalla G, Dossat C, Segurens B, Dasilva C, Salanoubat M, Levy M, Boudet N, Castellano S, Anthouard V, Jubin C, Castelli V, Katinka M, Vacherie B, Biemont C, Skalli Z, Cattolico L, Poulain J, De Berardinis V, Cruaud C, Duprat S, Brottier P, Coutanceau JP, Gouzy J, Parra G, Lardier G, Chapple C, McKernan KJ, McEwan P, Bosak S, Kellis M, Volff JN, Guigo R, Zody MC, Mesirov J, Lindblad-Toh K, Birren B, Nusbaum C, Kahn D, Robinson-Rechavi M, Laudet V, Schachter V, Quetier F, Saurin W, Scarpelli C, Wincker P, Lander ES, Weissenbach J, Roest CH (2004) Genome duplication in the teleost fish Tetraodon nigroviridis reveals the early vertebrate proto-karyotype. Nature 431:946–957

Kawata T, Shevchenko A, Fukuzawa M, Jermyn KA, Totty NF, Zhukovskaya NV, Sterling AE, Mann M, Williams JG (1997) SH2 signaling in a lower eukaryote: a STAT protein that regulates stalk cell differentiation in dictyostelium. Cell 89:909–916

Kisseleva MV, Cao L, Majerus PW (2002) Phosphoinositide-specific inositol polyphosphate 5-phosphatase IV inhibits Akt/protein kinase B phosphorylation and leads to apoptotic cell death. J Biol Chem 277:6266–6272

Levy DE, Darnell JE Jr (2002) Stats: transcriptional control and biological impact. Nat Rev Mol Cell Biol 3:651–662

Meyer A, VdP Y (2005) From 2R to 3R: evidence for a fish-specific genome duplication (FSGD). Bioessays 27:937–945

Miyoshi K, Cui Y, Riedlinger G, Robinson P, Lehoczky J, Zon L, Oka T, Dewar K, Hennighausen L (2001) Structure of the mouse Stat 3/5 locus: evolution from *Drosophila* to zebrafish to mouse. Genomics 71:150–155

Na J, Tunggal B, Eichinger L (2007) STATc is a key regulator of the transcriptional response to hyperosmotic shock. BMC Genomics 8:123

O'Shea JJ, Gadina M, Schreiber RD (2002) Cytokine signaling in 2002: new surprises in the Jak/Stat pathway. Cell 109(Suppl):S121–S131

Pesu M, Laurence A, Kishore N, Zwillich SH, Chan G, O'Shea JJ (2008) Therapeutic targeting of Janus kinases. Immunol Rev 223:132–142

Sackton TB, Lazzaro BP, Schlenke TA, Evans JD, Hultmark D, Clark AG (2007) Dynamic evolution of the innate immune system in *Drosophila*. Nat Genet 39:1461–1468

Schindler C, Shuai K, Prezioso VR, Darnell JE Jr (1992) Interferon-dependent tyrosine phosphorylation of a latent cytoplasmic transcription factor. Science 257:809–813

Schindler C, Levy DE, Decker T (2007) JAK-STAT signaling: from interferons to cytokines. J Biol Chem 282:20059–20063

Sharman AC, Holland PWH (1996) Conservation, duplication and divergence of developmental genes during chordate evolution. Neth J Zool 46:47–67

Shuai K, Schindler C, Prezioso VR, Darnell JE Jr (1992) Activation of transcription by IFN-gamma: tyrosine phosphorylation of a 91-kD DNA binding protein. Science 258:1808–1812

Sidow A (1996) Gen(om)e duplications in the evolution of early vertebrates. Curr Opin Genet Dev 6:715–722

Van Valen L (1973) A new evolutionary law. Evol Theor 1:1–30

Vinkemeier U, Moarefi I, Darnell JE Jr, Kuriyan J (1998) Structure of the amino-terminal protein interaction domain of STAT-4. Science 279:1048–1052

Wang Y, Levy DE (2006) C. elegans STAT: evolution of a regulatory switch. FASEB J 20:1641–1652

Wang X, Watt PM, Louis EJ, Borts RH, Hickson ID (1996) Pat1: a topoisomerase II-associated protein required for faithful chromosome transmission in *Saccharomyces cerevisiae*. Nucleic Acids Res 24:4791–4797

Weiler S, Mou S, DeBerry CS, Keller JR, Ruscetti F, Ferris DK, Longo DL, Linnekin D (2011) JAK2 is associated with the c-kit proto-oncogene product and is phosphorylated in response to stem cell factor. Blood 87:3688–3693

Wittbrodt JMA, Schartl M (1998) More genes in fish? Bioessays 20:511–515

Wojciak JM et al (2009) Structural basis for recruitment of CBP/p300 coactivators by STAT1 and STAT2 transactivation domains. EMBO J 28(7):948–958

Yao Z, Cui Y, Watford WT, Bream JH, Yamaoka K, Hissong BD, Li D, Durum SK, Jiang Q, Bhandoola A, Hennighausen L, O'Shea JJ (2006) Stat5a/b are essential for normal lymphoid development and differentiation. Proc Natl Acad Sci USA 103:1000–1005

Zhukovskaya NV et al (2004) Dd-STATb, a Dictyostelium STAT protein with a highly aberrant SH2 domain, functions as a regulator of gene expression during growth and early development. Development 131(2):447–458

JAK/STAT and Chromatin Regulation in *Drosophila*

Kriti Gaur and Willis X. Li

Abstract
The formation of distinct chromatin domains, such as euchromatin and heterochromatin, in eukaryotic cells is a critical mechanism by which proper gene expression and development are controlled. Covalent DNA and histone modifications that establish stable heterochromatin states, such as cytosine methylation, histone hypoacetylation and histone H3-Lys9 methylation are among the best understood. The specific pattern of these modifications provides a mechanism for the spread and maintenance of heterochromatin in conjunction with the recruitment of additional factors, a key step in epigenetic gene silencing. In this review, we will summarize the current understanding of the modulation of chromatin structure dynamics, its implications for disease and development, especially recent work highlighting the contribution of the JAK/STAT signaling pathway in controlling heterochromatin stability.

Introduction

The DNA inside a eukaryotic nucleus is packaged with proteins to form chromatin. By weight, chromatin structure is roughly comprised of one-third DNA, one third-histones, and one-third nonhistone chromosomal proteins, along with a little RNA. The nucleosome core is the basic repeating unit of chromatin, composed of 147 bp

K. Gaur
Department of Biomedical Genetics, University of Rochester Medical Center, 601 Elmwood Avenue, Rochester, NY 14642, USA

W.X. Li (✉)
Department of Biomedical Genetics, University of Rochester Medical Center, Rochester, NY 14642, USA

Department of Medicine, University of California, San Diego, La Jolla, CA, USA
e-mail: willisli@ucsd.edu

Th. Decker and M. Müller (eds.), *Jak-Stat Signaling: From Basics to Disease*,
DOI 10.1007/978-3-7091-0891-8_8, © Springer-Verlag Wien 2012

of DNA wrapped 1.65 turns around the histone octamer, forming a 10 nm nucleosomal fiber (Luger et al. 1997). This nucleosomal fiber is folded helically into a 30 nm fiber and further into a 60–130 nm chromonema fiber to allow for maximum DNA compaction (Luger et al. 1997). Although the precise geometry of higher-order chromatin folding has not been resolved beyond this point, one level of organization evident by cytological studies of interphase nuclei, is the separation of chromatin into euchromatin and heterochromatin (Sadoni et al. 1999). While euchromatin decondenses during interphase, heterochromatin remains relatively more condensed, showing intense staining. Although largely an oversimplification, chromatin folding has been assumed to act as a barrier, limiting access of regulatory factors to condensed chromatin domains (Dillon and Festenstein 2002) (Fig. 1).

The partitioning of chromatin into heterochromatin and euchromatin states and their respective characteristics initially revealed the connection between gene activity and chromatin structure, as active genes are often found in largely decondensed euchromatin and silenced genes in condensed heterochromatin (Li et al. 2007). Heterochromatic regions, on the other hand, have relatively fewer genes, albeit not devoid of genes. For instance, in *Drosophila melanogaster*,

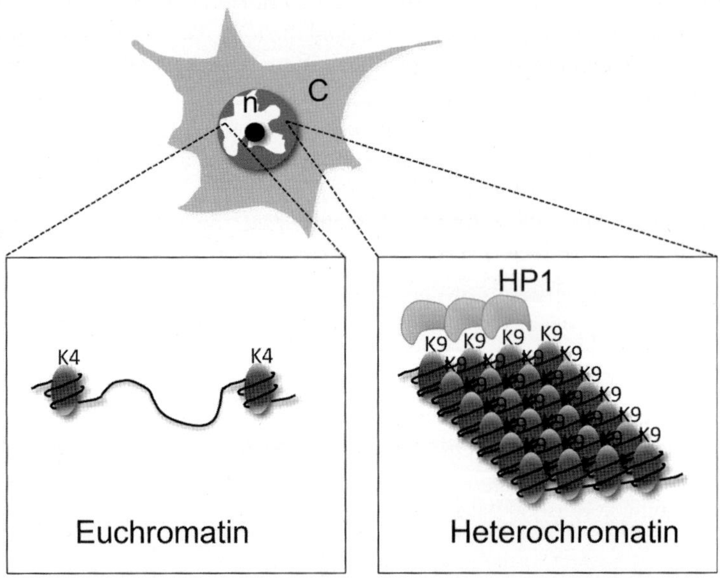

Fig. 1 Chromatin domains. A schematic showing of the euchromatin and heterochromatin domains and their simplified organization. A cell consists of cytoplasmic (C) and nuclear (n) regions. Within the nucleus, chromatin is organized as electron-dense (*dark gray*) and – light (*white*) regions, surrounding the nucleolus (*filled circle*). Lower panels represent amplified euchromatin and heterochromatin regions. DNA wraps around core histone molecules creating a "beads-on-a-string" arrangement. In euchromatin and transcribed genes, in which the chromosomal DNA assumes an "open" structure, histone H3 is usually methylated at lys4 (K4). In heterochromatin, histone H3 is methylated on lys9 (K9), which recruits HP1 and other non-histone proteins (not shown), resulting in compaction of the chromosome (Alberts 2002)

the entire Y chromosome, most of the fourth chromosome, the proximal 40% of the X chromosome and the pericentric 20% of the major autosomes are heterochromatic; these heterochromatin regions, nonetheless, contain approximately 40–50% of genes in the fly genome (Weiler and Wakimoto 1995). Additional heterochromatic traits include methylated DNA, highly repetitive sequences, low rate of meiotic recombination and a tendency to replicate throughout S phase. Conversely, euchromatic DNA has a high proportion of genes and unique sequences and tends to replicate throughout S phase (Weiler and Wakimoto 1995).

Euchromatin and heterochromatin appear to differ functionally as well. In a phenomenon termed position effect varigeation (PEV) genes normally active in a euchromatic domain will typically be silenced, often showing a variegating phenotype when placed adjacent to or within a heterchromatic domain, by transposition or chromosomal rearrangement (Grewal and Elgin 2002). The variegated pattern or mottled appearance results when these genes are appropriately expressed (time and place) in some cells but not in others. For example, in the *Drosophila* chromosomal inversion $In(1)w^{m-4h}$, the white gene fails to express in some eye cells, leading to white patches in the eye (Festenstein et al. 1999). Such loss of normal expression, apparently the consequence of heterochromatic packaging, is described as silencing.

Extensive research efforts focused on understanding the regulation of chromatin structure in the past decade have underscored the significant impact of chromatin organization on virtually all DNA-related metabolic processes including transcription, recombination, DNA repair, replication, kinetochore and centromere formation, to name a few (Li et al. 2007). Given that epigenetic mechanisms are essential in establishing and maintaining chromatin states, thus controlling development by defining complex gene expression patterns, and that this epigenetic information contained in chromatin can be inherited, chromatin research has therefore become central to modern epigenetics.

Epigenetic modulation of the genome occurs through histone modifications, chromatin remodeling and DNA methylation. All three processes use different machineries to regulate chromatin structure, and cross talk to ensure proper gene expression (Richards and Elgin 2002). In this chapter, we will summarize the prevailing view of how chromatin structure is regulated in *Drosophila*, the emerging function of JAK/STAT signaling in chromatin regulation, and then discuss how modulation of chromatin can exert effects on disease states like cancer and aging.

Histone Modifications

The nucleosome's properties are dynamically regulated by a panoply of posttranslational modifications to the histone tails and globular domains (Liu et al. 2005). These include methylation of arginine (R) residues and methylation, acetylation, ubiquitination, ADP-ribosylation, and sumolation of lysines (K); and phosphorylation of serines and threonines on histones H3 and H4 (Methylation of R residues, ubiquitination, ADP-ribosylation, and phosphorylation have not been detected in the fly genome) (Richards and Elgin 2002).

Among these epigenetic marks, acetylation or deactylation and methylation or demethylation of lysine residues in the conserved amino termini of histone 3 and histone 4 (H3 and H4) are prominently associated with changes in gene expression and chromosome structure (Giordano and Avantaggiati 1999) (Fig. 1). Acetylation of H3 and H4 is associated with transcriptional activation or euchromatin, whereas decreased acteylation is correlated with transcriptional repression or heterochromatin. The degree of histone acetylation is modulated by the activities of histone acetyltransferases (HATs) and histone deactylases (HDACs)(Giordano and Avantaggiati 1999).

In addition to histone hypoacetylation, di or tri-methylation (me) of H3 on lysine 4 (H3 K4me) functions as a euchromatic mark whereas, H3 K9me, H3 K27me and H4 K20me is associated with heterochromatin assembly (Shaffer et al. 2006) (Fig. 1). In *Drosophila,* methylation of H3K9 is carried out by a family of conserved histone methyltransferase (HMTase) named Su(var)3–9, which directly associates with another conserved non-chromosomal protein called Heterochromatin Protein 1 (HP1, sometimes referred to as HP1a), encoded by the gene *Su(var)2–5* (Shaffer et al. 2006).

Role of H3 K9me/HP1 System in Chromatin Regulation

HP1

HP1 is 23 kDa in size, and has an amino-terminus chromodomain (CD) and a carboxy-terminal chromoshadow (CSD) domain, separated by a flexible hinge (Dialynas et al. 2008). The two domains enable dimer formation with other HP1 molecules and modification proteins including *Su(var)3–9*. Heterochromatin formation and spreading occurs when H3K9me binds HP1 through either the N- or C-terminal domains, which then recruits *Su(var)3–9* and orthologues that go on to specifically methylate the newly arrived histone, further perpetuating heterochromatin assembly (Dialynas et al. 2008). The H3 K9me/HP1 system is crucial for maintaining the silenced epigenetic state in pericentric heterochromatin, but of less importance at the telomeres where other mechanisms maybe involved (Shaffer et al. 2006).

In *Drosophila* the HP1 family of proteins includes HP1a–e, and while each individual isoform possesses the characteristic domain organization, they display varying functional, expression and/or chromosomal domain localization profiles (Dialynas et al. 2008). In the mouse genome three HP1 family members, HP1alpha, M31 (HP1beta) and M32 (HP1gamma) exhibit functional similarity to *Drosophila* HP1 (Wasenius et al. 2003). Similarly, the human genome encodes three HP1-like proteins, $HP1^{Hs\alpha}$, $HP1^{Hs\beta}$ and $HP1^{Hs\gamma}$, whose chromosomal localization and functions parallel that of flies to some degree (Pomeroy et al. 2002). Conservation of HP1 function across species was demonstrated when human $HP1^{Hs\alpha}$ was reported to rescue lethality associated with mutations of the *Drosophila* gene encoding HP1, *Su(var)2–5* (Pomeroy et al. 2002).

Null alleles of *Su(var)2–5,* while homozygous lethal, dominantly suppress position effect variegation (PEV) in heterozygous flies, and mutations that increase the levels of HP1 promote gene silencing and thus enhance variegation (Eissenberg et al. 1990). Likewise, by increasing the dosage of mouse HP1 in murine T cells, silencing of a variegating reporter gene inserted near centric heterochromatin was achieved (Festenstein et al. 2003). These reports implicate HP1 as a key factor in the creation of a stable and inaccessible heterochromatin state, with increases in HP1 levels enhancing heterochromatin and vice versa, thus allowing HP1 to behave as a negative regulator of gene expression. In accordance with these data, targeting HP1 to upstream regions of reporter genes integrated at euchromatic sites within the *Drosophila* genome caused heterochromatization thereby silencing reporter genes (Danzer and Wallrath 2004). Targeting HP1 proteins upstream of reporter genes on transiently transfected plasmids in mammalian cell culture also repressed gene expression (Lehming et al. 1998; van der Vlag et al. 2000). Armed with the ability to nucleate heterochromatin, HP1 proteins can thus silence genes at both centric regions and ectopic sites within euchromatin.

In contrast to a negative role in gene expression, there are several instances where HP1 plays a positive role in transcription. First, *Drosophila* HP1 is essential for proper expression of genes that reside within heterochromatin (Yasuhara and Wakimoto 2006). Second, HP1 is required in the expression of some euchromatic genes that bind to HP1 and possess H3K9me (Cryderman et al. 2005). Surprisingly, these genes exhibit decreased expression in a *Su(var)2–5* mutant background, supporting a positive role for HP1 in expression. Finally, the role of HP1 as a positive transcriptional enforcer derives from studies of activated genes in *Drosophila* (Piacentini et al. 2003). HP1 localizes to heat shock and developmental "puffs" in polytene chromosomes that are centers of intense gene activity. In the absence of HP1, these genes are no longer induced (Piacentini et al. 2003). Collectively these reports point to the versatility of HP1 function in transcription, including both repression and activation, thus demonstrating that the presence of H3K9 methylation along with HP1 association does not solely work as marker of gene repression.

HP1 and Cancer

Since cancer progression involves perturbation of gene expression on a global scale and given that defects in a chromatin packaging protein or an enzyme that modifies histones can alter the chromatin status of multiple genomic regions, simultaneously affecting the expression of hundreds of genes, it comes as no surprise that HP1 with its role in transcription in conjunction with its distribution throughout the genome has been implicated in cancer formation. Although currently no human diseases are associated with mutations in the genes encoding $HP1^{Hs\alpha}$, $HP1^{Hs\beta}$ or $HP1^{Hs\gamma}$, changes in the expression levels of these genes, however, have been reported for several cancers, including breast, colon and ovarian cancers (Dialynas et al. 2008). Gene expression profiling studies showed that $HP1^{Hs\alpha}$ mRNA levels are lower in papillary thyroid carcinomas compared to normal thyroid tissue, while increased levels of $HP1^{Hs\alpha}$ correlate with limited metastasis in colon cancer (Wasenius et al. 2003). Similarly, decreased amounts of $HP1^{Hs\alpha}$ mRNA were observed in patients with embryonal brain tumors with a poor prognosis, compared to those with more positive outcomes (Pomeroy et al. 2002). In fact, reduction in $HP1^{Hs\alpha}$ mRNA levels was successfully used as a predictor of treatment failure for individuals battling embryonal brain cancer (Pomeroy et al. 2002).

Although numerous studies have pointed to a link between alterations in HP1 protein levels and cancer progression, studies in breast cancer have surpassed mere correlation and provided a causal role for $HP1^{Hs}\alpha$ in determining breast cancer cell invasiveness. Downregulation of $HP1^{Hs\alpha}$ was seen in invasive/metastatic cells relative to poorly invasive/non-metastatic cells (Kirschmann et al. 2000). Consistent with these results, $HP1^{Hs\alpha}$ levels in metastatic tissues from breast cancer patients were lower by about 95%, compared to levels present in primary breast cancer tumors (Kirschmann et al. 2000). Knocking-down $HP1^{Hs\alpha}$ in poorly invasive/metastatic cells enhanced *in vitro* invasion by 50% relative to controls (without altering cellular growth rates), illustrating a causal role for $HP1^{Hs\alpha}$ in metastasis. Similarly, induction of exogenous $HP1^{Hs\alpha}$ in highly invasive/metastatic cells also inhibited invasiveness by 30% relative to controls, with no effect on growth (Norwood et al. 2006). Thus, HP1 proteins appear to play a role in inhibiting cancer cell invasiveness. It was in *Drosophila* that the role of the JAK/STAT pathway in regulating HP1 localization and maintaining heterochromatic gene silencing were demonstrated (Shi et al. 2006, 2008).

The JAK/STAT Signaling Pathway

Basic Elements of the JAK/STAT Pathway in *Drosophila*

The Janus Kinase (JAK) and Signal transducer and activator of transcription (STAT) pathway is a signaling module first identified in vertebrates as mediating the response to some cytokines (Shuai et al. 1993) and was subsequently found to be conserved in invertebrates (Aaronson and Horvath 2002; Hou et al. 1996; Yan et al. 1996). In the canonical pathway, the final effector of this pathway is the transcription factor STAT that localizes to the cytoplasm in its inactive state. STAT translocates to the nucleus upon ligand induced activation of the receptor, followed by binding to target gene promoters and activating gene expression. The JAK/STAT machinery consists of transmembrane receptors that upon recognition of extracellular signals (cytokines, growth factors and some peptides) recruit tyrosine kinases of the JAK family to their intracellular domains. The receptor-associated JAKs then phosphorylate the receptor as well as themselves creating STAT docking sites, that is also phosphorylated by JAK, enabling it to dimerize and transpose to

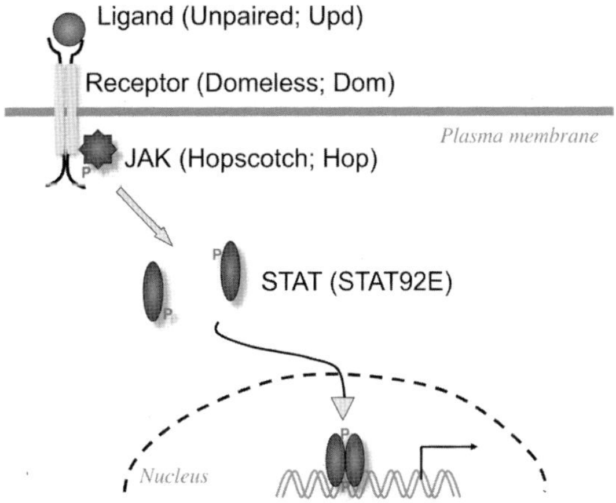

Fig. 2 Canonical JAK/STAT signaling in *Drosophila*. A schematic representation of JAK/STAT signaling in *Drosophila*, including confirmed elements of the cascade. The ligand (Unpaired) binding to the receptor (Domeless) activates Hopscotch bound to the intracellular domain, which in turn phosphorylates itself, the receptor and inactive STAT92E. Unphosphorylated STAT92E normally resides in the cytoplasm and upon phosphorylation by Hopscotch, STAT92E dimerizes and translocates to the nucleus and activates transcription. Adapted from Li (2008)

the nucleus and activate transcription (Fig. 2) (Aaronson and Horvath 2002; Levy and Darnell 2002).

Unlike vertebrates where there are several ligands, receptors, JAK kinases and STAT proteins, the *Drosophila* JAK/STAT pathway is less redundant and well characterized comprised of a receptor, Domeless (Dome); a JAK kinase, Hopscotch (Hop); a transcription factor STAT92E; and a ligand, Unpaired (Upd) (Binari and Perrimon 1994; Brown et al. 2001; Harrison et al. 1998; Hou et al. 1996; Yan et al. 1996). Regulators of the JAK/STAT pathway include *Su(var)2–10*, (*dPIAS*), *SOCS* and *STAM* (Hombria and Brown 2002).

Functions of the JAK/STAT Signaling Pathway and Modulation of Disease States

The JAK/STAT pathway has a multitude of roles in the regulation of animal development, growth control and homeostasis. In *Drosophila* the pathway is required for cell growth, eye development, segmentation, tracheal development, spermatogenesis and hematopoiesis (Hombria and Brown 2002; Tulina and Matunis 2001).

There are four JAK kinases and seven STAT isoforms in mammals and their diversity in their distribution and amino acid sequences provide an ability to respond to a variety of extracellular signals (Li 2008). Defects in JAK/STAT signaling are associated with severe developmental problems. In mammals, Jak1 deficient animals die prenatally, due to the failure of cytokine signaling during neurogenesis. Jak2 deficiency causes embryonic lethality due to failure of erythropoiesis. Mice lacking Jak3 develop SCID (Severe Combined Immunodeficiency) and defects in Tyk2, the fourth mammalian JAK, leads to an impaired immune response. Consistent with these findings, mice with null STAT mutations manifest various kinds of impairment in growth control and host defense (Bromberg 2001; Levy and Darnell 2002; Ward et al. 2000).

Dysregulation of the JAK/STAT pathway has also been implicated in a variety of cancers (Ward et al. 2000; Yu and Jove 2004). The most direct evidence for this comes from JAK2 fusion proteins found in lymphoid and myeloid leukemia cells. In both cases, fusion of the oligomerization domain of proteins such as Tel and BCR to the catalytic domain of JAK2 was responsible for the constitutive association and activation of JAK and activation of STAT (Ward et al. 2000). In fact, two constitutively activated mutants of the *Drosophila* JAK have been found to induce a leukemia-like defect. Tumorous lethal (*Tum-L* or hop^{Tum-L}) is a *Drosophila* JAK (*hopscotch* or *hop*) mutant with a glycine to glutamic acid substitution at position 341 (Binari and Perrimon 1994; Hanratty and Dearolf 1993; Harrison et al. 1995; Luo et al. 1995; Ward et al. 2000). In this mutant, hemocyte over-proliferation of particular blood-cell types (plasmatocytes and lamellocytes) and abnormal differentiation of prohemocytes results in the formation of melanotic tumors in the larval and adult body cavity, and is lethal at higher temperatures (Fig. 3a, b) (Li 2008). The other mutation (hop^{T42}), a lysine substitution for glutamic acid 695 has an even more severe phenotype (Hanratty and Dearolf 1993; Luo et al. 1995).

Although the constitutive activation of STATs has been detected in many types of cancers, the role of STATs in oncogenesis is not well understood. Constitutive activation of STAT3 has been reported in ovarian carcinomas and constitutive activation of STAT5 led to the development of mammary tumors in transgenic mice versus ones expressing wild-type and truncated forms. However, activated STAT1 was shown to inhibit angiogenesis, tumorigenicity and metastasis when tumor cells derived from a fibrosarcoma of a Stat1 knockout mice were reconstituted with a Stat1 expression vector (Yu and Jove 2004). Given that both signal transduction and epigenetic regulation are implicated in tumor progression and that the JAK/STAT pathway plays important roles in both leukemogenesis and lymphomagenesis, it was not long before researchers began examining JAK/STAT mediated effects on epigenome and their synergism in tumor formation.

JAK/STAT Signaling and Chromatin Regulation

JAK Mediates Global Heterochromatic Gene Silencing

Shi et al. (2006) utilized a genetic screen to find loci important for the *Tum-L* phenotype, induced by the overactivation of the JAK/STAT pathway resulting in hematopoietic tumors (Shi et al. 2006) (Fig. 3c). The authors chose to focus on genes that altered chromatin structure and had widespread effects on transcription.

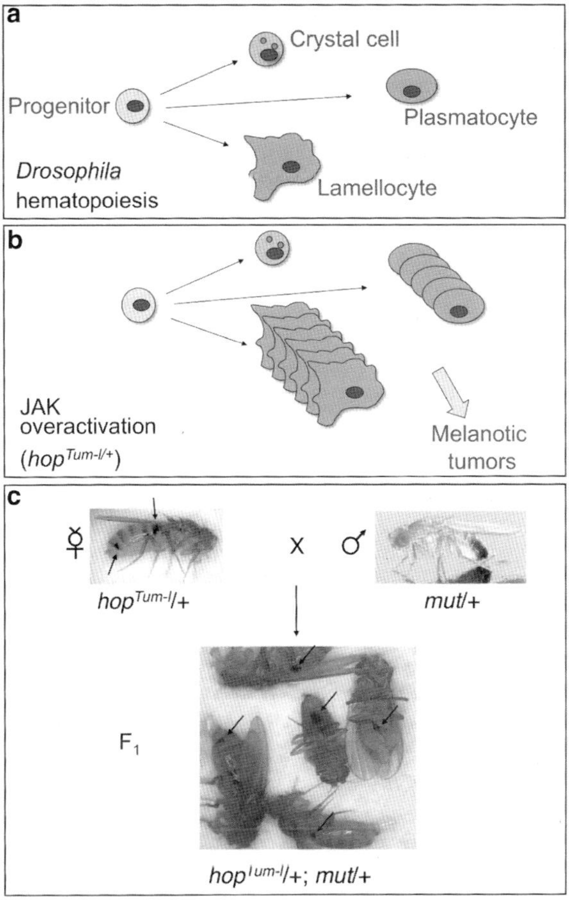

Fig. 3 ***Drosophila*** **JAK in hematopoiesis and blood-tumor formation, and the genetic screen.** (**a**) Progenitor cells give rise to three distinct blood cell-types in the larval hemolymph (blood) – the small crystal cells, the large intermediate plasmatocytes that perform phagocytosis and the terminally differentiated lamellocytes that are also phagocytic. (**b**) The hyperactive JAK kinase allele hop^{Tum-l} causes overproliferation of plasmatocytes and lamellocytes, that manifest as melanotic tumors. (**c**) Schematic representation of a genetic screen for genes that modify the blood tumor phenotype associated with hop^{Tum-l}. Examples of blood cell tumors in the body cavity (*arrows*), visible without staining, in heterozygous hop^{Tum-l} adult flies. The frequency and size these tumors in the F1 generation were compared to parents to identify new modifier mutations (a and b were adapted from Li 2008)

Of the several Hop^{Tuml} modifiers that were uncovered, HP1 and the histone H3 methyltransferase *Su(var)3–9* were of particular interest as they were essential heterochromatin components and were required for heterochromatin mediated gene silencing (Grewal and Elgin 2002).

As mentioned previously, heterochromatic gene silencing causes the phenomenon of position effect variegation (PEV), and mutations of HP1 and *Su(var)3–9* dominantly suppress PEV (Grewal and Elgin 2002). Shi et al. employed several transgenic fly lines with easily observable reporter genes inserted in chromosomal sections to detect changes in chromatin condensation, as these correlated with reporter gene activity. They demonstrated that the *Tum-l* mutation decreases PEV, whereas *hop* loss-of-function mutants increase PEV. The authors inferred that changes in chromatin structure likely leads to global regulation of genes, most (or many) of which may not be downstream of targets of the STAT92E transcription factor. To further demonstrate that JAK/STAT signaling functions to counteract heterochromatic gene silencing, Shi et al. investigated changes in heterochromatin markers HP1 and H3mK9. Results from immunostaining and western blot analysis convincingly showed that levels of heterochromatin increased in loss-of-function and reduced in gain-of-function *hop* mutants. In fact, a moderate increase in HP1 levels was sufficient in completely suppressing Hop^{tuml} induced hematopoietic tumors. Additionally, the authors reported a functional interaction between the JAK/STAT pathway ligand *Unpaired* (Upd) and HP1, as reducing or increasing HP1 levels enhanced or suppressed the characteristic large eye phenotype of *Upd* flies, while modifying HP1 levels did not have any effect on eye development in a wild-type genetic background. Taken together, the findings of Shi et al., established that the collaboration between effectors of heterochromatin formation and *hop*, is critical in epigenetically mediating gene silencing, which is an important tumor suppressive mechanism.

STAT-Chromatin Connection

An obvious and important issue raised by Shi et al. (2006) concerned the role of *Drosophila* STAT in heterochromatin formation. While phosphorylation of *STAT92E* by the JAK Kinase *Hop* led to disruption of heterochromatin and caused changes in gene expression (Shi et al. 2006) and a *STAT92E* hypomorphic allele partially suppressed the *Tuml* tumor phenotype (Yan et al. 1996), however, a direct role for STAT in controlling heterochromatin formation remained unclear.

Using the same approach as their previous study (Shi et al. 2006), Shi et al. (2008) found that reducing $Stat92E^{+}$ dosage decreased PEV and thus heterochromatin formation, an effect similar to that of *Hop* over-activation (Shi et al. 2008). This came as a surprise since in the canonical model of JAK/STAT signaling *Hop* activates *STAT92E*, that functions as a positive effector of JAK signaling. Therefore, reducing STAT92E levels was expected to increase heterochromatin, just as lowering Hop levels did. The authors argued that loss of STAT92E having the same effect as Hop over-activation on PEV suggested that the effects of JAK-STAT signaling on heterochromatin could not be mediated via the canonical JAK-STAT pathway. Furthermore, they used immunostaining to show that overexpression of $Stat92E^{+}$ resulted in higher levels of heterochromatin, and that in $Stat92E^{-/-}$ cells, heterochromatin formation was markedly reduced. To further confirm that STAT92E was required for HP1 localization on heterochromatin, the authors conducted chromatin immunoprecipitation (ChIP) analysis using cells expressing

STAT92E RNAi and reported reduced association of HP1 with heterochromatin sequences following STAT92E RNAi knockdown. These results indicated that STAT92E was an essential component for the association of HP1 with heterochromatin.

To explain their paradoxical observation that modifying levels of STAT92E had the opposite effects to *hop* gain- or loss-of-function on heterochromatin formation, particularly in light of the canonical mode of JAK-STAT signaling, the authors proposed the existence of a non-canonical model of JAK-STAT signaling as evidenced by the following results. First, using immunostaining Shi et al. (2008) reported the co-localization of STAT92E with HP1 in the nucleus in unstimulated *Drosophila* cells, against the conventional view of latent STAT proteins residing primarily in the cytoplasm. Interestingly unphosphorylated mammalian STAT3 and STAT5 have been demonstrated previously (Liu et al. 2005) to predominantly localize in the nucleus as well (Harrison et al. 1995; Iyer and Reich 2008; Liu et al. 2005). Additionally, STAT92E co-localization with HP1 in heterochromatinized sections was also evident in S2 cells using a STAT92E-Green fluorescent protein (GFP) transgene and by ChIP, showing that STAT92E is indeed present on heterochromatin. Moreover, STAT92E and HP1 co-localized in multiple regions of heterochromatin, including the chromocenter and telomeres, as seen in stained polytene chromosome squash prepared from wild-type salivary glands. Unlike the distribution of total STAT92E, a proportion of which localizes in the nucleus on heterochromatin and the rest in the cytoplasm, the distribution of phosphorylated STAT92E was uniform in the nucleus and did not co-localize with HP1 or heterochromatin. These findings indicated that while some of the latent/unphosphorylated portion of STAT92E resides in the nucleus on heterochromatin, active/phosphorylated STAT92E is excluded from heterochromatic regions.

Second, consistent with the results of the co-localization studies, STAT92E and HP1 were found to physically interact. Shi et al. (2008), also delineated the importance of two HP1 binding sites contained in STAT92E – the conserved sequence Pro-X-Val-X-Leu (X denotes any amino acid). Mutating both sites effectively abolished the HP1-STAT92E interaction. Modifying the phosphorylation state of STAT92E also influenced the binding between STAT92E and HP1. This was observed when co-immunoprecipitation of STAT92E and HP1 was reduced upon an increase of STAT92E phosphorylation in $hop^{tum-l/+}$ embryos or when Hop was overexpressed.

Third, the authors employed a time course microscopic analysis, before and after stimulation of STAT92E phosphorylation to investigate how STAT92E localization changes and destabilizes heterochromatin. In *ex vivo* cultured salivary glands, Shi et al. (2008) reported the association of unphosphorylated STAT92E with heterochromatin before stimulation. Stimulation and phosphorylation of STAT92E caused phospho-STAT92E dispersal and movement to euchromatic segments (where it likely bound to cognate promoters) along with diffusion of HP1 from heterochromatin (Fig. 4). These findings indicated that STAT92E translocation after phosphorylation precedes and is required for removal of HP1 from heterochromatin.

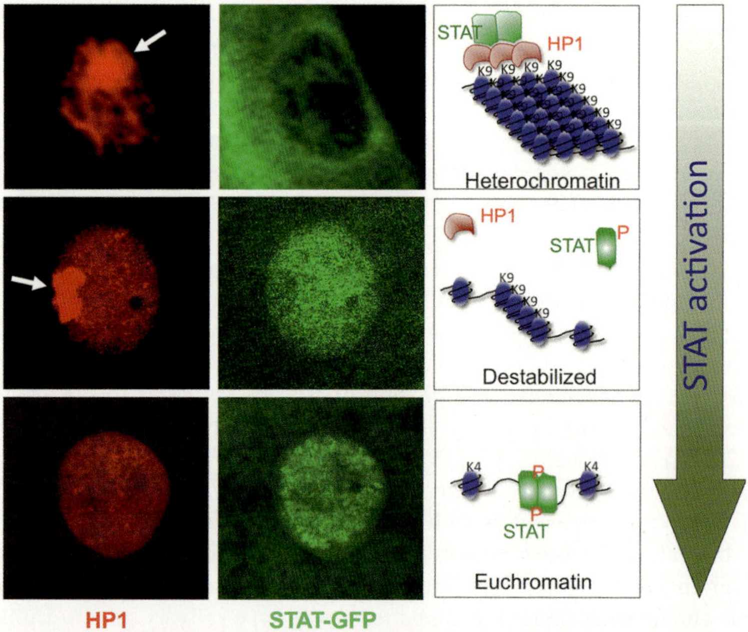

Fig. 4 Non-canonical JAK/STAT signaling in *Drosophila*. *Drosophila* larval salivary glands expressing UAS-STAT92E-GFP were cultured *ex vivo* and treated with pervanadate to activate STAT. Distribution of HP1 (*red*) and STAT92E-GFP (*green*) was examined before (*top row*), 20 min (*middle row*), or 1 h after treatment. Note that following pervanadate stimulation, STAT92E-GFP moves away from heterochromatin (HP1 foci) (20 min) and then binds to chromosomes as distinct bands (60 min). *Right panels* are schematic showing of the non-canonical mode of JAK/STAT signaling: a portion of unphosphorylated STAT resides in the nucleus on heterochromatin in association with HP1. Increased phosphorylation of STAT by JAK or other tyrosine kinases, lowers the amount of STAT localized on heterochromatin. Thus causes dissociation of HP1 from heterochromatin resulting in heterochromatin disruption. Dispersed phospho-STAT binds to cognate sequences in euchromatin and activates transcription of target genes. Genes normally located within heterchromatin thus become accessible to STAT and other transcription factors (Images are from Shi et al. 2008. With permission)

Finally, Shi et al. (2008) demonstrated that STAT92E activation does not destabilize heterochromatin indirectly by inducing transcription of other genes. In salivary glands treated with protein synthesis inhibitor cyclohexamide (CHX) prior to activation, phosphorylated STAT92E-induced dispersal of HP1 was unaffected. These results suggested that since STAT-activation mediated heterochromatin destabilization does not need new protein synthesis therefore it cannot be caused by induction of transcriptional targets of STAT92E. Collectively, these findings suggest that unphosphorylated STAT92E normally functions to stabilize HP1 localization at heterochromatin, while its activation by phosphorylation results in STAT92E diffusing away from heterochromatin, thereby causing HP1 displacement and heterochromatin destabilization, independent of transcription.

Non-canonical JAK-STAT Signaling

The conventional model of JAK/STAT signaling posits that inactive STAT resides in the cytoplasm, however, in the non-canonical model proposed by Shi et al. (2008), a portion of the unphosphorylated-STAT protein pool was found to localize in the nucleus on heterochromatin and associate with HP1 (Fig. 4b) (Shi et al. 2008). This heterochromatin bound inactive-STAT is critical for maintaining HP1 localization and heterochromatin stability. Upon activation, STAT is phosphorylated, causing it to disperse from heterochromatin, thereby promoting diffusion of HP1 and destruction of heterochromatin. This mechanism functions independently of STAT transcriptional induction of target genes (Shi et al. 2008).

It is not clear whether JAK enters the nucleus to phosphorylate STAT, or whether the translocation of unphosphorylated nuclear STAT occurs in response to the altered equilibrium between nuclear and cytoplasmic or phosphorylated and unphosphorylated pools of STAT, induced upon JAK activation. In fact, JAK translocation to the nucleus has previously been shown for mammalian JAK2, which was enriched in nuclear liver extracts and liver cells (Ram and Waxman 1997). Recently, it has also been reported that in human hematopoietic cells, JAK2 directly phosphorylates Tyr41 on histone H3 in the nucleus, thereby disrupting HP1 binding and destabilizing heterochromatin (Dawson et al. 2009). These studies highlight the similarities and differences between mammals and *Drosophila* in the involvement of JAK/STAT signaling in heterochromatin regulation.

Heterochromatin and Tumor Suppression

Heterochromatin consists of highly condensed chromosomal DNA, which is generally believed to be inaccessible to transcription factors and is "transcriptionally silent". While constitutive heterochromatin remains condensed throughout the cell cycle and is regularly found at pericentric and telomere regions, heterochromatization of euchromatic regions results in facultative heterochromatin (Grewal and Elgin 2002; Grewal and Jia 2007). Heterochromatin, especially facultative heterochromatin was thought to be inert until recently, when the fast exchange rate of heterochromatin stability factor HP1 indicated that constitutive heterochromatin possesses dynamic properties and should be readily accessible by regulatory factors for remodeling (Cheutin et al. 2003; Festenstein et al. 2003).

Heterochromatin has implications for disease states such as cancer. For instance, SUV39H1 catalyzes methylation of H3K9 and H4K20, and is important for heterochromatin establishment and gene silencing (Rea et al. 2000). Deletion of Suv39h1 renders hematopoietic cells susceptible to Ras-induced lymphomas, by circumventing cellular senescence (Braig et al. 2005). SUV39H1 also binds to the tumor suppressor Rb (Retinoblastoma protein), which in turn, recruits HP1 to promote heterochromatin formation and thus mediate silencing of cyclin E and cyclin A2, and suppress tumor formation (Nielsen et al. 2001). Another chromatin modifier, the histone methyltransferase, Mll1 (Myeloid/Lymphoid Leukemia 1), when activated, interacts with the SWI/SNF chromatin remodeling complex and methylates H3K4 to activate gene expression (Rozenblatt-Rosen et al. 1998). In fact MLL1 mutations induced by chromosomal translocation or duplication cause acute lymphoblastic leukemia (Rozenblatt-Rosen et al. 1998). In contrast, oncogenic c-Myc, inhibits heterochromatin formation by maintaining active chromatin regions in the genome (Fernandez et al. 2003; Frank et al. 2003; Oster et al. 2002). Increased heterochromatin formation also serves as an indicator of cellular senescence, which in turn regulates cell proliferation and provides protection from oncogene-induced tumor formation in mammals (Braig et al. 2005; Mathon and Lloyd 2001). Consistent with these findings, in *Drosophila*, heterochromatin formation prevents JAK-STAT activation-induced tumorigensis and heterochromatin destruction enhances tumorigenesis (Shi et al. 2006). These results were also corroborated in mammalians models, where impaired heterochromatin formation through loss of the key heterochromatin interacting factors such as HP1 or the Suv39h1 methyltransferase, or through amplification of JAK2 and the JMJD2C H3 demethylase, has been shown to contribute to cancer formation and/or progression (Braig et al. 2005; Cloos et al. 2006; Harrison et al. 1995; Norwood et al. 2006; Rui et al. 2010).

Concluding Remarks

It is increasingly clear that cancer development involves both genetic mutations and epigenetic dysregulation. While genetic mutations such as gain-of-function mutations in oncogenes and loss-of-function mutations in tumor suppressors have been extensively studied, the mechanisms by which epigenetic dysregulation arises and leads to cancers remain obscure. Recent studies have indicated that, in addition to controlling expression of STAT target genes, the JAK/STAT signaling pathway regulates heterochromatin stability, which represents a novel molecular mechanism by which signaling pathways can influence cellular epigenetic status and lead to tumorigenesis. Future research should focus on unraveling the molecular mechanisms by which the canonical and noncanonical mode of JAK/STAT signaling cooperate especially in the context of cancer development.

References

Aaronson DS, Horvath CM (2002) A road map for those who don't know JAK-STAT. Science 296:1653–1655

Alberts B (2002) Molecular biology of the cell, 4th edn. Garland Science, New York

Binari R, Perrimon N (1994) Stripe-specific regulation of pair-rule genes by hopscotch, a putative Jak family tyrosine kinase in *Drosophila*. Genes Dev 8:300–312

Braig M, Lee S, Loddenkemper C, Rudolph C, Peters AH, Schlegelberger B, Stein H, Dorken B, Jenuwein T, Schmitt CA (2005) Oncogene-induced senescence as an initial barrier in lymphoma development. Nature 436:660–665

Bromberg JF (2001) Activation of STAT proteins and growth control. Bioessays 23:161–169

Brown S, Hu N, Hombria JC (2001) Identification of the first invertebrate interleukin JAK/STAT receptor, the *Drosophila* gene domeless. Curr Biol 11:1700–1705

Cheutin T, McNairn AJ, Jenuwein T, Gilbert DM, Singh PB, Misteli T (2003) Maintenance of stable heterochromatin domains by dynamic HP1 binding. Science 299:721–725

Cloos PA, Christensen J, Agger K, Maiolica A, Rappsilber J, Antal T, Hansen KH, Helin K (2006) The putative oncogene GASC1 demethylates tri- and dimethylated lysine 9 on histone H3. Nature 442:307–311

Cryderman DE, Grade SK, Li Y, Fanti L, Pimpinelli S, Wallrath LL (2005) Role of *Drosophila* HP1 in euchromatic gene expression. Dev Dyn 232:767–774

Danzer JR, Wallrath LL (2004) Mechanisms of HP1-mediated gene silencing in *Drosophila*. Development 131:3571–3580

Dawson MA, Bannister AJ, Gottgens B, Foster SD, Bartke T, Green AR, Kouzarides T (2009) JAK2 phosphorylates histone H3Y41 and excludes HP1alpha from chromatin. Nature 461:819–822

Dialynas GK, Vitalini MW, Wallrath LL (2008) Linking heterochromatin protein 1 (HP1) to cancer progression. Mutat Res 647:13–20

Dillon N, Festenstein R (2002) Unravelling heterochromatin: competition between positive and negative factors regulates accessibility. Trends Genet 18:252–258

Eissenberg JC, James TC, Foster-Hartnett DM, Hartnett T, Ngan V, Elgin SC (1990) Mutation in a heterochromatin-specific chromosomal protein is associated with suppression of position-effect variegation in *Drosophila* melanogaster. Proc Natl Acad Sci USA 87:9923–9927

Fernandez PC, Frank SR, Wang L, Schroeder M, Liu S, Greene J, Cocito A, Amati B (2003) Genomic targets of the human c-Myc protein. Genes Dev 17:1115–1129

Festenstein R, Sharghi-Namini S, Fox M, Roderick K, Tolaini M, Norton T, Saveliev A, Kioussis D, Singh P (1999) Heterochromatin protein 1 modifies mammalian PEV in a dose- and chromosomal-context-dependent manner. Nat Genet 23:457–461

Festenstein R, Pagakis SN, Hiragami K, Lyon D, Verreault A, Sekkali B, Kioussis D (2003) Modulation of heterochromatin protein 1 dynamics in primary mammalian cells. Science 299:719–721

Frank SR, Parisi T, Taubert S, Fernandez P, Fuchs M, Chan HM, Livingston DM, Amati B (2003) MYC recruits the TIP60 histone acetyltransferase complex to chromatin. EMBO Rep 4:575–580

Giordano A, Avantaggiati ML (1999) p300 and CBP: partners for life and death. J Cell Physiol 181:218–230

Grewal SI, Elgin SC (2002) Heterochromatin: new possibilities for the inheritance of structure. Curr Opin Genet Dev 12:178–187

Grewal SI, Jia S (2007) Heterochromatin revisited. Nat Rev Genet 8:35–46

Hanratty WP, Dearolf CR (1993) The *Drosophila* tumorous-lethal hematopoietic oncogene is a dominant mutation in the hopscotch locus. Mol Gen Genet 238:33–37

Harrison DA, Binari R, Nahreini TS, Gilman M, Perrimon N (1995) Activation of a *Drosophila* Janus kinase (JAK) causes hematopoietic neoplasia and developmental defects. EMBO J 14: 2857–2865

Harrison DA, McCoon PE, Binari R, Gilman M, Perrimon N (1998) *Drosophila* unpaired encodes a secreted protein that activates the JAK signaling pathway. Genes Dev 12:3252–3263

Hombria JC, Brown S (2002) The fertile field of *Drosophila* Jak/STAT signalling. Curr Biol 12: R569–R575

Hou XS, Melnick MB, Perrimon N (1996) Marelle acts downstream of the *Drosophila* HOP/JAK kinase and encodes a protein similar to the mammalian STATs. Cell 84:411–419

Iyer J, Reich NC (2008) Constitutive nuclear import of latent and activated STAT5a by its coiled coil domain. FASEB J 22:391–400

Kirschmann DA, Lininger RA, Gardner LM, Seftor EA, Odero VA, Ainsztein AM, Earnshaw WC, Wallrath LL, Hendrix MJ (2000) Down-regulation of HP1Hsalpha expression is associated with the metastatic phenotype in breast cancer. Cancer Res 60:3359–3363

Lehming N, Le Saux A, Schuller J, Ptashne M (1998) Chromatin components as part of a putative transcriptional repressing complex. Proc Natl Acad Sci USA 95:7322–7326
Levy DE, Darnell JE Jr (2002) Stats: transcriptional control and biological impact. Nat Rev Mol Cell Biol 3:651–662
Li WX (2008) Canonical and non-canonical JAK-STAT signaling. Trends Cell Biol 18:545–551
Li B, Carey M, Workman JL (2007) The role of chromatin during transcription. Cell 128:707–719
Liu L, McBride KM, Reich NC (2005) STAT3 nuclear import is independent of tyrosine phosphorylation and mediated by importin-alpha3. Proc Natl Acad Sci USA 102:8150–8155
Luger K, Rechsteiner TJ, Flaus AJ, Waye MM, Richmond TJ (1997) Characterization of nucleosome core particles containing histone proteins made in bacteria. J Mol Biol 272:301–311
Luo H, Hanratty WP, Dearolf CR (1995) An amino acid substitution in the *Drosophila* hopTum-l Jak kinase causes leukemia-like hematopoietic defects. EMBO J 14:1412–1420
Mathon NF, Lloyd AC (2001) Cell senescence and cancer. Nat Rev Cancer 1:203–213
Nielsen SJ, Schneider R, Bauer UM, Bannister AJ, Morrison A, O'Carroll D, Firestein R, Cleary M, Jenuwein T, Herrera RE et al (2001) Rb targets histone H3 methylation and HP1 to promoters. Nature 412:561–565
Norwood LE, Moss TJ, Margaryan NV, Cook SL, Wright L, Seftor EA, Hendrix MJ, Kirschmann DA, Wallrath LL (2006) A requirement for dimerization of HP1Hsalpha in suppression of breast cancer invasion. J Biol Chem 281:18668–18676
Oster SK, Ho CS, Soucie EL, Penn LZ (2002) The myc oncogene: marvelous lY complex. Adv Cancer Res 84:81–154
Piacentini L, Fanti L, Berloco M, Perrini B, Pimpinelli S (2003) Heterochromatin protein 1 (HP1) is associated with induced gene expression in *Drosophila* euchromatin. J Cell Biol 161:707–714
Pomeroy SL, Tamayo P, Gaasenbeek M, Sturla LM, Angelo M, McLaughlin ME, Kim JY, Goumnerova LC, Black PM, Lau C et al (2002) Prediction of central nervous system embryonal tumour outcome based on gene expression. Nature 415:436–442
Ram PA, Waxman DJ (1997) Interaction of growth hormone-activated STATs with SH2-containing phosphotyrosine phosphatase SHP-1 and nuclear JAK2 tyrosine kinase. J Biol Chem 272:17694–17702
Rea S, Eisenhaber F, O'Carroll D, Strahl BD, Sun ZW, Schmid M, Opravil S, Mechtler K, Ponting CP, Allis CD et al (2000) Regulation of chromatin structure by site-specific histone H3 methyltransferases. Nature 406:593–599
Richards EJ, Elgin SC (2002) Epigenetic codes for heterochromatin formation and silencing: rounding up the usual suspects. Cell 108:489–500
Rozenblatt-Rosen O, Rozovskaia T, Burakov D, Sedkov Y, Tillib S, Blechman J, Nakamura T, Croce CM, Mazo A, Canaani E (1998) The C-terminal SET domains of ALL-1 and TRITHORAX interact with the INI1 and SNR1 proteins, components of the SWI/SNF complex. Proc Natl Acad Sci USA 95:4152–4157
Rui L, Emre NC, Kruhlak MJ, Chung HJ, Steidl C, Slack G, Wright GW, Lenz G, Ngo VN, Shaffer AL et al (2010) Cooperative epigenetic modulation by cancer amplicon genes. Cancer Cell 18:590–605
Sadoni N, Langer S, Fauth C, Bernardi G, Cremer T, Turner BM, Zink D (1999) Nuclear organization of mammalian genomes. Polar chromosome territories build up functionally distinct higher order compartments. J Cell Biol 146:1211–1226
Shaffer CD, Cenci G, Thompson B, Stephens GE, Slawson EE, Adu-Wusu K, Gatti M, Elgin SC (2006) The large isoform of *Drosophila* melanogaster heterochromatin protein 2 plays a critical role in gene silencing and chromosome structure. Genetics 174:1189–1204
Shi S, Calhoun HC, Xia F, Li J, Le L, Li WX (2006) JAK signaling globally counteracts heterochromatic gene silencing. Nat Genet 38:1071–1076
Shi S, Larson K, Guo D, Lim SJ, Dutta P, Yan SJ, Li WX (2008) *Drosophila* STAT is required for directly maintaining HP1 localization and heterochromatin stability. Nat Cell Biol 10:489–496

Shuai K, Ziemiecki A, Wilks AF, Harpur AG, Sadowski HB, Gilman MZ, Darnell JE (1993) Polypeptide signalling to the nucleus through tyrosine phosphorylation of Jak and Stat proteins. Nature 366:580–583

Tulina N, Matunis E (2001) Control of stem cell self-renewal in *Drosophila* spermatogenesis by JAK-STAT signaling. Science 294:2546–2549

van der Vlag J, den Blaauwen J, Sewalt RG, van Driel R, Otte AP (2000) Transcriptional repression mediated by polycomb group proteins and other chromatin-associated repressors is selectively blocked by insulators. J Biol Chem 275:697–704

Ward AC, Touw I, Yoshimura A (2000) The Jak-Stat pathway in normal and perturbed hematopoiesis. Blood 95:19–29

Wasenius VM, Hemmer S, Kettunen E, Knuutila S, Franssila K, Joensuu H (2003) Hepatocyte growth factor receptor, matrix metalloproteinase-11, tissue inhibitor of metalloproteinase-1, and fibronectin are up-regulated in papillary thyroid carcinoma: a cDNA and tissue microarray study. Clin Cancer Res 9:68–75

Weiler KS, Wakimoto BT (1995) Heterochromatin and gene expression in *Drosophila*. Annu Rev Genet 29:577–605

Yan R, Small S, Desplan C, Dearolf CR, Darnell JE Jr (1996) Identification of a Stat gene that functions in *Drosophila* development. Cell 84:421–430

Yasuhara JC, Wakimoto BT (2006) Oxymoron no more: the expanding world of heterochromatic genes. Trends Genet 22:330–338

Yu H, Jove R (2004) The STATs of cancer–new molecular targets come of age. Nat Rev Cancer 4:97–105

JAK/STAT Signaling and Invertebrate Immune Responses

Feng Zhou and Hervé Agaisse

Abstract
This review focuses on JAK/STAT signaling and its role in response to infection in invertebrates. Most of our knowledge comes from studies conducted in the model organism *Drosophila melanogaster*. However, we tentatively cover available information on JAK/STAT signaling in other invertebrates, including mosquitoes. Covered topics include the components of JAK/STAT signaling and their role in humoral, cellular and mucosal immunity. Finally, we summarize recent developments on the role of JAK/STAT signaling in the maintenance of homeostasis in response to intestinal challenge.

Components and Regulations of the JAK/STAT Signaling Pathway

Core Components of JAK/STAT Signaling in *Drosophila*

The *Drosophila* JAK/STAT pathway was discovered for its role in embryonic development (Binari and Perrimon 1994; Perrimon and Mahowald 1986). Genetic and biochemical studies have established that the core components of the pathway consist of the three known cytokine-like molecules of the Unpaired family (Upd1, Upd2 and Upd3), the GP130 receptor family member Domeless (Dome), the JAK kinase Hopscotch (Hop), and the transcriptional regulator Stat92E (Fig. 1).

F. Zhou • H. Agaisse (✉)
Section of Microbial Pathogenesis, Boyer Center for Molecular Medicine, Yale University School of Medicine, 295 Congress Avenue, New Haven, CT 06519, USA
e-mail: herve.agaisse@yale.edu

Th. Decker and M. Müller (eds.), *Jak-Stat Signaling: From Basics to Disease*,
DOI 10.1007/978-3-7091-0891-8_9,

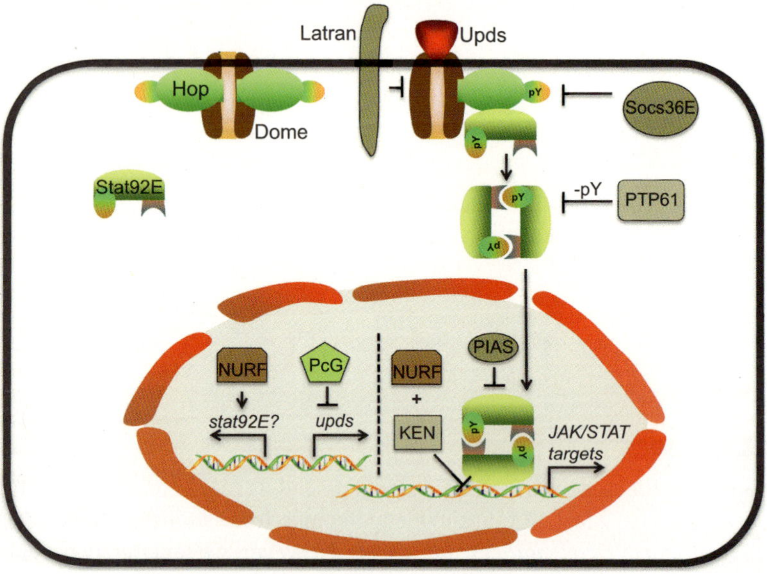

Fig. 1 Components and regulation of the *Drosophila* JAK/STAT pathway. The core components of the *Drosophila* JAK/STAT pathway consist of the Upd family of signaling cytokines (ligands), Dome (receptor), Hop (JAK kinase), and Stat92E (STAT). The pathway activity can be modulated at multiple points. Latran is a dominant-negative receptor that antagonizes Dome function. Socs36E inhibits Hop kinase activity. PTP61 probably dephosphorylates Stat92E. PIAS binds to and inhibits Stat92E activity. KEN competes with Stat92E at the promoters of a subset of JAK/STAT target genes and represses them via NURF recruitment. NURF and PcG regulate the transcriptions of the JAK/STAT components epigenetically

The Cytokine-Like Molecules

The *Drosophila* genome encodes three cytokine-like molecules of the Unpaired family, Upd1, Upd2, and Upd3, that are known activators of JAK/STAT signaling in flies (Agaisse et al. 2003; Gilbert et al. 2005; Harrison et al. 1998; Hombria et al. 2005). The founder of the *upd* family, *upd*, was discovered for its role in segmentation pattern formation during embryogenesis (Harrison et al. 1998). It codes for a secreted N-linked glycoprotein that interacts with the extracellular matrix (Harrison et al. 1998; Zeidler et al. 1999). In vitro experiment showed that recombinant Upd induces phosphorylation and activation of Hop (Harrison et al. 1998). A *upd2* mutant was generated and found to be viable and fertile, unlike the *upd* mutant (Gilbert et al. 2005; Hombria et al. 2005). Cell culture assay indicated that Upd2 is secreted as a diffusible ligand and activates JAK/STAT signaling in vitro and in vivo (Hombria et al. 2005). Similar to *upd2*, *upd3* expression is apparently dispensable for embryogenesis as determined by RNAi-mediated knock-down experiments (Hombria et al. 2005). However, no *upd3* mutant has been reported so far. To date, no function has been assigned to *upd* and *upd2* in response to infection. However, potential functions have been reported for *upd3* in lymph glands, where it is regulated in response to parasite infection (Jung et al. 2005);

in the adult circulating hemocytes, where it is expressed in response to septic injury (Agaisse et al. 2003); and in the adult intestine, where it is expressed in response to enteric infections (Buchon et al. 2009a, b; Jiang et al. 2009). These observations indicate that Upd and Upd3 are devoted to embryogenesis and immune/stress response, respectively, suggesting functional specialization among *Drosophila* cytokines involved in JAK/STAT signaling.

Computer-assisted analyses of current DNA sequence databases revealed that *upd*-like genes are present in the genome of several *Drosophila spp.*, but appear to be absent from other insects, including mosquitoes. The identity of the putative ligands involved in the activation of JAK/STAT signaling in mosquitoes is thus unknown. It is also unclear whether additional ligands may activate JAK/STAT signaling in *Drosophila spp.*

The Domeless Receptor (aka Mom)

Mutations in *domeless* are embryonic lethal and lead to patterning defects identical to the defects observed with mutations in the other core components of the JAK/STAT pathway. *domeless* encodes a protein of 1,282 a/a that displays limited homology to the mammalian GP130 family members (Brown et al. 2001; Chen et al. 2002). The protein contains the conserved cytokine receptor family domains that include four fibronectin type III (FN3) repeats and a STAT-binding YXXQ consensus. In vitro experiments demonstrated the physical interaction between Dome and Upd, Hop and Stat92E, as well as its ability to activate Hop (Chen et al. 2002).

The JAK Kinase (Hop)

hop encodes a protein of 1,177 a/a most similar to the mammalian JAK2 with 27% homology (Binari and Perrimon 1994). It contains seven JAK homology (JH1-JH7) domains, in which the tyrosine kinase catalytic domain (JH1) is located at the C terminus. The function of the JH2, kinase-like domain, is not well known. JH3-JH7 domains contain a FERM (*f*our-point-one, *e*zrin, *r*adixin, *m*oesin) domain that mediates association with receptors (Girault et al. 1999). Two gain-of-function mutations in Hop lead to the formation of melanotic tumors. The Tumorous-lethal (hop^{TumL}) mutant contains an a/a substitution at residue 341 (from G to E) in the JH4 region (Harrison et al. 1995; Luo et al. 1995). The hop^{T42} mutant contains an E to K substitution at position 695, a residue which is conserved in all known JAK kinases (Luo et al. 1997).

The Transcriptional Regulator Stat92E

Stat92E encodes a protein of 761 a/a with a molecular weight of 83 kDa. It is most similar to the mammalian STAT5 with 37% identity (Hou et al. 1996; Yan et al. 1996). It contains two domains that are conserved in other STAT proteins: a src homology 2 (SH2) domain, and a DNA-binding domain. The DNA-binding domain includes a highly conserved C-terminal tyrosine residue at position 704, which is phosphorylated by Hop (Yan et al. 1996). The consensus DNA recognition sequence by Stat92E is $TTCN_3GAA$, which resembles that of the mammalian

STATs (Hou et al. 1996; Yan et al. 1996). Although the DNA-binding domain is also the domain responsible for the transcriptional activity of Stat92E, a mutational study revealed that these two activities can be functionally uncoupled (Karsten et al. 2006).

JAK/STAT Components in Other Invertebrates

Genes coding for STAT-like proteins have been identified in the genome of a number of invertebrates, including the slime mould *Dictyostelium* (Araki et al. 1998), *C. elegans* (Dierking et al. 2011; Wang and Levy 2006), the tobacco hornworm moth *Manduca sexta* (Elliott and Zeidler 2008), the beetle *Tribolium* (Baumer et al. 2011), the human malaria vector *Anopheles gambiae* (Barillas-Mury et al. 1999; Christophides et al. 2002), and the brine shrimp *Artemia franciscana* (Cheng et al. 2010). However, not all of these organisms display the components of the canonical JAK/STAT pathway, as defined in *Drosophila*. Orthologues of *dome* and *hop* have been identified in *Anopheles gambiae* and *Aedes aegypti* (Waterhouse et al. 2007). The *Dictyostelium* STAT is activated by extracellular cAMP signaling and can function as either a repressor or activator in the absence of a JAK kinase (Araki et al. 1998; Fukuzawa and Williams 2000). The *C. elegans* genome encodes two STAT-like genes, *STA-1* and *STA-2*, but does not encode any JAKs (Dierking et al. 2011; Wang and Levy 2006). STA-2 was recently shown to be activated by the p38 MAPK pathway (Dierking et al. 2011).

Regulation of JAK/STAT Signaling in *Drosophila*

SOCS (Suppressors of Cytokine Signaling)

Mammalian SOCS down-regulate JAK/STAT signaling by inhibiting the kinase activity of JAKs. There are three SOCS-like genes in *Drosophila*, *Socs36E*, *Socs44A* and *Socs16D* (Callus and Mathey-Prevot 2002; Karsten et al. 2002). Only Socs36E has been shown to be a functional repressor of JAK/STAT signaling (Fig. 1). *Socs36E* is transcriptionally activated by Stat92E (Baeg et al. 2005; Callus and Mathey-Prevot 2002; Karsten et al. 2002; Rawlings et al. 2004) and the *10X-STAT-GFP* reporter displaying the STAT binding sites of *Socs36E* regulatory regions is used as a standard for analyzing the activation of JAK/STAT signaling in *Drosophila* (Bach et al. 2007). In agreement with its role as a negative regulator of the pathway, ectopic expression of Socs36E suppresses JAK/STAT signaling in the wing imaginal disc, mimics the outstretched wing phenotype displayed by *upd* hypormorphic alleles and the venation defects observed in animals displaying the $Stat92E^{HJ}$ mutation (Baeg et al. 2005; Callus and Mathey-Prevot 2002; Rawlings et al. 2004).

PIAS (Protein Inhibitors of Activated STAT)

A single PIAS-like protein, ZIMP, is encoded in the *Drosophila* genome. ZIMP binds to Stat92E and suppresses the melanotic tumor phenotype of flies displaying the hop^{TumL} mutation (Betz et al. 2001).

PTP61F

PTP61F is a phosphatase identified in two independent RNAi screens (Baeg et al. 2005; Muller et al. 2005). It is homologous to human PTPB1 (phosphor-Tyr phosphatase B1). *ptp61f* is transcriptionally activated by Stat92E. Knock-down of *ptp61* expression in vitro increases the levels of pTyr-Hop and pTyr-Stat92E (Baeg et al. 2005), and increases JAK/STAT activation. Epistasis analysis places *ptp61f* downstream of *hop* (Muller et al. 2005), suggesting it acts on Stat92E.

Ken and Barbie/BCL6

Ken and Barbie (KEN) is a homologue of the human BCL6 (B-cell lymphoma) and is a BTB/POZ domain-containing transcription repressor (Arbouzova et al. 2006). The KEN binding site partially overlaps with that of Stat92E, but is present only in the promoters of a subset of Stat92E target genes.

Latran

Latran (Lat) is related to the Dome receptor, but lacks the intracellular domain required for signal transduction. It acts by antagonizing the function of Dome in a dose-dependent manner (Kallio et al. 2010; Makki et al. 2010). A detailed description of the function of Latran in the context of wasp parasitization is presented in the hematopoiesis section of this review.

Epigenetic Regulation of JAK/STAT Signaling

Several recent studies revealed that JAK/STAT signaling is regulated by epigenetic modifications leading to repression or activation of the pathway. *upd* gene transcription is epigenetically repressed by the Polycomb Group (PcG) of epigenetic silencer proteins in the eye imaginal disc (Classen et al. 2009; Gonzalez et al. 2009). Mutations in any core PcG repressive complex 1 (PRC1) components result in the de-repression of *upd* gene expression, triggering the activation of JAK/STAT signaling, which leads to hyper-proliferation of the imaginal disc tissue. In *Drosophila* testis, the nucleosome-remodeling factor (NURF) positively regulates JAK/STAT signal to maintain the germline and somatic stem cells and prevent premature differentiation (Cherry and Matunis 2010). It is yet unknown which core components of the JAK/STAT pathway are epigenetically enhanced by the NURF, although the observation that the Stat92E protein level decreases in *nurf301*-null germline stem cell clones indicates that the *Stat92E* locus could be a NURF target. Interestingly, in hemocytes, NURF has been shown to repress JAK/STAT signaling, as *nurf301* mutation phenocopies hop^{TumL} mutation (Kwon et al. 2008), and can further increase melanotic tumor incidence in the hop^{TumL} background (Badenhorst et al. 2002). Genetic and biochemical evidence suggests the repressive activity of NURF is mediated by KEN-dependent recruitment of NURF to the promoters of a subset of JAK/STAT target genes (Kwon et al. 2008). The contrasting effects of NURF on JAK/STAT activity suggest that the epigenetic modification machinery can elicit different regulatory outcomes in different tissues.

JAK/STAT Signaling and Hematopoiesis

Blood Cell Types in *Drosophila*

There are three major types of hemocytes present in the *Drosophila* hemolymph: plasmatocytes, crystal cells and lamellocytes (Rizki 1978). Plasmatocytes make up 90–95% of the total blood cell population. They function primarily as professional phagocytes to engulf and degrade dead cells and bacteria (Franc et al. 1996; Rizki 1978; Tepass et al. 1994). In addition, they produce antimicrobial peptides (Imler and Bulet 2005), survey and respond to damaged tissues and tumor mass (Babcock et al. 2008; Pastor-Pareja et al. 2008), undergo autophagy to limit the growth of intracellular bacteria (Goto et al. 2010; Yano and Kurata 2008), and activate humoral immunity (Agaisse et al. 2003). Crystal cells represent the majority of the remaining population in embryo and larva, but are no longer present after metamorphosis (Rizki and Rizki 1980). They are slightly larger than plasmatocytes and contain the precursors to the prophenoloxidase cascade (Cerenius et al. 2008) involved in melanization, which is activated during wound healing and immune response (Tang 2009). Lamellocytes are the largest type of hemocytes and function primarily in the encapsulation of dead tissues during metamorphosis, and of foreign objects too large to be phagocytosed by plasmatocytes. They are found only in larva and are present in very low number under normal condition, but can be massively produced by rapid differentiation in response to parasitization by Hymenopteran wasp *Leptopilina boulardi* (Lanot et al. 2001; Markus et al. 2009; Nappi 1975; Rizki and Rizki 1992).

The Two Phases of *Drosophila* Hematopoiesis

Hematopoiesis in the fly is a biphasic developmental process that produces hemocytes in the embryo, larva, pupa and adult. The first phase of hematopoiesis occurs during embryogenesis, while the second phase takes place during larval stages. To date, there is no hematopoietic organ identified in adult. Therefore, hemocytes in adult are currently thought to be a mixture of embryonic and larval hemocytes. Embryonic hematopoiesis takes place from embryonic stage 5 to stage 12. An invariant number of hemocytes (around 700 for plasmatocytes and 30 for crystal cells) is derived from a population of prohemocytes (hemocyte precursors) at the head (procephalic) mesoderm and disperses throughout the embryo (Beer et al. 1987; Klapper et al. 1998; Tepass et al. 1994). There is no known role for JAK/STAT signaling during this phase. Larval hematopoiesis is described in detail in the next section.

Larval Hematopoiesis in the Lymph Gland

Larval hematopoiesis occurs in the lymph gland, a tissue derived from the lateral thoracic mesoderm and formed during embryogenesis (el Shatoury 1955; Rugendorff et al. 1994; Stark and Marshall 1930). The lymph gland persists through the onset of metamorphosis, when it ruptures to release the hemocytes to populate the pupa and the future adult (Holz et al. 2003; Robertson 1936). Prohemocytes in the lymph gland proliferate rapidly during the first half of larval development, enlarging the developing lymph gland to consist of, by the third in-star larva (L3), three to six mirrored paired lobes (one pair of primary lobes and usually two pairs of secondary lobes) separated by the dorsal vessel, the fly's rudimentary heart (Jung et al. 2005; el Shatoury 1955; Stark and Marshall 1930). Hemocyte differentiation occurs mainly in the primary lobes (Jung et al. 2005), which consist of three zones at L3. The posterior signaling center (PSC), recently emerged as the "niche" for the prohemocytes, maintains a pool of signaling prohemocytes (Krzemien et al. 2007; Mandal et al. 2007) that are distinguished by the markers Serrate (a Notch ligand) and Collier (a member of the family of COE transcription factors) (Crozatier et al. 1996; Lebestky et al. 2003). Adjacent to the PSC is the medullary zone, which contains densely packed prohemocytes (Jung et al. 2005; Krzemien et al. 2007). The cortical zone, located on the periphery of the medullary zone and derived from the medullary zone, contains less densely packed, proliferating and differentiating hemocytes and crystal cells (Jung et al. 2005). The JAK/STAT pathway components *dome* and *upd3* are expressed in the medullary zone (Jung et al. 2005; Krzemien et al. 2007). *upd3* is also expressed in the PSC (Jung et al. 2005). JAK/STAT signaling is required to maintain the undifferentiated state of prohemocytes in the medullary zone, probably through paracrine Upd3 signaling (Jung et al. 2005; Krzemien et al. 2007), or paracrine Hh signaling (Mandal et al. 2007), or a combination of both. Loss of JAK/STAT signaling in the medullary zone by removal of *Stat92E* expression results in the premature differentiation of prohemocytes and eventual disappearance of the medullary zone (Fig. 2).

JAK/STAT Signaling and Response to Wasp Parasitization

A hallmark of the response to wasp parasitization in *Drosophila* is the massive differentiation of hemocytes into lamellocytes in the lymph gland (Sorrentino et al. 2002). Aggregated lamellocytes encapsulate and eventually destroy wasp eggs by secreting cytotoxic substances such as quinoid intermediates and free radicals (Nappi and Ottaviani 2000; Russo et al. 1996). The potential role of JAK/STAT signaling activation in response to wasp infection was first inferred from the unique phenotype observed in gain-of-function mutations in the JAK/STAT pathway. At the restrictive temperature, gain-of-function hop^{TumL} and hop^{T42} mutations lead to a striking increase in lamellocyte numbers, a phenotype reminiscent of the response to wasp parasitization (Harrison et al. 1995; Luo et al. 1995). Lamellocytes can often take up more than half of the total circulating blood cells, leading to melanotic

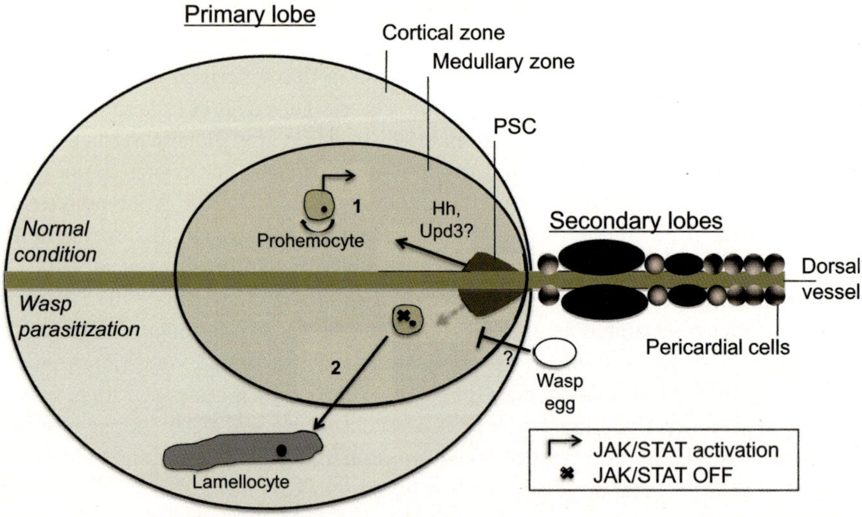

Fig. 2 JAK/STAT signaling in *Drosophila* hematopoiesis. The larval lymph gland consists of a pair of primary lobes and usually two pairs of secondary lobes. Three functional zones, the posterior signaling center (PSC), the medullary zone, and the cortical zone, delineate the primary lobe. JAK/STAT signaling is active in the medullary zone, probably via paracrine Hh or Upd3 signaling. It is required for prohemocyte maintenance (1). Upon wasp parasitization, rapid differentiation of lamellocytes requires a temporary shut-down of JAK/STAT signaling in the medullary zone (2)

masses in the mutant larvae. Reduction in *Stat92E* activity suppresses the mutant phenotype, indicating that JAK/STAT regulates lamellocyte production through *Stat92E* (Hou et al. 1996). In agreement with the notion that activation of JAK/STAT signaling is required for the differentiation of lamellocytes observed in response to wasp parasitization, loss-of-function mutations in *hop* lead to a significant reduction in the number of differentiated lamellocytes and encapsulation capacity (Sorrentino et al. 2004). However, recent development revealed that the requirement of the JAK/STAT pathway in cellular immune response after wasp parasitization is more complex than previously thought. It was demonstrated that the massive differentiation of lamellocytes in the lymph gland following wasp parasitization requires an acute *down-regulation* of the JAK/STAT pathway in the medullary zone (Makki et al. 2010) (Fig. 2). Upd3, the only *upd* gene expressed in the lymph gland, and required for maintaining JAK/STAT signaling, is down-regulated after wasp infection, resulting in a decreased level of Dome, which is transcriptionally controlled by JAK/STAT. The level of Lat, the short receptor related to Dome and antagonizes Dome, does not change. The resulting increase in the Lat/Dome ratio extinguishes the remaining JAK/STAT activity and facilitates lamellocyte differentiation. In agreement with this model, *lat* mutants fail to mount a cellular immune response after wasp parasitization. Thus, an apparent paradox emerges: the JAK/STAT pathway is required for lamellocyte differentiation following wasp parasitization (as demonstrated by the lack of cellular response in the *hop* mutant), but at the same time JAK/STAT signaling needs to be down-regulated to stimulate lamellocyte differentiation. As previously mentioned in the section on *Larval hematopoiesis in the lymph gland,* the JAK/STAT pathway is required to maintain prohemocytes and prevent them from premature differentiation in the medullary zone. Thus, the simplest explanation to reconcile these observations is that the loss of the cellular response in the *hop* mutant observed after wasp parasitization is most likely a secondary effect as a result of the premature loss of hemocyte progenitors. The mechanism leading to the down-regulation of *upd3* expression in lymph glands in response to wasp infection is unknown.

JAK/STAT Signaling and Humoral Immunity

A hallmark of the *Drosophila* response to systemic infection is the production of a collection of antimicrobial peptides (AMPs) by the fat body, a major immune-responsive tissue. There are seven known classes of AMPs, including attacin, cecropin, defensin, diptericin, drosocin, drosomycin, and metchnikowin (Imler and Bulet 2005). Collectively, AMPs confer a broad range of activities against bacteria and fungi. Two NF-kB/Rel-like transcription factors, DIF and Relish, regulate the expression of AMPs in response to infections by bacteria or fungi (Hedengren et al. 1999; Meng et al. 1999; Rutschmann et al. 2000). In addition to AMPs, several humoral factors, such as the complement-like protein Tep1, are produced in the fat body in response to infection (Lagueux et al. 2000).

The first evidence for the involvement of JAK/STAT in insect immune responses came from studies in the mosquito *Anopheles gambiae* (Barillas-Mury et al. 1999). Upon bacterial challenge, substantial clearance of cytoplasmic *Anopheles* STAT and a concurrent accumulation in the nucleus of fat body cells was observed, indicating the activation of the JAK/STAT pathway. Similar observations were made in *Drosophila*, where Stat92E translocates to the nucleus of fat body cells in response to immune challenge in a JAK-dependent manner (Agaisse et al. 2003). Gene expression profile experiments led to the identification of a subset of immune-responsive JAK/STAT-dependent humoral effectors, including Tep1 and the Turandot family member TotA (Agaisse et al. 2003; Boutros et al. 2002; Lagueux et al. 2000) (see below).

Tep1 belongs to a four-member family of thioester-containing proteins (TEP) with significant similarities to members of the complement C3/alpha2-macroglobulin super-family. Tep1 is strongly activated in the fat body upon immune challenge (Lagueux et al. 2000). Although the function of Tep1 in *Drosophila* is still unclear, studies conducted in *Anopheles gambiae* hemocytes revealed a role for aTEP1 in phagocytosis of Gram-positive and Gram-negative bacteria (Levashina et al. 2001). Similar to human complement factors, aTEP1 binds to bacterial surface and promotes their uptake by hemocytes. aTEP1 was also the first molecular determinant found to control the number of *Plasmodium* parasites in *Anopheles gambiae* by binding to the surface of ookinetes and mediating parasite killing via hitherto unknown mechanisms (Blandin et al. 2004). In mosquitoes, aTEP1 is constitutively secreted by hemocytes (Levashina et al. 2001) and its transcription does not rely on JAK/STAT signaling, but rather on NF-kB signaling (Frolet et al. 2006). Upon *Plasmodium* infection, however, inducible regulation of *aTEP1* appears to require JAK/STAT signaling (Gupta et al. 2009).

TotA was originally identified as a polypeptide secreted by the larval fat body in response to various stress conditions such as immune challenge and heat (Agaisse et al. 2003; Ekengren and Hultmark 2001; Ekengren et al. 2001). Gene expression profile studies revealed that *totA* is regulated by the JAK/STAT pathway in an immune inducible manner (Boutros et al. 2002). Studies revealed a signaling role of hemocytes in the activation of *totA*: in response to septic injury, hemocytes release Upd3, which in turn binds to Dome in the fat body cells and induces *totA* expression (Agaisse et al. 2003). To date, the exact function of TotA remains unknown.

In addition to activating a specific subset of humoral effectors, JAK/STAT signaling has also been shown to down-regulate humoral effectors activated by Relish (Kim et al. 2007). Using SL2 cells, it was shown that a Stat92E binding site present in the *attacin A* promoter, an AMP activated by Relish, recruits Stat92E, the JNK pathway target dAP-1, and the *Drosophila* HMG protein Dsp1 to form a repressosome. In vitro, the repressosome displaces Relish from the promoter and recruits histone deacetylase, thereby dampening the Relish-dependent output. Although the concept that excessive NF-kB signaling can be prevented by another immune signaling pathway could provide insight into how the host optimizes an immune reaction by integrating different signals, the in vivo relevance of these findings remains to be tested.

JAK/STAT Signaling and Mucosal Immunity

Local AMP Production

The *Drosophila* epithelial linings, such as the digestive and respiratory tracts, display the ability to produce AMPs that are potentially important to prevent local infection (Tzou et al. 2000). There are two modes of local AMP production. The first mode is constitutive AMP production that provides a battery of antimicrobial molecules in tissues including the salivary glands, the digestive tract (cardia and midgut) and the reproductive tract (the female spermatheca, oviduct and calyx; and the male ejaculatory duct) (Ferrandon et al. 1998; Tzou et al. 2000). This mode does not rely on NF-kB transcription factors, but requires tissue-specific transcription factors such as Homeobox gene *caudal* (Ryu et al. 2004). The second mode is inducible AMP production as evidenced in response to intestinal infection by *Erwinia carotovora carotovora* (*Ecc*) (Basset et al. 2000) or *Pseudomonas entomophila* (*Pe*) (Vodovar et al. 2005). Both *Ecc* and *Pe* infections in the digestive tract induce *diptericin (dipt)* expression (Buchon et al. 2009b; Liehl et al. 2006), while *Ecc* infection in the trachea (the respiratory tract) induces a strong expression of *drosomycin (drs)* (Tzou et al. 2000). The inducible expressions of *dipt*, and of *drs* in the trachea, depend on Relish (Buchon et al. 2009b; Liehl et al. 2006;

Onfelt Tingvall et al. 2001; Tzou et al. 2000). Global gene expression analysis revealed a set of immune genes specifically activated in the intestine in response to *Ecc* infection (Buchon et al. 2009b). The activation of the vast majority of the immune genes identified relied on the activity of Relish. However, a subset of genes, including *drosomycin 3* (*dro3*), was found to rely on JAK/STAT signaling. On the basis of the presence of putative STAT-binding sites in the vicinity of its regulatory regions, the authors hypothesized that *dro3* may be a direct target of Stat92E. However, this hypothesis has not yet been supported by direct experimental evidence. Nonetheless, these observations constitute the first evidence of a role for JAK/STAT signaling in *Drosophila* mucosal immunity.

Reactive Free Radical Production in the *Drosophila* and Anopheles Gut Epithelia

In addition to AMP production, a potent local ROS response has been observed in the *Drosophila* intestine in response to bacterial infection, and is responsible for the elimination of a majority of the ROS-sensitive microbes (Ryu et al. 2006). Two central players have been characterized that mediate this effective and potentially destructive oxidative stress response: the membrane-bound *Drosophila* dual oxidase (dDuox) protein that synthesizes ROS (Ha et al. 2005a), and the extracellular immune-responsive catalase (IRC), which removes oxygen radicals (Ha et al. 2005b). The ROS response is not mediated by the JAK/STAT pathway.

On the other hand, the production of nitric oxide (NO) in the midgut of *Anopheles gambiae* relies on JAK/STAT signaling (Gupta et al. 2009). *Plasmodium*, the causative agent for malaria, undergoes a series of developmental stages in the midgut of the mosquito vector *Anopheles gambiae* (Baton and Ranford-Cartwright 2005). Gametocytes are rapidly activated to produce gametes upon entering the mosquito midgut. Fertilization generates zygotes that develop to motile ookinetes 16–30 h after infection. Ookinetes cross the midgut epithelium and form protected capsules called oocysts on the basal side of the gut epithelium, where they multiply to give rise to a large number of sporozoites that are eventually released into the mosquito hemolymph upon rupture of the oocysts, thereby completing their life cycle in the mosquito. Recently, JAK/STAT signaling has been shown to mediate mosquito mucosal immunity against *Plasmodium* by regulating the expression of the nitric oxide synthase gene (NOS) (Gupta et al. 2009). Two STAT genes that arose from duplication reside in the genome of *Anopheles gambiae*, the intronless *AgSTAT-B*, and *AgSTAT-A*, which is the predominant form expressed in adult (Barillas-Mury et al. 1999; Christophides et al. 2002; Gupta et al. 2009). AgSTAT-B mediates the basal level of AgSTAT-A, and this STAT pathway (hereafter referred to as AgSTAT-A) was shown to control *NOS* expression in vivo (Gupta et al. 2009): *AgSTAT-A* silencing in female mosquitoes reduces the basal pre-invasion transcript level of *NOS*; *Plasmodium* infection induces the expression of *NOS* by fivefold, and this induction is completely abolished when *AgSTAT-A* is silenced. AgSTAT-A activation is important to limit *Plasmodium* infection, as

AgSTAT-A silencing increases the median number of *Plasmodium* oocysts by more than fourfold one week post-infection, while knock-down of *SOCS*, a suppressor of AgSTAT-A signaling, reduces the median number of oocysts by sevenfold. The effect of the AgSTAT-A pathway in *Plasmodium* infection is mediated by the NOS protein and NO production in the cytoplasm of midgut epithelial cells, as knock-down of both *SOCS* and *NOS* completely reverts the decrease in infection observed in the *SOCS*-silenced mosquito. Finally, it was demonstrated that NOS limits *Plasmodium* infection by decreasing parasite survival at the oocyst stage. Thus, the JAK/STAT pathway plays a crucial role in mosquito mucosal response to parasitization.

JAK/STAT Signaling and Anti-Viral Immunity

Although RNAi is the main immune mechanism to combat viral infection in *Drosophila* (Galiana-Arnoux et al. 2006; van Rij et al. 2006; Wang et al. 2006; Zambon et al. 2006), the JAK/STAT pathway has also been shown to contribute to anti-viral immunity. Infection by *Drosophila* C virus stimulates the DNA binding activity of Stat92E (Dostert et al. 2005). *hop* is involved in the control of virus replication and is required but not sufficient for the induction of some virus-regulated genes. Another recent study provided evidence that JAK/STAT affects the replication of Sindbis virus: flies heterozygous for a *Stat92E* mutation display increased viral load (Avadhanula et al. 2009).

In *Aedes aegypti*, the major vector for dengue virus, the JAK/STAT pathway is required to control virus infection (Souza-Neto et al. 2009). Susceptibility to the virus increases when either *dome* or *hop* expression was silenced in vitro. Microarray experiments identified five JAK/STAT-regulated and infection-responsive dengue virus restriction factors, including DVRF1 (dengue virus restriction factor 1) and DVRF2. The function of these two factors is unresolved. Thus, the involvement of JAK/STAT signaling in controlling viral replications in both *Drosophila* and mosquitoes suggests an evolutionarily conserved function of the pathway in anti-viral immunity.

JAK/STAT Signaling and the Maintenance of Homeostasis in Response to Intestinal Challenge

The *Drosophila* alimentary canal consists of the foregut (oesophagus and crop), the midgut and the hindgut (pylorus, ileum and rectum). The midgut, functionally equivalent to the mammalian small intestine, performs the function of digestion and nutrient absorption. It is a single-cell epithelium consisting chiefly of large, cuboidal, polyploid enterocytes (ECs; 90% of the total cell population). The other terminally differentiated cell type, much fewer in number, is the secretive entero-endocrine cells (EEs; 10% of the total cell population). In 2006, using genetic mosaic analysis and lineage labeling, it was shown that multipotent *Drosophila* intestinal stem cells (ISCs) are present in the adult midgut (Micchelli and Perrimon 2006; Ohlstein and Spradling 2006). They are located at a basal position relative to EC and EE, are smaller than EC in size, and display a wedge-like morphology (Lee et al. 2009; Ohlstein and Spradling 2006, 2007). Under normal physiological conditions, an ISC is often located adjacent to a quiescent enteroblast (EB), the daughter cell from ISC division. Upon stimulation by unknown molecular cues, quiescent EBs differentiate into ECs or EEs. The four aforementioned resident cell types, EC, EE, ISC and EB, all situate atop of a meshwork of extra-cellular matrix that is basally bordered by visceral muscles (VM) (Lee et al. 2009; Ohlstein and Spradling 2006).

Recent studies on the stress- and enteric infection-induced regeneration of the midgut epithelium have provided a useful model to study ISC dynamics under conditions of high regenerative pressure and revealed the requirement of the JAK/STAT pathway in this process (Fig. 3). A number of studies showed that ISCs undergo rapid, compensatory proliferation following a variety of chemical and enteric challenges, including dextran sulfate sodium, reactive oxygen radicals, and bacterial infections (Amcheslavsky et al. 2009; Buchon et al. 2009a, b; Cronin et al. 2009; Jiang et al. 2009). In the current model, the damage to the epithelial lining up-regulates a variety of stress-response mechanisms, including apoptosis and the JNK pathway, which lead to the induction of the Upd family of cytokines, notably the Upd3 cytokine, in the stressed epithelium (Buchon et al. 2009a; Jiang et al. 2009). Paracrine Upd3 signaling in turn activates the JAK/STAT pathway in three cell types. First, it enhances JAK/STAT activity in ISCs and EBs, mediating

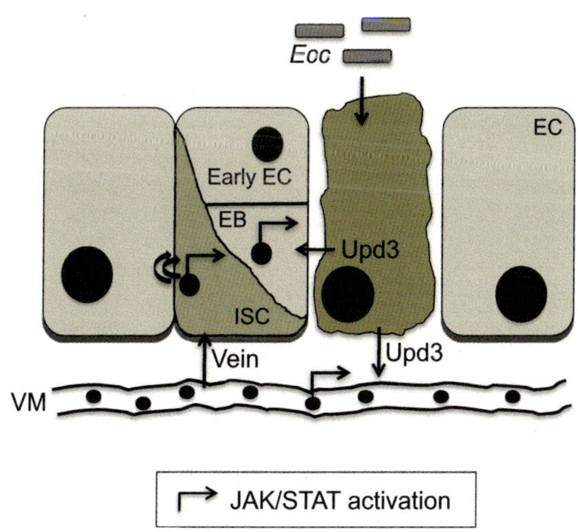

Fig. 3 JAK/STAT signaling in the homeostasis of the *Drosophila* midgut epithelium in response to intestinal challenge. Upon enteric infection, JAK/STAT activity is enhanced in intestinal stem cells (ISC) and enteroblasts (EB) via paracrine Upd3 signaling, mediating ISC proliferation and EB differentiation. JAK/STAT is also activated in visceral muscles (VM). This up-regulates Vein production, which enhances ISC proliferation

compensatory proliferation of ISCs and differentiation of EBs. In addition, JAK/STAT signaling is activated in VM, which partially contributes to the production of the EGF ligand, Vein (Buchon et al. 2010; Jiang et al. 2011). Vein in turn activates the mitogenic EGFR pathway in ISCs, further enhancing their proliferation.

Concluding Remarks

Since its discovery in *Drosophila* for its role in embryonic development, the JAK/STAT signaling pathway has been implicated in various biological processes. Although major progress has been made on the characterization of JAK/STAT signaling in the context of response to stress and infections, important questions remain unresolved. The same intracellular core components appear to be used in various biological contexts, but the functional specificity seems to be achieved through the regulation of cytokine expression in producing cells. For instance, the Upd1 cytokine appears to be devoted to embryonic development, whereas the Upd3 cytokine is specifically produced in response to stress and infections. An important objective in the future is therefore to uncover the mechanisms supporting the regulation of cytokine production in response to challenges. These studies should be conducted in cytokine producing cells as diverse as blood and intestinal cells and may bring different answers depending on the nature of the challenges considered. The discovery of the role of JAK/STAT signaling in mucosal immunity and tissue regeneration in response to intestinal infection in *Drosophila* has been a major development in the past few years. Similarly, future studies on the various roles of JAK/STAT signaling in the mosquito intestine should reveal important aspects of vector biology, thereby offering novel strategies to control malaria.

Acknowledgments We would like to thank Shaeri Mukherjee for her generous help in drawing the figures.

References

Agaisse H, Petersen UM, Boutros M, Mathey-Prevot B, Perrimon N (2003) Signaling role of hemocytes in *Drosophila* JAK/STAT-dependent response to septic injury. Dev Cell 5:441–450

Amcheslavsky A, Jiang J, Ip YT (2009) Tissue damage-induced intestinal stem cell division in *Drosophila*. Cell Stem Cell 4:49–61

Araki T, Gamper M, Early A, Fukuzawa M, Abe T, Kawata T, Kim E, Firtel RA, Williams JG (1998) Developmentally and spatially regulated activation of a *Dictyostelium* STAT protein by a serpentine receptor. EMBO J 17:4018–4028

Arbouzova NI, Bach EA, Zeidler MP (2006) Ken and Barbie selectively regulates the expression of a subset of Jak/STAT pathway target genes. Curr Biol 16:80–88

Avadhanula V, Weasner BP, Hardy GG, Kumar JP, Hardy RW (2009) A novel system for the launch of alphavirus RNA synthesis reveals a role for the Imd pathway in arthropod antiviral response. PLoS Pathog 5:e1000582

Babcock DT, Brock AR, Fish GS, Wang Y, Perrin L, Krasnow MA, Galko MJ (2008) Circulating blood cells function as a surveillance system for damaged tissue in *Drosophila* larvae. Proc Natl Acad Sci USA 105:10017–10022

Bach EA, Ekas LA, Ayala-Camargo A, Flaherty MS, Lee H, Perrimon N, Baeg GH (2007) GFP reporters detect the activation of the *Drosophila* JAK/STAT pathway in vivo. Gene Exp Pattern 7:323–331

Badenhorst P, Voas M, Rebay I, Wu C (2002) Biological functions of the ISWI chromatin remodeling complex NURF. Genes Dev 16:3186–3198

Baeg GH, Zhou R, Perrimon N (2005) Genome-wide RNAi analysis of JAK/STAT signaling components in *Drosophila*. Genes Dev 19:1861–1870

Barillas-Mury C, Han Y-S, Seeley D, Kafatos FC (1999) *Anopheles gambiae* Ag-STAT, a new insect member of the STAT family, is activated in response to bacterial infection. EMBO J 18:959–967

Basset A, Khush RS, Braun A, Gardan L, Boccard F, Hoffmann JA, Lemaitre B (2000) The phytopathogenic bacteria *Erwinia carotovora* infects *Drosophila* and activates an immune response. Proc Natl Acad Sci USA 97:3376–3381

Baton LA, Ranford-Cartwright LC (2005) Spreading the seeds of million-murdering death: metamorphoses of malaria in the mosquito. Trends Parasitol 21:573–580

Baumer D, Trauner J, Hollfelder D, Cerny A, Schoppmeier M (2011) JAK-STAT signalling is required throughout telotrophic oogenesis and short-germ embryogenesis of the beetle *Tribolium*. Dev Biol 350:169–182

Beer J, Technau GM, Campos-Ortega JA (1987) Lineage analysis of transplanted individual cells in embryos of *Drosophila melanogaster*. IV. Commitment and proliferative capabilities of mesodermal cells. Roux's Arch Dev Biol 196:222–230

Betz A, Lampen N, Martinek S, Young MW, Darnell JE Jr (2001) A *Drosophila* PIAS homologue negatively regulates stat92E. Proc Natl Acad Sci USA 98:9563–9568

Binari R, Perrimon N (1994) Stripe-specific regulation of pair-rule genes by hopscotch, a putative Jak family tyrosine kinase in *Drosophila*. Genes Dev 8:300–312

Blandin S, Shiao SH, Moita LF, Janse CJ, Waters AP, Kafatos FC, Levashina EA (2004) Complement-like protein TEP1 is a determinant of vectorial capacity in the malaria vector *Anopheles gambiae*. Cell 116:661–670

Boutros M, Agaisse H, Perrimon N (2002) Sequential activation of signaling pathways during innate immune responses in *Drosophila*. Dev Cell 3:711–722

Brown S, Hu N, Hombria JC (2001) Identification of the first invertebrate interleukin JAK/STAT receptor, the *Drosophila* gene domeless. Curr Biol 11:1700–1705

Buchon N, Broderick NA, Chakrabarti S, Lemaitre B (2009a) Invasive and indigenous microbiota impact intestinal stem cell activity through multiple pathways in *Drosophila*. Genes Dev 23:2333–2344

Buchon N, Broderick NA, Poidevin M, Pradervand S, Lemaitre B (2009b) *Drosophila* intestinal response to bacterial infection: activation of host defense and stem cell proliferation. Cell Host Microbe 5:200–211

Buchon N, Broderick NA, Kuraishi T, Lemaitre B (2010) *Drosophila* EGFR pathway coordinates stem cell proliferation and gut remodeling following infection. BMC Biol 8:152

Callus BA, Mathey-Prevot B (2002) SOCS36E, a novel *Drosophila* SOCS protein, suppresses JAK/STAT and EGF-R signalling in the imaginal wing disc. Oncogene 21:4812–4821

Cerenius L, Lee BL, Soderhall K (2008) The proPO-system: pros and cons for its role in invertebrate immunity. Trends Immunol 29:263–271

Chen HW, Chen X, Oh SW, Marinissen MJ, Gutkind JS, Hou SX (2002) mom identifies a receptor for the *Drosophila* JAK/STAT signal transduction pathway and encodes a protein distantly related to the mammalian cytokine receptor family. Genes Dev 16:388–398

Cheng CH, Chen GD, Yeh MS, Chu CY, Hsu YL, Hwang PP, Huang FL, Huang CJ (2010) Expression and characterization of the JAK kinase and STAT protein from brine shrimp, *Artemia franciscana*. Fish Shellfish Immunol 28:774–782

Cherry CM, Matunis EL (2010) Epigenetic regulation of stem cell maintenance in the *Drosophila* testis via the nucleosome-remodeling factor NURF. Cell Stem Cell 6:557–567

Christophides GK, Zdobnov E, Barillas-Mury C, Birney E, Blandin S, Blass C, Brey PT, Collins FH, Danielli A, Dimopoulos G et al (2002) Immunity-related genes and gene families in *Anopheles gambiae*. Science 298:159–165

Classen AK, Bunker BD, Harvey KF, Vaccari T, Bilder D (2009) A tumor suppressor activity of *Drosophila* Polycomb genes mediated by JAK-STAT signaling. Nat Genet 41:1150–1155

Cronin SJ, Nehme NT, Limmer S, Liegeois S, Pospisilik JA, Schramek D, Leibbrandt A, Simoes Rde M, Gruber S, Puc U et al (2009) Genome-wide RNAi screen identifies genes involved in intestinal pathogenic bacterial infection. Science 325:340–343

Crozatier M, Valle D, Dubois L, Ibnsouda S, Vincent A (1996) Collier, a novel regulator of *Drosophila* head development, is expressed in a single mitotic domain. Curr Biol 6:707–718

Dierking K, Polanowska J, Omi S, Engelmann I, Gut M, Lembo F, Ewbank JJ, Pujol N (2011) Unusual regulation of a STAT protein by an SLC6 family transporter in *C. elegans* epidermal innate immunity. Cell Host Microbe 9:425–435

Dostert C, Jouanguy E, Irving P, Troxler L, Galiana-Arnoux D, Hetru C, Hoffmann JA, Imler JL (2005) The Jak-STAT signaling pathway is required but not sufficient for the antiviral response of *Drosophila*. Nat Immunol 6:946–953

Ekengren S, Hultmark D (2001) A family of Turandot-related genes in the humoral stress response of *Drosophila*. Biochem Biophys Res Commun 284:998–1003

Ekengren S, Tryselius Y, Dushay MS, Liu G, Steiner H, Hultmark D (2001) A humoral stress response in *Drosophila*. Curr Biol 11:714–718

el Shatoury HH (1955) The structure of the lymph gland of *Drosophila* larvae. Roux Arch EntwMech Organ 147:489–495

Elliott GC, Zeidler MP (2008) MsSOCS expression indicates a potential role for JAK/STAT signalling in the early stages of *Manduca sexta* spermatogenesis. Insect Mol Biol 17:475–483

Ferrandon D, Jung AC, Criqui M, Lemaitre B, Uttenweiler-Joseph S, Michaut L, Reichhart J, Hoffmann JA (1998) A drosomycin-GFP reporter transgene reveals a local immune response in *Drosophila* that is not dependent on the Toll pathway. EMBO J 17:1217–1227

Franc NC, Dimarcq JL, Lagueux M, Hoffmann J, Ezekowitz RA (1996) Croquemort, a novel *Drosophila* hemocyte/macrophage receptor that recognizes apoptotic cells. Immunity 4:431–443

Frolet C, Thoma M, Blandin S, Hoffmann JA, Levashina EA (2006) Boosting NF-kappaB-dependent basal immunity of *Anopheles gambiae* aborts development of *Plasmodium berghei*. Immunity 25:677–685

Fukuzawa M, Williams JG (2000) Analysis of the promoter of the *cudA* gene reveals novel mechanisms of *Dictyostelium* cell type differentiation. Development 127:2705–2713

Galiana-Arnoux D, Dostert C, Schneemann A, Hoffmann JA, Imler JL (2006) Essential function in vivo for Dicer-2 in host defense against RNA viruses in *Drosophila*. Nat Immunol 7:590–597

Gilbert MM, Weaver BK, Gergen JP, Reich NC (2005) A novel functional activator of the *Drosophila* JAK/STAT pathway, unpaired2, is revealed by an in vivo reporter of pathway activation. Mech Dev 122:939–948

Girault JA, Labesse G, Mornon JP, Callebaut I (1999) The N-termini of FAK and JAKs contain divergent band 4.1 domains. Trends Biochem Sci 24:54–57

Gonzalez I, Simon R, Busturia A (2009) The Polyhomeotic protein induces hyperplastic tissue overgrowth through the activation of the JAK/STAT pathway. Cell Cycle 8:4103–4111

Goto A, Yano T, Terashima J, Iwashita S, Oshima Y, Kurata S (2010) Cooperative regulation of the induction of the novel antibacterial Listericin by peptidoglycan recognition protein LE and the JAK-STAT pathway. J Biol Chem 285:15731–15738

Gupta L, Molina-Cruz A, Kumar S, Rodrigues J, Dixit R, Zamora RE, Barillas-Mury C (2009) The STAT pathway mediates late-phase immunity against *Plasmodium* in the mosquito *Anopheles gambiae*. Cell Host Microbe 5:498–507

Ha EM, Oh CT, Bae YS, Lee WJ (2005a) A direct role for dual oxidase in *Drosophila* gut immunity. Science 310:847–850

Ha EM, Oh CT, Ryu JH, Bae YS, Kang SW, Jang IH, Brey PT, Lee WJ (2005b) An antioxidant system required for host protection against gut infection in *Drosophila*. Dev Cell 8:125–132

Harrison DA, Binari R, Nahreini TS, Gilman M, Perrimon N (1995) Activation of a *Drosophila* Janus kinase (JAK) causes hematopoietic neoplasia and developmental defects. EMBO J 14:2857–2865

Harrison DA, McCoon PE, Binari R, Gilman M, Perrimon N (1998) *Drosophila unpaired* encodes a secreted protein that activates the JAK signaling pathway. Genes Dev 12:3252–3263

Hedengren M, Asling B, Dushay MS, Ando I, Ekengren S, Wihlborg M, Hultmark D (1999) Relish, a central factor in the control of humoral but not cellular immunity in *Drosophila*. Mol Cell 4:827–837

Holz A, Bossinger B, Strasser T, Janning W, Klapper R (2003) The two origins of hemocytes in *Drosophila*. Development 130:4955–4962

Hombria JC, Brown S, Hader S, Zeidler MP (2005) Characterisation of Upd2, a *Drosophila* JAK/STAT pathway ligand. Dev Biol 288:420–433

Hou XS, Melnick MB, Perrimon N (1996) *marelle* acts downstream of the *Drosophila* HOP/JAK kinase and encodes a protein similar to the mammalian STATs. Cell 84:411–419

Imler JL, Bulet P (2005) Antimicrobial peptides in *Drosophila*: structures, activities and gene regulation. Chem Immunol Allergy 86:1–21

Jiang H, Patel PH, Kohlmaier A, Grenley MO, McEwen DG, Edgar BA (2009) Cytokine/Jak/Stat signaling mediates regeneration and homeostasis in the *Drosophila* midgut. Cell 137:1343–1355

Jiang H, Grenley MO, Bravo MJ, Blumhagen RZ, Edgar BA (2011) EGFR/Ras/MAPK signaling mediates adult midgut epithelial homeostasis and regeneration in *Drosophila*. Cell Stem Cell 8:84–95

Jung SH, Evans CJ, Uemura C, Banerjee U (2005) The *Drosophila* lymph gland as a developmental model of hematopoiesis. Development 132:2521–2533

Kallio J, Myllymaki H, Gronholm J, Armstrong M, Vanha-aho LM, Makinen L, Silvennoinen O, Valanne S, Ramet M (2010) Eye transformer is a negative regulator of *Drosophila* JAK/STAT signaling. FASEB J 24:4467–4479

Karsten P, Hader S, Zeidler MP (2002) Cloning and expression of *Drosophila* SOCS36E and its potential regulation by the JAK/STAT pathway. Mech Dev 117:343–346

Karsten P, Plischke I, Perrimon N, Zeidler MP (2006) Mutational analysis reveals separable DNA binding and trans-activation of *Drosophila* STAT92E. Cell Signal 18:819–829

Kim LK, Choi UY, Cho HS, Lee JS, Lee WB, Kim J, Jeong K, Shim J, Kim-Ha J, Kim YJ (2007) Down-regulation of NF-kappaB target genes by the AP-1 and STAT complex during the innate immune response in *Drosophila*. PLoS Biol 5:e238

Klapper R, Holz A, Janning W (1998) Fate map and cell lineage relationships of thoracic and abdominal mesodermal anlagen in *Drosophila melanogaster*. Mech Dev 71:77–87

Krzemien J, Dubois L, Makki R, Meister M, Vincent A, Crozatier M (2007) Control of blood cell homeostasis in *Drosophila* larvae by the posterior signalling centre. Nature 446:325–328

Kwon SY, Xiao H, Glover BP, Tjian R, Wu C, Badenhorst P (2008) The nucleosome remodeling factor (NURF) regulates genes involved in *Drosophila* innate immunity. Dev Biol 316:538–547

Lagueux M, Perrodou E, Levashina EA, Capovilla M, Hoffmann JA (2000) Constitutive expression of a complement-like protein in toll and JAK gain-of-function mutants of *Drosophila*. Proc Natl Acad Sci USA 97:11427–11432

Lanot R, Zachary D, Holder F, Meister M (2001) Postembryonic hematopoiesis in *Drosophila*. Dev Biol 230:243–257

Lebestky T, Jung SH, Banerjee U (2003) A Serrate-expressing signaling center controls *Drosophila* hematopoiesis. Genes Dev 17:348–353

Lee WC, Beebe K, Sudmeier L, Micchelli CA (2009) *Adenomatous polyposis coli* regulates *Drosophila* intestinal stem cell proliferation. Development 136:2255–2264
Levashina EA, Moita LF, Blandin S, Vriend G, Lagueux M, Kafatos FC (2001) Conserved role of a complement-like protein in phagocytosis revealed by dsRNA knockout in cultured cells of the mosquito, *Anopheles gambiae*. Cell 104:709–718
Liehl P, Blight M, Vodovar N, Boccard F, Lemaitre B (2006) Prevalence of local immune response against oral infection in a *Drosophila/Pseudomonas* infection model. PLoS Pathog 2:e56
Luo H, Hanratty WP, Dearolf CR (1995) An amino acid substitution in the *Drosophila* hopTum-l Jak kinase causes leukemia-like hematopoietic defects. EMBO J 14:1412–1420
Luo H, Rose P, Barber D, Hanratty WP, Lee S, Roberts TM, D'Andrea AD, Dearolf CR (1997) Mutation in the Jak kinase JH2 domain hyperactivates *Drosophila* and mammalian Jak-Stat pathways. Mol Cell Biol 17:1562–1571
Makki R, Meister M, Pennetier D, Ubeda JM, Braun A, Daburon V, Krzemien J, Bourbon HM, Zhou R, Vincent A et al (2010) A short receptor downregulates JAK/STAT signalling to control the *Drosophila* cellular immune response. PLoS Biol 8:e1000441
Mandal L, Martinez-Agosto JA, Evans CJ, Hartenstein V, Banerjee U (2007) A Hedgehog- and Antennapedia-dependent niche maintains *Drosophila* haematopoietic precursors. Nature 446:320–324
Markus R, Laurinyecz B, Kurucz E, Honti V, Bajusz I, Sipos B, Somogyi K, Kronhamn J, Hultmark D, Ando I (2009) Sessile hemocytes as a hematopoietic compartment in *Drosophila melanogaster*. Proc Natl Acad Sci USA 106:4805–4809
Meng X, Khanuja BS, Ip YT (1999) Toll receptor-mediated *Drosophila* immune response requires Dif, an NF-kappaB factor. Genes Dev 13:792–797
Micchelli CA, Perrimon N (2006) Evidence that stem cells reside in the adult *Drosophila* midgut epithelium. Nature 439:475–479
Muller P, Kuttenkeuler D, Gesellchen V, Zeidler MP, Boutros M (2005) Identification of JAK/STAT signalling components by genome-wide RNA interference. Nature 436:871–875
Nappi AJ (1975) Inhibition by parasites of melanotic tumour formation in *Drosophila melanogaster*. Nature 255:402–404
Nappi AJ, Ottaviani E (2000) Cytotoxicity and cytotoxic molecules in invertebrates. Bioessays 22:469–480
Ohlstein B, Spradling A (2006) The adult *Drosophila* posterior midgut is maintained by pluripotent stem cells. Nature 439:470–474
Ohlstein B, Spradling A (2007) Multipotent *Drosophila* intestinal stem cells specify daughter cell fates by differential Notch signaling. Science 315:988–992
Onfelt Tingvall T, Roos E, Engstrom Y (2001) The *imd* gene is required for local Cecropin expression in *Drosophila* barrier epithelia. EMBO Rep 2:239–243
Pastor-Pareja JC, Wu M, Xu T (2008) An innate immune response of blood cells to tumors and tissue damage in *Drosophila*. Dis Model Mech 1:144–154, discussion 153
Perrimon N, Mahowald AP (1986) *l(1)hopscotch*, a larval-pupal zygotic lethal with a specific maternal effect on segmentation in *Drosophila*. Dev Biol 118:28–41
Rawlings JS, Rennebeck G, Harrison SM, Xi R, Harrison DA (2004) Two *Drosophila* suppressors of cytokine signaling (SOCS) differentially regulate JAK and EGFR pathway activities. BMC Cell Biol 5:38
Rizki TM (1978) The circulatory system and associated cells and tissues. In: Ashburner M, Wright TRF (eds) The genetics and biology of *Drosophila*. Academic, New York
Rizki TM, Rizki RM (1980) Properties of the larval hemocytes of *Drosophila melanogaster*. Experientia 36:1223–1226
Rizki TM, Rizki RM (1992) Lamellocyte differentiation in *Drosophila* larvae parasitized by *Leptopilina*. Dev Comp Immunol 16:103–110
Robertson CW (1936) The metamorphosis of *Drosophila melanogaster*, including an accurately timed account of the principal morphological changes. J Morphol 59:351–399

Rugendorff AE, Younossi-Hartenstein A, Hartenstein V (1994) Embryonic origin and differentiation of the *Drosophila* heart. Roux's Arch Dev Biol 203:266–280
Russo J, Dupas S, Frey F, Carton Y, Brehelin M (1996) Insect immunity: early events in the encapsulation process of parasitoid (*Leptopilina boulardi*) eggs in resistant and susceptible strains of *Drosophila*. Parasitology 112(Pt 1):135–142
Rutschmann S, Jung AC, Hetru C, Reichhart JM, Hoffmann JA, Ferrandon D (2000) The Rel protein DIF mediates the antifungal but not the antibacterial host defense in *Drosophila*. Immunity 12:569–580
Ryu JH, Nam KB, Oh CT, Nam HJ, Kim SH, Yoon JH, Seong JK, Yoo MA, Jang IH, Brey PT et al (2004) The homeobox gene *Cauda*l regulates constitutive local expression of antimicrobial peptide genes in *Drosophila* epithelia. Mol Cell Biol 24:172–185
Ryu JH, Ha EM, Oh CT, Seol JH, Brey PT, Jin I, Lee DG, Kim J, Lee D, Lee WJ (2006) An essential complementary role of NF-kappaB pathway to microbicidal oxidants in *Drosophila* gut immunity. EMBO J 25:3693–3701
Sorrentino RP, Carton Y, Govind S (2002) Cellular immune response to parasite infection in the *Drosophila* lymph gland is developmentally regulated. Dev Biol 243:65–80
Sorrentino RP, Melk JP, Govind S (2004) Genetic analysis of contributions of dorsal group and JAK-Stat92E pathway genes to larval hemocyte concentration and the egg encapsulation response in *Drosophila*. Genetics 166:1343–1356
Souza-Neto JA, Sim S, Dimopoulos G (2009) An evolutionary conserved function of the JAK-STAT pathway in anti-dengue defense. Proc Natl Acad Sci USA 106:17841–17846
Stark MB, Marshall AK (1930) The blood-forming organ of the larva of *Drosophila* melanogaster. Bell Telephone Syst Tech Pub Monogr 1931:1204–1206
Tang H (2009) Regulation and function of the melanization reaction in *Drosophila*. Fly 3:105–111 (Austin)
Tepass U, Fessler LI, Aziz A, Hartenstein V (1994) Embryonic origin of hemocytes and their relationship to cell death in *Drosophila*. Development 120:1829–1837
Tzou P, Ohresser S, Ferrandon D, Capovilla M, Reichhart JM, Lemaitre B, Hoffmann JA, Imler JL (2000) Tissue-specific inducible expression of antimicrobial peptide genes in *Drosophila* surface epithelia. Immunity 13:737–748
van Rij RP, Saleh MC, Berry B, Foo C, Houk A, Antoniewski C, Andino R (2006) The RNA silencing endonuclease Argonaute 2 mediates specific antiviral immunity in *Drosophila melanogaster*. Genes Dev 20:2985–2995
Vodovar N, Vinals M, Liehl P, Basset A, Degrouard J, Spellman P, Boccard F, Lemaitre B (2005) *Drosophila* host defense after oral infection by an entomopathogenic *Pseudomonas* species. Proc Natl Acad Sci USA 102:11414–11419
Wang Y, Levy DE (2006) *C. elegans* STAT: evolution of a regulatory switch. FASEB J 20:1641–1652
Wang XH, Aliyari R, Li WX, Li HW, Kim K, Carthew R, Atkinson P, Ding SW (2006) RNA interference directs innate immunity against viruses in adult *Drosophila*. Science 312:452–454
Waterhouse RM, Kriventseva EV, Meister S, Xi Z, Alvarez KS, Bartholomay LC, Barillas-Mury C, Bian G, Blandin S, Christensen BM et al (2007) Evolutionary dynamics of immune-related genes and pathways in disease-vector mosquitoes. Science 316:1738–1743
Yan R, Small S, Desplan C, Dearolf CR, Darnell JE Jr (1996) Identification of a Stat gene that functions in *Drosophila* development. Cell 84:421–430
Yano T, Kurata S (2008) Induction of autophagy via innate bacterial recognition. Autophagy 4:958–960
Zambon RA, Vakharia VN, Wu LP (2006) RNAi is an antiviral immune response against a dsRNA virus in *Drosophila melanogaster*. Cell Microbiol 8:880–889
Zeidler MP, Perrimon N, Strutt DI (1999) Polarity determination in the *Drosophila* eye: a novel role for Unpaired and JAK/STAT signaling. Genes Dev 13:1342–1353

Stat5 as a Hematopoietic Master Regulator for Differentiation and Neoplasia Development

Harini Nivarthi, Katrin Friedbichler, and Richard Moriggl

Abstract

Stat5 transcription factors have a crucial role in hematopoiesis from hematopoietic stem cells to fully differentiated cells. The individual contributions of Stat5a and Stat5b genes to the generation of hematopoietic cells and to their malignant transformation are subject of the following review. Absence of Stat5 proteins causes lymphopenia and Stat5 was recognized to be indispensable for the development of B-, T- and NK-cells. The few peripheral T-cells that develop in Stat5-deficient mice have an activated phenotype and these T-cells contribute to the development of autoimmunity. Moreover, deletion of Stat5 in myeloid cells causes myelodysplasia (red cell anemia and thrombocytopenia). In addition, generation and function of mast cells and eosinophils depends on Stat5. Impoartantly, Stat5 was found to be highly expressed and constitutively activated in many human hematopoietic neoplasms, where it regulates expression of genes controlling cell survival and cell cycle progression. Expression of Stat5 in hematopoietic neoplasms was both found to be elevated at the mRNA and protein level. Interestingly, higher Stat5 levels were linked with tyrosine kinase inhibitor drug resistance.

Introduction

The Signal Transducer and Activator of Transcription 5 (Stat5) plays a crucial role in hematopoiesis (Bunting 2007). The Stat5 gene encodes two isoforms, Stat5a and Stat5b, which have certain overlapping and distinct functions (Ferbeyre and Moriggl 2011).

H. Nivarthi • K. Friedbichler • R. Moriggl (✉)
Ludwig Boltzmann Institute for Cancer Research (LBI-CR), Währinger Straße 13A, 1090 Vienna, Austria
e-mail: richard.moriggl@lbicr.lbg.ac.at

Th. Decker and M. Müller (eds.), *Jak-Stat Signaling: From Basics to Disease*, DOI 10.1007/978-3-7091-0891-8_10, © Springer-Verlag Wien 2012

Stat5 Functions in Hematopoietic Stem Cells (HSCs)

The first Stat5 knockout mice generated lacked the N-termini of both Stat5 proteins (Teglund et al. 1998) and are now recognized as Stat5$^{\Delta N}$ mice expressing hypomorphic alleles. The N-terminus of Stat5 is the docking platform for the glucocorticoid receptor and it provides the oligomerization domain for Stat dimer interaction on chromatin (Ferbeyre and Moriggl 2011). Stat5$^{\Delta N}$ mice have normal numbers of HSCs and the cells are capable of engrafting into lethally irradiated hosts (Bunting et al. 2002). However, their reconstitution ability is highly reduced in competitive transplantation experiments, suggesting a defect in the ability of the HSCs to 'self-renew' (Bradley et al. 2002). Once Stat5-deficient mice (Cui et al. 2004) (Stat5^{null}) were available, competitive reconstitution experiments with fetal liver cells showed a very drastic defect in the repopulation capacity in the absence of Stat5 (Li et al. 2007; Yao et al. 2006). The number of progenitor cells is also reduced in Stat5$^{\Delta N}$ mice, as assayed by the number and size of colony forming units (Bunting et al. 2002). The N-terminus of Stat5 is essential for the induction of bcl-2 and bcl-x_L and the suppression of the microRNAs miR15/16 (negative regulators of bcl-2 and bcl-x_L) required for survival of HSCs (Li et al. 2007) (Fig. 1). The above data clearly indicate a crucial role for Stat5 in the maintenance and renewal of HSCs. A role for activated Stat5 (pYStat5) is now emerging in hematopoietic cancer stem cells. Down regulation of Stat5 expression by RNA interference or deletion of Stat5 impairs the long-term expansion of leukemic stem/progenitor cells in primary acute myeloid leukemia (AML) or chronic myeloid leukemia (CML) (Schepers et al. 2007; Scherr et al. 2006). Stat5b activity has been shown to be linked to leukemia initiating cells in MN1 and HOXA9 expressing AML cell lines (Heuser et al. 2009). In AML patients, Stat5 activation by the mutant receptor tyrosine kinase, FMS-like tyrosine kinase-3 with internal tandem duplications (FLT3ITD), leads to high expression of the pro-survival gene Mcl-1 which promotes the survival of leukemic stem cells (LSCs) (Yoshimoto et al. 2009). Stat5 activation in the HSCs of patients with truncated granulocyte colony stimulating factor (G-CSF) receptor provides them with a clonal advantage that can lead to AML or myelodysplastic syndrome. These patients suffer from severe congenital neutropenia and are treated with exogenous G-CSF to boost neutrophil numbers (Liu et al. 2008). Interestingly, the activation of Stat5 by cytokines in the stem cells of AML patients shows a high degree of heterogeneity and does not correlate with the surface expression of the cytokine receptors (Han et al. 2009). Further studies are required to understand the differential activation of Stat5 by cytokines in LSCs compared to normal HSCs.

Stat5 as a Key Regulator for Myelo- and Erythropoiesis

Stat5 is the key signaling molecule downstream of a variety of myeloid cytokines and growth factors including interleukin (IL)-3, IL-5, thrombopoietin (TPO), erythropoietin (EPO), stem cell factor (SCF), fms-like tyrosine kinase-3 ligand

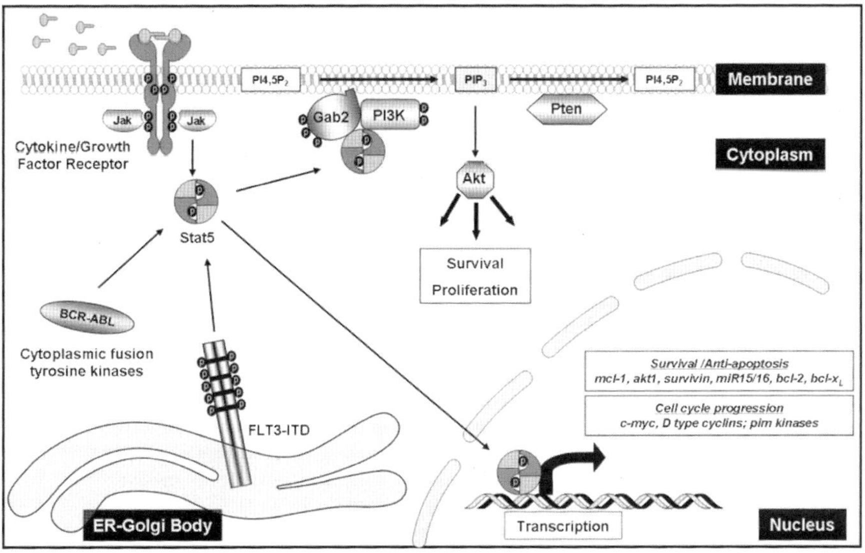

Fig. 1 Oncogenic activity of Stat5: Stat5 can be activated by many different mechanisms. It is phosphorylated by Jak kinases associated with the cytokine and growth factor receptors, at the membrane. In the cytoplasm, it can be activated by oncogenic fusion tyrosine kinases such as BCR-ABL. Other oncogenic receptor tyrosine kinases such as FLT3-ITD phosphorylate Stat5 at the endoplasmic reticulum (ER) – Golgi body. Stat5 targets include anti-apoptotic genes, such as *bcl-2, bcl-*X_L*, mcl-1* and *survivin.* Stat5 maintains the levels of bcl-2 by suppressing the expression of the microRNAs miR15/16, which negatively regulate the levels of bcl-2. It induces expression of *D type cyclins, c-myc* and *pim kinases*, genes that promote cell cycle progression. In the cytoplasm, activated Stat5 can interact with Gab2 which leads to the activation of the PI3K/Akt pathway, via generation of phosphatidyl inositol trisphosphate (PIP3). This pathway is negatively regulated by the tumor suppressor Pten phosphatase. The Akt pathway also results in the transcription of survival and proliferation genes, which further augments the transforming potential of Stat5

(Flt3L), G-CSF and granulocyte macrophage colony stimulating factor (GM-CSF). $Stat5^{\Delta N}$ embryos display anemia, that can be rescued by ectopic expression of the survival genes bcl-2 or bcl-X_L (Dolznig et al. 2006). However, most of the $Stat5^{null}$ embryos die during definitive erythropoiesis on pure C57Bl/6 or Balb/c backgrounds (Cui et al. 2004). Interestingly, in a mixed background, Stat3 activity can compensate for the function of Stat5 in erythropoiesis and a few mice survive. However, these mice are severely sick and display dwarfism, autoimmune disorders and neutrophil infiltration in organs. The defect in erythropoiesis in $Stat5^{null}$ erythroid progenitors is due to their inability to absorb iron efficiently; as they have reduced expression of the transferrin receptor (CD71) and iron regulatory protein-2, which are direct targets of Stat5 (Kerenyi et al. 2008). The key role of Stat5 in erythropoiesis is highlighted by the fact that the expression of constitutively activated Stat5 (cS5) in $Jak2^{-/-}$ and $EpoR^{-/-}$ fetal liver cells leads to the development of functional erythroblasts in transplantation and colony forming assays (Grebien et al. 2008).

The progenitors derived from the bone marrow of Stat5$^{\Delta N}$ mice are deficient in the ability to give rise to myeloid colonies (Bunting 2007; Teglund et al. 1998). The mast cells in these mice are not only drastically reduced (Shelburne et al. 2003), but they also show functional defects in degranulation upon IgE binding (Barnstein et al. 2006). This phenotype is aggravated in Stat5^{null} mice, and re-expression of Stat5a in fetal liver cells restores their capability to differentiate into functional mast cells in vitro (Li et al. 2007). The Stat5$^{\Delta N}$ mice also suffer from thrombocytopenia, due to defects in TPO signaling (Bradley et al. 2002; Bunting et al. 2002). Stat5 is also essential for the differentiation of eosinophils upon IL-5 treatment, in mice and in humans (Buitenhuis et al. 2003; Zhu et al. 2004). Stat5 plays an interesting ambivalent role in granulopoiesis. While Stat5 is required cell intrinsically for the survival of granulocytes it represses G-CSF production in liver endothelial cells (LECs). During inflammation, Stat5 is rapidly degraded in the LECs to induce G-CSF production (Fievez et al. 2007). Mice lacking Stat5 in hematopoietic cells have reduced numbers of neutrophils. Especially under myelosuppressive conditions, these mice are unable to produce higher numbers of neutrophils and to respond to GM-CSF (Kimura et al. 2009).

Stat5 Signaling in Lymphocytes

Stat5 is crucial for signaling by major lymphoid cytokines like IL-2, IL-4, IL-7, IL-9, IL-15 and IL-21 (Giliani et al. 2005). It is also activated by the cytokine thymic stromal lymphopoietin (TSLP) which plays a role in B-cell development and T-helper 2 cell (Th2) polarization (Kang and Der 2004). Stat5^{null} mice exhibit a severe combined immunodeficiency phenotype reminiscent of γ_c, Jak3 and IL-7Rα deficient mice (Yao et al. 2006).

Stat5-Regulated B-cell Development

B-cell development is dependent on IL-7 signaling, as indicated by the complete lack of mature B-cells in IL-7$^{-/-}$ and IL-7R$\alpha^{-/-}$ mice (Malin et al. 2010a). Expression of constitutively active Stat5b (Stat5b-CA) in the lymphoid cells can rescue B-cell development in IL-7R$\alpha^{-/-}$ mice. The expansion of the pro-B-cells is associated with the expression of the Stat5 target genes *cyclin D2, pim-1 and bcl-*x_L, suggesting a role for Stat5 in the survival and proliferation of pro-B-cells (Fig. 2) (Goetz et al. 2004). B-cell development in Stat5ab$^{\Delta N}$ mice is only mildly affected (Sexl et al. 2000), but Stat5^{null} mice show a drastic reduction in mature B-cell numbers due to a developmental block at the pre-pro-B-cell stage (Dai et al. 2007; Hoelbl et al. 2006; Yao et al. 2006). Earlier studies suggested a direct transcriptional regulation of Pax5 and Ebf1 by Stat5 (Dai et al. 2007; Hirokawa et al. 2003). However, this observation has been contradicted by later publications. In fact, the

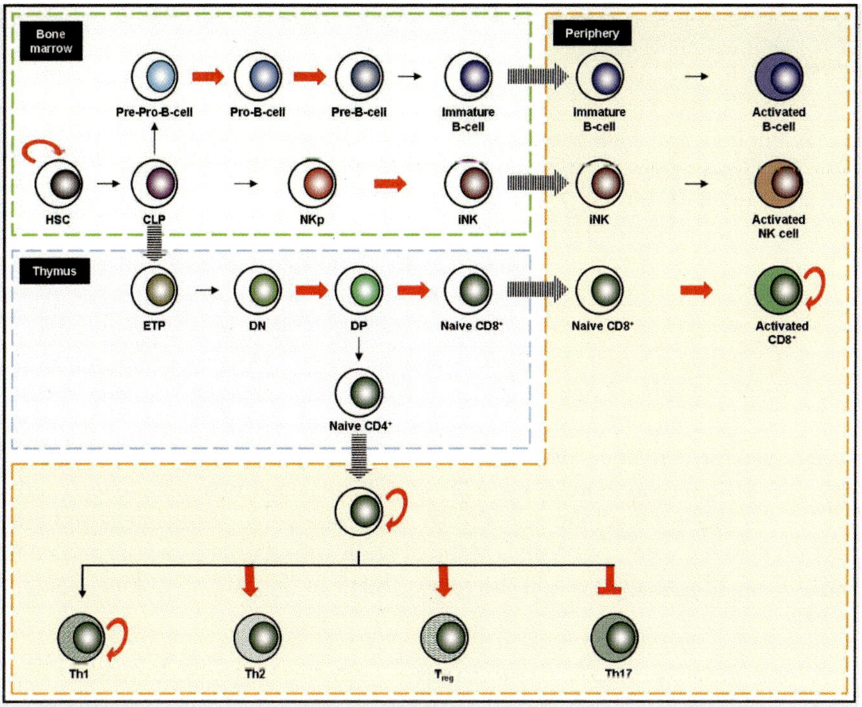

Fig. 2 Stat5 functions in lymphopoiesis: Stat5 plays an indispensable role at various stages of lymphoid differentiation (marked in *red arrows*). Stat5 is essential for the self-renewal of hematopoietic stem cells (HSCs). The HSCs give rise to common lymphoid progenitor (CLP) which is the precursor for the cells of the lymphoid lineage. In the bone marrow, the CLPs give rise to pre-pro-B-cells which differentiate into pre-B-cells; both of these cell types require Stat5 for their survival and proliferation. The pro-B-cells give rise to pre-B-cells which differentiate into immature B-cells. These cells migrate to the periphery, where they become activated upon seeing the antigen. The CLP can also give rise to natural killer cell progenitors (NKp). These cells require Stat5 to differentiate into inactive NK cells (iNK), which can be activated in the periphery to perform their cytolytic functions. The CLP is also the precursor of the early thymic progenitor (ETP) in the thymus. It differentiates into $CD4^{-}$ $CD8^{-}$ double negative (DN) cells, which then express CD4 and CD8 to become double positive (DP) cells. Stat5 signaling, induced by IL-7 or thymic stromal lymphopoietin (TSLP) is essential for the survival of DN and DP cells. The DP cells lose one of the markers to become either $CD4^{+}$ or $CD8^{+}$ single positive cells. Stat5, activated by IL-7, is essential for the differentiation of $CD8^{+}$ cells. IL-2 mediated activation of Stat5 is indispensable for the activation and proliferation of $CD8^{+}$ cells in the periphery. The $CD4^{+}$ cells also require Stat5 signaling for proliferation and expansion. They can differentiate into four different lineages. The Th1 cells require Stat5 signaling by IL-2 for proliferation. The concerted action of Stat5 and Gata-3 is essential for the differentiation of the Th2 subset. Stat5 regulates FoxP3 transcription which is the driver of regulatory T-cell (T_{regs}) differentiation. In contrast, Stat5 inhibits the differentiation of Th17 cells

levels of Pax5 and Ebf1 were found to be quite normal in Bcl-2 rescued Stat5 deficient pro-B-cells, suggesting a more permissive role of Stat5 in B-cell development. Indeed, the apoptosis of Stat5 deficient pro-B-cells is due to aberrant expression of the Igκ light chain, as Stat5 represses the recombination of the Igκ locus (Malin et al. 2010b).

TSLP induces pYStat5 and promotes the differentiation of fetal liver cells to immature B-cells. The TSLP receptor complex includes the IL-7Rα and early studies implied that Stat5 activation by TSLP in a pre-B-cell line is independent of the Jak kinases (Levin et al. 1999). However, recent publications have shown that activation of Stat5 by TSLP in $CD4^+$ T-cells and mouse embryonic fibroblasts requires the activation of Jak1 and Jak2 (Wohlmann et al. 2010).

Stat5 was reported to play an important role in the generation of human memory B-cells. The memory B-cells in the germinal centers of patient tonsils were shown to express pYStat5. Knockdown of Stat5 by shRNA decreased proliferation of a human Burkitt lymphoma cell line. Moreover, overexpression of constitutively active Stat5b in primary human B-cells dramatically increased their survival and expansion in culture (Scheeren et al. 2005). However, Stat5 was shown to be dispensable for activation of murine B-cells as mice with deletion of Stat5 in mature B-cells display normal numbers of follicular and marginal zone B-cells. The B-cells lacking Stat5 are also able to differentiate into plasma cells and give rise to functional memory B-cells (Malin et al. 2010b). Further work is needed to clarify these differences seen in the role of Stat5 in human and murine memory B-cells.

Stat5-Regulated T-cell Development

Stat5 plays an important role in the differentiation and function of T-cell subsets (Fig. 2). While the Th1 subset is essential for cell mediated immunity accomplished by the cytolytic activity of $CD8^+$ T-cells; the Th2 subset is crucial for mounting a humoral response against extracellular pathogens. The $Stat5^{\Delta N}$ mice show a mild reduction in the number of $CD8^+$ T-cells in the periphery, but normal thymocyte and γδ T-cell numbers (Moriggl et al. 1999). However, the analysis of $Stat5^{null}$ mice showed a massive reduction of thymocyte numbers, which results in a severe reduction in $CD8^+$ T-cells and a complete absence of γδ T-cells (Hoelbl et al. 2006; Yao et al. 2006). Deletion of Stat5 in $CD4^+CD8^+$ double positive thymocytes also leads to a severe reduction of $CD8^+$ T-cells (Hoelbl et al. 2006). Interestingly, Stat5 regulates differentiation of $CD8^+$ T-cells in a dose dependent manner (Ermakova et al. 2011). Indeed, the IL-7/Stat5 mediated signaling pathway, which leads to the induction of the transcription factor Runx3 and survival signals by bcl-2, can even circumvent the requirement for the T-cell receptor (TCR) for differentiation of $CD8^+$ T-cells (Park et al. 2010). Transgenic mice with ectopic expression of wild type Stat5b in lymphoid cells show an expansion of $CD8^+$ T-cells (Kelly et al. 2003), while those that express Stat5b-CA show an increase in the number of $CD8^+$ and γδ T-cells (Burchill et al. 2003). It has been shown that peripheral T-cells of $Stat5^{\Delta N}$ mice are highly deficient in proliferation upon stimulation with IL-2 or IL-4 despite normal TCR activation (Moriggl et al. 1999). Moreover, $CD8^+$ T-cells from untreated HIV^+ patients show decreased expression of Stat5 mRNA and protein and are also deficient in their ability to activate Stat5 upon IL-7 stimulation (Vranjkovic et al. 2010).

Naive $CD4^+$ T-cells can undergo at least four distinct cellular fates: Th1, Th2, T_{regs} (regulatory T-cells) and Th17 (Zhu and Paul 2008). Th2 differentiation is dependent on TCR stimulation, IL-4 and IL-2 signaling. IL-4 leads to the activation of the transcription factors Gata3, Stat6 and Stat5, while IL-2 signaling predominantly activates Stat5. These transcription factors interact in a complex manner to determine cell fate decisions of Th subsets (Zhu 2010). Early studies showed that IL-2 can induce the expression of the high affinity IL-2Rα chain, also known as CD25. This phenomenon is mediated by Stat5, along with a GATA family protein (John et al. 1996). Moreover, IL-2 signaling 'primes' T-cells to Th2 differentiation by inducing and maintaining the expression of the IL-4Rα chain via Stat5 (Liao et al. 2008). Profound Th2 defects were identified in $Stat5a^{null}$ mice (Kagami et al. 2001). Stat5a and Gata3 directly bind the promoter regions of *Il1rl1* (Guo et al. 2009) which encodes the receptor IL-33Rα. Upon binding its ligand, IL-33Rα leads to the production of IL-13, which further amplifies the Th2 responses (Oboki et al. 2010). Interestingly, Gata3 can up-regulate CD25 expression via c-maf (Hwang et al. 2002) and Stat5 maintains Gata3 expression in Th2 cells (Guo et al. 2009). Stat5 is also activated in naive $CD4^+$ T-cells upon TSLP treatment, and promotes their survival and proliferation (Rochman et al. 2010). Stat5 also assists Th2 differentiation by epigenetic modification of the *Il4/Il13* gene locus and is required for IL-4-induced Th2 priming (Zhu 2010).

Stat5 deficient mice lack $CD4^+CD25^+$ regulatory T-cells (T_{regs}). Stat5 is a direct transcriptional regulator of Foxp3 and CD25, the key molecules required for T_{regs} differentiation (Yao et al. 2007). A new subset of T-cells, Th17, has been recently discovered, which have been associated with various autoimmune disorders (Korn et al. 2009). Addition of IL-2 to in vitro differentiation culture systems, leads to inhibition of Th17 differentiation. IL-2 deficient mice also have higher number of Th17 cells. Moreover, mice lacking Stat5 in T-cells ($Stat5^{fl/fl}$ CD4-Cre) develop more Th17 cells, implying that IL-2 inhibits Th17 differentiation in a Stat5 dependent manner (Laurence et al. 2007). In fact, Stat5 displaces Stat3 from the promoter region of *Il17* and directly suppresses the transcription. The balance between the amount of Stat5 and Stat3 determines the differentiation of a cell to the Th17 lineage (Yang et al. 2011). Considering the opposing roles of Stat5 in the development of T_{regs} and Th17 cells, it has been suggested that the auto-immune phenotype seen in Stat5-deficient mice could be due to a skewed ratio of T_{regs} to Th17 cells (Yao et al. 2007).

The generation of memory T-cells has not been defined very well. However, it has been shown that both IL-7 and IL-15 are essential for the survival and maintenance of $CD8^+$ memory T-cells (Osborne and Abraham 2010). Recent experiments have shown that $CD8^+$ T-cells transduced with retroviruses expressing cS5 are able to expand dramatically more than the control cells in a LCMV infection model (Hand et al. 2010).

Stat5-Regulated NK Cell Development

The development of NK cells is regulated primarily by the cytokines IL-2, IL-15 and IL-21. IL-15$^{-/-}$ and IL-15R$\alpha^{-/-}$ mice show a drastic reduction in the number of peripheral NK cells (Zwirner and Domaica 2010). Stat5$^{\Delta N}$ mice display a massive decrease in the number of NK cells (Moriggl et al. 1999). Stat5b^{null} mice show reduced NK cell numbers and impaired cytolytic activity of whole splenocyte cultures (Imada et al. 1998). Moreover, conditional deletion of Stat5 in NK cell progenitors (using a novel Ncr1-cre mouse) abrogates their ability to differentiate into immature NK cells (Eckelhart et al. 2011). A Stat5 binding site has been identified in the promoter of the human perforin gene suggesting a role for Stat5 in regulating NK cell functions (Yu et al. 1999). Recently, a case study of two male siblings has been reported who have a four nucleotide deletion (CTCC, position 424–427) in the Stat5b mRNA. These patients suffer from growth hormone insensitivity, as Stat5b is indispensable for growth hormone signaling in the liver. Notably, the patients also suffer from lymphopenia and particularly from reduced number of NK cells (Pugliese-Pires et al. 2010).

Role of Stat5 for Hematopoietic Cancer Development and Progression

Stat5 has been shown to play a crucial role in the generation of a variety of hematopoietic neoplasms (Table 1). Aberrant activation of Stat5 can render the proliferation of many hematopoietic cells factor independent (Grebien et al. 2008; Moriggl et al. 2005), which is a hallmark of oncogenic transformation. Moreover, a persistently activated mutated Stat5a has been used to show its role in the generation of hematopoietic malignancies in mouse bone marrow transplant models (Li et al. 2010; Moriggl et al. 2005). Interestingly, the leukemogenic potential of oncogenic Stat5a is critically dependent on the phosphorylation of two serine residues in the C-terminus (Friedbichler et al. 2010). Intriguingly, most of the activating mutations in patients have been found in upstream kinases and receptors, but not in Stat5 itself. The best studied of these mutations is the BCR-ABL translocation product that leads to a persistent activation of Stat5. It has been shown that Stat5 is not only required for initiation of the leukemia but it is indispensable for leukemia maintenance, thereby identifying Stat5 as a target for leukemia therapy (Hoelbl et al. 2006; Warsch et al. 2011). Stat5 has been implicated in leukemia/lymphomas induced by a variety of fusion tyrosine kinases, such as TEL-JAK2, NPM-ALK, TEL-ABL and TEL-PDGFRβ (Table 1). A comprehensive study with these fusion tyrosine kinases showed that Stat5 mediated over-expression of Rad51 (involved in dsDNA break repair by homologous recombination repair mechanism) is one of the important contributors for the resistance of the malignant cells against DNA damage inducing drugs (Slupianek et al. 2002).

Table 1 Stat5 activation in hematopoietic malignancies

Disease	Cell type	Kinase	References
Leukemia			
Chronic myelocytic leukemia (CML)	Granulocytes	BCR-ABL BCR-JAK2	Lane et al. (2008), Slupianek et al. (2002)
Acute lymphocytic leukemia (ALL)	B- or T-lymphocytes	BCR-ABL TEL-JAK2 TEL-ABL EML1-ABL	De Keersmaecker et al. (2005), Slupianek et al. (2002)
Acute myelocytic leukemia (AML)	Myeloid cells	KITD816V FLT3ITD BCR-ABL JAK3A572V	Birkenkamp et al. (2001), Walters et al. (2006a)
Megakaryocytic leukemia	Megakaryocytes	JAK3A572V JAK2T875N	Scott et al. (2007), Walters et al. (2006b)
Myeloproliferative disorders			
(EMS)/Stem cell leukemia lymphoma syndrome	Myeloid progenitor cells	ZNF198-FGFR1	Heath and Cross (2004)
Polycythemia vera	Erythrocytes	JAK2V716F JAK2 Exon 12	Scott et al. (2007), Shide et al. (2008)
Essential thrombocythemia	Megakaryocytes	JAK2V716F	Shide et al. (2008)
Idiopathic myelofibrosis	Megakaryocytes	JAK2V716F JAK2 Exon 12	Scott et al. (2007), Shide et al. (2008)
Severe congenital neutropenia	Promyelocyte/ myelocyte	Truncated G-CSFR	Liu et al. (2008)
Chronic myelo-monocytic leukemia (CMML)	Monocytes	TEL-PDGFRβ	Tomasson et al. (2000)
Mastocytosis	Mast cells	KITD816V BCR-ABL FLP1L1-PDGFRα	Valent et al. (2003)
Lymphoma			
Cutaneous T-cell lymphoma	T-cells	IL-2R	Vermeer et al. (2008), Zhang et al. (1996)
Anaplastic large cell lymphoma	T-cells	TEL-JAK2 TEL-ABL NPM1-ALK	Zhang et al. (1996), Zhang et al. (2007)
B-cell lymphoma	B-cells	BCR-ABL V-ABL JAK1S646F JAK1V658F JAK2R683S/G	Kearney et al. (2009), Malin et al. (2010a), Mullighan et al. (2009)
HTLV-I-dependent T-cell neoplasia	T-cells	HTLV infection	Nicot et al. (2001)

Mutations in Jak kinases were identified in multiple hematopoietic neoplasms (Table 1). However, the activation of Stat5 is not limited to JAK kinases localized at the cell membrane, but certain mutated growth factor receptors such as KITD816V and FLT3ITD can activate Stat5 at the endoplasmic reticulum (ER)-Golgi membrane network (Fig. 2) (Choudhary et al. 2009). In a similar manner, the Stat5 pathway has also been successfully hijacked by the human T-cell leukemia virus (HTLV) to induce adult T-cell leukemia/lymphoma (ATL) (Nicot et al. 2001).

Stat5 activation was identified in cutaneous T-cell lymphoma and Sézary syndrome. IL-2 signaling pathway components are amplified in the genome of the malignant cells from these patients (Vermeer *e*t al. 2008). Recently, Stat5 has been reported to be present in the mitochondria of murine lymphoma cell line (LSTRA) and a murine pro-B-cell line (Ba/F3), where it binds the D-loop regulatory region of mitochondrial DNA. The authors hypothesize that this might be the mechanism for a shift in metabolism of these cancer cells, known as Warburg effect (Chueh et al. 2010); but further proof from other types of cancer is needed to draw the conclusion that Stat5 contributes to energy supply of cancer cells. Stat5 can transcribe genes implicated in mitochondrial function, such as c-myc, AKT1, GLUT1, Bcl-2 and Bcl-X_L, which indicates that Stat5 might also regulate the metabolism of cancer cells (Ferbeyre and Moriggl 2011).

Concluding Remarks

Stat5 is an important transcription factor for proper development of a functional hematopoietic system including proper differentiation of hematopoietic lineages. However, the mechanism of the shift from differentiation to increased self renewal of hematopoietic progenitors in context of pYStat5 remains enigmatic. Surprisingly, in leukemic cells pYStat5 is often localized in the cytoplasm, where it enhances the PI_3K-AKT signaling. Intriguingly, the expression of Stat5 mRNA and protein differs in the transformation stages of diseased myeloid cells in CML patients; and higher Stat5 levels contribute to resistance to tyrosine kinase inhibitors (Warsch et al. 2011). Today, pYStat5 is used as a biomarker in clinics for hematopoietic neoplasms and is associated with bad prognosis. However, Stat5 is a rather weak oncogene, and transformation of cells usually requires the activation of other core cancer pathways. Indeed, many of these pathways (such as PI3K-AKT, G1-S cell cycle progression, DNA repair processes and survival pathways) interact with Stat5 signaling (Fig. 2) (Ferbeyre and Moriggl 2011). One of the challenges in this area is the identification of these pathways, within the context of specific disorders. Sequencing of the human cancer genome has shown that about 90% of the mutations occur in tumor suppressor gene pathways. However, the specific tumor suppressor genes that contribute to pYStat5 induced leukemogenesis are unknown and will continue to provide interesting avenues for research.

Acknowledgements This work was supported by the Austrian Science Fund FWF grant SFB F28.

References

Barnstein BO, Li G, Wang Z, Kennedy S, Chalfant C, Nakajima H, Bunting KD, Ryan JJ (2006) Stat5 expression is required for IgE-mediated mast cell function. J Immunol 177:3421–3426

Birkenkamp KU, Geugien M, Lemmink HH, Kruijer W, Vellenga E (2001) Regulation of constitutive STAT5 phosphorylation in acute myeloid leukemia blasts. Leukemia 15:1923–1931

Bradley HL, Hawley TS, Bunting KD (2002) Cell intrinsic defects in cytokine responsiveness of STAT5-deficient hematopoietic stem cells. Blood 100:3983–3989

Buitenhuis M, Baltus B, Lammers JW, Coffer PJ, Koenderman L (2003) Signal transducer and activator of transcription 5a (STAT5a) is required for eosinophil differentiation of human cord blood-derived CD34+ cells. Blood 101:134–142

Bunting KD (2007) STAT5 signaling in normal and pathologic hematopoiesis. Front Biosci 12:2807–2820

Bunting KD, Bradley HL, Hawley TS, Moriggl R, Sorrentino BP, Ihle JN (2002) Reduced lymphomyeloid repopulating activity from adult bone marrow and fetal liver of mice lacking expression of STAT5. Blood 99:479–487

Burchill MA, Goetz CA, Prlic M, O'Neil JJ, Harmon IR, Bensinger SJ, Turka LA, Brennan P, Jameson SC, Farrar MA (2003) Distinct effects of STAT5 activation on CD4+ and CD8+ T cell homeostasis: development of CD4+CD25+ regulatory T cells versus CD8+ memory T cells. J Immunol 171:5853–5864

Choudhary C, Olsen JV, Brandts C, Cox J, Reddy PN, Bohmer FD, Gerke V, Schmidt-Arras DE, Berdel WE, Muller-Tidow C, Mann M, Serve H (2009) Mislocalized activation of oncogenic RTKs switches downstream signaling outcomes. Mol Cell 36:326–339

Chueh FY, Leong KF, Yu CL (2010) Mitochondrial translocation of signal transducer and activator of transcription 5 (STAT5) in leukemic T cells and cytokine stimulated cells. Biochem Biophys Res Commun 402:778–783

Cui Y, Riedlinger G, Miyoshi K, Tang W, Li C, Deng CX, Robinson GW, Hennighausen L (2004) Inactivation of Stat5 in mouse mammary epithelium during pregnancy reveals distinct functions in cell proliferation, survival, and differentiation. Mol Cell Biol 24:8037–8047

Dai X, Chen Y, Di L, Podd A, Li G, Bunting KD, Hennighausen L, Wen R, Wang D (2007) Stat5 is essential for early B cell development but not for B cell maturation and function. J Immunol 179:1068–1079

De Keersmaecker K, Graux C, Odero MD, Mentens N, Somers R, Maertens J, Wlodarska I, Vandenberghe P, Hagemeijer A, Marynen P, Cools J (2005) Fusion of EML1 to ABL1 in T-cell acute lymphoblastic leukemia with cryptic t(9,14) (q34;q32). Blood 105:4849–4852

Dolznig H, Grebien F, Deiner EM, Stangl K, Kolbus A, Habermann B, Kerenyi MA, Kieslinger M, Moriggl R, Beug H, Mullner EW (2006) Erythroid progenitor renewal versus differentiation: genetic evidence for cell autonomous, essential functions of EpoR, Stat5 and the GR. Oncogene 25:2890–2900

Eckelhart E, Warsch W, Zebedin E, Simma O, Stoiber D, Kolbe T, Rulicke T, Mueller M, Casanova E, Sexl V (2011) A novel Ncr1-Cre mouse reveals the essential role of STAT5 for NK cell survival and development. Blood 117:1565–1573

Ermakova O, Piszczek L, Luciani L, Cavalli FM, Ferreira T, Farley D, Rizzo S, Paolicelli RC, Al-Banchaabouchi M, Nerlov C, Moriggl R, Luscombe NM, Gross C (2011) Sensitized phenotypic screening identifies gene dosage sensitive region on chromosome 11 that predisposes to disease in mice. EMBO Mol Med 3:50–66

Ferbeyre G, Moriggl R (2011) The role of Stat5 transcription factors as tumor suppressors or oncogenes. Biochim Biophys Acta 1815:104–114

Fievez L, Desmet C, Henry E, Pajak B, Hegenbarth S, Garze V, Bex F, Jaspar F, Boutet P, Gillet L, Vanderplasschen A, Knolle PA, Leo O, Moser M, Lekeux P, Bureau F (2007) STAT5 is an ambivalent regulator of neutrophil homeostasis. PLoS One 2:e727

Friedbichler K, Kerenyi MA, Kovacic B, Li G, Hoelbl A, Yahiaoui S, Sexl V, Mullner EW, Fajmann S, Cerny-Reiterer S, Valent P, Beug H, Gouilleux F, Bunting KD, Moriggl R (2010) Stat5a serine 725 and 779 phosphorylation is a prerequisite for hematopoietic transformation. Blood 116:1548–1558

Giliani S, Mella P, Savoldi G, Mazzolari E (2005) Cytokine-mediated signalling and early defects in lymphoid development. Curr Opin Allergy Clin Immunol 5:519–524

Goetz CA, Harmon IR, O'Neil JJ, Burchill MA, Farrar MA (2004) STAT5 activation underlies IL7 receptor-dependent B cell development. J Immunol 172:4770–4778

Grebien F, Kerenyi MA, Kovacic B, Kolbe T, Becker V, Dolznig H, Pfeffer K, Klingmuller U, Muller M, Beug H, Mullner EW, Moriggl R (2008) Stat5 activation enables erythropoiesis in the absence of EpoR and Jak2. Blood 111:4511–4522

Guo L, Wei G, Zhu J, Liao W, Leonard WJ, Zhao K, Paul W (2009) IL-1 family members and STAT activators induce cytokine production by Th2, Th17, and Th1 cells. Proc Natl Acad Sci USA 106:13463–13468

Han L, Wierenga AT, Rozenveld-Geugien M, van de Lande K, Vellenga E, Schuringa JJ (2009) Single-cell STAT5 signal transduction profiling in normal and leukemic stem and progenitor cell populations reveals highly distinct cytokine responses. PLoS One 4:e7989

Hand TW, Cui W, Jung YW, Sefik E, Joshi NS, Chandele A, Liu Y, Kaech SM (2010) Differential effects of STAT5 and PI3K/AKT signaling on effector and memory CD8 T-cell survival. Proc Natl Acad Sci USA 107:16601–16606

Heath C, Cross NC (2004) Critical role of STAT5 activation in transformation mediated by ZNF198-FGFR1. J Biol Chem 279:6666–6673

Heuser M, Sly LM, Argiropoulos B, Kuchenbauer F, Lai C, Weng A, Leung M, Lin G, Brookes C, Fung S, Valk PJ, Delwel R, Lowenberg B, Krystal G, Humphries RK (2009) Modeling the functional heterogeneity of leukemia stem cells: role of STAT5 in leukemia stem cell self-renewal. Blood 114:3983–3993

Hirokawa S, Sato H, Kato I, Kudo A (2003) EBF-regulating Pax5 transcription is enhanced by STAT5 in the early stage of B cells. Eur J Immunol 33:1824–1829

Hoelbl A, Kovacic B, Kerenyi MA, Simma O, Warsch W, Cui Y, Beug H, Hennighausen L, Moriggl R, Sexl V (2006) Clarifying the role of Stat5 in lymphoid development and Abelson-induced transformation. Blood 107:4898–4906

Hwang ES, White IA, Ho IC (2002) An IL-4-independent and CD25-mediated function of c-maf in promoting the production of Th2 cytokines. Proc Natl Acad Sci USA 99:13026–13030

Imada K, Bloom ET, Nakajima H, Horvath-Arcidiacono JA, Udy GB, Davey HW, Leonard WJ (1998) Stat5b is essential for natural killer cell-mediated proliferation and cytolytic activity. J Exp Med 188:2067–2074

John S, Robbins CM, Leonard WJ (1996) An IL-2 response element in the human IL-2 receptor alpha chain promoter is a composite element that binds Stat5, Elf-1, HMG-I(Y) and a GATA family protein. EMBO J 15:5627–5635

Kagami S, Nakajima H, Suto A, Hirose K, Suzuki K, Morita S, Kato I, Saito Y, Kitamura T, Iwamoto I (2001) Stat5a regulates T helper cell differentiation by several distinct mechanisms. Blood 97:2358–2365

Kang J, Der SD (2004) Cytokine functions in the formative stages of a lymphocyte's life. Curr Opin Immunol 16:180–190

Kearney L, Gonzalez De Castro D, Yeung J, Procter J, Horsley SW, Eguchi-Ishimae M, Bateman CM, Anderson K, Chaplin T, Young BD, Harrison CJ, Kempski H, So CW, Ford AM, Greaves M (2009) Specific JAK2 mutation (JAK2R683) and multiple gene deletions in Down syndrome acute lymphoblastic leukemia. Blood 113:646–648

Kelly J, Spolski R, Imada K, Bollenbacher J, Lee S, Leonard WJ (2003) A role for Stat5 in CD8+ T cell homeostasis. J Immunol 170:210–217

Kerenyi MA, Grebien F, Gehart H, Schifrer M, Artaker M, Kovacic B, Beug H, Moriggl R, Mullner EW (2008) Stat5 regulates cellular iron uptake of erythroid cells via IRP-2 and TfR-1. Blood 112:3878–3888

Kimura A, Rieger MA, Simone JM, Chen W, Wickre MC, Zhu BM, Hoppe PS, O'Shea JJ, Schroeder T, Hennighausen L (2009) The transcription factors STAT5A/B regulate GM-CSF-mediated granulopoiesis. Blood 114:4721–4728

Korn T, Bettelli E, Oukka M, Kuchroo VK (2009) IL-17 and Th17 cells. Annu Rev Immunol 27:485–517

Lane SW, Fairbairn DJ, McCarthy C, Nandini A, Perry-Keene J, Kennedy GA (2008) Leukaemia cutis in atypical chronic myeloid leukaemia with a t(9;22) (p24;q11.2) leading to BCR-JAK2 fusion. Br J Haematol 142:503

Laurence A, Tato CM, Davidson TS, Kanno Y, Chen Z, Yao Z, Blank RB, Meylan F, Siegel R, Hennighausen L, Shevach EM, O'Shea JJ (2007) Interleukin-2 signaling via STAT5 constrains T helper 17 cell generation. Immunity 26:371–381

Levin SD, Koelling RM, Friend SL, Isaksen DE, Ziegler SF, Perlmutter RM, Farr AG (1999) Thymic stromal lymphopoietin: a cytokine that promotes the development of IgM+B cells in vitro and signals via a novel mechanism. J Immunol 162:677–683

Li G, Wang Z, Zhang Y, Kang Z, Haviernikova E, Cui Y, Hennighausen L, Moriggl R, Wang D, Tse W, Bunting KD (2007) STAT5 requires the N-domain to maintain hematopoietic stem cell repopulating function and appropriate lymphoid-myeloid lineage output. Exp Hematol 35:1684–1694

Li G, Miskimen KL, Wang Z, Xie XY, Tse W, Gouilleux F, Moriggl R, Bunting KD (2010) Effective targeting of STAT5-mediated survival in myeloproliferative neoplasms using ABT-737 combined with rapamycin. Leukemia 24:1397–1405

Liao W, Schones DE, Oh J, Cui Y, Cui K, Roh TY, Zhao K, Leonard WJ (2008) Priming for T helper type 2 differentiation by interleukin 2-mediated induction of interleukin 4 receptor alpha-chain expression. Nat Immunol 9:1288–1296

Liu F, Kunter G, Krem MM, Eades WC, Cain JA, Tomasson MH, Hennighausen L, Link DC (2008) Csf3r mutations in mice confer a strong clonal HSC advantage via activation of Stat5. J Clin Invest 118:946–955

Malin S, McManus S, Busslinger M (2010a) STAT5 in B cell development and leukemia. Curr Opin Immunol 22:168–176

Malin S, McManus S, Cobaleda C, Novatchkova M, Delogu A, Bouillet P, Strasser A, Busslinger M (2010b) Role of STAT5 in controlling cell survival and immunoglobulin gene recombination during pro-B cell development. Nat Immunol 11:171–179

Moriggl R, Topham DJ, Teglund S, Sexl V, McKay C, Wang D, Hoffmeyer A, van Deursen J, Sangster MY, Bunting KD, Grosveld GC, Ihle JN (1999) Stat5 is required for IL-2-induced cell cycle progression of peripheral T cells. Immunity 10:249–259

Moriggl R, Sexl V, Kenner L, Duntsch C, Stangl K, Gingras S, Hoffmeyer A, Bauer A, Piekorz R, Wang D, Bunting KD, Wagner EF, Sonneck K, Valent P, Ihle JN, Beug H (2005) Stat5 tetramer formation is associated with leukemogenesis. Cancer Cell 7:87–99

Mullighan CG, Zhang J, Harvey RC, Collins-Underwood JR, Schulman BA, Phillips LA, Tasian SK, Loh ML, Su X, Liu W, Devidas M, Atlas SR, Chen IM, Clifford RJ, Gerhard DS, Carroll WL, Reaman GH, Smith M, Downing JR, Hunger SP, Willman CL (2009) JAK mutations in high-risk childhood acute lymphoblastic leukemia. Proc Natl Acad Sci USA 106:9414–9418

Nicot C, Mulloy JC, Ferrari MG, Johnson JM, Fu K, Fukumoto R, Trovato R, Fullen J, Leonard WJ, Franchini G (2001) HTLV-1 p12(I) protein enhances STAT5 activation and decreases the interleukin-2 requirement for proliferation of primary human peripheral blood mononuclear cells. Blood 98:823–829

Oboki K, Ohno T, Kajiwara N, Saito H, Nakae S (2010) IL-33 and IL-33 receptors in host defense and diseases. Allergol Int 59:143–160

Osborne LC, Abraham N (2010) Regulation of memory T cells by gammac cytokines. Cytokine 50:105–113

Park JH, Adoro S, Guinter T, Erman B, Alag AS, Catalfamo M, Kimura MY, Cui YZ, Lucas PJ, Gress RE, Kubo M, Hennighausen L, Feigenbaum L, Singer A (2010) Signaling by intrathymic cytokines, not T cell antigen receptors, specifies CD8 lineage choice and promotes the differentiation of cytotoxic-lineage T cells. Nat Immunol 11:257, U210

Pugliese-Pires PN, Tonelli CA, Dora JM, Silva PC, Czepielewski M, Simoni G, Arnhold IJ, Jorge AA (2010) A novel STAT5B mutation causing GH insensitivity syndrome associated with hyperprolactinemia and immune dysfunction in two male siblings. Eur J Endocrinol 163:349–355

Rochman Y, Kashyap M, Robinson GW, Sakamoto K, Gomez-Rodriguez J, Wagner KU, Leonard WJ (2010) Thymic stromal lymphopoietin-mediated STAT5 phosphorylation via kinases JAK1 and JAK2 reveals a key difference from IL-7-induced signaling. Proc Natl Acad Sci USA 107:19455–19460

Scheeren FA, Naspetti M, Diehl S, Schotte R, Nagasawa M, Wijnands E, Gimeno R, Vyth-Dreese FA, Blom B, Spits H (2005) STAT5 regulates the self-renewal capacity and differentiation of human memory B cells and controls Bcl-6 expression. Nat Immunol 6:303–313

Schepers H, van Gosliga D, Wierenga AT, Eggen BJ, Schuringa JJ, Vellenga E (2007) STAT5 is required for long-term maintenance of normal and leukemic human stem/progenitor cells. Blood 110:2880–2888

Scherr M, Chaturvedi A, Battmer K, Dallmann I, Schultheis B, Ganser A, Eder M (2006) Enhanced sensitivity to inhibition of SHP2, STAT5, and Gab2 expression in chronic myeloid leukemia (CML). Blood 107:3279–3287

Scott LM, Tong W, Levine RL, Scott MA, Beer PA, Stratton MR, Futreal PA, Erber WN, McMullin MF, Harrison CN, Warren AJ, Gilliland DG, Lodish HF, Green AR (2007) JAK2 exon 12 mutations in polycythemia vera and idiopathic erythrocytosis. N Engl J Med 356:459–468

Sexl V, Piekorz R, Moriggl R, Rohrer J, Brown MP, Bunting KD, Rothammer K, Roussel MF, Ihle JN (2000) Stat5a/b contribute to interleukin 7-induced B-cell precursor expansion, but abl- and bcr/abl-induced transformation are independent of stat5. Blood 96:2277–2283

Shelburne CP, McCoy ME, Piekorz R, Sexl V, Roh KH, Jacobs-Helber SM, Gillespie SR, Bailey DP, Mirmonsef P, Mann MN, Kashyap M, Wright HV, Chong HJ, Bouton LA, Barnstein B, Ramirez CD, Bunting KD, Sawyer S, Lantz CS, Ryan JJ (2003) Stat5 expression is critical for mast cell development and survival. Blood 102:1290–1297

Shide K, Shimoda HK, Kumano T, Karube K, Kameda T, Takenaka K, Oku S, Abe H, Katayose KS, Kubuki Y, Kusumoto K, Hasuike S, Tahara Y, Nagata K, Matsuda T, Ohshima K, Harada M, Shimoda K (2008) Development of ET, primary myelofibrosis and PV in mice expressing JAK2 V617F. Leukemia 22:87–95

Slupianek A, Hoser G, Majsterek I, Bronisz A, Malecki M, Blasiak J, Fishel R, Skorski T (2002) Fusion tyrosine kinases induce drug resistance by stimulation of homology-dependent recombination repair, prolongation of G(2)/M phase, and protection from apoptosis. Mol Cell Biol 22:4189–4201

Teglund S, McKay C, Schuetz E, van Deursen JM, Stravopodis D, Wang D, Brown M, Bodner S, Grosveld G, Ihle JN (1998) Stat5a and Stat5b proteins have essential and nonessential, or redundant, roles in cytokine responses. Cell 93:841–850

Tomasson MH, Sternberg DW, Williams IR, Carroll M, Cain D, Aster JC, Ilaria RL Jr, Van Etten RA, Gilliland DG (2000) Fatal myeloproliferation, induced in mice by TEL/PDGFbetaR expression, depends on PDGFbetaR tyrosines 579/581. J Clin Invest 105:423–432

Valent P, Akin C, Sperr WR, Horny HP, Metcalfe DD (2003) Mast cell proliferative disorders: current view on variants recognized by the World Health Organization. Hematol Oncol Clin North Am 17:1227–1241

Vermeer MH, van Doorn R, Dijkman R, Mao X, Whittaker S, van Voorst Vader PC, Gerritsen MJ, Geerts ML, Gellrich S, Soderberg O, Leuchowius KJ, Landegren U, Out-Luiting JJ, Knijnenburg J, Ijszenga M, Szuhai K, Willemze R, Tensen CP (2008) Novel and highly recurrent chromosomal alterations in Sezary syndrome. Cancer Res 68:2689–2698

Vranjkovic A, Crawley AM, Patey A, Angel JB (2010) IL-7-dependent STAT-5 activation and CD8+ T cell proliferation are impaired in HIV infection. J Leukoc Biol 89:499–506

Walters DK, Goss VL, Stoffregen EP, Gu TL, Lee K, Nardone J, McGreevey L, Heinrich MC, Deininger MW, Polakiewicz R, Druker BJ (2006a) Phosphoproteomic analysis of AML cell lines identifies leukemic oncogenes. Leuk Res 30:1097–1104

Walters DK, Mercher T, Gu TL, O'Hare T, Tyner JW, Loriaux M, Goss VL, Lee KA, Eide CA, Wong MJ, Stoffregen EP, McGreevey L, Nardone J, Moore SA, Crispino J, Boggon TJ, Heinrich MC, Deininger MW, Polakiewicz RD, Gilliland DG, Druker BJ (2006b) Activating alleles of JAK3 in acute megakaryoblastic leukemia. Cancer Cell 10:65–75

Warsch W, Kollmann K, Eckelhart E, Fajmann S, Cerny-Reiterer S, Holbl A, Gleixner KV, Dworzak M, Mayerhofer M, Hoermann G, Herrmann H, Sillaber C, Egger G, Valent P, Moriggl R, Sexl V (2011) High STAT5 levels mediate imatinib resistance and indicate disease progression in chronic myeloid leukemia. Blood 117:3409–3420

Wohlmann A, Sebastian K, Borowski A, Krause S, Friedrich K (2010) Signal transduction by the atopy-associated human thymic stromal lymphopoietin (TSLP) receptor depends on Janus kinase function. Biol Chem 391:181–186

Yang XP, Ghoreschi K, Steward-Tharp SM, Rodriguez-Canales J, Zhu J, Grainger JR, Hirahara K, Sun HW, Wei L, Vahedi G, Kanno Y, O'Shea JJ, Laurence A (2011) Opposing regulation of the locus encoding IL-17 through direct, reciprocal actions of STAT3 and STAT5. Nat Immunol 12:247–254

Yao Z, Cui Y, Watford WT, Bream JH, Yamaoka K, Hissong BD, Li D, Durum SK, Jiang Q, Bhandoola A, Hennighausen L, O'Shea JJ (2006) Stat5a/b are essential for normal lymphoid development and differentiation. Proc Natl Acad Sci USA 103:1000–1005

Yao Z, Kanno Y, Kerenyi M, Stephens G, Durant L, Watford WT, Laurence A, Robinson GW, Shevach EM, Moriggl R, Hennighausen L, Wu C, O'Shea JJ (2007) Nonredundant roles for Stat5a/b in directly regulating Foxp3. Blood 109:4368–4375

Yoshimoto G, Miyamoto T, Jabbarzadeh-Tabrizi S, Iino T, Rocnik JL, Kikushige Y, Mori Y, Shima T, Iwasaki H, Takenaka K, Nagafuji K, Mizuno S, Niiro H, Gilliland GD, Akashi K (2009) FLT3-ITD up-regulates MCL-1 to promote survival of stem cells in acute myeloid leukemia via FLT3-ITD-specific STAT5 activation. Blood 114:5034–5043

Yu CR, Ortaldo JR, Curiel RE, Young HA, Anderson SK, Gosselin P (1999) Role of a STAT binding site in the regulation of the human perforin promoter. J Immunol 162:2785–2790

Zhang Q, Nowak I, Vonderheid EC, Rook AH, Kadin ME, Nowell PC, Shaw LM, Wasik MA (1996) Activation of Jak/STAT proteins involved in signal transduction pathway mediated by receptor for interleukin 2 in malignant T lymphocytes derived from cutaneous anaplastic large T-cell lymphoma and Sezary syndrome. Proc Natl Acad Sci USA 93:9148–9153

Zhang Q, Wang HY, Liu X, Wasik MA (2007) STAT5A is epigenetically silenced by the tyrosine kinase NPM1-ALK and acts as a tumor suppressor by reciprocally inhibiting NPM1-ALK expression. Nat Med 13:1341–1348

Zhu J (2010) Transcriptional regulation of Th2 cell differentiation. Immunol Cell Biol 88:244–249

Zhu J, Paul WE (2008) CD4 T cells: fates, functions, and faults. Blood 112:1557–1569

Zhu Y, Chen L, Huang Z, Alkan S, Bunting KD, Wen R, Wang D, Huang H (2004) Cutting edge: IL-5 primes Th2 cytokine-producing capacity in eosinophils through a STAT5-dependent mechanism. J Immunol 173:2918–2922

Zwirner NW, Domaica CI (2010) Cytokine regulation of natural killer cell effector functions. Biofactors 36:274–288

Regulation of Dendritic Cell Development by STATs

Haiyan S. Li and Stephanie S. Watowich

Abstract
Dendritic cells (DCs) serve as a critical link between the innate and adaptive immune responses due to their ability to sense pathogens and respond by activating adaptive immune cell types. Delineating the molecular control of DC development will provide important information about the generation of natural immunity as well as approaches to regulate DCs in clinical settings. DCs are generated from hematopoietic stem cells through specialized progenitor subsets in response to cytokine and transcriptional cues, with FMS-like tyrosine kinase 3 ligand (Flt3L) and Flt3L receptor (Flt3) signaling providing a major pathway supporting homeostatic DC generation. Recent work has indicated that granulocyte-macrophage colony-stimulating factor (GM-CSF) and type I interferons (IFNs) also play important roles in regulating DC subset production. Here we review new insight into the mechanisms by which cytokine-activated STAT proteins control the DC developmental process.

Introduction

Dendritic cells (DCs) are known as 'professional' antigen-presenting cells, unique in their capacity to maintain self-tolerance and initiate primary T- and B-cell responses. Diverse DC subsets have been reported based on phenotypic markers,

H.S. Li
Department of Immunology, The University of Texas M.D. Anderson Cancer Center, 301402, Unit 902, Houston, TX 77030-1402, USA

S.S. Watowich (✉)
Department of Immunology, The University of Texas M.D. Anderson Cancer Center, 301402, Unit 902, Houston, TX 77030-1402, USA

The University of Texas Graduate School of Biomedical Sciences, Houston, TX, USA
e-mail: swatowic@mdanderson.org

Th. Decker and M. Müller (eds.), *Jak-Stat Signaling: From Basics to Disease*,
DOI 10.1007/978-3-7091-0891-8_11, © Springer-Verlag Wien 2012

anatomical location and functionality, with resident, migratory and plasmacytoid DCs (pDC) as the major categories (reviewed in Watowich and Liu 2010; Merad and Manz 2009). Resident and migratory DCs can be further divided into distinct classes; for example, in mouse spleen resident CD11c^{+} CD11b^{+} CD4^{+} CD8α^{-} and CD11c^{+} CD11b^{-} CD4^{-} CD8α^{+} DC subsets exist, generally referred as CD8α^{-} and CD8α^{+} DCs, respectively. Resident DCs are located primarily in lymphoid tissue and have also been termed conventional DCs (cDCs) while migratory subsets, which travel to lymphoid organs from peripheral sites upon activation, are additionally referred to as tissue DCs. By contrast, pDCs are found mainly in bone marrow, blood, spleen and lymph node, are defined as CD11c^{lo} CD11b^{-} B220^{+} PDCA-1^{+} SiglecH^{+} plasma cell-like DCs, and are distinguished by their ability to produce abundant quantities of type I interferons (IFNs) upon viral or Toll-like receptor stimulation. In this chapter, we use the term "DC" to refer to the collective group of resident and migratory DCs, while excluding the pDC lineage, which we classify separately. Moreover, in cases where clear indications exist in the literature, we describe specific DC subsets by category such as cDC (referring to resident DCs) or tissue DC (referring to migratory DCs). It should be noted that delineation of DC subsets and molecular cues regulating their development is an intense area of investigation with continuous information emerging.

The origin of cDCs and pDCs is restricted to the lineage-negative (lin^{-}), Flt3^{+} compartment in bone marrow (D'Amico and Wu 2003), whereas migratory DCs appear to descend from hematopoietic progenitors or monocytes (reviewed in Merad and Manz 2009). Within the lin^{-} Flt3^{+} progenitor subset are found macrophage-DC precursors (MDPs, lin^{-} c-kit^{hi} CD115^{+} CX$_3$CR1^{+} Flt3^{+}), which give rise to cDCs, pDCs, monocytes and macrophages upon adoptive transfer (Fogg et al. 2006; Liu et al. 2009), and a more developmentally confined progenitor, the common DC progenitor (CDP, lin^{-} c-kit^{lo} CD115^{+} Flt3^{+}), which is able to generate cDCs and pDCs nearly exclusively (Naik et al. 2007; Onai et al. 2007; Liu et al. 2009). Unlike pDCs that are fully developed in the bone marrow, cDC precursors (pre-cDCs, lin^{-} CD11c^{+} MHC class II^{-} SIRP-α^{int} Flt3^{+}) are believed to exit bone marrow and migrate to lymphoid tissues for subsequent division and terminal differentiation (Liu et al. 2009).

As with other hematopoietic lineages, the molecular regulation of DC subset specification and differentiation from hematopoietic stem cells (HSCs) is complicated and involves cellular events driven by cytokines and lineage-restricted transcription factors. Herein we focus on factors regulating cDC and pDC development, as these subsets appear to share a common progenitor and in some cases similar developmental mechanisms (Naik et al. 2007; Onai et al. 2007; Liu et al. 2009) (reviewed in Watowich and Liu 2010; Merad and Manz 2009). For instance, engagement of Flt3 with its ligand Flt3L activates the most critical signaling cascade for pDC and cDC generation in vivo, as implicated by the phenotypes of Flt3 and Flt3L gene knockout and transgenic animals (McKenna et al. 2000; Manfra et al. 2003; Waskow et al. 2008; Kingston et al. 2009). The importance of Flt3L in DC development is further demonstrated by expansion of pDC and DC populations following its administration in vivo and by its ability to induce pDC and DC production ex vivo (Maraskovsky et al. 1996; Daro et al. 2000; O'Keeffe et al. 2002; Gilliet et al. 2002). By contrast, granulocyte/macrophage colony stimulating factor (GM-CSF) is commonly used in cell culture to generate DCs from bone marrow precursors and blood monocytes (Inaba et al. 1992), but its role in homeostatic cDC production in vivo appears to be largely dispensable or redundant with other factors, as genetic deficiency of GM-CSF or its receptor leads to only minimal reduction of cDCs (Vremec et al. 1997; Kingston et al. 2009). Moreover, GM-CSF favors DC generation yet inhibits the development of pDCs (Gilliet et al. 2002; Esashi et al. 2008). Interferon-α, a type I IFN secreted by pDCs and other blood cells upon viral infection, promotes the production of CDPs and pDCs in vivo while abrogating cDC generation (Li et al. 2011), similar to the inhibitory action of IFN-β upon cDCs (Hahm et al. 2005). Thus, Flt3L, GM-CSF and IFN-α/β have emerged as important DC regulators, albeit with distinct activities. These cytokines activate multiple signal transduction cascades, most prominently the STAT pathways. In this chapter, we focus on the mechanisms of DC development by STAT proteins.

Flt3L-STAT3

Flt3L promotes the proliferation, differentiation and survival of many hematopoietic lineages, in addition to serving as the most critical growth factor for pDCs and cDCs in vivo and in vitro (Stirewalt and Radich 2003; Watowich and Liu 2010). Flt3L is produced by several tissues, including lymphoid organs (bone marrow, thymus and spleen), prostate, kidney, small intestine and placenta (Stirewalt and Radich 2003). The most abundant isoform of Flt3L in humans is the full-length transmembrane protein, which can be cleaved to generate a soluble isoform containing only the extracellular domain (Hannum et al. 1994; Lyman et al. 1994). Structural analysis of soluble human Flt3L revealed that it forms a noncovalently linked homodimer with each monomer demonstrating an α-helical bundle configuration (Savvides et al. 2000). In mouse, Flt3L is present mainly as a 220 a.a. membrane-bound isoform (Lyman et al. 1995a). A soluble variant of Flt3L is also produced in humans and mice by virtue of a stop codon introduced near the end of exon 6 (Lyman and Jacobsen 1998). It is unclear whether the multiple Flt3L isoforms are functionally distinct, however all are biologically active and show no restriction in species specificity.

Flt3L signals through its receptor Flt3, also known as fetal liver kinase-2 (flk-2), stem cell kinase 1 (STK-1), or CD135. Flt3 belongs to the type III tyrosine kinase receptor family, presenting as a monomeric, membrane-bound receptor with two intracellular kinase moieties linked by a kinase-insert region (Stirewalt and Radich 2003) (Fig. 1). Flt3 is expressed primarily on early hematopoietic progenitor cells with myeloid or lymphoid potential (reviewed in Lyman and Jacobsen 1998); certain non-hematopoietic organs including placenta, gonads and brain have also been reported to express *Flt3* mRNA (Stirewalt and Radich 2003). The relatively restricted expression pattern of Flt3 within the hematopoietic system confines the

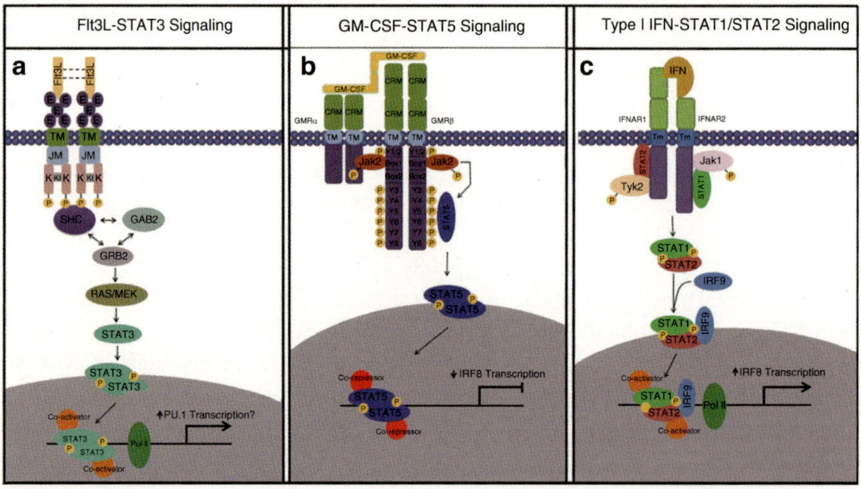

Fig. 1 Model for cytokine-STAT signaling cascades regulating DC development. Cytokine-activated STATs have been found to regulate genes encoding critical pDC/DC-related transcription factors, signaling pathways that may influence pDC and DC development. A model of proposed signaling mechanisms is shown. STATs may also control the expression of genes controlling pDC/DC progenitor proliferation and survival as well as mature pDC and DC functions (not shown). (**a**) Flt3-STAT3. Upon Flt3L binding, Flt3 dimerizes and undergoes autophosphorylation, subsequently activating effector molecules including SHC, GAB2, GRB2 and RAS/MEK, which lead to STAT3 tyrosine phosphorylation directly or indirectly. Activated STAT3 dimerizes and translocates to the nucleus, inducing signals that mediate pDC and DC generation, potentially including direct induction of *Sfpi1* expression. Flt3 comprises five extracellular immunoglobulin-like domains (E), a transmembrane domain (TM), a juxtamembrane domain (JM) and two tyrosine-kinase domains (K) linked via a kinase insert region (KI). (**b**) GM-CSF-STAT5. GM-CSF stimulates the formation of hexameric (2GMRα:2βc:2GM-CSF) or dodecameric (2x hexameric) ligand:receptor complexes, which activate JAK2 and, subsequently, STAT5. STAT5 dimers translocate to the nucleus where they suppress *Irf8* gene expression, inhibiting pDC development. GMRα contains a CRM domain, a TM domain and a cytoplasmic domain while βc consists of two CRMs, a TM domain and a cytoplasmic domain that associates with Jak2. (**c**) Type I IFN-STAT1/2. Type I IFN stimulation activates Tyk2 and Jak1, which are associated with IFNAR1 and IFNAR2, respectively. Jak activation leads to phosphorylation of STAT2 and STAT1, which may induce STAT1 homodimerization as well as STAT2 and STAT1 interaction with IRF9 to form STAT1:STAT2:IRF9 complexes (known as ISGF3). ISGF3 and/or STAT1 homodimers (not shown) accumulate in the nucleus and induce *Irf8* expression, potentially enhancing pathways that stimulate/support pDC differentiation. The signaling network involved in type I IFN-mediated DC repression is unclear

action of Flt3L, which is essential for normal hematopoiesis and immune system function. Low amounts of circulating Flt3L (Lyman et al. 1995b) and the proximity of Flt3L-producing cells suggest that paracrine stimulation of Flt3^{+} hematopoietic progenitors may occur in the bone marrow microenvironment; autocrine mechanisms have also been proposed, however this remains to be established (Stirewalt and Radich 2003).

Flt3L binding leads to dimerization of Flt3 and concomitant juxtaposition of the cytoplasmic tyrosine kinase domains, inducing rapid autophosphorylation (within minutes) of the receptor (Turner et al. 1996). The activated Flt3L-Flt3 complex, which is rapidly internalized (Turner et al. 1996), stimulates multiple signaling cascades that seem to be tissue and cell type-specific (Dosil et al. 1993; Zhang et al. 1999). In the Flt3$^+$ Ba/F3 cell line, activated Flt3 stimulates tyrosine phosphorylation of SH2-containing sequence protein (SHC), the inositol phosphatase SHIP, the protein phosphatase SHP2 and GRB2-binding protein (GAB2), which subsequently initiates multiple signal transduction cascades including the phosphatidylinositol 3-kinase (PI3K)/Akt and RAS/mitogen-activated protein kinase (MAPK) pathways (Marchetto et al. 1999; Zhang and Broxmeyer 2000). STATs are thought to be downstream intermediates following activation of RAS, RAF or MAPK/ERK kinases (MEK), without involvement of JAK activity (Zhang et al. 2000) (Fig. 1). By contrast, in circumstances in which Flt3 is overexpressed or constitutively activated, STAT5 is stimulated through interaction with phosphorylated tyrosines 589 and 591 within the Flt3 juxtamembrane region (Spiekermann et al. 2003; Rocnik et al. 2006). Phosphorylation of these tyrosine residues, however, does not occur upon physiological Flt3L stimulation (Rocnik et al. 2006), and STATs may be activated directly via recruitment to the receptor or indirectly through Src kinases (Heiss et al. 2006). In lin$^-$ Flt3$^+$ pDC/DC progenitors, STAT3 is the major STAT activated, becoming tyrosine phosphorylated within minutes of Flt3L addition (Esashi et al. 2008), suggesting direct stimulation via Flt3 (Fig. 1).

Flt3L was first described as an effective DC growth factor in mice receiving daily injection of the recombinant protein, a regimen that stimulates a significant increase in pDCs and DC subsets in bone marrow, blood, lymphoid organs and other tissues (e.g. thymus, spleen, gastro-intestinal lymphoid tissue and liver) (Maraskovsky et al. 1996). Subsequently, Flt3L was found to significantly boost DC generation from human blood ex vivo in the presence of GM-CSF and IL-4 (Hubert et al. 1998), or to promote pDC and DC production in human and mouse bone marrow cultures (Blom et al. 2000; Gilliet et al. 2002). Inducible expression of Flt3L in vivo also results in massive expansion of pDCs and DCs in multiple organs (Manfra et al. 2003). DCs that develop in response to Flt3L in vivo and ex vivo appear to correspond to pDC and DC populations that are generated in homeostatic conditions in vivo (Naik 2008). Importantly, Flt3L treatment also stimulates the development and/or accumulation of MDPs and CDPs from HSCs (Waskow et al. 2008) (Li and Watowich, unpublished data), highlighting its critical function throughout the DC developmental process. Correspondingly, ablation of Flt3L-Flt3 signaling by genetic disruption (McKenna et al. 2000; Waskow et al. 2008; Kingston et al. 2009) or treatment with Flt3 inhibitors (Tussiwand et al. 2005) leads to severe reduction (>90%) in pDC and DC amounts, and a mild decrease (30–40%) of bone marrow MDPs (Kingston et al. 2009). By contrast, enforced expression of Flt3 in Flt3$^-$ hematopoietic progenitors, which normally lack DC potential, enables their development into pDCs and DCs (Onai et al. 2006), suggesting Flt3 signaling may have instructive function in DC lineage development.

In a first attempt to explore the molecular mechanisms involved in Flt3L-induced DC development, Laouar et al. reported that conditional deletion of STAT3 in the bone marrow caused a profound loss of splenic $CD11c^+$ cells in C57BL/6 mice and impaired the generation of both pDC and DC subsets from Flt3L-supplemented bone marrow cultures ex vivo (Laouar et al. 2003). In addition, hematopoietic STAT3 deficiency abrogated Flt3L-stimulated induction of $CD11c^+$ DCs as well as common lymphoid progenitors (CLPs), which can serve as DC precursors, yet had no effect on HSCs or common myeloid progenitors (CMPs) (Laouar et al. 2003). These data suggested a role for STAT3 in homeostatic maintenance/development of DCs as well as Flt3L-driven generation of DCs, pDCs and their progenitors. Subsequently, our laboratory demonstrated that STAT3 is essential for the proliferation of lin^- $Flt3^+$ bone marrow progenitors in response to Flt3L (Esashi et al. 2008), consistent with its critical role in Flt3L-dependent generation of pDCs and DCs in vitro (Laouar et al. 2003; Esashi et al. 2008). Moreover, we found that STAT3-deficient pDCs retain their ability to produce type I IFNs upon TLR9 stimulation (Esashi et al. 2008). Collectively, these results suggest that STAT3 mediates growth-promoting signals in pDC/DC progenitors elicited by Flt3L stimulation, yet is dispensable for terminal pDC differentiation. Under steady state conditions, however, pDC/DC progenitors are found at similar amounts in hematopoietic STAT3-deficient mice relative to littermate controls (Li et al. 2011), indicating STAT3 is not required for homeostatic maintenance of the progenitor compartment. Furthermore, recent data indicates that STAT3 deletion in the hematopoietic system (Li et al. 2011) or in $CD11c^+$ cells (Melillo et al. 2010) leads to a significant reduction in pDCs in bone marrow and spleen while $CD11c^+$ $CD11b^+$ cDCs are present at near normal numbers and proportions; these results agree with the concept that STAT3 is necessary for proliferation of pDC progenitors (Esashi et al. 2008) yet indicate that cDCs or their precursors may be regulated by STAT3-independent mechanisms under homeostatic conditions in vivo. Flt3 controls cDC proliferation/survival in spleen in steady state (Waskow et al. 2008), thus the fact that $CD11c^+$ $CD11b^+$ cDCs remain at normal amounts in STAT3-deficient mice suggests that Flt3-driven cDC proliferation in the periphery employs signal cascades independent of STAT3. The reason for the discrepancy in the role for STAT3 in DC development between different reports (Laouar et al. 2003; Melillo et al. 2010) remains unclear, however the recent work underscores the critical role for STAT3 in pDC homeostasis in vivo (Melillo et al. 2010) (Li et al. 2011). Significantly, ectopic expression of STAT3 in $Flt3^-$ progenitors not only initiates pDC development from this subset, which is normally unable to generate pDCs, but also upregulates Flt3 expression (Onai et al. 2006). These results suggest that STAT3 may mediate instructive signals for pDC lineage development and potentially reinforce this signaling network via Flt3 induction.

Flt3L stimulates the expression of several DC-related transcription factors, including *Sfpi1* (encoding PU.1), *Ifr8* and *Spib* (Esashi et al. 2008). An important question centers on whether STATs serve a role in activating and/or maintaining the DC lineage-specific transcriptional network. STAT3 has been shown to regulate

Sfpi1 expression by binding the distal promoter region in primary erythroid progenitors or myeloid tissue culture cells (Hegde et al. 2009; Yoon and Watowich, unpublished data). These results, together with the observation that *Sfpi1* mRNA is induced by STAT3 overexpression in $Flt3^{-}$ bone marrow progenitors (Onai et al. 2006), suggest that *Sfpi1* may be a direct target of the Flt3-STAT3 signaling cascade in pDC/DC progenitors/precursors (Fig. 1). Recently, PU.1 was shown to induce Flt3 expression in pDC/DC progenitors in a dose-dependent manner (Carotta et al. 2010), suggesting the presence of a self-regulatory loop of Flt3 transcription mediated by STAT3 and PU.1 following Flt3 signaling.

GM-CSF-STAT5

GM-CSF is the most commonly used growth factor in the laboratory and clinic to generate DCs from peripheral blood and/or bone marrow. GM-CSF is normally undetectable in circulation, but is readily induced in many cell types in response to multiple stimuli or disease conditions (Hamilton and Anderson 2004). As revealed by X-ray crystallography, human GM-CSF is characterized by a highly compact four α-helix bundle structure containing a hydrophobic core (Rozwarski et al. 1996). Human and mouse GM-CSF display a high degree of homology in sequence and predicted structure, yet there is little species cross-reactivity due to the fact that distinct residues are required for receptor binding (Shanafelt et al. 1991).

GM-CSF receptor (GMR) is widely expressed on hematopoietic cells at low density (100–1,000 receptors/cell) (Guthridge et al. 1998). The heteromeric receptor is composed of a cytokine-specific α chain (GMRα) and a common β (βc) chain that is shared with receptors for IL-3 and IL-5 (Gearing et al. 1989; Hayashida et al. 1990; Kitamura et al. 1991). Both GMRα and βc chains belong to the class I cytokine receptor superfamily, and are characterized by the presence of the cytokine-receptor homology module (CRM) in the extracellular domain, which contains four conserved cysteine residues, a Trp-Ser-X-Trp-Ser motif (WSXWS motif) and tandem fibronectin type III domains (Bazan 1990). An "elbow" region formed by a fold between the fibronectin type III domains in GMRα serves as the critical ligand binding interface (Hansen et al. 2008), similar to growth hormone binding to the growth hormone receptor extracellular region (de Vos et al. 1992). The GMRα consists of one CRM and binds to GM-CSF at low affinity (Gearing et al. 1989). By contrast, the βc chain contains 2 CRMs and was reported to present as a membrane-bound homodimer in the absence of ligand stimulation (Fig. 1) (Hayashida et al. 1990; Murphy and Young 2006). While βc is the principal signaling component in cells expressing GM-CSF, IL-3 or IL-5 receptors, and is required for high affinity ligand binding, it is clear that different α subunits are required for cytokine-specific signal transduction (Hayashida et al. 1990; Kitamura et al. 1991; Tavernier et al. 1991; Geijsen et al. 2001). Recently, a hexameric complex consisting of 2 βc: 2 GMRα: 2 GM-CSF was identified by crystal structure analysis; this structure may further assemble into a higher order dodecamer (12-mer) (Hansen et al. 2008). The dodecameric ligand-receptor complex provides a structural basis for clustering the signaling βc subunits, and enables association of GMRα cytoplasmic domains, which may also participate in signal transduction (Lia et al. 1996).

The βc chain constitutively associates with the tyrosine kinase Jak2, which is critical for GMR signal transduction (Parganas et al. 1998), via a conserved Box 1 motif in the membrane-proximal region of the cytoplasmic domain (Quelle et al. 1994). Jak2 clustering upon ligand binding activates the kinase by a trans-phosphorylation mechanism; activated Jak2 subsequently phosphorylates tyrosine residues in the intracellular region of the GMR. Early studies with Ba/F3 cells expressing a dominant negative Jak2 isoform demonstrated that mutant Jak2 suppressed phosphorylation of βc and inhibited GM-CSF-induced activation of immediate response genes (Watanabe et al. 1996), indicating the importance of Jak2 for GM-CSF signaling. This was confirmed by studies with fetal liver progenitors from *Jak2*$^{-/-}$ mice, which failed to respond to GM-CSF and IL-3 (Parganas et al. 1998), demonstrating a nonredundant role for Jak2 in response to cytokine signals employing βc. Activated Jak2 and βc recruit SH2 and PTB domain-containing proteins to the GMR complex and initiate multiple signaling cascades, including STAT, Ras/MAPK and PI3K/Atk pathways (Hercus et al. 2009). STAT5A and STAT5B are the predominant STATs activated by GM-CSF. Sakurai et al. have indicated that each of the eight phosphorylated tyrosine residues in βc can serve as a docking site for STAT5 in a GM-CSF-dependent in vitro system (Sakurai et al. 2000), although it is unclear whether all βc tyrosines are functionally similar in vivo. Studies in Ba/F3 cells show that Y577 in βc mediates GM-CSF-dependent phosphorylation of SHC, while Y577, Y612, or Y695 appear to be similar in their ability to induce phosphorylation of SHP2 (Okuda et al. 1997). βc also associates with other tyrosine kinases, such as Lyn, Btk and Fyn, although the function of these interactions remain largely unknown (Geijsen et al. 2001).

GM-CSF is a potent growth factor for DCs, however it has a striking suppressive activity on pDC generation in cultures with total bone marrow or purified progenitor cells, indicating its potential to block pDCs at an early developmental stage (Gilliet et al. 2002; Esashi et al. 2008). Elevated circulating amounts of GM-CSF in mice carrying a GM-CSF-encoding transgene (Vremec et al. 1997) or in animals treated with recombinant GM-CSF (Daro et al. 2000; O'Keeffe et al. 2002) leads to increases in both CD8α^{+} and CD8α^{-} cDC numbers in lymphoid organs, demonstrating that GM-CSF can drive cDC generation in vivo. By contrast, mice carrying targeted null mutations of the GM-CSF or βc genes show marginal reduction in cDC populations in thymus, spleen and lymph node (Vremec et al. 1997; Kingston et al. 2009). βc$^{-/-}$ mice lack responsiveness to GM-CSF and IL-5 yet retain the ability to respond to IL-3, due to the presence of an alternative murine IL-3 receptor (Hara and Miyajima 1992; Nicola et al. 1996). Taken together, these results indicate that GM-CSF has a limited role in cDC homeostasis, and this may be due to the presence of compensatory cytokine signals including IL-3 and Flt3L. However, during conditions in which GM-CSF is expressed at elevated amounts, such as inflammation or infection, GM-CSF may boost the production of DCs from pDC/DC progenitors and/or monocytes. The DCs that arise in these conditions do not appear to correspond to DCs present in steady state in mouse and are considered to be a distinct "inflammatory" population (reviewed in Naik 2008). In fact, evidence for a monocyte to DC transition in vivo during inflammatory conditions has been reported. For example, $CD11c^+$ DCs differentiated from adoptively-transferred monocytes can be recovered from the spleens of recipient mice with systemic inflammation (Naik et al. 2006) or from inflamed peritoneum induced by thioglycollate treatment (Geissmann et al. 2003). Moreover, a novel inflammatory DC population secreting high levels of TNF-α and iNOS (Tip DCs) was reported during *L. moncytogenes* infection (Serbina et al. 2003), which is accompanied by elevated cytokine production. Hence, a principal role for GM-CSF may be to enhance the amount of antigen-presenting cells during inflammation or infection by stimulating production of inflammatory DCs.

The numerous signaling pathways elicited by GM-CSF posed challenges for specifying the roles of individual factors. As STAT5 is strongly activated in pDC/DC progenitors stimulated with GM-CSF, our group determined whether STAT5 is involved in regulating DC development. We found that STAT5 is essential for the suppressive activity of GM-CSF on pDC generation in vitro (Esashi et al. 2008). Moreover, bone marrow chimeric $Stat5a^{-/-}$ $Stat5b^{-/-}$ mice (termed here $Stat5^{-/-}$) demonstrate increased pDC proportions and decreased cDC frequencies in bone marrow and spleen compared to $Stat5^{+/+}$ chimeras, suggesting that STAT5 inhibits homeostatic pDC development in vivo (Esashi et al. 2008). Further analysis within the $CD11c^+$ splenic population showed an increase in the $CD11b^-$ $CD8\alpha^+$ subset proportion and a reduction in the $CD11b^+$ $CD4^+$ frequency, suggesting STAT5 signaling may also influence cDC differentiation in the spleen. By contrast, STAT5 is not required for Flt3L-driven pDC or DC development from lin^- $Flt3^+$ bone marrow progenitors (Esashi et al. 2008). GM-CSF is believed to exert its suppressive function on pDC development at a pDC/DC progenitor stage since it blocks pDC generation from the lin^- $Flt3^+$ subset, while terminally differentiated pDCs do not convert or develop into DCs in response to GM-CSF. In addition, GM-CSF alone is able to stimulate pDC production from $Stat5^{-/-}$ lin^- $Flt3^+$ progenitors suggesting that GM-CSF can induce pDC development in the absence of STAT5 (Esashi et al. 2008). Consistent with our ex vivo results, GM-CSF delivery by hydrodynamic gene transfer markedly inhibits pDC generation in vivo, and this inhibition is partially abrogated in hematopoietic STAT5-deficient mice [i.e. *Tg (Tek-cre)12Flv,* $Stat5^{f/f}$] (Li et al. 2011) (Li and Watowich, unpublished data). Taken together, these results suggest that STAT5 functions as a negative signal for pDC development in steady state conditions as well as "emergency" situations with high circulating amounts of GM-CSF.

To explore how GM-CSF employs STAT5 to exert its suppressive activity on pDCs, the expression of DC-related transcription factors was compared in $Stat5^{-/-}$ and $Stat5^{+/+}$ lin^- $Flt3^+$ progenitors stimulated with Flt3L or Flt3L+GM-CSF. As reported (Esashi et al. 2008), GM-CSF inhibited the expression of *Irf7*, *Irf8* and *Spib* induced by Flt3L, but promoted *Irf4* expression, in accordance with the function of these transcription factors in pDC versus cDC production (i.e. IRF7, IRF8 and SpiB are important for pDC development while IFR4 is required for $CD11c^+$ $CD8\alpha^-$ cDCs). Inspection of proximal promoter regions revealed consensus STAT sites in the *Irf4* and *Irf8* genes, and further analysis indicated that *Irf8* transcription is directly inhibited by STAT5 upon GM-CSF stimulation (Esashi et al. 2008). IRF8 is essential for pDC development and maturation, and deletion of *Irf8* leads to a profound loss of pDCs and CD8α^+ cDCs (Schiavoni et al. 2002). Hence, the significant suppression of *Irf8* expression by GM-CSF-STAT5 signaling is likely to contribute to GM-CSF-mediated inhibition of pDC development (Esashi et al. 2008) (Fig. 1).

Type I IFN-STAT1/STAT2

Type I IFNs comprise a large family consisting of multiple IFN-α subtypes, a single IFN-β and other members such as IFN-κ, IFN-ε and IFN-ω. The IFNs exhibit a wide spectrum of activities in the immune system, including anti-viral effects, pro-apoptotic activity and APC-stimulating capability (Stark et al. 1998; Pestka 2000; Biron 2001; Theofilopoulos et al. 2005; Trinchieri 2010). While pDCs are considered to be a primary source of type I IFN secretion upon viral infection (Siegal et al. 1999), many other cells types, including leukocytes, T cells and NK cells, produce type I IFN upon activation. All type I IFNs bind and signal through a single receptor composed of two subunits, IFNAR1 and IFNAR2, which is expressed by many cell types (reviewed in Stark et al. 1998; Theofilopoulos et al. 2005). IFNAR1 exists primarily as a single isoform, but alternative processing of IFNAR2 transcripts produces multiple isoforms that possess identical IFN-binding sequences in the extracellular domain (Domanski et al. 1995; Lutfalla et al. 1995). Moreover, the two subunits have distinct three-dimensional structures as the extracellular domain of IFNAR1 contains two putative cytokine binding sites formed by four fibronectin type III repeats, while IFNAR2 contains a single putative cytokine binding site (Fig. 1) (Uze et al. 1995). Interaction of type I IFNs is proposed to involve two-step process whereby ligand engages IFNAR2 initially and subsequently IFNAR1 is recruited to the complex (reviewed in Stark et al. 1998; Theofilopoulos et al. 2005). Full-length IFNAR2 is required for activation of JAK-STAT signal transduction (Lutfalla et al. 1995).

Like other cytokine receptors, IFNAR1 and IFNAR2 lack intrinsic enzymatic activity, however their cytoplasmic domains noncovalently associate with the JAK kinases Tyk2 and Jak1, respectively (Colamonici et al. 1994a, b; Gauzzi et al. 1996). In vitro binding assays revealed a minimum sequence in IFNAR1, comprising residues 479–511, as the Tyk2 interaction site; this region is proximal to the inducible STAT2 docking site at Y466. By contrast, the Jak1 binding region in IFNAR2 was localized to residues 300–346, which are nearby the constitutive STAT2-docking region at residues 404–462 (Mogensen et al. 1999). In addition, IFNAR2 also binds constitutively to STAT1 in the presence of STAT2 (Li et al. 1997). Engagement of IFNAR by ligand results in phosphorylation of Jak1, Tyk2 and both receptor subunits. Subsequently, STAT2 interacts with phosphorylated Y466 in IFNAR1 via its SH2 domain. This is thought to position STAT2 and STAT1 in the correct configuration to become tyrosine phosphorylated, leading to their heterodimerization and dissociation from the receptor complex (reviewed in Stark et al. 1998). STAT1:STAT2 heterodimers bind to p48/IRF9 to form the multimeric protein complex ISGF3, which stimulates transcription of type I IFN genes (Stark et al. 1998; Platanias 2005). STAT1 homodimers may also form in response to IFNAR activation (Platanias 2005). Of note, IFNAR1 was reported to bind to STAT3 at residues 525–544 (Yang et al. 1996), however the involvement of STAT3 in the type I IFN signaling pathway is unclear. In addition to the JAK-STAT pathway, type I IFNs can activate other signaling cascades including the MAPK cascade and PI3K (Platanias 2005).

Many investigators have reported that type I IFNs promote terminal differentiation and maturation of DCs by stimulating expression of MHC and costimulatory molecules (Trinchieri 2010). More recently, it was shown that IFN-α stimulates HSCs to exit quiescence and to enter the cell cycle via a STAT1-dependent mechanism (Essers et al. 2009); these results highlight the potential for IFN-α to induce the development of hematopoietic lineages. Consistent with this, we found that IFN-α hydrodynamic gene transfer (IFN-α HGT), which elicits elevated IFN-α amounts in the blood similar to viral infection (e.g. 2–3 ng/mL), induced an approximate twofold increase in bone marrow pDC/DC progenitor numbers (i.e. lin^- $Flt3^+$ cells) after 4 days of treatment (Li et al. 2011). In addition, IFN-α stimulated pDC production from pDC/DC progenitors in vitro in the presence of a primary DC growth factor, such as Flt3L or GM-CSF. Thus, the enhanced production of pDCs in bone marrow and spleen of mice that received IFN-α HGT suggests two developmental processes: the generation of pDC/DC progenitors from HSC and the differentiation of pDCs from pDC/DC progenitors. On the other hand, high dose IFN-β treatment was shown to suppress development of MHC class II^+ $CD11c^+$ DCs in vivo and in vitro (Hahm et al. 2005). The suppression of DCs was also observed during infection with measles virus and LCMV (Hahm et al. 2005), while enhanced pDC production was found following vesicular stomatitis virus infection (Li and Watowich, unpublished data). Zuniga reported that LCMV infection drives differentiation of bone marrow pDCs to DCs in the presence of Flt3L by upregulating CD11b and suppressing B220 expression via IFNAR-dependent mechanisms (Zuniga et al. 2004), however IFN-α signaling does not appear to stimulate conversion between pDCs and DCs or vice versa (Li et al. 2011). Thus, type I IFNs exert multiple effects on the DC developmental pathway and may distinctly regulate pDC generation and DC suppression.

We found that pDCs that were generated in the presence of IFN-α (IFN-α-conditioned pDCs) share many features with pDCs that develop in response to Flt3L, however the IFN-α-conditioned pDCs fail to secrete type I IFN upon TLR7 and TLR9 stimulation, but produce enhanced amounts of the proinflammatory cytokines IL-6, IL-23 and TNFα. Furthermore, IFN-α-conditioned pDCs preferentially facilitate the differentiation of Th17 cells compared to pDCs that develop in steady state or in response to Flt3L (Li et al. 2011). Genetic deletion of STAT1 abrogates the development of pDCs elicited by IFN-α in vitro or IFN-α HGT in vivo, indicating an indispensable role for STAT1 in their production (Li et al. 2011). *Stat1*$^{-/-}$ mice display normal amounts of pDCs and DCs in bone marrow and spleen (Li et al. 2011), indicating STAT1 is not necessary for DC homeostasis in these organs. Strikingly, however, *Stat1*$^{-/-}$ mice are deficient in Peyer's patch pDCs, a DC population that closely resembles IFN-α-conditioned pDCs in terms of phenotypic markers, transcription factor expression profiles and functional features (i.e. lack of IFN secretion) (Contractor et al. 2007; Li et al. 2011). This finding suggests that IFN-STAT1 signals may play important roles in pDC development in the gut and under inflammatory conditions where type I IFNs are produced. By contrast, DC suppression induced by viruses or IFN-β was dependent on STAT2, but independent of STAT1 (Hahm et al. 2005). Given the fact that IFNAR2 associates with STAT1 in a STAT2-dependent manner (Li et al. 1997) and IFN-β binding to IFNAR2 alone can stimulate early transcription of IFN-responsive genes (Lewerenz et al. 1998), it is possible that STAT2-mediated initial transcripts are sufficient to block DC development in response to IFN-β. No alteration in DC subsets has been reported in *Stat2*$^{-/-}$ animals (Hahm et al. 2005), suggesting STAT2 is not essential for DC homeostasis.

Analysis of the expression of DC-specific transcription factors in response to IFN-α suggested that IFN-α induced *Irf8* transcription in pDC/DC progenitors by a mechanism involving direct interaction of IFN-α-activated STAT1 with the *Irf8* promoter (Li et al. 2011). Since IRF8 is important for pDC development, this signaling cascade may account, at least in part, for increased pDC production in response to IFN-α (Fig. 1) (Li et al. 2011). IFN-α was also reported to enhance *IRF8* expression in human NK and T cell cultures (Lehtonen et al. 2003), as well as in patients with chronic myelogenous leukemia (Schmidt et al. 1998), but the underlying mechanisms have remained unresolved until recently. In terms of DC development, virus-induced type I IFNs do not seem to enhance DC apoptosis, but rather inhibit cell proliferation to limit DC expansion (Hahm et al. 2005). This is consistent with the anti-proliferative role for IFN-α and IFN-β (reviewed in Stark et al. 1998). Detailed analysis of how type I IFN-responsive STAT1 and/or STAT2 regulate the cell cycle machinery in DCs will extend our understanding of type I IFN-mediated DC suppression. In addition, IFN-α was suggested to inhibit IL-2-stimulated STAT5 DNA binding in T cells (Erickson et al. 2002); GM-CSF, on the other hand, blocks IFN-α-induced STAT1 activation (Kasper et al. 2007). This cross-inhibition by STAT1 and STAT5 signals are in accordance with their unique roles in pDC versus DC development, and may factor into their disparate activities in these lineages.

Concluding Remarks and Future Perspectives

Flt3L, GM-CSF and type I IFNs are three well-studied cytokines that signal through their cognate receptors to regulate pDC and DC lineage commitment and differentiation. In each system, STATs are activated and interact with the proximal promoter regions of certain pDC- and DC-specific transcription factor genes (Fig. 1). These signals may instruct and/or reinforce the developmental decision of pDC/DC progenitors. A major focus for the future is uncovering genome-wide STAT targets in pDCs, DCs and their progenitors, and determining how STAT-regulated genes participate in pDC/DC lineage development. In addition, other cytokines, such as thrombopoietin, M-CSF and IL-3, can independently regulate DC generation, or cooperate with other DC growth factors to influence DC development, yet the intracellular signaling pathways they employ remain unknown. The interplay and cross-regulation among multiple signaling pathways and numerous transcription factors are complex, and mechanisms by which these events regulate DCs remain as a challenging question. Compared to pDCs, less is known about the signaling networks involved in cDC subset (e.g. CD8α^+ vs. CD8α^- cDC) diversification, an area that should be explored in future studies. Unraveling these mechanisms is important for learning how to manipulate pDC and cDC amounts, as well as their effects on innate and adaptive immune functions, and may aid in our ability to target pDC and DC lineages in clinical therapies for cancer or immune disease.

References

Bazan JF (1990) Structural design and molecular evolution of a cytokine receptor superfamily. Proc Natl Acad Sci USA 87:6934–6938

Biron CA (2001) Interferons alpha and beta as immune regulators – a new look. Immunity 14:661–664

Blom B, Ho S, Antonenko S, Liu YJ (2000) Generation of interferon alpha-producing predendritic cell (Pre-DC)2 from human CD34(+) hematopoietic stem cells. J Exp Med 192:1785–1796

Carotta S, Dakic A, D'Amico A, Pang SH, Greig KT, Nutt SL, Wu L (2010) The transcription factor PU.1 controls dendritic cell development and Flt3 cytokine receptor expression in a dose-dependent manner. Immunity 32:628–641

Colamonici O, Yan H, Domanski P, Handa R, Smalley D, Mullersman J, Witte M, Krishnan K, Krolewski J (1994a) Direct binding to and tyrosine phosphorylation of the alpha subunit of the type I interferon receptor by p135tyk2 tyrosine kinase. Mol Cell Biol 14:8133–8142

Colamonici OR, Uyttendaele H, Domanski P, Yan H, Krolewski JJ (1994b) p135tyk2, an interferon-alpha-activated tyrosine kinase, is physically associated with an interferon-alpha receptor. J Biol Chem 269:3518–3522

Contractor N, Louten J, Kim L, Biron CA, Kelsall BL (2007) Cutting edge: Peyer's patch plasmacytoid dendritic cells (pDCs) produce low levels of type I interferons: possible role for IL-10, TGFbeta, and prostaglandin E2 in conditioning a unique mucosal pDC phenotype. J Immunol 179:2690–2694

D'Amico A, Wu L (2003) The early progenitors of mouse dendritic cells and plasmacytoid predendritic cells are within the bone marrow hemopoietic precursors expressing Flt3. J Exp Med 198:293–303

Daro E, Pulendran B, Brasel K, Teepe M, Pettit D, Lynch DH, Vremec D, Robb L, Shortman K, McKenna HJ, Maliszewski C, Maraskovsky E (2000) Polyethylene glycol-modified GM-CSF expands CD11b^{high}CD11c^{high} but not CD11b^{low}CD11c^{high} murine dendritic cells in vivo: a comparative analysis with Flt3 ligand. J Immunol 165:49–58

de Vos AM, Ultsch M, Kossiakoff AA (1992) Human growth hormone and extracellular domain of its receptor: crystal structure of the complex. Science 255:306–312

Domanski P, Witte M, Kellum M, Rubinstein M, Hackett R, Pitha P, Colamonici OR (1995) Cloning and expression of a long form of the beta subunit of the interferon alpha beta receptor that is required for signaling. J Biol Chem 270:21606–21611

Dosil M, Wang S, Lemischka IR (1993) Mitogenic signalling and substrate specificity of the Flk2/Flt3 receptor tyrosine kinase in fibroblasts and interleukin 3-dependent hematopoietic cells. Mol Cell Biol 13:6572–6585

Erickson S, Matikainen S, Thyrell L, Sangfelt O, Julkunen I, Einhorn S, Grander D (2002) Interferon-alpha inhibits Stat5 DNA-binding in IL-2 stimulated primary T-lymphocytes. Eur J Biochem 269:29–37

Esashi E, Wang YH, Perng O, Qin XF, Liu YJ, Watowich SS (2008) The signal transducer STAT5 inhibits plasmacytoid dendritic cell development by suppressing transcription factor IRF8. Immunity 28:509–520

Essers MA, Offner S, Blanco-Bose WE, Waibler Z, Kalinke U, Duchosal MA, Trumpp A (2009) IFNalpha activates dormant haematopoietic stem cells in vivo. Nature 458:904–908

Fogg DK, Sibon C, Miled C, Jung S, Aucouturier P, Littman DR, Cumano A, Geissmann F (2006) A clonogenic bone marrow progenitor specific for macrophages and dendritic cells. Science 311:83–87

Gauzzi MC, Velazquez L, McKendry R, Mogensen KE, Fellous M, Pellegrini S (1996) Interferon-alpha-dependent activation of Tyk2 requires phosphorylation of positive regulatory tyrosines by another kinase. J Biol Chem 271:20494–20500

Gearing DP, King JA, Gough NM, Nicola NA (1989) Expression cloning of a receptor for human granulocyte-macrophage colony-stimulating factor. EMBO J 8:3667–3676

Geijsen N, Koenderman L, Coffer PJ (2001) Specificity in cytokine signal transduction: lessons learned from the IL-3/IL-5/GM-CSF receptor family. Cytokine Growth Factor Rev 12:19–25

Geissmann F, Jung S, Littman DR (2003) Blood monocytes consist of two principal subsets with distinct migratory properties. Immunity 19:71–82

Gilliet M, Boonstra A, Paturel C, Antonenko S, Xu XL, Trinchieri G, O'Garra A, Liu YJ (2002) The development of murine plasmacytoid dendritic cell precursors is differentially regulated by FLT3-ligand and granulocyte/macrophage colony-stimulating factor. J Exp Med 195:953–958

Guthridge MA, Stomski FC, Thomas D, Woodcock JM, Bagley CJ, Berndt MC, Lopez AF (1998) Mechanism of activation of the GM-CSF, IL-3, and IL-5 family of receptors. Stem Cells 16:301–313

Hahm B, Trifilo MJ, Zuniga EI, Oldstone MB (2005) Viruses evade the immune system through type I interferon-mediated STAT2-dependent, but STAT1-independent, signaling. Immunity 22:247–257

Hamilton JA, Anderson GP (2004) GM-CSF biology. Growth Factors 22:225–231

Hannum C, Culpepper J, Campbell D, McClanahan T, Zurawski S, Bazan JF, Kastelein R, Hudak S, Wagner J, Mattson J et al (1994) Ligand for FLT3/FLK2 receptor tyrosine kinase regulates growth of haematopoietic stem cells and is encoded by variant RNAs. Nature 368:643–648

Hansen G, Hercus TR, McClure BJ, Stomski FC, Dottore M, Powell J, Ramshaw H, Woodcock JM, Xu Y, Guthridge M, McKinstry WJ, Lopez AF, Parker MW (2008) The structure of the GM-CSF receptor complex reveals a distinct mode of cytokine receptor activation. Cell 134:496–507

Hara T, Miyajima A (1992) Two distinct functional high affinity receptors for mouse interleukin-3 (IL-3). EMBO J 11:1875–1884

Hayashida K, Kitamura T, Gorman DM, Arai K, Yokota T, Miyajima A (1990) Molecular cloning of a second subunit of the receptor for human granulocyte-macrophage colony-stimulating factor (GM-CSF): reconstitution of a high-affinity GM-CSF receptor. Proc Natl Acad Sci USA 87:9655–9659

Hegde S, Ni S, He S, Yoon D, Feng GS, Watowich SS, Paulson RF, Hankey PA (2009) Stat3 promotes the development of erythroleukemia by inducing Pu.1 expression and inhibiting erythroid differentiation. Oncogene 28:3349–3359

Heiss E, Masson K, Sundberg C, Pedersen M, Sun J, Bengtsson S, Ronnstrand L (2006) Identification of Y589 and Y599 in the juxtamembrane domain of Flt3 as ligand-induced autophosphorylation sites involved in binding of Src family kinases and the protein tyrosine phosphatase SHP2. Blood 108:1542–1550

Hercus TR, Thomas D, Guthridge MA, Ekert PG, King-Scott J, Parker MW, Lopez AF (2009) The granulocyte-macrophage colony-stimulating factor receptor: linking its structure to cell signaling and its role in disease. Blood 114:1289–1298

Hubert P, Greimers R, Franzen-Detrooz E, Doyen J, Delanaye P, Boniver J, Delvenne P (1998) In vitro propagated dendritic cells from patients with human-papilloma virus-associated preneoplastic lesions of the uterine cervix: use of Flt3 ligand. Cancer Immunol Immunother 47:81–89

Inaba K, Inaba M, Romani N, Aya H, Deguchi M, Ikehara S, Muramatsu S, Steinman RM (1992) Generation of large numbers of dendritic cells from mouse bone marrow cultures supplemented with granulocyte/macrophage colony-stimulating factor. J Exp Med 176:1693–1702

Kasper S, Kindler T, Sonnenschein S, Breitenbuecher F, Bohmer FD, Huber C, Fischer T (2007) Cross-inhibition of interferon-induced signals by GM-CSF through a block in Stat1 activation. J Interferon Cytokine Res 27:947–959

Kingston D, Schmid MA, Onai N, Obata-Onai A, Baumjohann D, Manz MG (2009) The concerted action of GM-CSF and Flt3-ligand on in vivo dendritic cell homeostasis. Blood 114:835–843

Kitamura T, Sato N, Arai K, Miyajima A (1991) Expression cloning of the human IL-3 receptor cDNA reveals a shared beta subunit for the human IL-3 and GM-CSF receptors. Cell 66:1165–1174

Laouar Y, Welte T, Fu XY, Flavell RA (2003) STAT3 is required for Flt3L-dependent dendritic cell differentiation. Immunity 19:903–912

Lehtonen A, Lund R, Lahesmaa R, Julkunen I, Sareneva T, Matikainen S (2003) IFN-alpha and IL-12 activate IFN regulatory factor 1 (IRF-1), IRF-4, and IRF-8 gene expression in human NK and T cells. Cytokine 24:81–90

Lewerenz M, Mogensen KE, Uze G (1998) Shared receptor components but distinct complexes for alpha and beta interferons. J Mol Biol 282:585–599

Li X, Leung S, Kerr IM, Stark GR (1997) Functional subdomains of STAT2 required for preassociation with the alpha interferon receptor and for signaling. Mol Cell Biol 17:2048–2056

Li HS, Gelbard A, Martinez GJ, Esashi E, Zhang H, Nguyen-Jackson H, Liu Y-J, Overwijk WW, Watowich SS (2011) Cell intrinsic role for interferon-α-STAT1 signals in regulating murine Peyer's Patch plasmacytoid dendritic cells and conditioning an inflammatory response. Blood 118(14):3879–3889

Lia F, Rajotte D, Clark SC, Hoang T (1996) A dominant negative granulocyte-macrophage colony-stimulating factor receptor alpha chain reveals the multimeric structure of the receptor complex. J Biol Chem 271:28287–28293

Liu K, Victora GD, Schwickert TA, Guermonprez P, Meredith MM, Yao K, Chu FF, Randolph GJ, Rudensky AY, Nussenzweig M (2009) In vivo analysis of dendritic cell development and homeostasis. Science 324:392–397

Lutfalla G, Holland SJ, Cinato E, Monneron D, Reboul J, Rogers NC, Smith JM, Stark GR, Gardiner K, Mogensen KE et al (1995) Mutant U5A cells are complemented by an interferon-alpha beta receptor subunit generated by alternative processing of a new member of a cytokine receptor gene cluster. EMBO J 14:5100–5108

Lyman SD, Jacobsen SE (1998) c-kit ligand and Flt3 ligand: stem/progenitor cell factors with overlapping yet distinct activities. Blood 91:1101–1134

Lyman SD, James L, Johnson L, Brasel K, de Vries P, Escobar SS, Downey H, Splett RR, Beckmann MP, McKenna HJ (1994) Cloning of the human homologue of the murine flt3 ligand: a growth factor for early hematopoietic progenitor cells. Blood 83:2795–2801

Lyman SD, James L, Escobar S, Downey H, de Vries P, Brasel K, Stocking K, Beckmann MP, Copeland NG, Cleveland LS et al (1995a) Identification of soluble and membrane-bound isoforms of the murine flt3 ligand generated by alternative splicing of mRNAs. Oncogene 10:149–157

Lyman SD, Seaberg M, Hanna R, Zappone J, Brasel K, Abkowitz JL, Prchal JT, Schultz JC, Shahidi NT (1995b) Plasma/serum levels of flt3 ligand are low in normal individuals and highly elevated in patients with Fanconi anemia and acquired aplastic anemia. Blood 86:4091–4096

Manfra DJ, Chen SC, Jensen KK, Fine JS, Wiekowski MT, Lira SA (2003) Conditional expression of murine Flt3 ligand leads to expansion of multiple dendritic cell subsets in peripheral blood and tissues of transgenic mice. J Immunol 170:2843–2852

Maraskovsky E, Brasel K, Teepe M, Roux ER, Lyman SD, Shortman K, McKenna HJ (1996) Dramatic increase in the numbers of functionally mature dendritic cells in Flt3 ligand-treated mice: multiple dendritic cell subpopulations identified. J Exp Med 184:1953–1962

Marchetto S, Fournier E, Beslu N, Aurran-Schleinitz T, Dubreuil P, Borg JP, Birnbaum D, Rosnet O (1999) SHC and SHIP phosphorylation and interaction in response to activation of the FLT3 receptor. Leukemia 13:1374–1382

McKenna HJ, Stocking KL, Miller RE, Brasel K, De Smedt T, Maraskovsky E, Maliszewski CR, Lynch DH, Smith J, Pulendran B, Roux ER, Teepe M, Lyman SD, Peschon JJ (2000) Mice lacking flt3 ligand have deficient hematopoiesis affecting hematopoietic progenitor cells, dendritic cells, and natural killer cells. Blood 95:3489–3497

Melillo JA, Song L, Bhagat G, Blazquez AB, Plumlee CR, Lee C, Berin C, Reizis B, Schindler C (2010) Dendritic cell (DC)-specific targeting reveals Stat3 as a negative regulator of DC function. J Immunol 184:2638–2645

Merad M, Manz MG (2009) Dendritic cell homeostasis. Blood 113:3418–3427

Mogensen KE, Lewerenz M, Reboul J, Lutfalla G, Uze G (1999) The type I interferon receptor: structure, function, and evolution of a family business. J Interferon Cytokine Res 19:1069–1098

Murphy JM, Young IG (2006) IL-3, IL-5, and GM-CSF signaling: crystal structure of the human beta-common receptor. Vitam Horm 74:1–30

Naik SH (2008) Demystifying the development of dendritic cell subtypes, a little. Immunol Cell Biol 86:439–452

Naik SH, Metcalf D, van Nieuwenhuijze A, Wicks I, Wu L, O'Keeffe M, Shortman K (2006) Intrasplenic steady-state dendritic cell precursors that are distinct from monocytes. Nat Immunol 7:663–671

Naik SH, Sathe P, Park HY, Metcalf D, Proietto AI, Dakic A, Carotta S, O'Keeffe M, Bahlo M, Papenfuss A, Kwak JY, Wu L, Shortman K (2007) Development of plasmacytoid and conventional dendritic cell subtypes from single precursor cells derived in vitro and in vivo. Nat Immunol 8:1217–1226

Nicola NA, Robb L, Metcalf D, Cary D, Drinkwater CC and Begley CG (1996) Functional inactivation in mice of the gene for the interleukin-3 (IL-3)-specific receptor beta-chain: implications for IL-3 function and the mechanism of receptor transmodulation in hematopoietic cells. Blood 87:2665–2674

O'Keeffe M, Hochrein H, Vremec D, Pooley J, Evans R, Woulfe S, Shortman K (2002) Effects of administration of progenipoietin 1, Flt-3 ligand, granulocyte colony-stimulating factor, and pegylated granulocyte-macrophage colony-stimulating factor on dendritic cell subsets in mice. Blood 99:2122–2130

Okuda K, Smith L, Griffin JD, Foster R (1997) Signaling functions of the tyrosine residues in the betac chain of the granulocyte-macrophage colony-stimulating factor receptor. Blood 90:4759–4766

Onai N, Obata-Onai A, Tussiwand R, Lanzavecchia A, Manz MG (2006) Activation of the Flt3 signal transduction cascade rescues and enhances type I interferon-producing and dendritic cell development. J Exp Med 203:227–238

Onai N, Obata-Onai A, Schmid MA, Ohteki T, Jarrossay D, Manz MG (2007) Identification of clonogenic common Flt3(+)M-CSFR(+) plasmacytoid and conventional dendritic cell progenitors in mouse bone marrow. Nat Immunol 8:1207–1216

Parganas E, Wang D, Stravopodis D, Topham DJ, Marine JC, Teglund S, Vanin EF, Bodner S, Colamonici OR, van Deursen JM, Grosveld G, Ihle JN (1998) Jak2 is essential for signaling through a variety of cytokine receptors. Cell 93:385–395

Pestka S (2000) The human interferon alpha species and receptors. Biopolymers 55:254–287

Platanias LC (2005) Mechanisms of type-I- and type-II-interferon-mediated signalling. Nat Rev Immunol 5:375–386

Quelle FW, Sato N, Witthuhn BA, Inhorn RC, Eder M, Miyajima A, Griffin JD, Ihle JN (1994) JAK2 associates with the beta c chain of the receptor for granulocyte-macrophage colony-stimulating factor, and its activation requires the membrane-proximal region. Mol Cell Biol 14:4335–4341

Rocnik JL, Okabe R, Yu JC, Lee BH, Giese N, Schenkein DP, Gilliland DG (2006) Roles of tyrosine 589 and 591 in STAT5 activation and transformation mediated by FLT3-ITD. Blood 108:1339–1345

Rozwarski DA, Diederichs K, Hecht R, Boone T, Karplus PA (1996) Refined crystal structure and mutagenesis of human granulocyte-macrophage colony-stimulating factor. Proteins 26:304–313

Sakurai Y, Arai K, Watanabe S (2000) In vitro analysis of STAT5 activation by granulocyte-macrophage colony-stimulating factor. Genes Cells 5:937–947

Savvides SN, Boone T, Andrew Karplus P (2000) Flt3 ligand structure and unexpected commonalities of helical bundles and cystine knots. Nat Struct Biol 7:486–491

Schiavoni G, Mattei F, Sestili P, Borghi P, Venditti M, Morse HC III, Belardelli F, Gabriele L (2002) ICSBP is essential for the development of mouse type I interferon-producing cells and for the generation and activation of CD8alpha(+) dendritic cells. J Exp Med 196:1415–1425

Schmidt M, Nagel S, Proba J, Thiede C, Ritter M, Waring JF, Rosenbauer F, Huhn D, Wittig B, Horak I, Neubauer A (1998) Lack of interferon consensus sequence binding protein (ICSBP) transcripts in human myeloid leukemias. Blood 91:22–29

Serbina NV, Salazar-Mather TP, Biron CA, Kuziel WA, Pamer EG (2003) TNF/iNOS-producing dendritic cells mediate innate immune defense against bacterial infection. Immunity 19:59–70

Shanafelt AB, Johnson KE, Kastelein RA (1991) Identification of critical amino acid residues in human and mouse granulocyte-macrophage colony-stimulating factor and their involvement in species specificity. J Biol Chem 266:13804–13810

Siegal FP, Kadowaki N, Shodell M, Fitzgerald-Bocarsly PA, Shah K, Ho S, Antonenko S, Liu YJ (1999) The nature of the principal type 1 interferon-producing cells in human blood. Science 284:1835–1837

Spiekermann K, Bagrintseva K, Schwab R, Schmieja K, Hiddemann W (2003) Overexpression and constitutive activation of FLT3 induces STAT5 activation in primary acute myeloid leukemia blast cells. Clin Cancer Res 9:2140–2150

Stark GR, Kerr IM, Williams BR, Silverman RH, Schreiber RD (1998) How cells respond to interferons. Ann Rev Biochem 67:227–264

Stirewalt DL, Radich JP (2003) The role of FLT3 in haematopoietic malignancies. Nat Rev Cancer 3:65–665

Tavernier J, Devos R, Cornelis S, Tuypens T, Van der Heyden J, Fiers W, Plaetinck G (1991) A human high affinity interleukin-5 receptor (IL5R) is composed of an IL5-specific alpha chain and a beta chain shared with the receptor for GM-CSF. Cell 66:1175–1184

Theofilopoulos AN, Baccala R, Beutler B, Kono DH (2005) Type I interferons (alpha/beta) in immunity and autoimmunity. Annu Rev Immunol 23:307–336

Trinchieri G (2010) Type I interferon: friend or foe? J Exp Med 207:2053–2063

Turner AM, Lin NL, Issarachai S, Lyman SD, Broudy VC (1996) FLT3 receptor expression on the surface of normal and malignant human hematopoietic cells. Blood 88:3383–3390

Tussiwand R, Onai N, Mazzucchelli L, Manz MG (2005) Inhibition of natural type I IFN-producing and dendritic cell development by a small molecule receptor tyrosine kinase inhibitor with Flt3 affinity. J Immunol 175:3674–3680

Uze G, Lutfalla G, Mogensen KE (1995) Alpha and beta interferons and their receptor and their friends and relations. J Interferon Cytokine Res 15:3–26

Vremec D, Lieschke GJ, Dunn AR, Robb L, Metcalf D, Shortman K (1997) The influence of granulocyte/macrophage colony-stimulating factor on dendritic cell levels in mouse lymphoid organs. Eur J Immunol 27:40–44

Waskow C, Liu K, Darrasse-Jeze G, Guermonprez P, Ginhoux F, Merad M, Shengelia T, Yao K, Nussenzweig M (2008) The receptor tyrosine kinase Flt3 is required for dendritic cell development in peripheral lymphoid tissues. Nat Immunol 9:676–683

Watanabe S, Itoh T, Arai K (1996) JAK2 is essential for activation of c-fos and c-myc promoters and cell proliferation through the human granulocyte-macrophage colony-stimulating factor receptor in BA/F3 cells. J Biol Chem 271:12681–12686

Watowich SS, Liu YJ (2010) Mechanisms regulating dendritic cell specification and development. Immunol Rev 238:76–92

Yang CH, Shi W, Basu L, Murti A, Constantinescu SN, Blatt L, Croze E, Mullersman JE, Pfeffer LM (1996) Direct association of STAT3 with the IFNAR-1 chain of the human type I interferon receptor. J Biol Chem 271:8057–8061

Zhang S, Broxmeyer HE (2000) Flt3 ligand induces tyrosine phosphorylation of gab1 and gab2 and their association with shp-2, grb2, and PI3 kinase. Biochem Biophys Res Commun 277:195–199

Zhang S, Mantel C, Broxmeyer HE (1999) Flt3 signaling involves tyrosyl-phosphorylation of SHP-2 and SHIP and their association with Grb2 and Shc in Baf3/Flt3 cells. J Leukoc Biol 65:372–380

Zhang S, Fukuda S, Lee Y, Hangoc G, Cooper S, Spolski R, Leonard WJ, Broxmeyer HE (2000) Essential role of signal transducer and activator of transcription (Stat)5a but not Stat5b for Flt3-dependent signaling. J Exp Med 192:719–728

Zuniga EI, McGavern DB, Pruneda-Paz JL, Teng C, Oldstone MB (2004) Bone marrow plasmacytoid dendritic cells can differentiate into myeloid dendritic cells upon virus infection. Nat Immunol 5:1227–1234

STAT Transcription Factors: Controlling All Aspects of NK Cell Biology

Eva Maria Putz, Eva Zebedin, and Veronika Sexl

Abstract
Besides B and T cells, Natural Killer (NK) cells constitute the third lymphocytic population with a broad spectrum of skills and functions. For several decades NK cells have been portrayed as first line defense against virally infected and malignant cells. But recent reports unraveled far more diverse properties of NK cells, e.g. their involvement in reproductive immunology and in mucosal defense of pathogens in the gut, and especially their ability to retain memory over several months. This chapter combines well established paradigms of NK cell biology with recent findings and special emphasis on the JAK/STAT signaling pathway. NK cell development, activation and cytotoxic function are tightly regulated by a plethora of cytokines – prominent inducers of the JAK/STAT signaling cascade. The availability and detailed analysis of gene-targeted mice underscores the importance of STATs controlling all aspects of NK cell biology.

Being part of the innate immune system NK cells kill rapidly and without great selectivity. Once stimulated, NK cells produce large amounts of cytokines and chemokines such as interferon-γ (IFN-γ), tumor necrosis factor-α (TNF-α), granulocyte-macrophage colony-stimulating factor (GM-CSF) as well as macrophage inflammatory proteins (MIP-1α and MIP-1β). Thereby NK cells interfere with and regulate other cellular components of the innate and adaptive immune system. It was Eva Klein in the year 1975 who coined the phrase of "Natural Killers" when she and Rolf Kiessling first discovered a novel lymphocyte population with the ability to kill leukemic tumor cells without prior sensitization (Kiessling et al. 1975). Since their discovery extensive research has been focused on these killers and it became evident that NK cells – always

E.M. Putz • V. Sexl (✉)
University of Veterinary Medicine Vienna, Veterinärplatz 1, 1210 Vienna, Austria
e-mail: veronika.sexl@vetmeduni.ac.at

E. Zebedin
Medical University of Vienna, Waehringerstrasse 13a, 1090 Vienna, Austria

Th. Decker and M. Müller (eds.), *Jak-Stat Signaling: From Basics to Disease*,
DOI 10.1007/978-3-7091-0891-8_12, © Springer-Verlag Wien 2012

regarded as simple and innate – have been underestimated for a long time. Changes of old paradigms are on the way. Recent provocative findings assigned an adaptive-like feature to NK cells namely the ability to retain a memory over a certain period of time. The capability to memorize was shown in two different murine models, in cytomegalovirus (CMV) infection (Sun et al. 2009) and in hapten-induced contact hypersensitivity response (O'Leary et al. 2006). A recent report extended these findings to human NK cells: in accordance with the murine models also humans infected with hantavirus show a rapid expansion and long-term persistence of "memory NK cells" far beyond what is considered "normal" for a classical innate immune response (Bjorkstrom et al. 2010). Thus, NK cells are now discussed as an "evolutionary bridge between innate and adaptive immunity" (Sun and Lanier 2009).

Moreover, functional aspects of NK cells have been linked to distinct NK cell subsets. Different murine and human NK cell subsets were defined, such as "gut NK" cells (Leon et al. 2003; Tagliabue et al. 1982), which recently were re-named as NK-22 cells according to their potential to produce IL-22 (Satoh-Takayama et al. 2008). NK-22 cells have attracted considerable interest because of strong evidence for a protective role in mucosal defense against pathogens (Di Santo et al. 2010). Further, a unique subset of non-destructive NK cells was discovered in the uterus prior to and during pregnancy, which seems to have a major impact in tissue-remodeling during the menstrual cycle and mainly in the first trimester of pregnancy (Manaster and Mandelboim 2010).

Thus, Natural Killer cells exhibit a stunningly broad spectrum of skills and functions. They may be on the one hand considered as very powerful weapons against infections and tumors. One the other hand they represent a peaceful cellular compartment involved in reproductive immunology.

The Transcription Factor Network Regulating NK Cell Development: STAT5 in the Center of Attention

The development of conventional NK cells mainly takes place in the bone marrow (BM). However, their maturation and education can proceed in any kind of secondary lymphoid organs, as for example in lymph nodes, liver and spleen (Huntington et al. 2007). It all starts from hematopoietic stem cells (HSCs) in the BM, which give rise to lineage negative (Lin-) NK cell precursors (NKPs) that are characterized by the expression of the IL-2 and IL-15 receptor common ß subunit (also known as CD122) and the complete lack of NK lineage markers such as NK1.1, integrin α2 (DX5) and Ly49 receptors (Rosmaraki et al. 2001). In the presence of IL-2 and/or IL-15 NKPs can differentiate at first into immature NK cells (iNKs, defined as Lin-CD122+NK1.1+DX5−), which sequentially acquire the NK cells' characteristic receptor repertoire (Kim et al. 2002; Roth et al. 2000) and finally reach the fully mature NK cell state (mNKs), as defined by NK1.1+DX5+NKp46+CD3− expression. Even though IL-2 is an important cytokine for NK cell survival, proliferation and cytolytic activity in vitro, *Il2*-deficient mice harbor normal NK cell numbers (Vosshenrich et al. 2005). In contrast, *Il15*-deficient mice are completely devoid of peripheral NK cells supporting the central role of IL-15 for NK cell development (Kennedy et al. 2000). Interestingly, IL-2 and IL-15 induce largely overlapping signaling pathways. Signaling of IL-2/IL-15 via the IL-2/IL-15 receptor βγ complex results in tyrosine phosphorylation and activation of STAT3 and STAT5 by JAK1 and JAK3, respectively (Miyazaki et al. 1994). Phosphorylated STAT3 and STAT5 then homo-dimerize and translocate to the nucleus, where they act as transcription factors. Consistent with the critical role of STAT5 in IL-2/IL-15 signal transduction, *Jak3* knockout (Nosaka et al. 1995), *Stat5aΔN*, *Stat5bΔN* (Imada et al. 1998) and *Stat5a/bΔN* double knockout mice (Moriggl et al. 1999) display NK cell defects. Mice that lack *Stat5* exclusively in the NK cell compartment are arrested in NK cell development at an early stage (Eckelhart et al. 2011), determining an intrinsic key role for STAT5 in NK cell development. Additional transcription factors, which have been shown to be indispensable for NK cell development, are Ikaros (Boggs et al. 1998), PU.1 (Colucci et al. 2001) and ID2 (Ikawa et al. 2001). Recently, the transcription factor E4BP4 was implicated in the development of NK cells and is now discussed as critical factor determining NK lineage fate decision (Gascoyne et al. 2009; Kamizono et al. 2009). It is attractive to speculate that STAT5 is the "missing" link that closes the circle from IL-15 to E4BP4 thus driving NK cell development (Fig. 1). This hypothesis relies on the documentation of a developmental block occurring in *Stat5*-deficient NK cells, which is superimposable to the phenotype observed in *E4bp4*- and *Il15r*-deficient mice. Further studies are required to finally delineate the transcriptional network around STAT5 allowing NK cell development.

In addition, the T-box transcription factor and STAT downstream target T-bet is required for the final (homeostatic) maturation and function of NK cells (Townsend et al. 2004). During NK cell differentiation T-bet expression is induced by IL-12 and IL-15, cytokines known to control NK development and NK effector functions via STAT4 (Bacon et al. 1995) and STAT5 (Miyazaki et al. 1994). T-bet interacts with various other transcription factors, such as Blimp-1, which is also induced by IL-15 early in NK cell development and together with T-bet mediates NK cell maturation and homeostasis (Kallies et al. 2011). A different, but highly homologous T-box transcription factor is Eomesodermin (Eomes), which shows a selective expression pattern for NK cells and cytotoxic T cells (Pearce et al. 2003). T-bet and Eomes are both required for the expression of the IL-2/IL-15 receptor common ß subunit – thus closing the circle. T-bet and Eomes have largely redundant functions, only the monoallelic loss of Eomes and the complete deletion of T-bet in combination ($Eomes^{+/-}Tbx21^{-/-}$) results in the loss of IL-15-dependent lineages (Intlekofer et al. 2005) and phenocopies *Il15*-deficient mice (Kennedy et al. 2000).

How to Educate an NK Cell: STAT1 as Determinator

NK cell development includes a process called "education" which finally allows the establishment of fully mature and functional NK cells. "Education" represents an important step to prevent autoreactivity and is dominated by the MHC class I

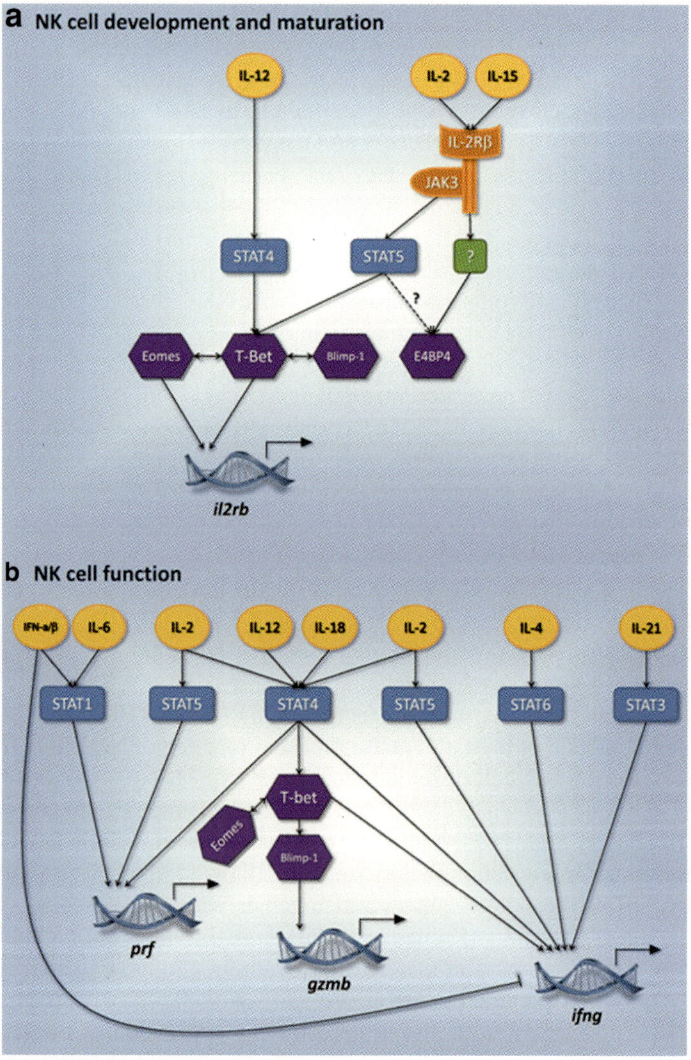

Fig. 1 JAK/STAT signaling network in NK cells

repertoire of the individual (Orr and Lanier 2010). NK cell-mediated autoreactivity is of patho-physiological relevance and bears consequences for diseases such as type 1 diabetes (Gur et al. 2010), psoriasis (Martin et al. 2002) and rheumatoid arthritis (Yen et al. 2001). Despite its importance the molecular mechanism of NK cell education is not yet fully understood. Several models such as the "licensing", "arming", "disarming" and "tuning" models have been proposed, nevertheless the details of NK cell education remain enigmatic (Brodin and Hoglund 2008). The common denominator of all models is the involvement and engagement of the self-major histocompatibility complex (MHC) class I. Only if NK cells express specific inhibitory receptors for MHC I molecules characteristic for the particular individual, full NK cell maturation and competence will be achieved by initiating a yet undefined process (Anfossi et al. 2006; Joncker and Raulet 2008; Raulet and Vance 2006; Yokoyama and Kim 2006). Interestingly, neither the exposure to MHC class I ligands during NK cell development in the BM nor endogenous MHC class I expression by NK cells themselves seem to be absolutely required for this educational process (Elliott et al. 2010). Thus, it remains a challenging open question, how, when and where NK cells are "educated" in terms of generating "better" or "worse" killers. Understanding the molecular details of NK cell education may have important consequences. Once we do understand the process, we may interfere and either enhance the clearance of undesired target cells or block NK cells that are causing autoimmunity. STAT1 is a key player in this regard: *Stat1*-deficient mice show a severe reduction in MHC class I levels (Marques et al. 2004; Meraz et al. 1996). Not unexpectedly, *Stat1*-deficient NK cells show a highly impaired cytolytic capacity (Kovacic et al. 2006; Lee et al. 2000). Thus one may speculate, that *Stat1*-deficient NK cells are not "well-educated" and display an immature phenotype. This may explain the reduced cytolytic ability and the reduced levels of self-MHC class I-recognizing receptors on *Stat1*-deficient NK cells (Robbins et al. 2005).

STATs United: Natural Killer Cells Live Up to their Name

Once fully mature, NK cells kill their targets rapidly and without prior sensitization. This ability demands that killing is under tight control to avoid self-destruction and autoimmunity. A complex integration of signals provided by a plethora of activating and inhibitory NK cell surface receptors acquired during education serves this purpose (Lanier 2008; Long 2008; Zompi and Colucci 2005). NK cells can kill their targets either by direct cell-mediated cytotoxicity or by secretion of soluble factors like IFN-γ and TNF-α (Smyth et al. 2002). The most prominent mechanism is the release of cytoplasmic granules filled with toxic enzymes like perforin, which accounts for the formation of pores in target cell membranes, and granzymes, lytic enzymes that induce apoptosis (Froelich et al. 2009; Trambas and Griffiths 2003). The family of human granzymes consists of five members, of which granzyme A and B have been studied most extensively. Interestingly, mice express twice as many granzymes compared to humans (Bovenschen and Kummer 2010). Since NK cells have to be ready to kill instantaneously and at any time, they are equipped with basal levels of perforin and granzymes that are further increased upon stimulation.

Several cytokines induce perforin expression in NK cells via the JAK/STAT signaling cascade, including IL-2 via STAT5 (Salcedo et al. 1993), IL-6 and IFN-α via STAT1 (Liang et al. 2003; Yu et al. 1999) and IL-12 via STAT4 (Yamamoto et al. 2002) (Fig. 1). In addition, also granzyme B expression underlies the control of the latter cytokines (DeBlaker-Hohe et al. 1995). However, to date not much is known about the involvement of STAT proteins in the regulation of granzyme B. In murine NK cells Kallies et al. (2011) demonstrated the importance of T bet and

Blimp-1 in IL-12-mediated granzyme B expression – suggesting that granzyme B up-regulation is mediated by STAT4. Apart from that, no published data link granzyme expression and STAT signaling in NK cells conclusively.

Besides the granule-exocytosis pathway, NK cells kill target cells by the induction of apoptosis using TNF superfamily members, also known as death ligands and receptors. Ligation of Fas or TNF-related apoptosis-inducing ligand (TRAIL)-receptor on the target cell surface to NK cells' Fas ligand (FasL) and TRAIL induces a machinery of death-bringing signals resulting in target cell lysis (Arase et al. 1995; Zamai et al. 1998). Several crossroads with JAK/STAT signaling exist: TRAIL expression is markedly induced by IFN-α/β via TYK2 (Rani et al. 2007). Moreover, interleukins like IL-2 and IL-12 inducing STAT1, STAT4 and STAT5 signaling are potent inducers of Fas and FasL (Medvedev et al. 1997; Trinchieri et al. 1984). Similarly, IFN-γ fulfils its duties as major NK cell product and death-bringing protein by inducing Fas and FasL expression via STAT1 in tumor and immune cells, respectively, thereby triggering apoptosis (Xu et al. 1998).

Being a dangerous and effective weapon, IFN-γ expression itself is controlled and induced in a cytokine-dependent manner. IL-2 (Wang et al. 1999), IL-12 (Szabo et al. 2002; Uemura et al. 2010) and IL-18 (Lauwerys et al. 1999) are potent inducers of IFN-γ via STAT4 and T-bet. Additionally, IL-21 and IL-4 are able to induce IFN-γ gene transcription in a strictly STAT3- or STAT6-dependent manner, respectively (Morris et al. 2006; Strengell et al. 2003). In contrast, type I IFNs not only fail to trigger, but even block IFN-γ production via STAT1 (Nguyen et al. 2000). Thereby, STAT1 is an important break for IFN-γ production and seems to represent a key node in the signaling network that balances IFN-γ production. The central inhibitory role of STAT1 for IFN-γ production is underscored by observations in *Stat1*-deficient NK cells. In these cells IFN-α/β induce IFN-γ expression (Nguyen et al. 2000), whereas the IL-12-induced IFN-γ production is impaired (Lee et al. 2000).

For the control of viral infections NK cells and interferons play a central role. Viral infections such as cytomegalovirus (CMV) or lymphocytic choriomeningitis virus (LCMV) infections are paralleled by a rapid increase in IFN-γ expression, which accompanies the proliferation, accumulation and cytotoxic response of NK cells (Biron et al. 1996). Several studies defined IFN-α/β and IL-12 as the major immunoregulatory cytokines during CMV infections (Orange and Biron 1996a, b). IFN-α/β strongly induce the activation of NK cells (Biron 1998; Gidlund et al. 1978; Liang et al. 2003) leading to NK cell proliferation and accumulation at the site of infection. Nevertheless, it is highly unlikely that IFN-α/β stimulates NK cell proliferation directly, as IFN-α/β fail to elicit NK cell proliferation in vitro and even exert anti-proliferative effects at high concentrations (Loza and Perussia 2004). In spite of this, type I IFNs induce STAT1-mediated upregulation of IL-15, which is genuinely responsible for the observed NK cell expansion during viral infection (Nguyen et al. 2002). In addition to STAT1 and STAT2, type I IFNs have been reported to conditionally activate all of the STATs, including STAT4 (Matikainen et al. 2001). NK cells harbor high basal levels of STAT4, which is bound to the type I IFN receptor (IFNAR). One key characteristic of viral infections is the rapid and potent induction of STAT1, which instantaneously displaces STAT4 from the IFN receptor (Miyagi et al. 2007).

Summing up, natural killer cells live up to their name by triggering apoptosis in target cells by different mechanisms, many of which are induced by interleukins and interferons that thereby involve all members of the JAK/STAT signaling cascade. It is not surprising that these complex biological responses are orchestrated by a multitude of different signaling molecules. We are just at the very beginning to shed light on their challenging interplay.

NK Cells Versus Tumor Growth: STATs Battle on Both Sides

At the beginning of the last century Paul Ehrlich proposed the idea that nascent transformed cells arise continuously in our bodies and that an intact immune system scans for and clears them from the system (Ehrlich 1908). Thus, whether or not a tumor develops critically depends on the competence of the immune system. In the interaction of host and tumor cells, "three E's" have been defined as key steps of tumor surveillance: "elimination", "equilibrium" and "escape" (Dunn et al. 2002). The immune system is eager to eliminate transformed cells. In some situations elimination may not be complete, but the immune system still controls expansion of the transformed cells. This situation is referred to as "equilibrium". However, this state of equilibrium exerts selective pressure on the tumor cells, which try to evade immune recognition and to decrease immunogenicity. This is frequently achieved by down-regulating MHC class I in order to "escape" T cell-mediated tumor surveillance. A decrease in MHC class I surface expression on the other hand renders tumor cells highly susceptible to NK cell-mediated eradication. More than 25 years ago, Klas Kärre suggested that NK cells are able to recognize aberrant cells not only by their expression of "stress" or "danger" signals, but rather by the "absence of the expected" (MHC class I). He formulated the hypothesis of "recognition of missing self" (Karre 2008; Karre et al. 1986), a concept that was supported by multiple subsequent studies (Hoglund et al. 1991; Liao et al. 1991; Pende et al. 1998). Several components of the major histocompatibility complex are upregulated by IFNs and direct transcriptional targets of STAT1 (as e.g. MHC class I heavy and light chain, and the accessory molecules TAP1, LMP2, LMP7) (Lee et al. 1999). Thus, it is not surprising that *Stat1*-deficiency severely impairs MHC class I expression (Kamiya et al. 2004; Lee et al. 1999; Meraz et al. 1996). Accordingly, tumor cells devoid of *Stat1* and thus low in MHC class I expression are highly susceptible to NK cell-mediated lysis (Kovacic et al. 2006). At the flip side of the coin, *Stat1*-deficiency impairs NK cell maturation and leads to a severe reduction in NK cell cytotoxicity (Lee et al. 2000; Robbins et al. 2005). As a consequence tumor surveillance by *Stat1*-deficient NK cells is highly impaired (Lee et al. 2000). It is currently unclear whether this is the consequence of cell intrinsic effects within the NK cell compartment or if STAT1 is required for NK cell education, which cannot take place efficiently in *Stat1*-deficient animals due to reduced MHC class I levels. The generation of mice lacking *Stat1* in NK cells only will help to clarify this issue and to shed light on the topic of NK cell maturation. Interestingly, in syngenic *Stat1*-deficient leukemia models, the combination of impaired NK cell cytotoxicity and reduced MHC class I expression on leukemic cells resulted in increased disease latency of *Stat1*-deficient mice (Kovacic et al. 2006). Thus, the loss of MHC class I on the leukemic cells dominated tumor surveillance and led to an increased clearance of the leukemic cells despite the reduced cytotoxic ability of *Stat1*-deficient NK cells.

Similarly and as expected, genetically modified mice targeting the prominent NK cell effector molecules perforin (van den Broek et al. 1996) and IFN-γ (Street et al. 2001; Wendel et al. 2008) or the NK cell activating receptor NKG2D (Guerra et al. 2008) show a severe reduction in NK cell-mediated tumor surveillance of metastasizing prostate and mammary carcinomas and methylcholanthrene-induced sarcomas. Furthermore, it is plausible that mouse strains harboring reduced NK cell numbers fail to induce proper NK cell-mediated tumor clearance, as it was shown for *Irf1*- (Duncan et al. 1996) and *Irf2*-deficient mice (Lohoff et al. 2000) (Table 1).

According to their important roles in regulating cytokine-dependent immunity, the components of the JAK/STAT pathway play distinct roles in the fate of immune responses in the tumor microenvironment, either promoting or inhibiting cancer (Yu et al. 2009). STAT3 has recently been assigned a key role in this process with rather unexpected outcomes. Since STAT3 mediates the expression of cytokines, growth factors and angiogenic factors, and the corresponding receptors in turn activate STAT3, it is one of the central transcriptional regulators that mediate the crosstalk between tumor and immune cells (Wang et al. 2004). Tumor-infiltrating NK cells show constitutively activated STAT3, which is thought to be triggered by interferons and IL-21 and in turn leads to enhanced NK cell cytotoxicity in vivo (Eriksen et al. 2009). Inhibitor studies and mouse models targeting STAT3 have given controversial results in respect of the role of STAT3 signaling on immune surveillance of tumors. To bypass the requirement of STAT3 during embryogenesis (Takeda et al. 1997), experiments by Kortylewski et al. (2005) were conducted in conditional knockout mice using the *Mx1-Cre-loxP* system. Intriguingly, the cytotoxic activity of NK cells per se is not affected by deletion of *Stat3*, but *Stat3*-deficient NK cells show enhanced cytotoxicity after in vivo challenge with B16F10 tumor cells. This indicates the complexity of intercellular communication between immune and tumor cells.

Presumably the most potent inducer of NK cell activity is IL-12, which mainly signals via TYK2/JAK2, STAT4 and T-bet and results in robust IFN-γ production (Shimoda et al. 2002; Thierfelder et al. 1996). Therefore one would expect that upon deletion of one of these signaling components NK cell-mediated tumor surveillance is diminished. Indeed, mice deficient in *Tyk2* developed NK cell-surveilled leukemia with a higher incidence and shortened latency compared with wild type controls (Stoiber et al. 2004). *Tbx21*-deficient NK cells show reduced in vivo survival, inefficient tumor cell killing and poor IFN-γ production, rendering them highly inefficient in establishing a proper anti-tumor activity against B16F10 melanoma cells (Werneck et al. 2008).

Table 1 Lessons from knockout mice

Gene	Gene targeted mouse phenotype	General observations	Gene targeted NK cell phenotype
Jak1	Die perinatally (Rodig et al. 1998)	Neurological and lymphoid deficits, signaling failure (Rodig et al. 1998)	*n.d.*
Jak2	Embryonically lethal (Neubauer et al. 1998)	Anemic, lack erythrocytes (Neubauer et al. 1998)	*n.d.*
Jak3	Viable (Park et al. 1995)	Severe defects in lymphoid cells, lack of lymph nodes (Park et al. 1995)	Absent (Park et al. 1995)
Tyk2	Viable (Karaghiosoff et al. 2000)	Impaired cytokine signaling to IFNs and IL-12 (Karaghiosoff et al. 2000)	Impaired IL-12/IL-18 signaling and STAT4 activation (Shimoda et al. 2002); reduced cytotoxicity and Leishmania clearance (Schleicher et al. 2004); impaired tumor surveillance (Stoiber et al. 2004)
Stat1	Viable (Durbin et al. 1996; Meraz et al. 1996)	Highly sensitive to infections, cells are unresponsive to IFNs, reduced MHC class I (Kennedy et al. 2000)	Impaired cytotoxicity, reduced IFN-γ production, defect in tumor surveillance (Lee et al. 2000); maturation deficiency (Robbins et al. 2005)
Stat2	Viable (Park et al. 2000)	Sensitive to viral infections (Park et al. 2000)	*n.d.*
Stat3	Embryonically lethal (Takeda et al. 1997)		Increased cytotoxicity (YAC-1) (Kortylewski et al. 2005); co-stimulation of IL-21/INF-α activates STAT3 and cytotoxicity (Eriksen et al. 2009)
Stat4	Viable (Kaplan et al. 1996)		Failure to enhance NK cytotoxicity in response to IL-12 (Kaplan et al. 1996)
Stat5	Embryonically lethal (Cui et al. 2004)		Developmental block, reduced number of NK cells (Eckelhart et al. 2011)
Stat6	Viable (Wang et al. 2004)	Impaired IL-4 signaling, defect in Th2 cell differentiation (Wang et al. 2004)	*n.d.*
Irf1	Viable (Matsuyama et al. 1993)	BM deficient to produce IL-15 (Ogasawara et al. 1998)	Reduced NK cell numbers, no killing capacity, no IFN-γ production, defective tumor surveillance (RMA-S) (Duncan et al. 1996)
Irf2	Viable (Matsuyama et al. 1993)		Reduced NK cell numbers, defective development and RMA-S tumor rejection (Lohoff et al. 2000); selective loss of mature NK cells due to premature death (Taki et al. 2005)

(continued)

Table 1 (continued)

Gene	Gene targeted mouse phenotype	General observations	Gene targeted NK cell phenotype
Tbx21 (T-bet)	Viable (Szabo et al. 2002)	Defective differentiation and IFN-γ production of CD4+ T cells, but not CD8+ T cells (Szabo et al. 2002)	Reduced IFN-γ production, impaired cytotoxicity (Szabo et al. 2002)
Eomes	Embryonically lethal (Russ et al. 2000)		Severe reduction of peripheral NK cells, fail to mature terminally, intrinsic increased proliferation and apoptosis, lack granzyme B and perforin (Townsend et al. 2004); reduced tumor surveillance (B16F10) (Werneck et al. 2008)

n.d. not determined

A consolidated view of all this data reveals the central role of JAKs and STATs in NK cell-mediated tumor surveillance. Both, NK and cytotoxic T cells are ultimately responsible for the defense against malignant cells. Regrettably, the knowledge about distinct functions of signaling molecules within NK cells is limited, owing to the lack of tools for NK cell specific gene deletion. Recently Eckelhart et al. reported on a *Ncr1*-iCreTg mouse, which for the first time allows the conditional mutagenesis specifically in NK cells (Eckelhart et al. 2011). Additionally, they showed that by crossing the *Ncr1*-iCreTg to *Stat5fl/fl* mice, NK cells were almost completely absent causing the virtual abrogation of NK cell-mediated tumor surveillance without affecting T cell-controlled immunity. This mouse model will be very helpful in the future to determine the impact of NK cells in the surveillance of different tumor types. Moreover, it represents a powerful tool to assess the function of individual members of the JAK/STAT signaling pathway in the process of NK cell-mediated tumor surveillance.

STATs as Therapeutic Targets: The Good and the Evil

In contrast to traditional anticancer therapy, which targets all rapidly dividing cells, novel drugs are designed to intervene in signal transduction pathways. Those signaling interceptors aim to ensure specificity and to reduce side effects associated with classical anticancer therapy. As cytotoxic drugs may compromise the immune response, special attention should be paid to the interference of newly developed drugs with the immune system. An example in this regard was published by Zebedin et al. (2008) and deals with the adverse side effects coming along with the inhibition of PI3Kδ. The lesson we learned from this study was that inhibiting PI3Kδ in mice suffering from leukemia has no beneficial effects in vivo. Drugs interfering with this kinase may act as a double-edged sword by inhibiting both, leukemic cells as well as the NK cell compartment.

The involvement of JAK/STAT signaling in the pathophysiology of various diseases opens a window of opportunity for therapeutic intervention. Particularly in cancer therapy, STAT3 and STAT5 are obvious targets for novel therapeutic strategies (Hoelbl et al. 2006, 2010; Turkson and Jove 2000; Yue and Turkson 2009) and the pharmaceutical industry is trying to develop specific drugs interfering with this pathway. Considering recent publications, drugs targeting STAT5 would not only affect the viability of the leukemic cells (Hoelbl et al. 2010) but they also would profoundly impair the NK cell compartment (Eckelhart et al. 2011). This certainly represents a significant disadvantage for leukemia treatment and is of particular importance when treating patients with minimal residual disease where NK cells are considered important players in the clearance of remaining tumor cells. In this regard it is interesting to note that signal interceptors used to treat patients with BCR/ABL-induced leukemia such as imatinib or dasatinib may also enhance or stimulate immune control (Borg et al. 2004; Kreutzman et al. 2010; Rohon et al. 2010). The underlying mechanisms are currently unclear. On the other hand, NK cell activity may also represent the prime target of desire and JAK/STAT signaling could be blocked on purpose to inhibit over-boarding and self-destructive effects of the NK compartment. JAK2, JAK3 and TYK2 inhibitors are available and are currently intensively studied as treatment options for diseases such as rheumatoid arthritis (Riese et al. 2010) and psoriasis (Strange et al. 2010).

Concluding Remarks

We are only beginning to understand the many ways how NK cells shape and contribute to immune responses and pathophysiological processes. Understanding the contribution of JAKs and STATs will be an important aspect of further insight into the role, regulation and function of NK cells.

References

Anfossi N, Andre P, Guia S, Falk CS, Roetynck S, Stewart CA, Breso V, Frassati C, Reviron D, Middleton D, Romagne F, Ugolini S, Vivier E (2006) Human NK cell education by inhibitory receptors for MHC class I. Immunity 25:331–342

Arase H, Arase N, Saito T (1995) Fas-mediated cytotoxicity by freshly isolated natural killer cells. J Exp Med 181:1235–1238

Bacon CM, Petricoin EF III, Ortaldo JR, Rees RC, Larner AC, Johnston JA, O'Shea JJ (1995) Interleukin 12 induces tyrosine phosphorylation and activation of STAT4 in human lymphocytes. Proc Natl Acad Sci USA 92:7307–7311

Biron CA (1998) Role of early cytokines, including alpha and beta interferons (IFN-alpha/beta), in innate and adaptive immune responses to viral infections. Semin Immunol 10:383–390

Biron CA, Su HC, Orange JS (1996) Function and regulation of natural killer (NK) cells during viral infections: characterization of responses in vivo. Methods 9:379–393

Bjorkstrom NK, Lindgren T, Stoltz M, Fauriat C, Braun M, Evander M, Michaelsson J, Malmberg KJ, Klingstrom J, Ahlm C, Ljunggren HG (2010) Rapid expansion and long-term persistence of elevated NK cell numbers in humans infected with hantavirus. J Exp Med 208:13–21

Boggs SS, Trevisan M, Patrene K, Geogopoulos K (1998) Lack of natural killer cell precursors in fetal liver of ikaros knockout mutant mice. Nat Immun 16:137–145

Borg C, Terme M, Taieb J, Menard C, Flament C, Robert C, Maruyama K, Wakasugi H, Angevin E, Thielemans K, Le Cesne A, Chung-Scott V, Lazar V, Tchou I, Crepineau F, Lemoine F, Bernard J, Fletcher JA, Turhan A, Blay JY, Spatz A, Emile JF, Heinrich MC, Mecheri S, Tursz T, Zitvogel L (2004) Novel mode of action of c-kit tyrosine kinase inhibitors leading to NK cell-dependent antitumor effects. J Clin Invest 114:379–388

Bovenschen N, Kummer JA (2010) Orphan granzymes find a home. Immunol Rev 235:117–127

Brodin P, Hoglund P (2008) Beyond licensing and disarming: a quantitative view on NK-cell education. Eur J Immunol 38:2934–2937

Colucci F, Samson SI, DeKoter RP, Lantz O, Singh H, Di Santo JP (2001) Differential requirement for the transcription factor PU.1 in the generation of natural killer cells versus B and T cells. Blood 97:2625–2632

Cui Y, Riedlinger G, Miyoshi K, Tang W, Li C, Deng CX, Robinson GW, Hennighausen L (2004) Inactivation of Stat5 in mouse mammary epithelium during pregnancy reveals distinct functions in cell proliferation, survival, and differentiation. Mol Cell Biol 24:8037–8047

DeBlaker-Hohe DF, Yamauchi A, Yu CR, Horvath-Arcidiacono JA, Bloom ET (1995) IL-12 synergizes with IL-2 to induce lymphokine-activated cytotoxicity and perforin and granzyme gene expression in fresh human NK cells. Cell Immunol 165:33–43

Di Santo JP, Vosshenrich CA, Satoh-Takayama N (2010) A 'natural' way to provide innate mucosal immunity. Curr Opin Immunol 22:435–441

Duncan GS, Mittrucker HW, Kagi D, Matsuyama T, Mak TW (1996) The transcription factor interferon regulatory factor-1 is essential for natural killer cell function in vivo. J Exp Med 184:2043–2048

Dunn GP, Bruce AT, Ikeda H, Old LJ, Schreiber RD (2002) Cancer immunoediting: from immunosurveillance to tumor escape. Nat Immunol 3:991–998

Durbin JE, Hackenmiller R, Simon MC, Levy DE (1996) Targeted disruption of the mouse Stat1 gene results in compromised innate immunity to viral disease. Cell 84:443–450

Eckelhart E, Warsch W, Zebedin E, Simma O, Stoiber D, Kolbe T, Rulicke T, Mueller M, Casanova E, Sexl V (2011) A novel Ncr1-Cre mouse reveals the essential role of STAT5 for NK cell survival and development. Blood 117:1565–1573

Ehrlich P (1908) Über den jetzigen Stand der Karzinomforschung. Presented at the Vereinigung für Wissenschaftliche Arbeit, University of Amsterdam, Amsterdam, 1 June 1908

Elliott JM, Wahle JA, Yokoyama WM (2010) MHC class I-deficient natural killer cells acquire a licensed phenotype after transfer into an MHC class I-sufficient environment. J Exp Med 207:2073–2079

Eriksen KW, Sondergaard H, Woetmann A, Krejsgaard T, Skak K, Geisler C, Wasik MA, Odum N (2009) The combination of IL-21 and IFN-alpha boosts STAT3 activation, cytotoxicity and experimental tumor therapy. Mol Immunol 46:812–820

Froelich CJ, Pardo J, Simon MM (2009) Granule-associated serine proteases: granzymes might not just be killer proteases. Trends Immunol 30:117–123

Gascoyne DM, Long E, Veiga-Fernandes H, de Boer J, Williams O, Seddon B, Coles M, Kioussis D, Brady HJ (2009) The basic leucine zipper transcription factor E4BP4 is essential for natural killer cell development. Nat Immunol 10:1118–1124

Gidlund M, Orn A, Wigzell H, Senik A, Gresser I (1978) Enhanced NK cell activity in mice injected with interferon and interferon inducers. Nature 273:759–761

Guerra N, Tan YX, Joncker NT, Choy A, Gallardo F, Xiong N, Knoblaugh S, Cado D, Greenberg NM, Raulet DH (2008) NKG2D-deficient mice are defective in tumor surveillance in models of spontaneous malignancy. Immunity 28:571–580

Gur C, Porgador A, Elboim M, Gazit R, Mizrahi S, Stern-Ginossar N, Achdout H, Ghadially H, Dor Y, Nir T, Doviner V, Hershkovitz O, Mendelson M, Naparstek Y, Mandelboim O (2010) The activating receptor NKp46 is essential for the development of type 1 diabetes. Nat Immunol 11:121–128

Hoelbl A, Kovacic B, Kerenyi MA, Simma O, Warsch W, Cui Y, Beug H, Hennighausen L, Moriggl R, Sexl V (2006) Clarifying the role of Stat5 in lymphoid development and Abelson-induced transformation. Blood 107:4898–4906

Hoelbl A, Schuster C, Kovacic B, Zhu B, Wickre M, Hoelzl MA, Fajmann S, Grebien F, Warsch W, Stengl G, Hennighausen L, Poli V, Beug H, Moriggl R, Sexl V (2010) Stat5 is indispensable for the maintenance of bcr/abl-positive leukaemia. EMBO Mol Med 2:98–110

Hoglund P, Glas R, Ohlen C, Ljunggren HG, Karre K (1991) Alteration of the natural killer repertoire in H-2 transgenic mice: specificity of rapid lymphoma cell clearance determined by the H-2 phenotype of the target. J Exp Med 174:327–334

Huntington ND, Vosshenrich CA, Di Santo JP (2007) Developmental pathways that generate natural-killer-cell diversity in mice and humans. Nat Rev Immunol 7:703–714

Ikawa T, Fujimoto S, Kawamoto H, Katsura Y, Yokota Y (2001) Commitment to natural killer cells requires the helix-loop-helix inhibitor Id2. Proc Natl Acad Sci USA 98:5164–5169

Imada K, Bloom ET, Nakajima H, Horvath-Arcidiacono JA, Udy GB, Davey HW, Leonard WJ (1998) Stat5b is essential for natural killer cell-mediated proliferation and cytolytic activity. J Exp Med 188:2067–2074

Intlekofer AM, Takemoto N, Wherry EJ, Longworth SA, Northrup JT, Palanivel VR, Mullen AC, Gasink CR, Kaech SM, Miller JD, Gapin L, Ryan K, Russ AP, Lindsten T, Orange JS, Goldrath AW, Ahmed R, Reiner SL (2005) Effector and memory CD8+ T cell fate coupled by T-bet and eomesodermin. Nat Immunol 6:1236–1244

Jonсker NT, Raulet DH (2008) Regulation of NK cell responsiveness to achieve self-tolerance and maximal responses to diseased target cells. Immunol Rev 224:85–97

Kallies A, Carotta S, Huntington ND, Bernard NJ, Tarlinton DM, Smyth MJ, Nutt SL (2011) A role for Blimp-1 in the transcriptional network controlling natural killer cell maturation. Blood 117:1869–1879

Kamiya S, Owaki T, Morishima N, Fukai F, Mizuguchi J, Yoshimoto T (2004) An indispensable role for STAT1 in IL-27-induced T-bet expression but not proliferation of naive CD4+ T cells. J Immunol 173:3871–3877

Kamizono S, Duncan GS, Seidel MG, Morimoto A, Hamada K, Grosveld G, Akashi K, Lind EF, Haight JP, Ohashi PS, Look AT, Mak TW (2009) Nfil3/E4bp4 is required for the development and maturation of NK cells in vivo. J Exp Med 206:2977–2986

Kaplan MH, Sun YL, Hoey T, Grusby MJ (1996) Impaired IL-12 responses and enhanced development of Th2 cells in Stat4-deficient mice. Nature 382:174–177

Karaghiosoff M, Neubauer H, Lassnig C, Kovarik P, Schindler H, Pircher H, McCoy B, Bogdan C, Decker T, Brem G, Pfeffer K, Muller M (2000) Partial impairment of cytokine responses in Tyk2-deficient mice. Immunity 13:549–560

Karre K (2008) Natural killer cell recognition of missing self. Nat Immunol 9:477–480

Karre K, Ljunggren HG, Piontek G, Kiessling R (1986) Selective rejection of H-2-deficient lymphoma variants suggests alternative immune defence strategy. Nature 319:675–678

Kennedy MK, Glaccum M, Brown SN, Butz EA, Viney JL, Embers M, Matsuki N, Charrier K, Sedger L, Willis CR, Brasel K, Morrissey PJ, Stocking K, Schuh JC, Joyce S, Peschon JJ (2000) Reversible defects in natural killer and memory CD8 T cell lineages in interleukin 15-deficient mice. J Exp Med 191:771–780

Kiessling R, Klein E, Wigzell H (1975) "Natural" killer cells in the mouse. I. Cytotoxic cells with specificity for mouse Moloney leukemia cells. Specificity and distribution according to genotype. Eur J Immunol 5:112–117

Kim S, Iizuka K, Kang HS, Dokun A, French AR, Greco S, Yokoyama WM (2002) In vivo developmental stages in murine natural killer cell maturation. Nat Immunol 3:523–528

Kortylewski M, Kujawski M, Wang T, Wei S, Zhang S, Pilon-Thomas S, Niu G, Kay H, Mule J, Kerr WG, Jove R, Pardoll D, Yu H (2005) Inhibiting Stat3 signaling in the hematopoietic system elicits multicomponent antitumor immunity. Nat Med 11:1314–1321

Kovacic B, Stoiber D, Moriggl R, Weisz E, Ott RG, Kreibich R, Levy DE, Beug H, Freissmuth M, Sexl V (2006) STAT1 acts as a tumor promoter for leukemia development. Cancer Cell 10:77–87

Kreutzman A, Juvonen V, Kairisto V, Ekblom M, Stenke L, Seggewiss R, Porkka K, Mustjoki S (2010) Mono/oligoclonal T and NK cells are common in chronic myeloid leukemia patients at diagnosis and expand during dasatinib therapy. Blood 116:772–782

Lanier LL (2008) Up on the tightrope: natural killer cell activation and inhibition. Nat Immunol 9:495–502

Lauwerys BR, Renauld JC, Houssiau FA (1999) Synergistic proliferation and activation of natural killer cells by interleukin 12 and interleukin 18. Cytokine 11:822–830

Lee CK, Gimeno R, Levy DE (1999) Differential regulation of constitutive major histocompatibility complex class I expression in T and B lymphocytes. J Exp Med 190:1451–1464

Lee CK, Rao DT, Gertner R, Gimeno R, Frey AB, Levy DE (2000) Distinct requirements for IFNs and STAT1 in NK cell function. J Immunol 165:3571–3577

Leon F, Roldan E, Sanchez L, Camarero C, Bootello A, Roy G (2003) Human small-intestinal epithelium contains functional natural killer lymphocytes. Gastroenterology 125:345–356

Liang S, Wei H, Sun R, Tian Z (2003) IFNalpha regulates NK cell cytotoxicity through STAT1 pathway. Cytokine 23:190–199

Liao NS, Bix M, Zijlstra M, Jaenisch R, Raulet D (1991) MHC class I deficiency: susceptibility to natural killer (NK) cells and impaired NK activity. Science 253:199–202

Lohoff M, Duncan GS, Ferrick D, Mittrucker HW, Bischof S, Prechtl S, Rollinghoff M, Schmitt E, Pahl A, Mak TW (2000) Deficiency in the transcription factor interferon regulatory factor (IRF)-2 leads to severely compromised development of natural killer and T helper type 1 cells. J Exp Med 192:325–336

Long EO (2008) Negative signaling by inhibitory receptors: the NK cell paradigm. Immunol Rev 224:70–84

Loza MJ, Perussia B (2004) Differential regulation of NK cell proliferation by type I and type II IFN. Int Immunol 16:23–32

Manaster I, Mandelboim O (2010) The unique properties of uterine NK cells. Am J Reprod Immunol 63:434–444

Marques L, Brucet M, Lloberas J, Celada A (2004) STAT1 regulates lipopolysaccharide- and TNF-alpha-dependent expression of transporter associated with antigen processing 1 and low molecular mass polypeptide 2 genes in macrophages by distinct mechanisms. J Immunol 173:1103–1110

Martin MP, Nelson G, Lee JH, Pellett F, Gao X, Wade J, Wilson MJ, Trowsdale J, Gladman D, Carrington M (2002) Cutting edge: susceptibility to psoriatic arthritis: influence of activating killer Ig-like receptor genes in the absence of specific HLA-C alleles. J Immunol 169:2818–2822

Matikainen S, Paananen A, Miettinen M, Kurimoto M, Timonen T, Julkunen I, Sareneva T (2001) IFN-alpha and IL-18 synergistically enhance IFN-gamma production in human NK cells: differential regulation of Stat4 activation and IFN-gamma gene expression by IFN-alpha and IL-12. Eur J Immunol 31:2236–2245

Matsuyama T, Kimura T, Kitagawa M, Pfeffer K, Kawakami T, Watanabe N, Kundig TM, Amakawa R, Kishihara K, Wakeham A et al (1993) Targeted disruption of IRF-1 or IRF-2 results in abnormal type I IFN gene induction and aberrant lymphocyte development. Cell 75:83–97

Medvedev AE, Johnsen AC, Haux J, Steinkjer B, Egeberg K, Lynch DH, Sundan A, Espevik T (1997) Regulation of Fas and Fas-ligand expression in NK cells by cytokines and the involvement of Fas-ligand in NK/LAK cell-mediated cytotoxicity. Cytokine 9:394–404

Meraz MA, White JM, Sheehan KC, Bach EA, Rodig SJ, Dighe AS, Kaplan DH, Riley JK, Greenlund AC, Campbell D, Carver-Moore K, DuBois RN, Clark R, Aguet M, Schreiber RD (1996) Targeted disruption of the Stat1 gene in mice reveals unexpected physiologic specificity in the JAK-STAT signaling pathway. Cell 84:431–442

Miyagi T, Gil MP, Wang X, Louten J, Chu WM, Biron CA (2007) High basal STAT4 balanced by STAT1 induction to control type 1 interferon effects in natural killer cells. J Exp Med 204:2383–2396

Miyazaki T, Kawahara A, Fujii H, Nakagawa Y, Minami Y, Liu ZJ, Oishi I, Silvennoinen O, Witthuhn BA, Ihle JN et al (1994) Functional activation of Jak1 and Jak3 by selective association with IL-2 receptor subunits. Science 266:1045–1047

Moriggl R, Topham DJ, Teglund S, Sexl V, McKay C, Wang D, Hoffmeyer A, van Deursen J, Sangster MY, Bunting KD, Grosveld GC, Ihle JN (1999) Stat5 is required for IL-2-induced cell cycle progression of peripheral T cells. Immunity 10:249–259

Morris SC, Orekhova T, Meadows MJ, Heidorn SM, Yang J, Finkelman FD (2006) IL-4 induces in vivo production of IFN-gamma by NK and NKT cells. J Immunol 176:5299–5305

Neubauer H, Cumano A, Muller M, Wu H, Huffstadt U, Pfeffer K (1998) Jak2 deficiency defines an essential developmental checkpoint in definitive hematopoiesis. Cell 93:397–409

Nguyen KB, Cousens LP, Doughty LA, Pien GC, Durbin JE, Biron CA (2000) Interferon alpha/beta-mediated inhibition and promotion of interferon gamma: STAT1 resolves a paradox. Nat Immunol 1:70–76

Nguyen KB, Salazar-Mather TP, Dalod MY, Van Deusen JB, Wei XQ, Liew FY, Caligiuri MA, Durbin JE, Biron CA (2002) Coordinated and distinct roles for IFN-alpha beta, IL-12, and IL-15 regulation of NK cell responses to viral infection. J Immunol 169:4279–4287

Nosaka T, van Deursen JM, Tripp RA, Thierfelder WE, Witthuhn BA, McMickle AP, Doherty PC, Grosveld GC, Ihle JN (1995) Defective lymphoid development in mice lacking Jak3. Science 270:800–802

Ogasawara K, Hida S, Azimi N, Tagaya Y, Sato T, Yokochi-Fukuda T, Waldmann TA, Taniguchi T, Taki S (1998) Requirement for IRF-1 in the microenvironment supporting development of natural killer cells. Nature 391:700–703

O'Leary JG, Goodarzi M, Drayton DL, von Andrian UH (2006) T cell- and B cell-independent adaptive immunity mediated by natural killer cells. Nat Immunol 7:507–516

Orange JS, Biron CA (1996a) An absolute and restricted requirement for IL-12 in natural killer cell IFN-gamma production and antiviral defense. Studies of natural killer and T cell responses in contrasting viral infections. J Immunol 156:1138–1142

Orange JS, Biron CA (1996b) Characterization of early IL-12, IFN-alphabeta, and TNF effects on antiviral state and NK cell responses during murine cytomegalovirus infection. J Immunol 156:4746–4756

Orr MT, Lanier LL (2010) Natural killer cell education and tolerance. Cell 142:847–856

Park SY, Saijo K, Takahashi T, Osawa M, Arase H, Hirayama N, Miyake K, Nakauchi H, Shirasawa T, Saito T (1995) Developmental defects of lymphoid cells in Jak3 kinase-deficient mice. Immunity 3:771–782

Park C, Li S, Cha E, Schindler C (2000) Immune response in Stat2 knockout mice. Immunity 13:795–804

Pearce EL, Mullen AC, Martins GA, Krawczyk CM, Hutchins AS, Zediak VP, Banica M, DiCioccio CB, Gross DA, Mao CA, Shen H, Cereb N, Yang SY, Lindsten T, Rossant J, Hunter CA, Reiner SL (2003) Control of effector CD8+ T cell function by the transcription factor Eomesodermin. Science 302:1041–1043

Pende D, Accame L, Pareti L, Mazzocchi A, Moretta A, Parmiani G, Moretta L (1998) The susceptibility to natural killer cell-mediated lysis of HLA class I-positive melanomas reflects the expression of insufficient amounts of different HLA class I alleles. Eur J Immunol 28:2384–2394

Rani MR, Pandalai S, Shrock J, Almasan A, Ransohoff RM (2007) Requirement of catalytically active Tyk2 and accessory signals for the induction of TRAIL mRNA by IFN-beta. J Interferon Cytokine Res 27:767–779

Raulet DH, Vance RE (2006) Self-tolerance of natural killer cells. Nat Rev Immunol 6:520–531

Riese RJ, Krishnaswami S, Kremer J (2010) Inhibition of JAK kinases in patients with rheumatoid arthritis: scientific rationale and clinical outcomes. Best Pract Res Clin Rheumatol 24:513–526

Robbins SH, Tessmer MS, Van Kaer L, Brossay L (2005) Direct effects of T-bet and MHC class I expression, but not STAT1, on peripheral NK cell maturation. Eur J Immunol 35:757–765

Rodig SJ, Meraz MA, White JM, Lampe PA, Riley JK, Arthur CD, King KL, Sheehan KC, Yin L, Pennica D, Johnson EM Jr, Schreiber RD (1998) Disruption of the Jak1 gene demonstrates obligatory and nonredundant roles of the Jaks in cytokine-induced biologic responses. Cell 93:373–383

Rohon P, Porkka K, Mustjoki S (2010) Immunoprofiling of patients with chronic myeloid leukemia at diagnosis and during tyrosine kinase inhibitor therapy. Eur J Haematol 85:387–398

Rosmaraki EE, Douagi I, Roth C, Colucci F, Cumano A, Di Santo JP (2001) Identification of committed NK cell progenitors in adult murine bone marrow. Eur J Immunol 31:1900–1909

Roth C, Carlyle JR, Takizawa H, Raulet DH (2000) Clonal acquisition of inhibitory Ly49 receptors on developing NK cells is successively restricted and regulated by stromal class I MHC. Immunity 13:143–153

Russ AP, Wattler S, Colledge WH, Aparicio SA, Carlton MB, Pearce JJ, Barton SC, Surani MA, Ryan K, Nehls MC, Wilson V, Evans MJ (2000) Eomesodermin is required for mouse trophoblast development and mesoderm formation. Nature 404:95–99

Salcedo TW, Azzoni L, Wolf SF, Perussia B (1993) Modulation of perforin and granzyme messenger RNA expression in human natural killer cells. J Immunol 151:2511–2520

Satoh-Takayama N, Vosshenrich CA, Lesjean-Pottier S, Sawa S, Lochner M, Rattis F, Mention JJ, Thiam K, Cerf-Bensussan N, Mandelboim O, Eberl G, Di Santo JP (2008) Microbial flora drives interleukin 22 production in intestinal NKp46+ cells that provide innate mucosal immune defense. Immunity 29:958–970

Schleicher U, Mattner J, Blos M, Schindler H, Rollinghoff M, Karaghiosoff M, Muller M, Werner-Felmayer G, Bogdan C (2004) Control of Leishmania major in the absence of Tyk2 kinase. Eur J Immunol 34:519–529

Shimoda K, Tsutsui H, Aoki K, Kato K, Matsuda T, Numata A, Takase K, Yamamoto T, Nukina H, Hoshino T, Asano Y, Gondo H, Okamura T, Okamura S, Nakayama K, Nakanishi K, Niho Y, Harada M (2002) Partial impairment of interleukin-12 (IL-12) and IL-18 signaling in Tyk2-deficient mice. Blood 99:2094–2099

Smyth MJ, Hayakawa Y, Takeda K, Yagita H (2002) New aspects of natural-killer-cell surveillance and therapy of cancer. Nat Rev Cancer 2:850–861

Stoiber D, Kovacic B, Schuster C, Schellack C, Karaghiosoff M, Kreibich R, Weisz E, Artwohl M, Kleine OC, Muller M, Baumgartner-Parzer S, Ghysdael J, Freissmuth M, Sexl V (2004) TYK2 is a key regulator of the surveillance of B lymphoid tumors. J Clin Invest 114:1650–1658

Strange A, Capon F, Spencer CC, Knight J, Weale ME, Allen MH, Barton A, Band G, Bellenguez C, Bergboer JG, Blackwell JM, Bramon E, Bumpstead SJ, Casas JP, Cork MJ, Corvin A, Deloukas P, Dilthey A, Duncanson A, Edkins S, Estivill X, Fitzgerald O, Freeman C, Giardina E, Gray E, Hofer A, Huffmeier U, Hunt SE, Irvine AD, Jankowski J, Kirby B, Langford C, Lascorz J, Leman J, Leslie S, Mallbris L, Markus HS, Mathew CG, McLean WH, McManus R, Mossner R, Moutsianas L, Naluai AT, Nestle FO, Novelli G, Onoufriadis A, Palmer CN, Perricone C, Pirinen M, Plomin R, Potter SC, Pujol RM, Rautanen A, Riveira-Munoz E, Ryan AW, Salmhofer W, Samuelsson L, Sawcer SJ, Schalkwijk J, Smith CH, Stahle M, Su Z, Tazi-Ahnini R, Traupe H, Viswanathan AC, Warren RB, Weger W, Wolk K, Wood N, Worthington J, Young HS, Zeeuwen PL, Hayday A, Burden AD, Griffiths CE, Kere J, Reis A, McVean G, Evans DM, Brown MA, Barker JN, Peltonen L, Donnelly P, Trembath RC (2010) A genome-wide association study identifies new psoriasis susceptibility loci and an interaction between HLA-C and ERAP1. Nat Genet 42:985–990

Street SE, Cretney E, Smyth MJ (2001) Perforin and interferon-gamma activities independently control tumor initiation, growth, and metastasis. Blood 97:192–197

Strengell M, Matikainen S, Siren J, Lehtonen A, Foster D, Julkunen I, Sareneva T (2003) IL-21 in synergy with IL-15 or IL-18 enhances IFN-gamma production in human NK and T cells. J Immunol 170:5464–5469

Sun JC, Lanier LL (2009) Natural killer cells remember: an evolutionary bridge between innate and adaptive immunity? Eur J Immunol 39:2059–2064

Sun JC, Beilke JN, Lanier LL (2009) Adaptive immune features of natural killer cells. Nature 457:557–561

Szabo SJ, Sullivan BM, Stemmann C, Satoskar AR, Sleckman BP, Glimcher LH (2002) Distinct effects of T-bet in TH1 lineage commitment and IFN-gamma production in CD4 and CD8 T cells. Science 295:338–342

Tagliabue A, Befus AD, Clark DA, Bienenstock J (1982) Characteristics of natural killer cells in the murine intestinal epithelium and lamina propria. J Exp Med 155:1785–1796
Takeda K, Noguchi K, Shi W, Tanaka T, Matsumoto M, Yoshida N, Kishimoto T, Akira S (1997) Targeted disruption of the mouse Stat3 gene leads to early embryonic lethality. Proc Natl Acad Sci USA 94:3801–3804
Taki S, Nakajima S, Ichikawa E, Saito T, Hida S (2005) IFN regulatory factor-2 deficiency revealed a novel checkpoint critical for the generation of peripheral NK cells. J Immunol 174:6005–6012
Thierfelder WE, van Deursen JM, Yamamoto K, Tripp RA, Sarawar SR, Carson RT, Sangster MY, Vignali DA, Doherty PC, Grosveld GC, Ihle JN (1996) Requirement for Stat4 in interleukin-12-mediated responses of natural killer and T cells. Nature 382:171–174
Townsend MJ, Weinmann AS, Matsuda JL, Salomon R, Farnham PJ, Biron CA, Gapin L, Glimcher LH (2004) T-bet regulates the terminal maturation and homeostasis of NK and Valpha14i NKT cells. Immunity 20:477–494
Trambas CM, Griffiths GM (2003) Delivering the kiss of death. Nat Immunol 4:399–403
Trinchieri G, Matsumoto-Kobayashi M, Clark SC, Seehra J, London L, Perussia B (1984) Response of resting human peripheral blood natural killer cells to interleukin 2. J Exp Med 160:1147–1169
Turkson J, Jove R (2000) STAT proteins: novel molecular targets for cancer drug discovery. Oncogene 19:6613–6626
Uemura A, Takehara T, Miyagi T, Suzuki T, Tatsumi T, Ohkawa K, Kanto T, Hiramatsu N, Hayashi N (2010) Natural killer cell is a major producer of interferon gamma that is critical for the IL-12-induced anti-tumor effect in mice. Cancer Immunol Immunother 59:453–463
van den Broek ME, Kagi D, Ossendorp F, Toes R, Vamvakas S, Lutz WK, Melief CJ, Zinkernagel RM, Hengartner H (1996) Decreased tumor surveillance in perforin-deficient mice. J Exp Med 184.1781–1790
Vosshenrich CA, Ranson T, Samson SI, Corcuff E, Colucci F, Rosmaraki EE, Di Santo JP (2005) Roles for common cytokine receptor gamma-chain-dependent cytokines in the generation, differentiation, and maturation of NK cell precursors and peripheral NK cells in vivo. J Immunol 174:1213–1221
Wang KS, Ritz J, Frank DA (1999) IL-2 induces STAT4 activation in primary NK cells and NK cell lines, but not in T cells. J Immunol 162:299–304
Wang T, Niu G, Kortylewski M, Burdelya L, Shain K, Zhang S, Bhattacharya R, Gabrilovich D, Heller R, Coppola D, Dalton W, Jove R, Pardoll D, Yu H (2004) Regulation of the innate and adaptive immune responses by Stat-3 signaling in tumor cells. Nat Med 10:48–54
Wendel M, Galani IE, Suri-Payer E, Cerwenka A (2008) Natural killer cell accumulation in tumors is dependent on IFN-gamma and CXCR3 ligands. Cancer Res 68:8437–8445
Werneck MB, Lugo-Villarino G, Hwang ES, Cantor H, Glimcher LH (2008) T-bet plays a key role in NK-mediated control of melanoma metastatic disease. J Immunol 180:8004–8010
Xu X, Fu XY, Plate J, Chong AS (1998) IFN-gamma induces cell growth inhibition by Fas-mediated apoptosis: requirement of STAT1 protein for up-regulation of Fas and FasL expression. Cancer Res 58:2832–2837
Yamamoto K, Shibata F, Miyasaka N, Miura O (2002) The human perforin gene is a direct target of STAT4 activated by IL-12 in NK cells. Biochem Biophys Res Commun 297:1245–1252
Yen JH, Moore BE, Nakajima T, Scholl D, Schaid DJ, Weyand CM, Goronzy JJ (2001) Major histocompatibility complex class I-recognizing receptors are disease risk genes in rheumatoid arthritis. J Exp Med 193:1159–1167
Yokoyama WM, Kim S (2006) How do natural killer cells find self to achieve tolerance? Immunity 24:249–257
Yu CR, Ortaldo JR, Curiel RE, Young HA, Anderson SK, Gosselin P (1999) Role of a STAT binding site in the regulation of the human perforin promoter. J Immunol 162:2785–2790
Yu H, Pardoll D, Jove R (2009) STATs in cancer inflammation and immunity: a leading role for STAT3. Nat Rev Cancer 9:798–809

Yue P, Turkson J (2009) Targeting STAT3 in cancer: how successful are we? Expert Opin Investig Drugs 18:45–56

Zamai L, Ahmad M, Bennett IM, Azzoni L, Alnemri ES, Perussia B (1998) Natural killer (NK) cell-mediated cytotoxicity: differential use of TRAIL and Fas ligand by immature and mature primary human NK cells. J Exp Med 188:2375–2380

Zebedin E, Simma O, Schuster C, Putz EM, Fajmann S, Warsch W, Eckelhart E, Stoiber D, Weisz E, Schmid JA, Pickl WF, Baumgartner C, Valent P, Piekorz RP, Freissmuth M, Sexl V (2008) Leukemic challenge unmasks a requirement for PI3Kdelta in NK cell-mediated tumor surveillance. Blood 112:4655–4664

Zompi S, Colucci F (2005) Anatomy of a murder–signal transduction pathways leading to activation of natural killer cells. Immunol Lett 97:31–39

Function of JAKs and STATs in Lymphocytes: Bench to Bedside

Alejandro Villarino*, Xiang-Ping Yang*, Kiyoshi Hirahara, Yuka Kanno, John J. O' Shea, and Arian Laurence

Abstract

There are four members of Janus Kinase family (JAK-1, -2, -3 and TYK2) and seven members of STAT family (STAT-1, -2, -3, -4, -5a, -5b and -6) in the mammalian genome, each with unique functions in immune cells. Consistent with studies in mice, genetic evidence in humans has strongly linked the JAK/STAT pathway to primary immunodeficiencies, infection, autoimmunity and cancers. The following chapter will discuss the key role played by JAKs and STATs in the lymphocytes, with special emphasis on helper T cells, which are not only critical mediators of pathogenic inflammation, but also an outstanding model system for investigating JAK/STAT biology. Recent conceptual and technological advances will be highlighted, particularly those relating to human disease and the generation of JAK/STAT based therapeutics.

Role of JAK Kinases in Immune Cell Development and Differentiation

From their discovery in the early 1990s, JAKs and STATs have been indelibly linked to immune cells. The following chapter will summarize key immune-related functions of individual JAK/STAT family members while providing a genome-wide perspective of their mechanisms (i.e. how they influence target gene expression). Special emphasis will be placed on recent work implicating JAK/STAT signaling in human disease and on translational studies aimed at developing JAK/STAT based therapeutics.

---

*These authors contributed equally.

A. Villarino (✉) • X.-P. Yang* (✉) • K. Hirahara • Y. Kanno • J.J. O' Shea • A. Laurence
Molecular Immunology and Inflammation Branch, National Institute of Arthritis, Musculoskeletal and Skin Diseases, National Institutes of Health, 9000 Rockville Pike, Bethesda, MD 20892, USA
e-mail: villarinoav@mail.nih.gov; yangx2@mail.nih.gov

Th. Decker and M. Müller (eds.), *Jak-Stat Signaling: From Basics to Disease*,
DOI 10.1007/978-3-7091-0891-8_13, © Springer-Verlag Wien 2012

The JAKs, which include TYK2, JAK1, JAK2, and JAK3, were initially identified using PCR-based strategies and low-stringency hybridization (Firmbach-Kraft et al. 1990; Riedy et al. 1996). Despite the roughly 60 type I/II cytokines discovered, there are only four members in the JAK kinase family. Since the sequencing of other vertebrate genomes has been completed, we also know now that there are indeed only four JAKs in mammals, birds and fish. In *D. melanogaster* there is only one JAK kinase (hopscotch).

JAKs contain seven distinct JAK homologous regions (JH1-7). The kinase domain with enzymatic activity is located at the C-terminal JH1 region. It is preceded by a kinase-like domain (JH2) that lacks conserved sequence necessary for enzymatic activity. The N-terminus of JAKs including JH3-7 regions is involved in receptor association. Upon ligand binding, the receptors oligomerize and change the conformations which brings the JAKs into close proximity and leads to the activation of the JAKs via auto- and trans-phosphorylation of the tyrosine motif within the kinase domain. Subsequently, activated JAKs phosphorylate the receptors and create the docking sites for STATs. STATs are phosphorylated by JAKs and then form homodimer or heterodimer through the interaction between SH2 domain and phospho-tyrosine motif. These dimerized STATs translocate into the nucleus to regulate target gene expression.

Immunological Functions of Tyk2

Although TYK2 (*Ty*rosine *K*inase 2) was discovered shortly after JAK1 and JAK2, it was the advent of Tyk2 mutant cells which provided the first direct link between JAK Kinases and cytokine signaling (Firmbach-Kraft et al. 1990; Velazquez et al. 1992). Tyk2 is closely associated with interferons and, consistent with work in mutant cell lines, *Tyk2*-deficient mice exhibit reduced type I and type II interferon responses in vivo, leading to increased susceptibility to viral and bacterial infections (Karaghiosoff et al. 2000). A similar phenotype has been observed in humans, where a single nucleotide polymorphism has been shown to cause *TYK2*-deficiency, and in B10.Q/J mice, which express low levels of Tyk2 due to an analogous missense mutation (Minegishi et al. 2006; Shaw et al. 2003). In both cases, the loss of functional Tyk2 leads to diminished interferon responses and increased susceptibility to intracellular pathogens. Tyk2-deficient mice also exhibit increased Th2-type inflammation (i.e. eosinophilia and IgE production) in models of allergic lung disease, which highlights the more regulatory, anti-Th2 properties of Tyk2-activating cytokines (Seto et al. 2003).

Aside from interferons, Tyk2 can be activated by a variety of cytokines, including members of the IL-6 and IL-10 families. Consequently, cells from Tyk2-deficient humans and mice show impaired signaling by IL-6, IL-10, IL-12, IL-22, IL-23 and IL-27, among others. In particular, the ability of IL-12 to induce IFN-γ-producing T helper 1 (Th1) cells, which are critical for resistance to intracellular pathogens, is severely compromised in the absence of *Tyk2* (Karaghiosoff et al. 2000).

The ability of IL-6 and IL-23 to induce Th17 responses, which are critical for resistance to extracellular and fungal pathogens, is also diminished (Ishizaki et al. 2011; Oyamada et al. 2009) and, in humans, genetic susceptibility to inflammatory bowel disease and Behcet's disease has been linked to polymorphisms in both the TYK2 and IL-23 receptor (*IL23R*) loci (Lees et al. 2011; Remmers et al. 2010). Thus, acting downstream of multiple cytokines, Tyk2 is a key mediator of cytokine responses in settings of infection and inflammatory disease.

JAK3 in Lymphoid Cell Development and Function

JAK3 was the last JAK Kinase to be identified and is unique in that it is only known to associate with a single cytokine receptor chain, IL-2Rγ, also known as the common gamma chain (γc). By pairing with ligand-specific subunits, the γc functions downstream of multiple cytokines, including IL-2, IL-4, IL-7, IL-9, IL-15 and IL-21, collectively termed the γc family (Rochman et al. 2009). In both humans and mice, deletion or mutation of *JAK3* or γc results in severe combined immunodeficiency (SCID), which is characterized by profound loss of T cells and B cells, in large part, due to a lack of IL-7 receptor signaling (Buckley 2004; Russell et al. 1995). Consistent with this latter point, mice lacking IL-7R or STAT5, which is the principal signaling moiety employed by IL-7R, also exhibit severe defects in T and B cell development (Yao et al. 2006). Beyond IL-7, it is also known that other members of the γc family play important roles in shaping the lymphoid repertoire, such as IL-15, which is critical for NK and NKT cell development, and IL-2, which is critical for T regulatory cell (Treg) development and survival (discussed below). In addition, there is a growing list of innate immune cells, including lymphoid tissue inducer cells (LTi), IL-22-producing $NKp46^+$ cells, and IL-17-producing $CD4^-CD117^-$ $NKp46^-$ cells (Spits and Di Santo 2011), which are known to express IL-7R and γc (and presumably JAK3), but their developmental requirements, whether cytokine- and JAK3-dependent, have yet to be determined.

Hematopoietic stem cell transplant is currently the treatment of choice for JAK3–SCID and other SCIDs. However, though bone marrow transplantation from human leukocyte antigen (HLA)-matched siblings tends to have dramatic results, mismatched donors are less predictable; sometimes they are beneficial but, more often, they exhibit graft versus-host disease and have reduced life expectancy. Consequently, there has been great interest in developing alternative treatments for JAK-SCID and X-SCID and among these, gene therapy emerged as one of the most promising (Hacein-Bey-Abina et al. 2002). The goal of this technique is to correct the genetic lesions underlying the lymphopenic phenotype, in this case JAK3 or γc mutations, by introducing synthetic DNA, typically using retroviral vectors, into the genome of affected individuals. On one hand, the approach has been successful, with some patients experiencing complete recovery of T and B cell numbers, but there are also severe side effects, with an alarming number developing leukemia (Hacein-Bey-Abina et al. 2003). Many of these patients exhibited aberrant *LMO2* gene expression, which is known to be oncogenic, and in some cases there was evidence of retroviral *IL2RG* gene insertion in proximity to the *LMO2* gene (Hacein-Bey-Abina et al. 2003). However, JAK3 activation can influence the *LMO2* locus in other ways, such as by histone phosphorylation (Dawson et al. 2009), so it remains unclear whether the leukemias were due to aberrant insertion of the *IL2RG* gene or whether *IL2RG* gene itself is leukemogenic (Woods et al. 2006). 'Next generation' retroviral vectors have now been developed in order to minimize the risk of leukemia in SCID patients, for example lacking enhancer elements or using weak cellular promoters, but whether they are safe and efficacious remains to be determined (Almarza et al. 2011).

In contrast to those lacking IL-7/IL-7R, JAK3 or γc, IL-2 and IL-2-receptor deficient mice are not lymphopenic and develop a multi-organ autoimmune disease characterized by uncontrolled expansion of effector T cells. This phenotype is largely due to the loss of $CD4^+$ Treg cells, which specialize in dampening inflammation, particularly when directed at the body's own tissues (i.e. self-antigens). Consistent with this latter point, mice lacking STAT5, the principal signal downstream of IL-2R, also lack Tregs and develop autoimmune disease with slightly accelerated kinetics, likely due to the fact that other STAT5-activating, γc family cytokines, like IL-7, IL-9, IL-15 and IL-21, can also impact Treg cells. Thus, while IL-2 is not necessary for T cell development, it is crucial for immune tolerance due to its unique role in Treg homeostasis. Consequently, it can be surmised that JAK3 or γc-deficient individuals do not develop autoimmunity because of their profound lymphopenia and that, since STAT5-deficiency closely mirrors IL-2-deficiency in term of gross phenotype, STAT5-independent signals must be responsible for some the developmental/survival cues delivered by γc cytokines.

Aside from their obligatory role in development and survival, JAK3-activating, γc family cytokines are also critical for immune cell effector functions. IL-21, which primarily signals through STAT3 instead of STAT5, is one pertinent example. It is not required for T cell development but does promote the generation of IL-4-producing Th2 cells, IL-17-producing Th17 cells and T follicular helper cells (Tfh), which are essential for T cell-dependent antibody responses (Spolski and Leonard 2010). Due to this latter finding, and the fact that IL-21 is known to have direct effects on B cells, it has been proposed that alterations of this cytokine or its receptor may underlie antibody-mediated pathologies, such as systemic lupus erythromatosis, and that it may contribute to B cell abnormalities in SCID patients. IL-4 is also critical for differentiation of Th2 and Tfh effector cells and for providing T cell help for B cell antibody responses but, unlike IL-21, it inhibits Th17-type inflammation (Harrington et al. 2005; Park et al. 2005; Zaretsky et al. 2009). Likewise, IL-2 promotes differentiation of Th1- and Th2-type effector T cells but limits that of Th17 and Tfh cells (Malek 2008). IL-9 and TSLP, a related cytokine that does not use the γc but does activate STAT5, both promote Th2 development, particularly in settings of allergy and asthma, but their effects on other T cell subsets remain uncertain (Ziegler and Artis 2010).

Broad Functions of JAK1 and JAK2

In contrast with JAK3, studies in humans and mice have shown that JAK1/JAK2 are broadly expressed, found in both immune and non-immune cells, and that they are promiscuous, activated by most of the >40 known type II cytokine receptors. Consistent with this latter point, *Jak1*-deficient mice die soon after birth and, due to the essential role of IL-6 family cytokines in embryogenesis, exhibit severe developmental and immunological abnormalities (Rodig et al. 1998; Witthuhn et al. 1993). In particular, they exhibit a SCID phenotype that is grossly similar to that of *Jak3*-deficient mice, which suggests that JAK1 participates in γc signaling and STAT5 activation. In fact, some have proposed JAK1 is as important as JAK3 in this capacity, though further studies are needed to substantiate this claim (Haan et al. 2011).

The principal function of JAK2 is the mediation of signaling downstream of the Erythropoetin receptor, and mice that lack JAK2 die in utero of anaemia. Its importance is further highlighted in myeloproliferative diseases (MPD) characterized by erythroid over activity, including primary polycythaemia (PV), primary thrombocythemia (ET) and myelofibrosis (MF). Depending on the subtype, patients with these diseases have a 50% (MF, ET) – 90% (PV) incidence of an acquired mutation in the JAK2 gene, most commonly V617F (Baxter et al. 2005; James et al. 2005; Kralovics et al. 2005; Levine et al. 2005). Consequently a number of pharmaceutical companies are attempting to produce small molecule inhibitors of JAK2 as a treatment for myeloproliferative disease.

Nuclear Functions of JAK Kinases

Traditionally, JAK kinases are thought to act in the cytoplasm, where they phosphorylate cytokine receptor-associated molecules (i.e. STATs). However, recent work has shown that JAK2 can also be found in the nucleus, where it can phosphorylate Histone 3 at position tyrosine 41, thereby influencing the epigenetic accessibility and transcription of multiple target loci including several known oncogenes (i.e. *Myc* and *lmo2*) (Dawson et al. 2009). Consistent with the latter point, elevated JAK2 activity is associated with a genome-wide decrease in the repressive epigenetic mark H3K9me3 in patients with Hodgkin's lymphoma (Rui et al. 2010). Thus, while it is unquestionable that JAK kinases have critical cytoplasmic functions, it is now apparent that they also have important nuclear effects, though further work is needed to fully understand the mechanisms underlying this phenomenon.

Development of JAK Inhibitors as Therapeutics for Autoimmunity and Malignancy

Constitutive activation of JAKs and STATs has long been associated with cancers in humans, mice and lower model organism such as Drosophila (Luo et al. 1995; Migone et al. 1995). One landmark discovery in this field was the recognition that gain-of-function mutations in the pseudokinase domain of JAK2 underlie polycythemia vera and a spectrum of myeloproliferative disorders (Levine et al. 2007; Vainchenker et al. 2008). In addition, a number of fusion proteins involving JAK2 have been linked to hematological malignancies (Lacronique et al. 1997; Peeters et al. 1997; Tirado et al. 2010). JAK3 mutations are also strongly linked to Acute Myeloid Leukemia (AML), Acute Lymphocyte Leukemia (ALL) and lymphomas (Dien Bard et al. 2009; Malamut et al. 2010; Malinge et al. 2008; Mullighan et al. 2009; Tyner et al. 2008; Walters et al. 2006). JAK1 has also been linked with acute leukemias, though this subject remains controversial (Flex et al. 2008; Mullighan et al. 2009).

Genetic and biochemical evidence linking JAKs to autoimmunity and blood malignancies has served as the impetus for a new class of immunosuppressive drugs aimed at limiting JAK kinase activity (Ghoreschi et al. 2009; Leonard and O'Shea 1998). Due to the pleiotropic nature of JAK signaling, and the lethal or SCID phenotype of JAK-deficient mice, there was initial concern about the potential side effects of such compounds. However, though clinical trials have uncovered some adverse effects, including infections, anemia and neutropenia, these small molecule JAK inhibitors have proven to be remarkably well tolerated and, more importantly, appear to be highly efficacious for the treatment of immune- and cancer-related pathologies (Ghoreschi et al. 2009). Consequently, there are now several JAK inhibitors currently under development, many of which have advanced beyond stage I clinical trials (Table 1). Among these, Tofacitinib and Ruxolitinib are two of the most promising and best understood.

Tofacitinib is the first clinically useful oral JAK inhibitor. It blocks JAK3 and JAK1 with nanomolar potency and JAK2 to a lesser extent, but has little effect on other unrelated kinases (Ghoreschi et al. 2011). Consistent with known JAK functions, Tofacitinib affects the downstream signaling of multiple cytokines, including members of the γc family (i.e. IL-2, IL-4 and IL-7), IFN-γ and IL-12, making it a powerful inhibitor of both Th1- and Th2-type inflammation. It also blocks IL-6 and IL-23 signaling, thereby limiting the development/survival of pathogenic Th17 cells (Ghoreschi et al. 2010, 2011; McGeachy et al. 2009), and has effects on innate immune cell activation (Ghoreschi et al. 2011; Jiang et al. 2008). The clinical efficacy of Tofacitinib was first investigated in models of transplantation and autoimmunity, where it showed remarkable ability to limit pathogenic inflammation when delivered either during onset or peak of disease (Borie et al. 2005; Changelian et al. 2003; Conklyn et al. 2004; Kudlacz et al. 2004, 2008; Milici et al. 2008; Paniagua et al. 2005). It is now undergoing human clinical trials for rheumatoid arthritis, psoriasis and Sicca syndrome, as well as for renal transplant. Preliminary data from these trials have been promising; a phase II study for rheumatoid arthritis reported that 70–81% of patients experienced positive results compared to 29% in the placebo group and, of note, many of these had previously received no benefit from conventional therapies like methotrexate or tumor necrosis factor antagonists (Coombs et al. 2010; Kremer et al. 2009). It has also been shown to have significant, dose-dependent effects in patients with psoriasis, where it can reduce both inflammation and scaling (Boy et al. 2009), and in the

Table 1 Selected JAK inhibitors in clinical development

Kinase inhibited	Compound	Comments (updated on August 2011)
Janus kinase (JAK) 3	Tofacitinib	Efficacy shown in animal models of solid organ transplantation and chronic graft versus host disease
		In phase II trials for IBD, Crohn's, UC and transplant rejection; phase III trials underway in the treatment of rheumatoid arthritis (RA) and psoriasis
	VX-509	Phase II trials underway for RA
	R-348	In phase I trials for RA and psoriasis; in the preclinical phase for Sjogren's syndrome
	PS-608504	In preclinical stage as treatment for RA and psoriasis
	R-333	In preclinical stage as treatment for cutaneous lupus erythematosus
JAK 1 and JAK 2	Ruxolitinib	Has been submitted for FDA-approval for idiopathic myelofibrosis; phase II and III trials for cancer and myeloproliferative disorders, respectively; no longer being studied for RA
	INCB-28050	In phase II trials for RA
	CYT-387	In phase II trials for myelofibrosis; in preclinical trials for pulmonary hypertension and polycythemia vera
	GLPG-0634	Shown to reduce joint-destroying enzymes in mouse RA model
		Phase I trial complete; phase II proof of concept trial results for RA expected in Dec 2011
JAK 2	LY-2784544	In phase II clinical trials for myeloproliferative disorders
	AC-430	In phase I clinical trials for RA and lymphoma
CDK, Flt3, JAK2	SB-1317	In phase I clinical trials for acute lymphoblastic leukemia (ALL), acute myeloid leukemia (AML), chronic myeloid leukemia (CML) and myelodysplastic syndrome
JAK2, Flt3, Ret	SAR-302503	In phase II clinical trials for polycythemia vera, idiopathic myelofibrosis, thrombocytopenia
JAK1, JAK3, and aurora A/B kinase	AT-9283	In phase II trials for ALL, AML, CML and myelodysplastic syndrome
FLT3, TrkA, and JAK2	Lestaurtinib	In phase III clinical trials for treatment of acute myeloid leukemia patients who have an FLT3-activating mutation at first relapse from standard induction chemotherapy. In phase II trials for psoriasis and pancreatic cancer

setting of kidney transplantation, where it can help to prevent graft rejection. At present, there are more than a dozen trials for Tofacitinib, most of which involve inflammatory or auto-immune diseases. It has also emerged as a candidate for treating adult T-cell leukemia (ATL) and for neurological disorders like HTLV-I-associated myelopathy/tropical spastic paraparesis (HAM/TSP) (Ju et al. 2011). In addition, despite being a relatively weak inhibitor of JAK2, Tofacitinib has been shown to preferentially suppress oncogenic JAK2 mutants, which suggest that it may be useful for JAK2-dependendent myeloproliferative disorders (MPD)

(Manshouri et al. 2008). A number of other JAK3 inhibitors, all of which are less 'JAK-specific' than Tofacitinib, are also under development and will likely be applied to a similar panel of diseases (Geron et al. 2008; Hedvat et al. 2009; Ioannidis et al. 2011; Wernig et al. 2008).

Until the discovery that activating mutations underlie myeloproliferative disorders (MPD) (James et al. 2005), it was generally believed that manipulating JAK2 would have limited therapeutic value, particularly in light of its pleiotropic nature and the embryonic lethal phenotype of JAK2-deficient mice (Leonard and O'Shea 1998; Neubauer et al. 1998; Parganas et al. 1998). However, the link between JAK2 and MPD has since provided a logical rationale for targeting this kinase and data from clinical trials has shown that Ruxolitinib, which blocks JAK1 and JAK2 (Quintas-Cardama et al. 2010), may be a viable treatment option for both hematological and solid organ malignancies. Phase 1 and 2 trials have shown this compound to have remarkable efficacy for MPD, regardless of whether patients carried activating *JAK2* mutations or not, which suggests that some of its effects are, in fact, JAK2-independent (Verstovsek et al. 2010). Consistent with this latter point, CEP-701, a more selective oral JAK2 inhibitor, has displayed only modest effects in MPD (Santos et al. 2010). Based on these observations, Ruxolitinib is now being tested for several malignancies that are not associated with aberrant JAK kinase activity, such as prostate cancer, multiple myeloma, AML, and CML, and for autoimmune disorders like rheumatoid arthritis (Mesa 2010; Quintas-Cardama et al. 2010; Verstovsek et al. 2010). In addition, a compound related to Ruxolitinib, INCB028050, is also being pursued as a potential therapeutic for rheumatoid arthritis (Fridman et al. 2010).

The identification of patients with either inherited or acquired mutations in JAK family members has done much to identify the importance of these kinases in a range of immune and neoplastic diseases. This has in turn led to the development of a number of small molecule inhibitors of JAK kinases and their investigation as therapeutic agents. The wide spread use of clinical inhibitors may shed more light on the role of JAK's in cytokine biology over the next decade.

A Genome-Wide Perspective of STAT Function

As with JAK kinases, much of the pioneering work on STATs was done in cell lines. Primary lymphocytes, like T cells, macrophages and dendritic cells, have since become preferred model systems, largely because recent developments in cell biology and genetics, including flow cytometry and transgenic mice, have allowed for loss- and gain-of function studies to be performed both in vitro and in vivo. Advances in molecular biology, like the advent of genome-wide technologies (i.e. microarrays and next generation sequencing), have also been a major driving force in the study of STAT signaling. Thus while initial research tended to focus on a select few target genes (Darnell et al. 1994), it is now possible to get a panoramic, 'bird's eye view', thereby ushering what can be viewed as a new golden age of STAT research.

Technological Advances in STAT Biology

The canonical JAK/STAT pathway is initiated by extracellular association of cytokines (or growth factors) with their corresponding transmembrane receptors. That interaction results in the apposition of receptor-associated JAKs which, in turn, phosphorylate each other and the intracellular tail of the receptors, creating requisite docking sites for latent, cytoplasmic STAT monomers. JAK-mediated phosphorylation of STAT tyrosine residues then proceeds as the major activating event, though other JAK-independent modifications, like serine phosphorylation, acetylation and sumoylation, are also known to play a role (Begitt et al. 2011; Decker and Kovarik 2000; Yuan et al. 2005). Measurement of STAT tyrosine phosphorylation has become the standard method for assessing STAT activation. Initially, this was done by western blot using whole cell lysates and phospho-STAT antibodies, which has the advantage of sensitivity but the disadvantage of being a bulk assay that does not distinguish between different sub-populations of cells (Gronowski and Rotwein 1994). In contrast, recently developed flow cytometry-based techniques allow for detection of tyrosine phosphorylation at the single cell level, thus permitting STAT activity to be measured in small samples and to be directly linked to immune cell function (i.e. cytokine production) (Krutzik et al. 2008).

Upon phosphorylation, receptor-associated STATs form homo- and heterodimers (or tetramers) which then migrate to the nucleus and bind to consensus DNA elements, thereby regulating gene transcription. Traditionally, downstream targets of activated STATs were identified using a hypothesis-driven approach, meaning that one had to predict which genes might be affected prior to experimentation. Once candidates were selected, a number of methodologies could be used to determine whether they are influenced by STAT-mediated signals. The most common was to treat cells with STAT-activating cytokines and then measure putative targets using standard cellular and molecular assays, such as western blot, ELISA, flow cytometry or PCR. Given that all cytokines activate multiple signaling pathways and, in fact, most activate multiple STAT family members, the use of STAT-deficient cells (derived from STAT 'knock-out' mice) has become a vital component in this pipeline, making it possible to distinguish between STAT-dependent and -independent events. More recently, DNA microarrays have taken much of the guess work out of the process because they allow one to measure expression of thousands of genes at once and, thus, do not require presumptive knowledge. By combining genetic and microarray technologies, thousands of candidate targets have now been identified for each STAT (Alvarez and Frank 2004), though it also bears noting that this approach does not discriminate between *cis* effects (i.e. direct STAT binding to target genes) and *trans* effects (i.e. STATs influencing secondary targets, like transcription factors, which then regulate expression of a putative target gene).

There are several ways to identify primary targets of STAT signaling. One long established practice is to generate plasmid 'reporters' where candidate gene promoters are cloned upstream of a fluorescent marker, typically luciferase or GFP. These constructs are transfected or retrovirally transduced into cells that are exposed to cytokines or other STAT-activating agents, then fluorescence is measured as a function of STAT activity (Giguere 1991). Another approach, called electrophoretic mobility shift assay (EMSA), also involves the generation of synthetic DNA fragments containing putative STAT-binding sites. These are incubated together with purified STAT proteins (or whole cell lysates) and STAT-specific antibodies, then loaded onto a polyacrylamide or agarose gel. The binding of STAT/antibody complexes to the DNA results in a large multi-molecular complex which runs slowly in the gel and, thus, can easily be distinguished from unbound fragments (Fried and Crothers 1981). Each of these techniques has the advantage of flexibility – DNA constructs can be easily modified – but also several caveats relating to the synthetic nature of the constructs, which tend to be relatively short and, thus, lack distal regulatory elements (i.e. enhancers), and which ignore the role of epigenetics and DNA structure on gene expression. Moreover, they are predictive in nature; one must first know the putative binding sites in order to generate the corresponding DNA constructs.

Chromatin immuno-precipitation (ChIP) circumvents many of the issues associated with reporter constructs and EMSA because it allows for the measurement of STAT binding to genomic DNA in its native conformation and without extensive manipulation of cells (i.e. transfection/transduction) (Orlando 2000). Here, STATs are chemically fixed to DNA within cytokine treated (or untreated) cells. These cells are then lysed, the DNA sheared and anti-STAT antibodies used to precipitate DNA/protein complexes. PCR is then used to determine if a given DNA sequence falls within the STAT-bound fraction, thereby indicating whether or not there is a direct interaction (Fig. 1). ChIP is now the gold standard for measuring STAT binding to genomic DNA but, as with previous approaches, it is limited in scope, focusing only on pre-determined regions that are either based on rigid parameters, such as proximity to coding exons (i.e. promoters), or computational predictions of STAT binding sites, which are notorious for high false positive rates. This caveat can now be overcome by performing anti-STAT immuno-precipitation followed either by microarray detection (ChIP-on-Chip), which allows for thousands of genomic regions to be probed, or by high-throughput sequencing (ChIP-seq), which provides a comprehensive, genome-wide view of STAT binding. Both allow for STAT/DNA interactions to be probed on a vast scale but ChIP-seq has several advantages, including higher resolution, greater dynamic range, and fewer artifacts. Most importantly, it provides complete and unbiased coverage, which allows for STAT binding to be measured in parts of the genome that are typically ignored by array-based technologies, such as 'gene deserts', which are distal to protein coding genes but often rich in non-coding transcripts, like microRNAs and long non-coding RNAs, that could be subject to STAT-dependent regulation. Genome-wide binding of all STATs (save STAT2) has now been assayed by ChIP-seq and these data sets, which are publicly available through the Gene

Fig. 1 Schematic method of chromatin immunoprecipitation followed by massive parallel sequencing (CHIP-seq)

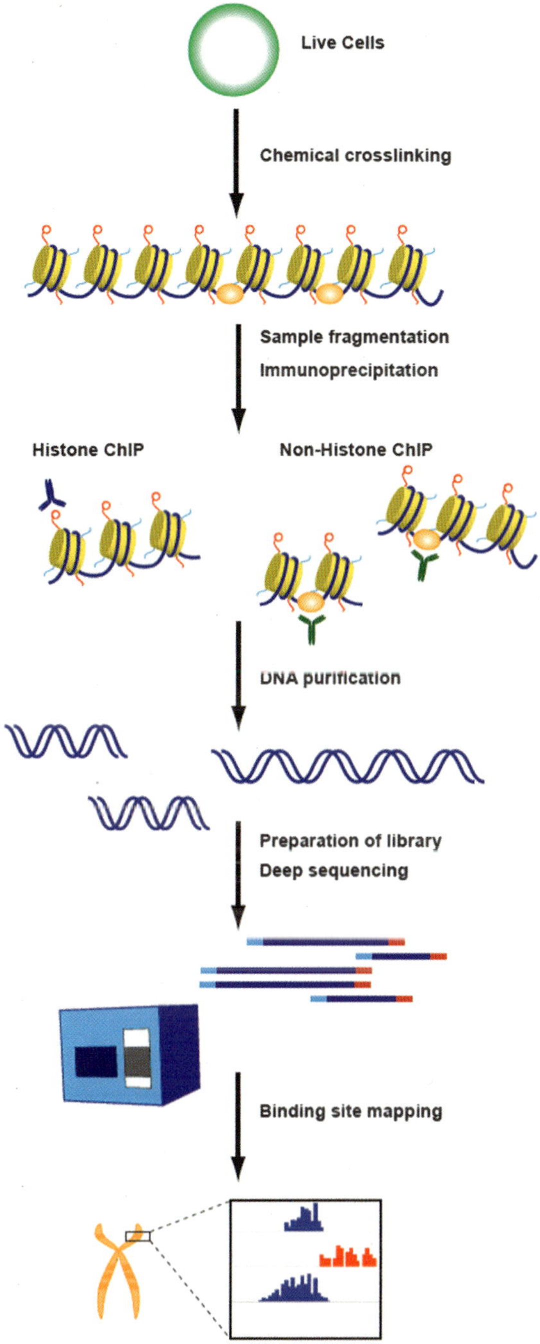

Expression Omnibus (GEO) repository (O'Shea et al. 2011), have become an invaluable resource for the cytokine research community.

STATs and Helper T Cell Differentiation

Upon encountering cognate antigen, naive CD4$^+$ T cells differentiate from a quiescent state into multiple effector subsets, each defined by the transcription factors they employ, the cytokines they secrete and, ultimately, the functions they execute. Such specification was first recognized within the Th1 and Th2 subsets, the former characterized by the transcription factor T-bet, production of IFN-γ, and the ability to combat intracellular pathogens; the latter by GATA-3, production of IL-4, and their role in resistance to helminth infection (Murphy et al. 2000; Zhu et al. 2010). Building on that original Th1/Th2 paradigm, a number of additional subsets have since been described including: (1) Th17 cells, which are characterized by the transcription factor retinoic acid orphan receptor gamma (RORγ), production of IL-17, and their ability to protect from fungal and mucosal pathogens, (2) T follicular helper cells (Tfh), which are characterized by the transcription factor Bcl6, production of IL-21, and their ability to provide 'help' for B cell antibody responses, and (3) T regulatory cells (Tregs), which are characterized by the transcription factor FoxP3, production of anti-inflammatory cytokines and their ability to suppress T cell responses (Crotty 2011; Korn et al. 2009; Sakaguchi et al. 2008). Other putative helper T cell subsets have also emerged, including IL-9-, and IL-22-producing T cells, but their differentiation requirements and effector functions are not yet fully understood, making it difficult to assess whether they represent bona fide subsets or more transient differentiation 'states' (Eyerich et al. 2009; Veldhoen et al. 2008b). What is clear is that functional specification of helper T cells is essential because it allows immune responses to be tailored for a given stimulus, be it microbial or environmental. However, though all T cell responses are intended for vital homeostatic processes, ranging from control of infection to wound repair and cancer surveillance, they can also promote host tissue damage when misdirected or overactive, leading to the idea that 'helper' T cells are key mediators of both protective and pathogenic adaptive immune responses (Jager and Kuchroo 2010).

Aside from being the end product, cytokines are key fate determinants during helper T cell differentiation. Not surprisingly, as mediators of cytokine signaling, JAKs and STATs are critical in this process, with particular JAK/STAT moieties known to promote particular T cell subsets. However, though it is true that each STAT can be associated with at least one lineage, there is significant overlap, meaning that multiple STATs can influence one subset and, conversely, that one STAT can influence multiple subsets. For instance, STAT4 is considered the prototypical Th1-type transcription factor but it is also recognized that STAT1 and STAT5 are important for driving Th1 commitment (Gollob et al. 1998). Likewise, thought STAT6 is the signature Th2-type transcription factor, it is known that other STATs, namely STAT3 and STAT5, contribute to Th2 commitment (Stritesky et al. 2011; Zhu et al. 2003). The pleiotropic nature of STAT signaling is also well illustrated by STAT3, which is required for differentiation of the Th2, Th17 and Tfh subsets (Nurieva et al. 2008; Stritesky et al. 2011; Yang et al. 2007), and by STAT5, which is required for differentiation of the Th1, Th2 and Treg subsets (Gollob et al. 1998; Yao et al. 2007; Zhu et al. 2003). Thus, while some STATs are strongly linked to a particular T cell subset, as is the case for STAT4 and STAT6, it is now appreciated that helper T cell differentiation results from the integration of multiple STAT signals, some of which are not subset-specific.

STATs and Epigenetics

STATs influence transcription by directly binding to consensus DNA elements (originated named as IFN-γ-activated site, GAS motifs) found proximal to target genes. However, at any given time, these binding sites may be inaccessible due to the dense three-dimensional structure of DNA, which is packaged together with histone proteins into (in ascending order) nucleosomes, chromatin and chromosomes. Accordingly, STAT function is heavily influenced by epigenetics, which are heritable (but often reversible) changes in gene activity imposed by conformational changes in chromatin structure. Within target genes, favorable, or 'open', epigenetic states are thought to promote STAT binding while unfavorable, or 'closed', epigenetic states are thought to preclude STAT binding. In addition, epigenetic modifications can be the downstream effect of STAT binding, meaning that STATs are sometimes required for epigenetic changes to occur at target loci, presumably because they can guide the epigenetic machinery towards a particular section of the genome (Robertson et al. 2008; Wei et al. 2010).

Epigenetic modifications come in two basic forms: permissive marks, which are associated with active gene transcription and include histone acetylation and tri-methylation of histone H3 at lysine 4 (H3K4), lysine 36 (H3K36) or lysine 79 (H3K79), and repressive marks, which are associated with transcriptional silencing and include DNA methylation and bi- or tri-methylation of histone H3 at lysine 9 (H3K9) or lysine 27 (H3K27) (Kouzarides 2007). These can be added or removed quickly in response to external cues (i.e. cytokines), thereby creating a dynamic epigenetic landscape based on the developmental/differentiation status of a cell. Not surprisingly, epigenetic regulation is critical for helper T cell differentiation, essential for both the induction and suppression of lineage-restricted gene products. In general, cytokine loci are 'closed' in naive T cells due to an abundance of repressive marks, particularly H3K27 methylation, and a relative lack of permissive marks like H3K4 and H3K36 methylation. However, when exposed to polarizing stimuli, there is a radical shift in this epigenetic profile such that, in effector T cells, repressive marks are removed and permissive marks added to some cytokine genes while further deposition of repressive marks leads to heritable silencing of others. For example, in Th1 cells, the *Ifng* locus has an 'open' conformation (high H3K4 and low H3K27) which allows for production of this Th1-defining cytokine, while the *Il-4* locus has a 'closed' conformation (low H3K4 methylation and high H3K27 methylation) which prevents elaboration of this Th2-type cytokine under Th1 conditions. The converse is true for Th2 cells – the

Ifng locus is closed and the *Il-4* locus is open, again matching the restricted cytokine pattern of this subset (Wei et al. 2009). On the other hand, the loci of 'master regulator' transcription factors like *Tbx21*, *Gata-3*, *Rorc* and *FoxP3* (among others), which cooperate with STATs to drive lineage specification, exhibit both permissive and repressive marks (termed 'bivalent' marks) in Th1, Th2, Th17, and Treg cells, suggesting that they can be expressed within multiple subsets and providing a molecular explanation for effector T cell plasticity (Lee et al. 2009; Wei et al. 2009). There is also strong evidence that 'master regulators' can influence the epigenetic status of cytokine genes, though it remains unclear how much this function contributes to their overall lineage-specifying capacity (Miller et al. 2008).

Function and Targets of Individual STATs in Helper T Cells

The importance of $CD4^+$ helper T cells is clearly illustrated by the opportunistic infections seen in humans lacking T cells, like those suffering from primary or secondary immunodeficiencies (i.e. X-SCID and AIDS, respectively), and by the causal relationship between self-reactive $CD4^+$ T cells and auto-immune diseases like lupus, diabetes and multiple sclerosis, among others (Leonard and O'Shea 1998) (Jager and Kuchroo 2010). Because they posses few direct effector mechanisms (i.e. target cell killing, phagocytosis), helper T cells are thought to control inflammatory responses by instructing the behavior of other immune cells, thereby acting as nodes for inter-cellular communication. This is achieved, in large part, by their ability to sense and produce cytokines, most of which operate through JAK/STAT signaling pathways. The following section will detail the key role played by each STAT in the differentiation and effector function of helper T cells, with special emphasis on recent conceptual and technological advances in the field.

STAT1

Initial work with mutagenized cell lines demonstrated that STAT1 is critical for mediating signaling transduction by Type I and II IFNs (Bromberg et al. 1996). Targeted disruption of *Stat1* gene in mice confirmed the importance of STAT1 in IFN-α/β and IFN-γ signaling: *Stat1*-deficient mice are highly susceptible to microbial and viral infection and tumor formation due to severely impaired IFN responses in innate immune cells (Durbin et al. 1996; Meraz et al. 1996). In T cells, IFN-γ is not only the signature cytokine of Th1 cells; it also induces STAT1-dependent expression T-bet and IL-12Rβ2, thereby mediating a positive feedback loop which amplifies Th1 differentiation. As a result, STAT1-deficient mice fail to generate robust Th1 responses when challenged with intracellular pathogens (i.e. Toxoplasma gondii), leading to reduced induction of host-protectve and anti-microbial proteins, including MHC-I, MHC-II, inducible nitric oxide synthase, and inducible GTPases, all of which are required to control parasite replication (Gavrilescu et al. 2004; Lieberman et al. 2004). Another STAT1-activating cytokine, IL-27 can also induce expression of T-bet and IL-12Rβ2 to potentiate the IL-12-induced Th1 differentiation, but is only required for the generation of IFN-γ-producing Th1 cells under select circumstances (Takeda et al. 2003).

In addition to its role in promoting Th1 differentiation, STAT1 is a negative regulator of Th17 differentiation. Accordingly, IFN-γ and IL-27 can both suppress Th17 responses via STAT1-dependent mechanisms (Harrington et al. 2005; Park et al. 2005; Stumhofer et al. 2006). In fact, recent work has shown that both of these cytokines can suppress expression of multiple Th17-associated genes, including IL-17A, IL-17F, IL-21, IL-23R and RORγ, and that they do this through both T-bet-dependent and -independent means, which indicates that STAT1 influences Th17 responses through both direct and indirect mechanisms (Villarino et al. 2010). STAT1 has also been shown to be a negative regulator of Treg cells. For instance, in an allogenic transplantation model, *Stat1*-deficient donor spleen cells have less potential to induce GVHD this correlates with a greater expansion of $CD4^+CD25^+FoxP3^+$ Treg cells (Ma et al. 2011).

Genome-wide STAT1 binding and its relationship to H3K4m3 have been investigated in Hela S3 cells (Robertson et al. 2008). There are ~20,600 and ~70,300 STAT1 binding sites in unstimuated and IFN-γ-stimulated cells respectively. STAT1 binding sites correspond to 25% of all H3K4 regions in the IFN-γ-stimulated Hela cells, suggesting that STAT1 may interact with an unexpected large fraction of regulatory elements genome-wide. Surprisingly, for the majority of the binding sites of STAT1 after IFN-γ stimulation, the histone active mark H3K4m1 and H3K4m3 are present even in the unstimulated cells, suggesting STAT1-independent histone modifications are common.

STAT3

T_H17 cells are defined by production of the cytokine IL-17, expression of ROR family transcription factors and by their pro-inflammatory functions, which protect again extracellular bacteria and fungi, but also contribute to pathology in numerous autoimmune diseases. STAT3-activating cytokines, including IL-6, IL-21 and IL-23, are critical for the Th17 differentiation program and, accordingly, T cell-specific deletion of STAT3 results in a profound Th17 defect (Mathur et al. 2007; Yang et al. 2007), thereby increasing susceptibility to certain pathogens and decreasing susceptibility to Th17-mediated autoimmunity (Durant et al. 2010; Harris et al. 2007). The converse is also true – mice lacking SOCS3, a potent inhibitor of STAT3 activity, exhibit an accumulation of T_H17 cells (Chen et al. 2006).

Thousands of direct STAT3 target genes have been identified by combining ChIP-seq and microarray technologies. Not surprisingly, many of these are involved in Th17 differentiation, including the cytokine receptors IL23R and IL6Ra, the cytokines IL-17A, IL-17F, IL-21 and IL-22, and the transcription factors RORγt, RORα, Ahr (aryl hydrocarbon receptor), Batf, IRF4 (interferon regulatory factor 4) and Maf (Bauquet et al. 2009; Ivanov et al. 2006; Schraml et al. 2009; Veldhoen et al. 2008a; Yang et al. 2008). It is also notable that STAT3 was shown to occupy multiple sites within the adjacent *Il17a* and *Il17f* loci (Yang et al. 2011), the most prominent of which are within the intergenic regions that coincides with conserved non-coding sequences (CNS) and histone acetyltransferase p300 binding sites (Akimzhanov et al. 2007). In addition, STAT3 was shown to bind genes associated with cellular proliferation and survival, including the anti-apoptotic proteins *Bcl2* and *Ier3* (immediate early response 3), and the proto-oncogenes *Fos*, *Jun* and *Fosl2*, which suggests a broad role in T cell fitness. Supporting this latter point, *Stat3*$^{-/-}$ T cells exhibit poor clonal expansion in mouse models of colitis and multiple sclerosis (Durant et al. 2010; Harris et al. 2007).

Beyond Th17 cells, STAT3-activating cytokines are also known to influence other effector subsets. For instance, IL-6 and IL-21 can each drive Tfh differentiation and genome-wide analysis of STAT3 binding downstream of these cytokines has revealed that Tfh-promoting genes, like *Bcl6* and *Il-21* itself, are direct STAT3 targets. The importance of those events is well illustrated by conditional STAT3-deficient mice, which have severe defects in Tfh development and T cell-dependent antibody responses (Nurieva et al. 2008). STAT3 has also been implicated in Th2 differentiation. Recent studies have shown that STAT3-deficient T cells produce less Th2-type cytokines (i.e. IL-4, IL5, IL-13) and Th2-type transcription factors (i.e. *Gata3*, *Batf*, and *Maf*) than WT controls (Stritesky et al. 2011). They also demonstrate that STAT3 binds directly to many of these loci and that, in the absence of STAT3, the ability of STAT6 to bind Th2-type genes is greatly reduced, which suggest that STAT3 may be required for STAT6 to gain access and/or induce transcription.

The role of STAT3 in the regulation of epigenetic modifications has been investigated under Th17 condition (Durant et al. 2010). In the absence of STAT3, permissive histone active marks (i.e. H3K4m3) are either absent or reduced at the gene loci of *Il17a*, *Il17f*, *Il21* and *Il6ra*, suggesting that STAT3 regulates the chromatin accessibility of these genes during the Th17 differentiation process. STAT3 is also required for the acquisition of permissive histone active marks at the gene loci of *Rorc*, *Rora*, and *Batf*, which encode key transcriptional factors for Th17 linage specification. However, although IRF4, Ahr and Maf have also been shown to play important roles in Th17 differentiation, the presence or absence of STAT3 did not change permissive marks at those genes.

STAT4

Activated downstream of IL-12, IL-23 and type I IFNs, STAT4 is a major driving force for IFN-γ production in innate and adaptive immune cells, making it an essential component of the inflammatory response against intracellular pathogens like Mycobacteria, Lysteria, Toxoplasma and Leishmania species (among many others). Prior to the advent of genome-wide assays, only a few direct STAT4 target genes were known, including *Ifng*, *Il18r1* (IL-18 receptor 1), *Hlx* (H2.0-like homeobox), *Map3k8* (mitogen-activated protein kinase kinase kinase 8) and *Furin* (Pesu et al. 2006; Thieu et al. 2008; Yu et al. 2007). ChIP-on-chip studies expanded this list to contain hundred genes, many of which are not closely associated with IFN-γ production or Th1 differentiation, such as *Gadd45g* (growth arrest and DNA-damage-inducible 45γ) and *Lcp2* (lymphocyte cytosolic protein 2) (Good et al. 2009). They also show that STAT4 binding does not always correlate to changes in gene expression, suggesting that its binding profile is broader than previously appreciated, and that binding alone may not be sufficient to drive transcription, which is consistent with other genome-wide surveys of transcription factor binding. Those findings have now been confirmed and expanded by ChIP-seq analysis of in vitro generated Th1 cells. These studies have uncovered at least 10,000 STAT4 binding sites throughout the genome, with about 40% localized to the promoters or gene bodies of approximately 4,000 annotated genes. The remaining 60% are located within intergenic regions, encompassing potential distal regulatory elements (i.e. enhancers) or non-coding loci (i.e. microRNAs or long non-coding RNAs) (Wei et al. 2010).

Aside from validating its role in Th1 differentiation, genome-wide profiling of STAT4 binding has revealed that it can also regulate other target genes, particularly cytokines, which are not traditionally associated with the Th1 program. For instance, though IL-10 has long been considered a Th2-type cytokine, and is a known target of STAT6 in Th2 cells, it is now understood that STAT4 can associate with the *Il-10* locus and that Th1 cells can produce IL-10 (Saraiva and O'Garra 2010). Another pertinent example is IL-21, a known target of STAT3 typically associated with the Th17 and Tfh subsets. STAT4 was found to bind upstream of the *Il-21* locus under Th1 conditions, which is consistent with recent work showing that IL-12 can induce IL-21 production in helper T cells, and with the emerging idea that IL-21 production is not restricted to STAT3-dependent lineages. For both IL-10 and IL-21, STAT4 binding was associated with positive regulation (i.e. STAT4 dependent induction), but for several other 'non-Th1' genes, it was associated with negative regulation (i.e. STAT4-dependent inhibition). Given that many of these are closely linked to other T cell subsets, including IL-4 (Th2) and IL-17 (Th17), these data argue that, beyond its role as a transcriptional activator, STAT4 also promotes Th1 responses by limiting the differentiation of alternative T cell fates.

Previous work has suggested that the deposition of histone modifications precedes STAT1 binding, which suggests that STAT1 has limited influence on the epigenetic landscape (Robertson et al. 2008). In contrast, comparative analysis of histone methylation in wild type and STAT4-deficient Th1 cells has revealed that >1,000 of the genes bound by STAT4 also have STAT4-dependent epigenetic modifications (Wei et al. 2010). About 200 of these were found to have STAT4-dependent, activating marks (i.e. H3K4me3) and to be positivity regulated by STAT4, therefore, representing a core signature of STAT4 targets which include not only signature Th1 cell genes, such as *Ifng* and *Tbx21*, but also many that are not considered to be subset specific, including *Il18rap* (IL-18 receptor accessory protein), *Icos* (inducible T cell co-stimulator), *Lilrb4* (leukocyte immunoglobulin-like receptor B4) and *Nkg7* (NK group 7). Another group of genes, including several hallmark Th2-type genes (i.e. IL-4, GATA-3), was found to contain STAT4-dependent, repressive marks (i.e. H3K27me3), which is consistent with a role for STAT4 as a transcriptional inhibitor, and with the idea that each STAT can influence multiple T cell subsets.

STAT5

The mammalian genome contains four copies of STAT5 encoded by two highly homologous genes, *STAT5a* and *STAT5b*, derived from a primordial duplication event and located adjacently on the same chromosome (11 in mouse and 17 in humans). These paralogs share multiple functions, many relating to fundamental cellular processes like proliferation and survival. Consequently, germline deletion of both results in perinatal lethality due to combined defects in hormone and cytokine receptor signaling. However, mice lacking either *Stat5a* or *Stat5b* are viable, which suggests a level of redundancy, but have distinct phenotypes, which suggests that, despite >96% sequence homology, each also has unique functions (Hennighausen and Robinson 2008). This latter point is supported by recent genome-wide studies showing that STAT5a and STAT5b can have distinct binding profiles in helper T cells, though it must also be noted that there was significant overlap between the two and that, in general, the differences were rather subtle, often spaced by a few hundred nucleotides and rarely demonstrating an absolute, locus-specific preference for one or the other (Liao et al. 2011).

Cell type-specific ablation of STAT5a and STAT5b has revealed a number of key immune functions, including an obligatory role in lymphoid lineage development and in the differentiation of effector T cell subsets. ChIP-seq analysis of STAT5 binding has confirmed and provided key mechanistic insights about these findings. For instance, though it has long been known that IL-2 (and other STAT5-activating cytokines) can promote Th1-type responses, a direct interaction between STAT5 and hallmark Th1 loci, including *Ifng*, *Tbx21* and *Il-12rβ2*, has only recently been established (Liao et al. 2011). Likewise, though γc cytokines have traditionally been associated with Th2-type responses, STAT5 has only recently been shown to act directly on Th2 loci, specifically *Il-4rα*, a key mediator of IL-4/STAT6 signaling (Liao et al. 2011). STAT5 binding to Th17 loci, including *Il-17a* and *Il-17f*, has also been shown to underlie the well known ability of IL-2 and γc cytokines to suppress Th17-type responses (Yang et al. 2011). Based on extensive overlap between STAT3 and STAT5 binding sites in Th17 cells, these latter studies propose a competitive model where STAT5 prevents inductive signals (i.e. STAT3) from accessing or activating regulatory elements in the *Il-17a/f* locus (Yang et al. 2011). Thus, in helper T cells, STAT5 binding appears to be pervasive, affecting a multitude of targets throughout the genome, and determinative, capable of exerting both positive and negative control over key subset-defining genes.

Mice lacking STAT5 in T cells are initially lymphopenic, which is consistent with the known role of JAK3 and γc cytokines in lymphocyte development, but as they age, these animals develop a lethal, T cell-dependent inflammatory disease characterized by massive outgrowth of Th1- and Th17-type effector T cells. Because this phenotype closely resembles that of IL-2 deficient mice, it is widely believed that the autoimmunity in STAT5-deficient mice is due to a lack of IL-2 receptor signaling, specifically, in IL-2-dependent Treg cells. Although ChIP-seq analysis of STAT5 binding in Treg cells has yet to be reported, conventional ChIP studies have shown that STAT5 does bind the promoter and first intron of *Foxp3*, a transcription factor that is both necessary and sufficient for specifying the Treg fate

(Burchill et al. 2007; Yao et al. 2007; Zorn et al. 2006). STAT5 has also been shown to bind directly to other Treg-associated genes, including the IL-2Rα (CD25) and the anti-apoptotic protein BCL2, making it a central node in the network of signals required for development and/or maintenance of this subset. Thus, despite exhibiting several important pro-inflammatory features, it has become apparent that the major, non-redundant function of STAT5 in T cells has to do with immune tolerance and, specifically, with its ability to suppress of Th17 differentiation while, at the same time, promoting that of Treg cells.

STAT6

Traditionally, Th2 differentiation is thought to be driven by IL-4 and its ability to activate STAT6, a potent inducer of hallmark Th2-type genes like *Il-4*, *Il-13*, *c-Maf* and *Gata-3*, though recent evidence suggests that other signals, including STAT3 and STAT5, can also participate (Kaplan et al. 1996; Shimoda et al. 1996; Takeda et al. 1996; Zhu et al. 2001). ChIP-seq analysis of STAT6 binding has confirmed the central importance of this pathway for Th2 differentiation, with many of its targets representing known Th2-type factors, and has demonstrated its dynamic effects on gene expression, with some genes exhibiting sustained binding and others a more transient or biphasic pattern. These studies also revealed substantial overlap with the binding profile of STAT5, which suggests a degree of cooperation, and with that of STAT4, which speaks towards antagonisms between these two transcription factors. Thus, STAT6 appears to promote Th2-type responses not only through direct, positive effects, but also by limiting alternative subsets (i.e. Th1 and Th17) via direct, inhibitory effects.

Similar to STAT4 in Th1 cells, STAT6 is a major influence on the epigenetic landscape of Th2 cells. Although it has long been known that STAT6 bindings are associated with activating histone marks at certain genes, particularly those most closely associated with the Th2 phenotype like *Il-4* and *Gata-3*, genome-wide analysis of histone methylation in WT and *Stat6*-deficient T cells has revealed that this may not be its principal function. Instead, STAT6 seems to inhibit the deposition of repressive histone marks, thereby preventing the ‘closing’ of target genes (Wei et al. 2010). Thus, STAT6 appears to be fundamentally different from STAT4 in its mechanism of action though, overall, the goal of each appears to be similar: to induce and or maintain an ‘open’ chromatin state at lineage-restricted genes. It is also notable that a subset of genes bound by STAT6 in Th2 cells are also bound by STAT4 in Th1 cells and that, in these cases, the two transcription factors seem to oppose one another, which is not surprising given the well known antagonism between the Th1 and Th2 subsets and, more specifically, the ability of IL-12/STAT4 and IL-4/STAT6 to oppose one another.

Counter-Regulation of STAT Family Members

Although STATs generally act as transcriptional activators, they can also limit gene expression. That dichotomy is clearly evident during helper T cell differentiation, where STATs are known to both promote and suppress lineage-restricted products (i.e. cytokines, transcription factors). Each STAT is known to inhibit at least one subset so that, when a particular STAT is deleted, there is a reduction of the subset(s) it is meant to promote and a concurrent expansion of the subset(s) it is meant to suppress. For instance, STAT1-deficient T cells exhibit reduced Th1-type responses due to the loss of STAT1-driven lineage induction and exaggerated Th2- and Th17-type responses due to the loss of STAT1-driven lineage repression. A similar pattern is observed for other STATs: STAT3-deficient T cells have reduced Th2-, Th17- and Tfh-type responses and enhanced Th1- and Treg-type responses; STAT4-deficient T cells have reduced Th1- and Tfh-type responses and enhanced Th2- and Th17-type responses; STAT5-deficient T cells have reduced Th1-, Th2- and Treg-type responses and enhanced Th17- and Tfh-type responses; STAT6-deficient T cells have reduced Th2-type responses and enhanced Th1- and Th17-type responses (Adamson et al. 2009). Thus, by exerting both positive and negative influences on lineage-restricted genes, a given STAT promotes the outgrowth of some subsets at the expense of others, eventually leading to the polarization of effector T cells from a heterogeneous mix at the onset of an immune response to a more homogeneous population during the peak and resolution phases.

It has long been known that all STATs (save STAT6) can bind to the same consensus DNA sequence, known as the GAS element (Ehret et al. 2001; Seidel et al. 1995). Given the prevalence of GAS elements throughout the genome, and the high degree of overlap between the binding profiles of different STATs (Ehret et al. 2001; Jothi et al. 2008), it has become apparent that STATs binding is more widespread than previously appreciated and that, in many instances, multiple STATs can bind to the same sites. This promiscuity may explain why different STATs can have similar effects, as is the case for induction of IL-10, which occurs via STAT3, STAT4 or STAT6, for suppression of IL-2, which occurs via STAT3, STAT4, STAT5 or STAT6, and for induction of CD25, which occurs via STAT3, STAT4 or STAT5 (Akaishi et al. 1998; Kim et al. 2001, 2006; O'Sullivan et al. 2004; Saraiva and O'Garra 2010; Villarino et al. 2006, 2007). It has also led to the idea that STATs may antagonize one another by competing for access to the same GAS elements. One example of such competition involves the role of STAT3 and STAT5 in Th17 differentiation. Studies have shown that STAT3, acting downstream of IL-6, IL-21 or IL-23, can promote IL-17 production and that STAT5, acting downstream of IL-2, can suppress it (Laurence et al. 2007). Recent work has provided a mechanism for this antagonism, demonstrating that STAT3 and STAT5 have remarkably similar binding profiles in Th17 cells and that, in fact, they compete for binding to the same GAS elements within the *Il17* locus, resulting in either repressive (STAT5) or activating (STAT3) epigenetic modifications (Yang et al. 2011). It remains unclear why one STAT acts as a positive regulator and the other as a negative regulator but these data are the first to link opposing cellular functions to a shared affinity for DNA binding. Future work should address the prevalence of this effect, whether STAT3 and STAT5 compete at multiple loci, and whether or not it applies to other 'antagonistic' STATs, such as STAT4/STAT6 or STAT1/STAT3.

As with IL-17 and Th17 cells, STAT3 and STAT5 are know to have opposite effects on FoxP3 and Treg cells. In this case, the genetic evidence is strong – $FoxP3^+$ Tregs cells are expanded in the absence of STAT3 and diminished in the absence of STAT5. However, though it is known that STAT3 and STAT5 can each interact with the *FoxP3* locus, it remains unknown whether there is direct competition between the two (Bettelli et al. 2006; Xu et al. 2010; Yao et al. 2007). Likewise, though STAT1, STAT4 and STAT6 have been shown to limit Treg differentiation, and STAT6 is able to bind the *FoxP3* locus (Ma et al. 2011; O'Malley et al. 2009; Takaki et al. 2008), competition with STAT5 has not been addressed. In addition, studies have shown that, when STAT3 is deleted specifically in $FoxP3^+$ cells, their regulatory capacity becomes impaired which, paradoxically, suggests that STAT3 is required for Treg function (Chaudhry et al. 2009). These latter findings suggest that the role of STAT3 in Treg biology is complex, though it also remains possible, and perhaps even likely, that there is some form of competition between STAT3 and STAT5 at *FoxP3* loci and other key Treg-associated loci.

Clinical Manifestation of STAT Mutants

Traditionally, gene-deficient or transgenic mice have been used to investigate STAT function in vivo. However, recent advances in DNA sequencing technologies have made it possible for genome-wide association studies (GWAS) to be performed in humans with high speed and at relatively low cost, thereby prompting the discovery of several single-nucleotide polymorphisms (SNPs) that affect STATs and manifest in clinical disease (Table 2). These human studies have confirmed many of the findings in mice, particularly in terms of disease etiology, and have provided valuable insights about mechanism, with several of the mutations found to influence STAT activity rather than expression levels.

The pioneering work of Dupuis et al. identified loss-of-function *STAT1* mutations in patients with heritable viral and mycobacterial susceptibility and demonstrate that this genetic lesion results in defective in interferon signaling, thus, mirroring studies in *Stat1*-deficient mice (Dupuis et al. 2001, 2003; Durbin et al. 1996; Meraz et al. 1996). Not surprisingly, patients with germline mutations in type I and type II interferon receptors exhibit a similar phenotype (Allende et al. 2001; Jouanguy et al. 1997), again, illustrating the critical role of the interferon/STAT1 pathway in resistance to infection. More recently, gain-of-function *STAT1* mutations have also been described. These are associated with increased susceptibility to *Candida* and fungal infections, not due to defects in interferon signaling but, rather, due to enhanced STAT1-mediated suppression of Th17-type responses (Liu et al. 2011; van de Veerdonk et al. 2011), thus, highlighting the antagonistic nature of the Th1 and Th17 subsets, as well as the broad functions of STAT1 in host defense against multiple classes of pathogens.

As in mice, genetic evidence in humans points towards STAT3 as a critical factor in resistance to fungi and extracellular bacteria. This effect is well illustrated

Table 2 Human diseases associated with mutations of the JAK/STAT pathway

JAKs/STATs	Immunological diseases	References
JAK2	Polycythemia vera (PV)	James et al. (2005)
	Essential thrombocytosis (ET)	Kralovics et al. (2005)
	Primary myelofibrosis (PMF)	Baxter et al. (2005) Levine et al. (2005)
JAK3	Severe combined immunodeficiency (SCID)	Russell et al. (1995) Macchi (1995)
TYK2	Primary immunodeficiency syndrome	Minegishi et al. (2006)
	Multiple sclerosis (MS)	Bahlo (2009)
	Inflammatory bowl disease (IBD)	Franke (2010)
STAT1	Atypical mycobacteria (*M. avium* etc.) infection	Liu et al. (2011)
	Chronic mucocutaneous candidiasis (CMC)	van de Veerdonk et al. (2011)
STAT3	Hyper IgE syndrome (HIES; Job's syndrome)	Minegishi (2007)
	Multiple sclerosis (MS)	Holland (2007)
	Inflammatory bowl disease (IBD)	Jakkula (2010) Franke (2008)
	Spondyloarthritis	Danoy (2010)
STAT4	Rheumatoid arthritis (RA)	Kobayashi (2008)
	Systemic lupus erythematosus (SLE)	
STAT5b	Dwarfism	Kofoed et al. (2003)
	Recurrent Herpes virus infection	Bernasconi (2006)

in patients with hyper-IgE syndrome (also known as Job's syndrome), a classic primary immunodeficiency that has recently been linked to dominant-negative mutations of STAT3. Affected individuals exhibit severe signaling defects downstream of multiple STAT3-activating cytokines, including IL-6, IL-10, IL-21, and IL-23, leading to defects in the generation Th17-type inflammation and, consequently, increased susceptibility to *Candida* and *Staphylococcus* infections (Heimall et al. 2009; Ma et al. 2008; Milner et al. 2008; Minegishi et al. 2006). Given that germline deletion of STAT3 is embryonic lethal (Takeda et al. 1997), it is clear that individuals with hyper-IgE syndrome retain some STAT3 activity, though obviously not enough to protect from Th17-type pathogens.

It is also noteworthy that, in contrast to mice lacking *Stat3* in Treg cells (Chaudhry et al. 2009), these patients do not have serious autoimmune complications, likely due to residual STAT3 activity in this lineage and or inability to generate pathogenic Th17 cells. Consistent with this latter point, and echoing studies in mice, a number of genome-wide association studies have linked STAT3 (and its upstream activators) to human autoimmune disease. For example, polymorphisms in *IL-23R*, *JAK2* and *STAT3* have been associated with, among others, Crohn's disease, asthma, psoriasis and ankylosing spondylitis, which strongly suggests a role for STAT3 in the pathogenesis of all these disorders (O'Shea et al. 2011).

Similarly, a variant *STAT4* allele has been found in rheumatoid arthritis, Crohn's disease and Sjogren's syndrome (Glas et al. 2010; Korman et al. 2008; Remmers et al. 2007). Surprisingly, the same variant allele is associated with systemic lupus erythematosus (SLE) disease, which is not a prototypical Th1 cell-mediated disease (Han et al. 2009; Namjou et al. 2009). However, one should keep in mind that SLE patients have an interferon-related pathology and STAT4 is activated by type I IFNs as well (Cho et al. 1996). Polymorphisms of the *STAT4* gene are not located within the coding region and presumably influence the level of gene expression. Consistent with the role of *STAT4* in SLE, polymorphisms in *TYK2* (tyrosine kinase 2) and *IRF5*, which are both involved in signaling cascade of type I interferons, involved have also been associated with SLE (Hellquist et al. 2009).

Homozygous missense mutations of *STAT5b* are linked to growth hormone insensitivity phenotype (Kofoed et al. 2003). As expected, given the key role of STAT5 in regulating Treg cell development, patients with *STAT5b* mutations have autoimmunity and impaired T_{Reg} cell function (Cohen et al. 2006).

Conclusions

Due to technological advances in molecular biology, and particularly the advent 'next generation' DNA sequencing, our understanding of Jak/STAT signaling has improved exponentially over the past decade. It is now clear that Jaks and STATs are critical for the pathogenesis of many human diseases, which has made targeting this pathway, such as JAK kinase inhibitors, an increasingly attractive therapeutic avenue. However, though genome-wide studies have taught us much about the mechanisms of JAK/STAT signaling, revealing thousands of direct binding sites, and establishing that STATs can have broad influence on epigenetics, there is still much to be learned, specifically about the functional consequences of JAKs and STATs in human health and disease. Looking forward, there are multiple ongoing clinical trials for Jak/STAT-related compounds, which should foster great interest for basic research into this pathway and, perhaps, herald a notable translation from bench to clinic.

Acknowledgements We apologize to the colleagues whose work was not cited owing to space limitations. Work of the authors was supported by NIH/NIAMS Intramural Research Program (IRP).

References

Adamson AS, Collins K, Laurence A, O'Shea JJ (2009) The current STATus of lymphocyte signaling: new roles for old players. Curr Opin Immunol 21:161–166

Akaishi H, Takeda K, Kaisho T, Shineha R, Satomi S, Takeda J, Akira S (1998) Defective IL-2-mediated IL-2 receptor alpha chain expression in Stat3-deficient T lymphocytes. Int Immunol 10:1747–1751

Akimzhanov AM, Yang XO, Dong C (2007) Chromatin remodeling of interleukin-17 (IL-17)-IL-17F cytokine gene locus during inflammatory helper T cell differentiation. J Biol Chem 282:5969–5972

Allende LM, Lopez-Goyanes A, Paz-Artal E, Corell A, Garcia-Perez MA, Varela P, Scarpellini A, Negreira S, Palenque E, Arnaiz-Villena A (2001) A point mutation in a domain of gamma interferon receptor 1 provokes severe immunodeficiency. Clin Diagn Lab Immunol 8:133–137

Almarza E, Zhang F, Santilli G, Blundell MP, Howe SJ, Thornhill SI, Bueren JA, Thrasher AJ (2011) Correction of SCID-X1 using an enhancerless Vav promoter. Hum Gene Ther 22:263–270

Alvarez JV, Frank DA (2004) Genome-wide analysis of STAT target genes: elucidating the mechanism of STAT-mediated oncogenesis. Cancer Biol Ther 3:1045–1050

Bahlo M, Booth DR, Broadley SA, Brown MA, Foote SJ, Griffiths LR, Kilpatrick TJ, Lechner-Scott J, Moscato P, Perreau VM et al (2009) Genome-wide association study identifies new multiple sclerosis susceptibility loci on chromosomes 12 and 20. Nat Genet 41:824–828

Bauquet AT, Jin H, Paterson AM, Mitsdoerffer M, Ho IC, Sharpe AH, Kuchroo VK (2009) The costimulatory molecule ICOS regulates the expression of c-Maf and IL-21 in the development of follicular T helper cells and TH-17 cells. Nat Immunol 10:167–175

Baxter EJ, Scott LM, Campbell PJ, East C, Fourouclas N, Swanton S, Vassiliou GS, Bench AJ, Boyd EM, Curtin N et al (2005) Acquired mutation of the tyrosine kinase JAK2 in human myeloproliferative disorders. Lancet 365:1054–1061

Begitt A, Droescher M, Knobeloch KP, Vinkemeier U (2011) SUMO conjugation of STAT1 protects cells from hyperresponsiveness to IFN{gamma}. Blood 118:1002–1007

Bernasconi A, Marino R, Ribas A, Rossi J, Ciaccio M, Oleastro M, Ornani A, Paz R, Rivarola MA, Zelazko M et al (2006) Characterization of immunodeficiency in a patient with growth hormone insensitivity secondary to a novel STAT5b gene mutation. Pediatrics 118: e1584–e1592

Bettelli E, Carrier Y, Gao W, Korn T, Strom TB, Oukka M, Weiner HL, Kuchroo VK (2006) Reciprocal developmental pathways for the generation of pathogenic effector TH17 and regulatory T cells. Nature 441:235–238

Borie DC, Changelian PS, Larson MJ, Si MS, Paniagua R, Higgins JP, Holm B, Campbell A, Lau M, Zhang S et al (2005) Immunosuppression by the JAK3 inhibitor CP-690,550 delays rejection and significantly prolongs kidney allograft survival in nonhuman primates. Transplantation 79:791–801

Boy MG, Wang C, Wilkinson BE, Chow VF, Clucas AT, Krueger JG, Gaweco AS, Zwillich SH, Changelian PS, Chan G (2009) Double-blind, placebo-controlled, dose-escalation study to evaluate the pharmacologic effect of CP-690,550 in patients with psoriasis. J Invest Dermatol 129:2299–2302

Bromberg JF, Horvath CM, Wen Z, Schreiber RD, Darnell JE Jr (1996) Transcriptionally active Stat1 is required for the antiproliferative effects of both interferon alpha and interferon gamma. Proc Natl Acad Sci USA 93:7673–7678

Buckley RH (2004) Molecular defects in human severe combined immunodeficiency and approaches to immune reconstitution. Annu Rev Immunol 22:625–655

Burchill MA, Yang J, Vogtenhuber C, Blazar BR, Farrar MA (2007) IL-2 receptor beta-dependent STAT5 activation is required for the development of Foxp3+ regulatory T cells. J Immunol 178:280–290

Changelian PS, Flanagan ME, Ball DJ, Kent CR, Magnuson KS, Martin WH, Rizzuti BJ, Sawyer PS, Perry BD, Brissette WH et al (2003) Prevention of organ allograft rejection by a specific Janus kinase 3 inhibitor. Science 302:875–878

Chaudhry A, Rudra D, Treuting P, Samstein RM, Liang Y, Kas A, Rudensky AY (2009) CD4+ regulatory T cells control TH17 responses in a Stat3-dependent manner. Science 326:986–991

Chen Z, Laurence A, Kanno Y, Pacher-Zavisin M, Zhu BM, Tato C, Yoshimura A, Hennighausen L, O'Shea JJ (2006) Selective regulatory function of Socs3 in the formation of IL-17-secreting T cells. Proc Natl Acad Sci USA 103:8137–8142

Cho SS, Bacon CM, Sudarshan C, Rees RC, Finbloom D, Pine R, O'Shea JJ (1996) Activation of STAT4 by IL-12 and IFN-alpha: evidence for the involvement of ligand-induced tyrosine and serine phosphorylation. J Immunol 157:4781–4789

Cohen AC, Nadeau KC, Tu W, Hwa V, Dionis K, Bezrodnik L, Teper A, Gaillard M, Heinrich J, Krensky AM et al (2006) Cutting edge: decreased accumulation and regulatory function of CD4+ CD25(high) T cells in human STAT5b deficiency. J Immunol 177:2770–2774

Conklyn M, Andresen C, Changelian P, Kudlacz E (2004) The JAK3 inhibitor CP-690550 selectively reduces NK and CD8+ cell numbers in cynomolgus monkey blood following chronic oral dosing. J Leukoc Biol 76:1248–1255

Coombs JH, Bloom BJ, Breedveld FC, Fletcher MP, Gruben D, Kremer JM, Burgos-Vargas R, Wilkinson B, Zerbini CA, Zwillich SH (2010) Improved pain, physical functioning and health status in patients with rheumatoid arthritis treated with CP-690,550, an orally active Janus kinase (JAK) inhibitor: results from a randomised, double-blind, placebo-controlled trial. Ann Rheum Dis 69:413–416

Crotty S (2011) Follicular helper CD4 T cells (T(FH)). Annu Rev Immunol 29:621–663

Darnell JE Jr, Kerr IM, Stark GR (1994) Jak-STAT pathways and transcriptional activation in response to IFNs and other extracellular signaling proteins. Science 264:1415–1421

Danoy P, Pryce K, Hadler J, Bradbury LA, Farrar C, Pointon J, Australo-Anglo-American Spondyloarthritis Consortium, Ward M, Weisman M, Reveille JD et al (2010) Association of variants at 1q32 and STAT3 with ankylosing spondylitis suggests genetic overlap with Crohn's disease. PLoS Genet 6:e1001195

Dawson MA, Bannister AJ, Gottgens B, Foster SD, Bartke T, Green AR, Kouzarides T (2009) JAK2 phosphorylates histone H3Y41 and excludes HP1alpha from chromatin. Nature 461:819–822

Decker T, Kovarik P (2000) Serine phosphorylation of STATs. Oncogene 19:2628–2637

Dien Bard J, Gelebart P, Anand M, Zak Z, Hegazy SA, Amin HM, Lai R (2009) IL-21 contributes to JAK3/STAT3 activation and promotes cell growth in ALK-positive anaplastic large cell lymphoma. Am J Pathol 175:825–834

Diveu C, McGeachy MJ, Boniface K, Stumhofer JS, Sathe M, Joyce-Shaikh B, Chen Y, Tato CM, McClanahan TK, de Waal Malefyt R et al (2009) IL-27 blocks RORc expression to inhibit lineage commitment of Th17 cells. J Immunol 182:5748–5756

Dupuis S, Dargemont C, Fieschi C, Thomassin N, Rosenzweig S, Harris J, Holland SM, Schreiber RD, Casanova JL (2001) Impairment of mycobacterial but not viral immunity by a germline human STAT1 mutation. Science 293:300–303

Dupuis S, Jouanguy E, Al-Hajjar S, Fieschi C, Al-Mohsen IZ, Al-Jumaah S, Yang K, Chapgier A, Eidenschenk C, Eid P et al (2003) Impaired response to interferon-alpha/beta and lethal viral disease in human STAT1 deficiency. Nat Genet 33:388–391

Durant L, Watford WT, Ramos HL, Laurence A, Vahedi G, Wei L, Takahashi H, Sun HW, Kanno Y, Powrie F, O'Shea JJ (2010) Diverse targets of the transcription factor STAT3 contribute to T cell pathogenicity and homeostasis. Immunity 32:605–615

Durbin JE, Hackenmiller R, Simon MC, Levy DE (1996) Targeted disruption of the mouse *Stat1* gene results in compromised innate immunity to viral disease. Cell 84:443–450

Ehret GB, Reichenbach P, Schindler U, Horvath CM, Fritz S, Nabholz M, Bucher P (2001) DNA binding specificity of different STAT proteins. Comparison of in vitro specificity with natural target sites. J Biol Chem 276:6675–6688

Eyerich S, Eyerich K, Pennino D, Carbone T, Nasorri F, Pallotta S, Cianfarani F, Odorisio T, Traidl-Hoffmann C, Behrendt H et al (2009) Th22 cells represent a distinct human T cell subset involved in epidermal immunity and remodeling. J Clin Invest 119:3573–3585

Firmbach-Kraft I, Byers M, Shows T, Dalla-Favera R, Krolewski JJ (1990) tyk2, prototype of a novel class of non-receptor tyrosine kinase genes. Oncogene 5:1329–1336

Franke A, Balschun T, Karlsen TH, Hedderich J, May S, Lu T, Schuldt D, Nikolaus S, Rosenstiel P, Krawczak M et al (2008) Replication of signals from recent studies of Crohn's disease identifies previously unknown disease loci for ulcerative colitis. Nat Genet 40:713–715

Franke A, McGovern DP, Barrett JC, Wang K, Radford-Smith GL, Ahmad T, Lees CW, Balschun T, Lee J, Roberts R et al (2010) Genome-wide meta-analysis increases to 71 the number of confirmed Crohn's disease susceptibility loci. Nat Genet 42:1118–1125

Flex E, Petrangeli V, Stella L, Chiaretti S, Hornakova T, Knoops L, Ariola C, Fodale V, Clappier E, Paoloni F et al (2008) Somatically acquired JAK1 mutations in adult acute lymphoblastic leukemia. J Exp Med 205:751–758

Fridman JS, Scherle PA, Collins R, Burn TC, Li Y, Li J, Covington MB, Thomas B, Collier P, Favata MF et al (2010) Selective inhibition of JAK1 and JAK2 is efficacious in rodent models of arthritis: preclinical characterization of INCB028050. J Immunol 184:5298–5307

Fried M, Crothers DM (1981) Equilibria and kinetics of lac repressor-operator interactions by polyacrylamide gel electrophoresis. Nucleic Acids Res 9:6505–6525

Gavrilescu LC, Butcher BA, Del Rio L, Taylor GA, Denkers EY (2004) STAT1 is essential for antimicrobial effector function but dispensable for gamma interferon production during Toxoplasma gondii infection. Infect Immun 72:1257–1264

Geron I, Abrahamsson AE, Barroga CF, Kavalerchik E, Gotlib J, Hood JD, Durocher J, Mak CC, Noronha G, Soll RM et al (2008) Selective inhibition of JAK2-driven erythroid differentiation of polycythemia vera progenitors. Cancer Cell 13:321–330

Ghoreschi K, Laurence A, O'Shea JJ (2009) Janus kinases in immune cell signaling. Immunol Rev 228:273–287

Ghoreschi K, Laurence A, Yang XP, Tato CM, McGeachy MJ, Konkel JE, Ramos HL, Wei L, Davidson TS, Bouladoux N et al (2010) Generation of pathogenic T(H)17 cells in the absence of TGF-beta signalling. Nature 467:967–971

Ghoreschi K, Jesson MI, Li X, Lee JL, Ghosh S, Alsup JW, Warner JD, Tanaka M, Steward-Tharp SM, Gadina M et al (2011) Modulation of innate and adaptive immune responses by tofacitinib (CP-690,550). J Immunol 186:4234–4243

Giguere V (1991) Application of the firefly luciferase reporter gene. Methods Mol Biol 7:237–241

Glas J, Seiderer J, Nagy M, Fries C, Beigel F, Weidinger M, Pfennig S, Klein W, Epplen JT, Lohse P et al (2010) Evidence for STAT4 as a common autoimmune gene: rs7574865 is associated with colonic Crohn's disease and early disease onset. PLoS One 5:e10373

Gollob JA, Murphy EA, Mahajan S, Schnipper CP, Ritz J, Frank DA (1998) Altered interleukin-12 responsiveness in Th1 and Th2 cells is associated with the differential activation of STAT5 and STAT1. Blood 91:1341–1354

Good SR, Thieu VT, Mathur AN, Yu Q, Stritesky GL, Yeh N, O'Malley JT, Perumal NB, Kaplan MH (2009) Temporal induction pattern of STAT4 target genes defines potential for Th1 lineage-specific programming. J Immunol 183:3839–3847

Gronowski AM, Rotwein P (1994) Rapid changes in nuclear protein tyrosine phosphorylation after growth hormone treatment in vivo. Identification of phosphorylated mitogen-activated protein kinase and STAT91. J Biol Chem 269:7874–7878

Haan C, Rolvering C, Raulf F, Kapp M, Druckes P, Thoma G, Behrmann I, Zerwes HG (2011) Jak1 has a dominant role over Jak3 in signal transduction through gammac-containing cytokine receptors. Chem Biol 18:314–323

Hacein-Bey-Abina S, Le Deist F, Carlier F, Bouneaud C, Hue C, De Villartay JP, Thrasher AJ, Wulffraat N, Sorensen R, Dupuis-Girod S et al (2002) Sustained correction of X-linked severe combined immunodeficiency by ex vivo gene therapy. N Engl J Med 346:1185–1193

Hacein-Bey-Abina S, Von Kalle C, Schmidt M, McCormack MP, Wulffraat N, Leboulch P, Lim A, Osborne CS, Pawliuk R, Morillon E et al (2003) LMO2-associated clonal T cell proliferation in two patients after gene therapy for SCID-X1. Science 302:415–419

Han JW, Zheng HF, Cui Y, Sun LD, Ye DQ, Hu Z, Xu JH, Cai ZM, Huang W, Zhao GP et al (2009) Genome-wide association study in a Chinese Han population identifies nine new susceptibility loci for systemic lupus erythematosus. Nat Genet 41:1234–1237

Harrington LE, Hatton RD, Mangan PR, Turner H, Murphy TL, Murphy KM, Weaver CT (2005) Interleukin 17-producing CD4+ effector T cells develop via a lineage distinct from the T helper type 1 and 2 lineages. Nat Immunol 6:1123–1132

Harris TJ, Grosso JF, Yen HR, Xin H, Kortylewski M, Albesiano E, Hipkiss EL, Getnet D, Goldberg MV, Maris CH et al (2007) Cutting edge: an in vivo requirement for STAT3 signaling in TH17 development and TH17-dependent autoimmunity. J Immunol 179: 4313–4317

Hedvat M, Huszar D, Herrmann A, Gozgit JM, Schroeder A, Sheehy A, Buettner R, Proia D, Kowolik CM, Xin H et al (2009) The JAK2 inhibitor AZD1480 potently blocks Stat3 signaling and oncogenesis in solid tumors. Cancer Cell 16:487–497

Heimall J, Freeman A, Holland SM (2009) Pathogenesis of hyper IgE syndrome. Clin Rev Allergy Immunol 38:32–38

Hellquist A, Jarvinen TM, Koskenmies S, Zucchelli M, Orsmark-Pietras C, Berglind L, Panelius J, Hasan T, Julkunen H, D'Amato M et al (2009) Evidence for genetic association and interaction between the TYK2 and IRF5 genes in systemic lupus erythematosus. J Rheumatol 36: 1631–1638

Hennighausen L, Robinson GW (2008) Interpretation of cytokine signaling through the transcription factors STAT5A and STAT5B. Genes Dev 22:711–721

Holland SM, DeLeo FR, Elloumi HZ, Hsu AP, Uzel G, Brodsky N, Freeman AF, Demidowich A, Davis J, Turner ML, Anderson VL et al (2007) STAT3 mutations in the hyper-IgE syndrome. N Engl J Med 357:1608–1619

Ioannidis S, Lamb ML, Wang T, Almeida L, Block MH, Davies AM, Peng B, Su M, Zhang HJ, Hoffmann E et al (2011) Discovery of 5-Chloro-N(2)-[(1S)-1-(5-fluoropyrimidin-2-yl)ethyl]-N(4)-(5-methyl-1H-pyr azol-3-yl)pyrimidine-2,4-diamine (AZD1480) as a Novel Inhibitor of the Jak/Stat Pathway. J Med Chem 54:262–276

Ishizaki M, Akimoto T, Muromoto R, Yokoyama M, Ohshiro Y, Sekine Y, Maeda H, Shimoda K, Oritani K, Matsuda T (2011) Involvement of tyrosine kinase-2 in both the IL-12/Th1 and IL-23/Th17 axes in vivo. J Immunol 187:181–189

Ivanov II, McKenzie BS, Zhou L, Tadokoro CE, Lepelley A, Lafaille JJ, Cua DJ, Littman DR (2006) The orphan nuclear receptor RORgammat directs the differentiation program of proinflammatory IL-17+ T helper cells. Cell 126:1121–1133

Jager A, Kuchroo VK (2010) Effector and regulatory T-cell subsets in autoimmunity and tissue inflammation. Scand J Immunol 72:173–184

James C, Ugo V, Le Couedic JP, Staerk J, Delhommeau F, Lacout C, Garcon L, Raslova H, Berger R, Bennaceur-Griscelli A et al (2005) A unique clonal JAK2 mutation leading to constitutive signalling causes polycythaemia vera. Nature 434:1144–1148

Jakkula E, Leppä V, Sulonen AM, Varilo T, Kallio S, Kemppinen A, Purcell S, Koivisto K, Tienari P, Sumelahti ML et al (2010) Genome-wide association study in a high-risk isolate for multiple sclerosis reveals associated variants in STAT3 gene. Am J Hum Genet 86:285–291

Jiang JK, Ghoreschi K, Deflorian F, Chen Z, Perreira M, Pesu M, Smith J, Nguyen DT, Liu EH, Leister W et al (2008) Examining the chirality, conformation and selective kinase inhibition of 3-((3R,4R)-4-methyl-3-(methyl(7H-pyrrolo[2,3-d]pyrimidin-4-yl)amino)piperi din-1-yl)-3-oxopropanenitrile (CP 690,550). J Med Chem 51:8012–8018

Jothi R, Cuddapah S, Barski A, Cui K, Zhao K (2008) Genome-wide identification of in vivo protein-DNA binding sites from ChIP-Seq data. Nucleic Acids Res 36:5221–5231

Jouanguy E, Altare F, Lamhamedi-Cherradi S, Casanova JL (1997) Infections in IFNGR-1-deficient children. J Interferon Cytokine Res 17:583–587

Ju W, Zhang M, Jiang JK, Thomas CJ, Oh U, Bryant BR, Chen J, Sato N, Tagaya Y, Morris JC et al (2011) CP-690,550, a therapeutic agent, inhibits cytokine-mediated Jak3 activation and proliferation of T cells from patients with ATL and HAM/TSP. Blood 117:1938–1946

Kaplan MH, Schindler U, Smiley ST, Grusby MJ (1996) Stat6 is required for mediating responses to IL-4 and for development of Th2 cells. Immunity 4:313–319

Karaghiosoff M, Neubauer H, Lassnig C, Kovarik P, Schindler H, Pircher H, McCoy B, Bogdan C, Decker T, Brem G et al (2000) Partial impairment of cytokine responses in Tyk2-deficient mice. Immunity 13:549–560

Kim HP, Kelly J, Leonard WJ (2001) The basis for IL-2-induced IL-2 receptor alpha chain gene regulation: importance of two widely separated IL-2 response elements. Immunity 15:159–172

Kim HP, Imbert J, Leonard WJ (2006) Both integrated and differential regulation of components of the IL-2/IL-2 receptor system. Cytokine Growth Factor Rev 17:349–366

Kobayashi S, Ikari K, Kaneko H, Kochi Y, Yamamoto K, Shimane K, Nakamura Y, Toyama Y, Mochizuki T, Tsukahara S et al (2008) Association of STAT4 with susceptibility to rheumatoid arthritis and systemic lupus erythematosus in the Japanese population. Arthritis Rheum 58:1940–1946

Kofoed EM, Hwa V, Little B, Woods KA, Buckway CK, Tsubaki J, Pratt KL, Bezrodnik L, Jasper H, Tepper A et al (2003) Growth hormone insensitivity associated with a STAT5b mutation. N Engl J Med 349:1139–1147

Korman BD, Alba MI, Le JM, Alevizos I, Smith JA, Nikolov NP, Kastner DL, Remmers EF, Illei GG (2008) Variant form of STAT4 is associated with primary Sjogren's syndrome. Genes Immun 9:267–270

Korn T, Bettelli E, Oukka M, Kuchroo VK (2009) IL-17 and Th17 Cells. Annu Rev Immunol 27:485–517

Kouzarides T (2007) Chromatin modifications and their function. Cell 128:693–705

Kralovics R, Passamonti F, Buser AS, Teo SS, Tiedt R, Passweg JR, Tichelli A, Cazzola M, Skoda RC (2005) A gain-of-function mutation of JAK2 in myeloproliferative disorders. N Engl J Med 352:1779–1790

Kremer JM, Bloom BJ, Breedveld FC, Coombs JH, Fletcher MP, Gruben D, Krishnaswami S, Burgos-Vargas R, Wilkinson B, Zerbini CA, Zwillich SH (2009) The safety and efficacy of a JAK inhibitor in patients with active rheumatoid arthritis: results of a double-blind, placebo-controlled phase IIa trial of three dosage levels of CP-690,550 versus placebo. Arthritis Rheum 60:1895–1905

Krutzik PO, Crane JM, Clutter MR, Nolan GP (2008) High-content single-cell drug screening with phosphospecific flow cytometry. Nat Chem Biol 4:132–142

Kudlacz E, Perry B, Sawyer P, Conklyn M, McCurdy S, Brissette W, Flanagan M, Changelian P (2004) The novel JAK-3 inhibitor CP-690550 is a potent immunosuppressive agent in various murine models. Am J Transplant 4:51–57

Kudlacz E, Conklyn M, Andresen C, Whitney-Pickett C, Changelian P (2008) The JAK-3 inhibitor CP-690550 is a potent anti-inflammatory agent in a murine model of pulmonary eosinophilia. Eur J Pharmacol 582:154–161

Lacronique V, Boureux A, Valle VD, Poirel H, Quang CT, Mauchauffe M, Berthou C, Lessard M, Berger R, Ghysdael J, Bernard OA (1997) A TEL-JAK2 fusion protein with constitutive kinase activity in human leukemia. Science 278:1309–1312

Laurence A, Tato CM, Davidson TS, Kanno Y, Chen Z, Yao Z, Blank RB, Meylan F, Siegel R, Hennighausen L et al (2007) Interleukin-2 signaling via STAT5 constrains T helper 17 cell generation. Immunity 26:371–381

Lee YK, Turner H, Maynard CL, Oliver JR, Chen D, Elson CO, Weaver CT (2009) Late developmental plasticity in the T helper 17 lineage. Immunity 30:92–107

Lees CW, Barrett JC, Parkes M, Satsangi J (2011) New IBD genetics: common pathways with other diseases. Gut 60:1739–1753

Leonard WJ, O'Shea JJ (1998) Jaks and STATs: biological implications. Annu Rev Immunol 16:293–322

Levine RL, Wadleigh M, Cools J, Ebert BL, Wernig G, Huntly BJ, Boggon TJ, Wlodarska I, Clark JJ, Moore S et al (2005) Activating mutation in the tyrosine kinase JAK2 in polycythemia vera, essential thrombocythemia, and myeloid metaplasia with myelofibrosis. Cancer Cell 7:387–397

Levine RL, Pardanani A, Tefferi A, Gilliland DG (2007) Role of JAK2 in the pathogenesis and therapy of myeloproliferative disorders. Nat Rev Cancer 7:673–683

Liao W, Lin JX, Wang L, Li P, Leonard WJ (2011) Modulation of cytokine receptors by IL-2 broadly regulates differentiation into helper T cell lineages. Nat Immunol 12:551–559

Lieberman LA, Banica M, Reiner SL, Hunter CA (2004) STAT1 plays a critical role in the regulation of antimicrobial effector mechanisms, but not in the development of Th1-type responses during toxoplasmosis. J Immunol 172:457–463

Liu L, Okada S, Kong XF, Kreins AY, Cypowyj S, Abhyankar A, Toubiana J, Itan Y, Audry M, Nitschke P et al (2011) Gain-of-function human STAT1 mutations impair IL-17 immunity and underlie chronic mucocutaneous candidiasis. J Exp Med 208:1635–1648

Luo H, Hanratty WP, Dearolf CR (1995) An amino acid substitution in the Drosophila hopTum-l Jak kinase causes leukemia-like hematopoietic defects. EMBO J 14:1412–1420

Ma CS, Chew GY, Simpson N, Priyadarshi A, Wong M, Grimbacher B, Fulcher DA, Tangye SG, Cook MC (2008) Deficiency of Th17 cells in hyper IgE syndrome due to mutations in STAT3. J Exp Med 205:1551–1557

Ma H, Lu C, Ziegler J, Liu A, Sepulveda A, Okada H, Lentzsch S, Mapara MY (2011) Absence of Stat1 in donor CD4 T cells promotes the expansion of Tregs and reduces graft-versus-host disease in mice. J Clin Invest 121:2554–2569

Macchi P, Villa A, Giliani S, Sacco MG, Frattini A, Porta F, Ugazio AG, Johnston JA, Candotti F, O'Shea JJ et al (1995) Mutations of Jak-3 gene in patients with autosomal severe combined immune deficiency (SCID). Nature 377:65–68

Malamut G, El Machhour R, Montcuquet N, Martin-Lanneree S, Dusanter-Fourt I, Verkarre V, Mention JJ, Rahmi G, Kiyono H, Butz EA et al (2010) IL-15 triggers an antiapoptotic pathway in human intraepithelial lymphocytes that is a potential new target in celiac disease-associated inflammation and lymphomagenesis. J Clin Invest 120:2131–2143

Malek TR (2008) The biology of interleukin-2. Annu Rev Immunol 26:453–479

Malinge S, Ragu C, Della-Valle V, Pisani D, Constantinescu SN, Perez C, Villeval JL, Reinhardt D, Landman-Parker J, Michaux L et al (2008) Activating mutations in human acute megakaryoblastic leukemia. Blood 112:4220–4226

Manshouri T, Quintas-Cardama A, Nussenzveig RH, Gaikwad A, Estrov Z, Prchal J, Cortes JE, Kantarjian HM, Verstovsek S (2008) The JAK kinase inhibitor CP-690,550 suppresses the growth of human polycythemia vera cells carrying the JAK2V617F mutation. Cancer Sci 99:1265–1273

Mathur AN, Chang HC, Zisoulis DG, Stritesky GL, Yu Q, O'Malley JT, Kapur R, Levy DE, Kansas GS, Kaplan MH (2007) Stat3 and Stat4 direct development of IL-17-secreting Th cells. J Immunol 178:4901–4907

McGeachy MJ, Chen Y, Tato CM, Laurence A, Joyce-Shaikh B, Blumenschein WM, McClanahan TK, O'Shea JJ, Cua DJ (2009) The interleukin 23 receptor is essential for the terminal differentiation of interleukin 17-producing effector T helper cells in vivo. Nat Immunol 10:314–324

Meraz MA, White JM, Sheehan KC, Bach EA, Rodig SJ, Dighe AS, Kaplan DH, Riley JK, Greenlund AC, Campbell D et al (1996) Targeted disruption of the *Stat1* gene in mice reveals unexpected physiologic specificity in the JAK-STAT signaling pathway. Cell 84:431–442

Mesa RA (2010) Ruxolitinib, a selective JAK1 and JAK2 inhibitor for the treatment of myeloproliferative neoplasms and psoriasis. IDrugs 13:394–403

Migone TS, Lin JX, Cereseto A, Mulloy JC, O'Shea JJ, Franchini G, Leonard WJ (1995) Constitutively activated Jak-STAT pathway in T cells transformed with HTLV-I. Science 269:79–81

Milici AJ, Kudlacz EM, Audoly L, Zwillich S, Changelian P (2008) Cartilage preservation by inhibition of Janus kinase 3 in two rodent models of rheumatoid arthritis. Arthritis Res Ther 10: R14

Miller SA, Huang AC, Miazgowicz MM, Brassil MM, Weinmann AS (2008) Coordinated but physically separable interaction with H3K27-demethylase and H3K4-methyltransferase activities are required for T-box protein-mediated activation of developmental gene expression. Genes Dev 22:2980–2993

Milner JD, Brenchley JM, Laurence A, Freeman AF, Hill BJ, Elias KM, Kanno Y, Spalding C, Elloumi HZ, Paulson ML et al (2008) Impaired T(H)17 cell differentiation in subjects with autosomal dominant hyper-IgE syndrome. Nature 452:773–776

Minegishi Y, Saito M, Morio T, Watanabe K, Agematsu K, Tsuchiya S, Takada H, Hara T, Kawamura N, Ariga T et al (2006) Human tyrosine kinase 2 deficiency reveals its requisite roles in multiple cytokine signals involved in innate and acquired immunity. Immunity 25:745–755

Minegishi Y, Saito M, Tsuchiya S, Tsuge I, Takada H, Hara T, Kawamura N, Ariga T, Pasic S, Stojkovic O, Metin A, Karasuyama H (2007) Dominant-negative mutations in the DNA-binding domain of STAT3 cause hyper-IgE syndrome. Nature 448:1058–1062

Mullighan CG, Zhang J, Harvey RC, Collins-Underwood JR, Schulman BA, Phillips LA, Tasian SK, Loh ML, Su X, Liu W et al (2009) JAK mutations in high-risk childhood acute lymphoblastic leukemia. Proc Natl Acad Sci USA 106:9414–9418

Murphy KM, Ouyang W, Farrar JD, Yang J, Ranganath S, Asnagli H, Afkarian M, Murphy TL (2000) Signaling and transcription in T helper development. Annu Rev Immunol 18:451–494

Namjou B, Sestak AL, Armstrong DL, Zidovetzki R, Kelly JA, Jacob N, Ciobanu V, Kaufman KM, Ojwang JO, Ziegler J et al (2009) High-density genotyping of STAT4 reveals multiple haplotypic associations with systemic lupus erythematosus in different racial groups. Arthritis Rheum 60:1085–1095

Neubauer H, Cumano A, Muller M, Wu H, Huffstadt U, Pfeffer K (1998) Jak2 deficiency defines an essential developmental checkpoint in definitive hematopoiesis. Cell 93:397–409

Nurieva RI, Chung Y, Hwang D, Yang XO, Kang HS, Ma L, Wang YH, Watowich SS, Jetten AM, Tian Q, Dong C (2008) Generation of T follicular helper cells is mediated by interleukin-21 but independent of T helper 1, 2, or 17 cell lineages. Immunity 29:138–149

O'Malley JT, Sehra S, Thieu VT, Yu Q, Chang HC, Stritesky GL, Nguyen ET, Mathur AN, Levy DE, Kaplan MH (2009) Signal transducer and activator of transcription 4 limits the development of adaptive regulatory T cells. Immunology 127:587–595

O'Shea JJ, Lahesmaa R, Vahedi G, Laurence A, Kanno Y (2011) Genomic views of STAT function in CD4+ T helper cell differentiation. Nat Rev Immunol 11:239–250

O'Sullivan A, Chang HC, Yu Q, Kaplan MH (2004) STAT4 is required for interleukin-12-induced chromatin remodeling of the CD25 locus. J Biol Chem 279:7339–7345

Orlando V (2000) Mapping chromosomal proteins in vivo by formaldehyde-crosslinked-chromatin immunoprecipitation. Trends Biochem Sci 25:99–104

Oyamada A, Ikebe H, Itsumi M, Saiwai H, Okada S, Shimoda K, Iwakura Y, Nakayama KI, Iwamoto Y, Yoshikai Y, Yamada H (2009) Tyrosine kinase 2 plays critical roles in the pathogenic CD4 T cell responses for the development of experimental autoimmune encephalomyelitis. J Immunol 183:7539–7546

Paniagua R, Si MS, Flores MG, Rousvoal G, Zhang S, Aalami O, Campbell A, Changelian PS, Reitz BA, Borie DC (2005) Effects of JAK3 inhibition with CP-690,550 on immune cell populations and their functions in nonhuman primate recipients of kidney allografts. Transplantation 80:1283–1292

Parganas E, Wang D, Stravopodis D, Topham DJ, Marine JC, Teglund S, Vanin EF, Bodner S, Colamonici OR, van Deursen JM et al (1998) Jak2 is essential for signaling through a variety of cytokine receptors. Cell 93:385–395

Park H, Li Z, Yang XO, Chang SH, Nurieva R, Wang YH, Wang Y, Hood L, Zhu Z, Tian Q, Dong C (2005) A distinct lineage of CD4 T cells regulates tissue inflammation by producing interleukin 17. Nat Immunol 6:1133–1141

Peeters P, Raynaud SD, Cools J, Wlodarska I, Grosgeorge J, Philip P, Monpoux F, Van Rompaey L, Baens M, Van den Berghe H, Marynen P (1997) Fusion of TEL, the ETS-variant gene 6 (ETV6), to the receptor-associated kinase JAK2 as a result of t(9;12) in a lymphoid and t(9;15;12) in a myeloid leukemia. Blood 90:2535–2540

Pesu M, Muul L, Kanno Y, O'Shea JJ (2006) Proprotein convertase furin is preferentially expressed in T helper 1 cells and regulates interferon gamma. Blood 108:983–985

Quintas-Cardama A, Vaddi K, Liu P, Manshouri T, Li J, Scherle PA, Caulder E, Wen X, Li Y, Waeltz P et al (2010) Preclinical characterization of the selective JAK1/2 inhibitor INCB018424: therapeutic implications for the treatment of myeloproliferative neoplasms. Blood 115:3109–3117

Remmers EF, Plenge RM, Lee AT, Graham RR, Hom G, Behrens TW, de Bakker PI, Le JM, Lee HS, Batliwalla F et al (2007) STAT4 and the risk of rheumatoid arthritis and systemic lupus erythematosus. N Engl J Med 357:977–986

Remmers EF, Cosan F, Kirino Y, Ombrello MJ, Abaci N, Satorius C, Le JM, Yang B, Korman BD, Cakiris A et al (2010) Genome-wide association study identifies variants in the MHC class I, IL10, and IL23R-IL12RB2 regions associated with Behcet's disease. Nat Genet 42:698–702

Riedy MC, Dutra AS, Blake TB, Modi W, Lal BK, Davis J, Bosse A, O'Shea JJ, Johnston JA (1996) Genomic sequence, organization, and chromosomal localization of human JAK3. Genomics 37:57–61

Robertson AG, Bilenky M, Tam A, Zhao Y, Zeng T, Thiessen N, Cezard T, Fejes AP, Wederell ED, Cullum R et al (2008) Genome-wide relationship between histone H3 lysine 4 mono- and tri-methylation and transcription factor binding. Genome Res 18:1906–1917

Rochman Y, Spolski R, Leonard WJ (2009) New insights into the regulation of T cells by gamma (c) family cytokines. Nat Rev Immunol 9:480–490

Rodig SJ, Meraz MA, White JM, Lampe PA, Riley JK, Arthur CD, King KL, Sheehan KC, Yin L, Pennica D et al (1998) Disruption of the *Jak1* gene demonstrates obligatory and nonredundant roles of the Jaks in cytokine-induced biologic responses. Cell 93:373–383

Rui L, Emre NC, Kruhlak MJ, Chung HJ, Steidl C, Slack G, Wright GW, Lenz G, Ngo VN, Shaffer AL et al (2010) Cooperative epigenetic modulation by cancer amplicon genes. Cancer Cell 18:590–605

Russell SM, Tayebi N, Nakajima H, Riedy MC, Roberts JL, Aman MJ, Migone TS, Noguchi M, Markert ML, Buckley RH et al (1995) Mutation of Jak3 in a patient with SCID: essential role of Jak3 in lymphoid development. Science 270:797–800

Sakaguchi S, Yamaguchi T, Nomura T, Ono M (2008) Regulatory T cells and immune tolerance. Cell 133:775–787

Santos FP, Kantarjian HM, Jain N, Manshouri T, Thomas DA, Garcia-Manero G, Kennedy D, Estrov Z, Cortes J, Verstovsek S (2010) Phase 2 study of CEP-701, an orally available JAK2 inhibitor, in patients with primary or post-polycythemia vera/essential thrombocythemia myelofibrosis. Blood 115:1131–1136

Saraiva M, O'Garra A (2010) The regulation of IL-10 production by immune cells. Nat Rev Immunol 10:170–181

Schraml BU, Hildner K, Ise W, Lee WL, Smith WA, Solomon B, Sahota G, Sim J, Mukasa R, Cemerski S et al (2009) The AP-1 transcription factor Batf controls T(H)17 differentiation. Nature 460:405–409

Seidel HM, Milocco LH, Lamb P, Darnell JE Jr, Stein RB, Rosen J (1995) Spacing of palindromic half sites as a determinant of selective STAT (signal transducers and activators of transcription) DNA binding and transcriptional activity. Proc Natl Acad Sci USA 92:3041–3045

Seto Y, Nakajima H, Suto A, Shimoda K, Saito Y, Nakayama KI, Iwamoto I (2003) Enhanced Th2 cell-mediated allergic inflammation in Tyk2-deficient mice. J Immunol 170:1077–1083

Shaw MH, Boyartchuk V, Wong S, Karaghiosoff M, Ragimbeau J, Pellegrini S, Muller M, Dietrich WF, Yap GS (2003) A natural mutation in the Tyk2 pseudokinase domain underlies altered susceptibility of B10.Q/J mice to infection and autoimmunity. Proc Natl Acad Sci USA 100:11594–11599

Shimoda K, van Deursen J, Sangster MY, Sarawar SR, Carson RT, Tripp RA, Chu C, Quelle FW, Nosaka T, Vignali DA et al (1996) Lack of IL-4-induced Th2 response and IgE class switching in mice with disrupted *Stat6* gene. Nature 380:630–633

Spits H, Di Santo JP (2011) The expanding family of innate lymphoid cells: regulators and effectors of immunity and tissue remodeling. Nat Immunol 12:21–27

Spolski R, Leonard WJ (2010) IL-21 and T follicular helper cells. Int Immunol 22:7–12

Stritesky GL, Muthukrishnan R, Sehra S, Goswami R, Pham D, Travers J, Nguyen ET, Levy DE, Kaplan MH (2011) The transcription factor STAT3 is required for T helper 2 cell development. Immunity 34:39–49

Stumhofer JS, Laurence A, Wilson EH, Huang E, Tato CM, Johnson LM, Villarino AV, Huang Q, Yoshimura A, Sehy D et al (2006) Interleukin 27 negatively regulates the development of interleukin 17-producing T helper cells during chronic inflammation of the central nervous system. Nat Immunol 7:937–945

Takaki H, Ichiyama K, Koga K, Chinen T, Takaesu G, Sugiyama Y, Kato S, Yoshimura A, Kobayashi T (2008) STAT6 Inhibits TGF-beta1-mediated Foxp3 induction through direct binding to the Foxp3 promoter, which is reverted by retinoic acid receptor. J Biol Chem 283:14955–14962

Takeda K, Tanaka T, Shi W, Matsumoto M, Minami M, Kashiwamura S, Nakanishi K, Yoshida N, Kishimoto T, Akira S (1996) Essential role of Stat6 in IL-4 signalling. Nature 380:627–630

Takeda K, Noguchi K, Shi W, Tanaka T, Matsumoto M, Yoshida N, Kishimoto T, Akira S (1997) Targeted disruption of the mouse *Stat3* gene leads to early embryonic lethality. Proc Natl Acad Sci USA 94:3801–3804

Takeda A, Hamano S, Yamanaka A, Hanada T, Ishibashi T, Mak TW, Yoshimura A, Yoshida H (2003) Cutting edge: role of IL-27/WSX-1 signaling for induction of T-bet through activation of STAT1 during initial Th1 commitment. J Immunol 170:4886–4890

Thieu VT, Yu Q, Chang HC, Yeh N, Nguyen ET, Sehra S, Kaplan MH (2008) Signal transducer and activator of transcription 4 is required for the transcription factor T-bet to promote T helper 1 cell-fate determination. Immunity 29:679–690

Tirado CA, Chen W, Huang LJ, Laborde C, Hiemenz MC, Valdez F, Ho K, Winick N, Lou Z, Koduru P (2010) Novel JAK2 rearrangement resulting from a t(9;22) (p24;q11.2) in B-acute lymphoblastic leukemia. Leuk Res 34:1674–1676

Tyner JW, Walters DK, Willis SG, Luttropp M, Oost J, Loriaux M, Erickson H, Corbin AS, O'Hare T, Heinrich MC et al (2008) RNAi screening of the tyrosine kinome identifies therapeutic targets in acute myeloid leukemia. Blood 111:2238–2245

Vainchenker W, Dusa A, Constantinescu SN (2008) JAKs in pathology: role of Janus kinases in hematopoietic malignancies and immunodeficiencies. Semin Cell Dev Biol 19:385–393

van de Veerdonk FL, Plantinga TS, Hoischen A, Smeekens SP, Joosten LA, Gilissen C, Arts P, Rosentul DC, Carmichael AJ, Smits-van der Graaf CA et al (2011) STAT1 mutations in autosomal dominant chronic mucocutaneous candidiasis. N Engl J Med 365:54–61

Velazquez L, Fellous M, Stark GR, Pellegrini S (1992) A protein tyrosine kinase in the interferon alpha/beta signaling pathway. Cell 70:313–322

Veldhoen M, Hirota K, Westendorf AM, Buer J, Dumoutier L, Renauld JC, Stockinger B (2008a) The aryl hydrocarbon receptor links TH17-cell-mediated autoimmunity to environmental toxins. Nature 453:106–109

Veldhoen M, Uyttenhove C, van Snick J, Helmby H, Westendorf A, Buer J, Martin B, Wilhelm C, Stockinger B (2008b) Transforming growth factor-beta 'reprograms' the differentiation of T helper 2 cells and promotes an interleukin 9-producing subset. Nat Immunol 9:1341–1346

Verstovsek S, Kantarjian H, Mesa RA, Pardanani AD, Cortes-Franco J, Thomas DA, Estrov Z, Fridman JS, Bradley EC, Erickson-Viitanen S et al (2010) Safety and efficacy of INCB018424, a JAK1 and JAK2 inhibitor, in myelofibrosis. N Engl J Med 363:1117–1127

Villarino AV, Stumhofer JS, Saris CJ, Kastelein RA, de Sauvage FJ, Hunter CA (2006) IL-27 limits IL-2 production during Th1 differentiation. J Immunol 176:237–247

Villarino AV, Tato CM, Stumhofer JS, Yao Z, Cui YK, Hennighausen L, O'Shea JJ, Hunter CA (2007) Helper T cell IL-2 production is limited by negative feedback and STAT-dependent cytokine signals. J Exp Med 204:65–71

Villarino AV, Gallo E, Abbas AK (2010) STAT1-activating cytokines limit Th17 responses through both T-bet-dependent and -independent mechanisms. J Immunol 185:6461–6471

Walters DK, Mercher T, Gu TL, O'Hare T, Tyner JW, Loriaux M, Goss VL, Lee KA, Eide CA, Wong MJ et al (2006) Activating alleles of JAK3 in acute megakaryoblastic leukemia. Cancer Cell 10:65–75

Wei G, Wei L, Zhu J, Zang C, Hu-Li J, Yao Z, Cui K, Kanno Y, Roh TY, Watford WT et al (2009) Global mapping of H3K4me3 and H3K27me3 reveals specificity and plasticity in lineage fate determination of differentiating CD4+ T cells. Immunity 30:155–167

Wei L, Vahedi G, Sun HW, Watford WT, Takatori H, Ramos HL, Takahashi H, Liang J, Gutierrez-Cruz G, Zang C et al (2010) Discrete roles of STAT4 and STAT6 transcription factors in tuning epigenetic modifications and transcription during T helper cell differentiation. Immunity 32:840–851

Wernig G, Kharas MG, Okabe R, Moore SA, Leeman DS, Cullen DE, Gozo M, McDowell EP, Levine RL, Doukas J et al (2008) Efficacy of TG101348, a selective JAK2 inhibitor, in treatment of a murine model of JAK2V617F-induced polycythemia vera. Cancer Cell 13:311–320

Witthuhn BA, Quelle FW, Silvennoinen O, Yi T, Tang B, Miura O, Ihle JN (1993) JAK2 associates with the erythropoietin receptor and is tyrosine phosphorylated and activated following stimulation with erythropoietin. Cell 74:227–236

Woods NB, Bottero V, Schmidt M, von Kalle C, Verma IM (2006) Gene therapy: therapeutic gene causing lymphoma. Nature 440:1123

Xu L, Kitani A, Stuelten C, McGrady G, Fuss I, Strober W (2010) Positive and negative transcriptional regulation of the *Foxp3* gene is mediated by access and binding of the Smad3 protein to enhancer I. Immunity 33:313–325

Yang XO, Panopoulos AD, Nurieva R, Chang SH, Wang D, Watowich SS, Dong C (2007) STAT3 regulates cytokine-mediated generation of inflammatory helper T cells. J Biol Chem 282:9358–9363

Yang XO, Pappu BP, Nurieva R, Akimzhanov A, Kang HS, Chung Y, Ma L, Shah B, Panopoulos AD, Schluns KS et al (2008) T helper 17 lineage differentiation is programmed by orphan nuclear receptors ROR alpha and ROR gamma. Immunity 28:29–39

Yang XP, Ghoreschi K, Steward-Tharp SM, Rodriguez-Canales J, Zhu J, Grainger JR, Hirahara K, Sun HW, Wei L, Vahedi G et al (2011) Opposing regulation of the locus encoding IL-17 through direct, reciprocal actions of STAT3 and STAT5. Nat Immunol 12:247–254

Yao Z, Cui Y, Watford WT, Bream JH, Yamaoka K, Hissong BD, Li D, Durum SK, Jiang Q, Bhandoola A et al (2006) Stat5a/b are essential for normal lymphoid development and differentiation. Proc Natl Acad Sci USA 103:1000–1005

Yao Z, Kanno Y, Kerenyi M, Stephens G, Durant L, Watford WT, Laurence A, Robinson GW, Shevach EM, Moriggl R et al (2007) Nonredundant roles for Stat5a/b in directly regulating Foxp3. Blood 109:4368–4375

Yu Q, Thieu VT, Kaplan MH (2007) Stat4 limits DNA methyltransferase recruitment and DNA methylation of the IL-18Ralpha gene during Th1 differentiation. EMBO J 26:2052–2060

Yuan ZL, Guan YJ, Chatterjee D, Chin YE (2005) Stat3 dimerization regulated by reversible acetylation of a single lysine residue. Science 307:269–273

Zaretsky AG, Taylor JJ, King IL, Marshall FA, Mohrs M, Pearce EJ (2009) T follicular helper cells differentiate from Th2 cells in response to helminth antigens. J Exp Med 206:991–999

Zhu J, Guo L, Watson CJ, Hu-Li J, Paul WE (2001) Stat6 is necessary and sufficient for IL-4's role in Th2 differentiation and cell expansion. J Immunol 166:7276–7281

Zhu J, Cote-Sierra J, Guo L, Paul WE (2003) Stat5 activation plays a critical role in Th2 differentiation. Immunity 19:739–748

Zhu J, Yamane H, Paul WE (2010) Differentiation of effector CD4 T cell populations (*). Annu Rev Immunol 28:445–489

Ziegler SF, Artis D (2010) Sensing the outside world: TSLP regulates barrier immunity. Nat Immunol 11:289–293

Zorn E, Nelson EA, Mohseni M, Porcheray F, Kim H, Litsa D, Bellucci R, Raderschall E, Canning C, Soiffer RJ et al (2006) IL-2 regulates FOXP3 expression in human CD4+CD25+ regulatory T cells through a STAT-dependent mechanism and induces the expansion of these cells in vivo. Blood 108:1571–1579

The Role of Janus Kinases in Hematopoietic Malignancies

Damla Olcaydu and Robert Kralovics

Abstract
The Janus family tyrosine kinases are indispensible for cytokine signaling and play a crucial role in blood cell production. However, their excessive activity causes various hematological phenotypes associated with overproduction of terminally differentiated cells and/or blastic transformation. Here we review the somatic mutations in the Janus family kinases and the associated hematological phenotypes.

Introduction

The production of blood cells is orchestrated by growth factors and cytokines that tightly regulate survival, proliferation and differentiation of hematopoietic stem cells. Binding of these factors to their cognate receptors on the surface of blood cells activates intracellular signaling cascades that modulate gene expression and regulate cell fate. As many cytokine receptors lack intrinsic catalytic activity, the transduction of growth signals requires molecules that associate with receptors on the cell surface and, upon ligand binding, activate downstream effector proteins in the cytosol and nucleus. Proper interaction and coordination of these signaling pathways is the basis for efficient production of various blood cell types. Alterations of cytokine signaling pathways and impairment of their regulation have been implicated in various disorders of the hematopoietic system. The Janus kinase (JAK) family of non-receptor protein tyrosine kinases, with its four members JAK1, JAK2, JAK3 and TYK2, plays a key role in hematopoiesis by mediating

D. Olcaydu • R. Kralovics (✉)
Center for Molecular Medicine, Austrian Academy of Sciences, Lazarettgasse 14, 1090 Vienna, Austria
e-mail: robert.kralovics@cemm.oeaw.ac.at

Th. Decker and M. Müller (eds.), *Jak-Stat Signaling: From Basics to Disease*,
DOI 10.1007/978-3-7091-0891-8_14, © Springer-Verlag Wien 2012

the transduction of growth signals from cytokine receptors to their effectors. In consequence, genetic aberrations that impair the function of JAK kinases have been implicated in various malignancies of the hematopoietic system, most prominently in the pathogenesis of leukemic disorders and the myeloproliferative neoplasms.

Tumors of the Hematopoietic Tissue

In first line, neoplasms of the hematopoietic tissue are classified according to the lineage of the malignant cell being either of myeloid, lymphoid, histiocytic/dendritic or of ambiguous lineage. Further categorization is based on the differentiation stage of the neoplastic cell clone. For example, precursor cell neoplasms (e.g. acute myeloid leukemia, lymphoblastic leukemia, blastic plasmacytoid dendritic cell neoplasm) comprise of undifferentiated cells that do not express lineage-specific cell surface markers, or cells that have a mixed lineage phenotype (expressing more than one lineage marker). On the counterpart, hematopoietic neoplasms that comprise of more mature and differentiated cells include myeloproliferative neoplasms, myelodysplastic syndromes, mature B-cell or T/NK-cell lymphomas, and histiocytic/dendritic cell neoplasms. Further classification is applied to mature lymphoid neoplasms according to the stage of differentiation of the malignant cell (e.g. mantle cell lymphoma), based on morphology (e.g. diffuse large B cell lymphoma), clinical features (e.g. diffuse large B cell lymphoma associated with chronic inflammation) or several different parameters including immunophenotypic and genotypic features that, in combination, serve the definition of specific disease entities. In case of the myeloid neoplasms, sub-classification is mainly based on the maturation stage. Whereas no or minimal maturation of blasts is present in acute myeloid leukemia, effective (e.g. myeloproliferative neoplasms) or ineffective myeloid cell maturation (e.g. myelodysplastic syndromes), or both in combination (e.g. myelodysplastic/myeloproliferative neoplasms) are observed in other myeloid malignancies. In the last years, a significant number of genetic aberrations were discovered in hematopoietic neoplasms, some of which associate with subgroups or specific disease entities. Thus, cytogenetic and mutational analysis has been incorporated into the classification scheme and diagnostic criteria of various hematological malignancies (Vardiman 2010).

Janus Kinases in Hematopoiesis

The growth factors and cytokines of the hematopoietic system include interleukins (IL), interferons (IFN), colony-stimulating factors (CSF), erythropoietin (EPO) and thrombopoietin (TPO). Most of these molecules bind to type I cytokine receptors, a group of homologous transmembrane receptors that share the characteristic signature sequence of four conserved cysteine residues in the extracellular domain (the WSXWS motif, with X representing a non-conserved amino acid residue). Type I cytokine receptors either comprise single chains [e.g. granulocyte CSF receptor

(G-CSFR), EPO receptor (EPOR) and TPO receptor (TPOR)] or heterodimers with a unique ligand-binding chain and a common signaling subunit such as a β-chain (receptors of granulocyte-macrophage CSF [GM-CSFR], IL-3 and IL-5), a γ-chain (the IL-2 receptor family) or a gp130 subunit (the IL-6 receptor family). In contrast to the type I family, type II cytokine receptors (IFN receptors and IL-10 receptor family) lack the specific WSXWS motif, but share conserved regions of the membrane proximal intracytoplasmic domain.

The most prominent common feature of type I and type II cytokine receptors is that they are devoid of intrinsic enzymatic activity. Thus, activation of downstream signaling pathways is mediated by tyrosine kinases of the JAK family. The JAK kinases are constitutively associated to the cytokine receptors via interaction of their FERM domain with the Box1 membrane proximal intracytoplasmic region of the receptor. Subsequent binding of a corresponding ligand induces homodimerization (e.g. G-CSFR) or heterodimerization (e.g. GM-CSFR) of cytokine receptor subunits; or causes a conformational change in receptor dimers that are preformed at the cell surface (e.g. EPOR). Activation upon ligand-binding locates receptor-associated JAKs into close proximity and results in transphosphorylation and/or crossphosphorylation of tyrosine residues promoting increased JAK kinase activity. Activated JAKs, in turn, phorphorylate tyrosine sites of the cytoplasmic cytokine receptor domain, thereby generating docking sites for SH2-domain containing signaling molecules such as the signal transducers and activators of transcription (STAT). After recruitment to the cytokine receptor-JAK complex, STATs are tyrosine phosphorylated by the JAKs and form homo- or heterodimers via interactions of their SH2 domains. The phosphorylated STAT dimers are actively translocated to the nucleus, where they bind to specific DNA target sequences and drive gene expression in cooperation with other transcription factors and coactivators.

The target genes that are regulated by STATs largely depend on the cell type, the activating cytokine and the specific STAT protein. Amongst many others, the STATs regulate expression of genes that play a role in cell cycle progression, proliferation, survival and angiogenesis (e.g. cyclin-dependent kinase inhibitors, genes of the Bcl-2 family, cyclins D and E, caspases, VEGF and MMP-2). Negative regulation of the JAK-STAT signaling pathway is tightly controlled at several levels. Constitutive pathways of regulation include protein tyrosine phosphatases such as SHP1, SHP2, CD45 and PTP1B that inactivate JAKs and STATs in the cytoplasm, whereas STAT activity in the nucleus is controlled by protein inhibitor of activated STAT (PIAS) proteins, which interfere with the DNA binding activity of the STATs. The major inducible pathway that regulates JAK-STAT signaling consists of the suppressor of cytokine signaling (SOCS) protein family, which acts as a classical negative feedback loop. Inhibition of JAK-STAT signaling by SOCS proteins is achieved through several ways, namely (1) via competition for the binding sites of JAKs and STATs at the cytokine receptor, (2) by direct binding and inhibition of JAK activity, and (3) by promoting ubiquitination and degradation of the JAKs.

JAK Fusion Proteins

Somatically acquired chromosomal aberrations have been frequently detected in patients with hematologic malignancies. Amongst other defects, translocations of chromosomal material have been identified as major pathogenetic factors in a number of hematopoietic neoplasms. The most studied example is the reciprocal translocation between chromosomes 9 and 22 that encodes the constitutively active BCR-ABL1 fusion protein and has been shown to play a causative role in chronic myeloid leukemia (CML). Similarly, chromosomal translocations that fuse JAK genes to other genes encoding e.g. transcription factors promote the production of chimeric fusion proteins that are constitutively active and alter hematopoietic cell function. The first report of a fusion protein involving a JAK kinase identified the TEL-JAK2 fusion protein that results from a chromosomal translocation between the short arms of chromosomes 9 and 12, denoted as t(9;12)(p24;p13) (Fig. 1) (Lacronique et al. 1997). The TEL-JAK2 fusion was detected in a patient with T-cell childhood acute lymphoblastic leukemia (ALL). Determination of the breakpoints revealed that the translocation caused a fusion between the 3′ part of the *JAK2* gene and the 5′ region of the Translocation Ets Leukemia (*TEL*) gene. Noteworthy, *TEL* encodes a member of the Ets family of transcription factors that at the time was already known to be involved in other leukemic translocations. Expression of TEL-JAK2 in the murine hematopoietic cell line Ba/F3 was shown to result in cytokine-independent growth and indicated a pathogenetic effect of the fusion product on cytokine signaling and hematopoietic cell proliferation (Lacronique et al. 1997). Further investigations revealed that the translocation fuses the kinase domain (JH1) of JAK2 to the oligomerization domain of the TEL transcription factor (Fig. 1) and gives rise to an oncogenic protein with constitutive activity of the JAK2 tyrosine kinase and its downstream signaling effectors (Ho et al. 1999).

The TEL-JAK2 fusion is not exclusive to T-cell leukemia, but has also been detected in a child with early B-precursor ALL as well as in an adult with atypical CML in transformation. The occurrence of the same chromosomal translocation and its aberrant fusion product in myeloid and lymphoid neoplasms suggests pathogenetic relevance of this aberrant kinase in both, lympho- and myeloproliferative diseases (Peeters et al. 1997). Accordingly, bone marrow transplantation experiments in mice have shown that retrovirally transduced TEL-JAK2 establishes a fatal disease phenotype with mixed myeloproliferation and T-cell lymphoproliferation (Schwaller et al. 1998). In addition to constitutive activation of the JAK-STAT pathway, the TEL-JAK2 fusion protein further activates the phosphotidylinositol 3′-kinase (PI3K), Ras/ERK, p38 and NF-κB signaling pathways (Santos et al. 2001; Nguyen et al. 2001; Malinge et al. 2006).

Schwaller et al. addressed whether STAT5 activation is necessary for the development of TEL-JAK2 positive myelo- and lymphoproliferative disease. Mice transplanted with retrovirally transduced bone marrow cells expressing TEL-JAK2 rapidly developed a fatal myelo- and lymphoproliferative syndrome, whereas mice that were reconstituted with Stat5a/b-deficient bone marrow cells expressing TEL-JAK2 did not show a pathologic phenotype. These results imply

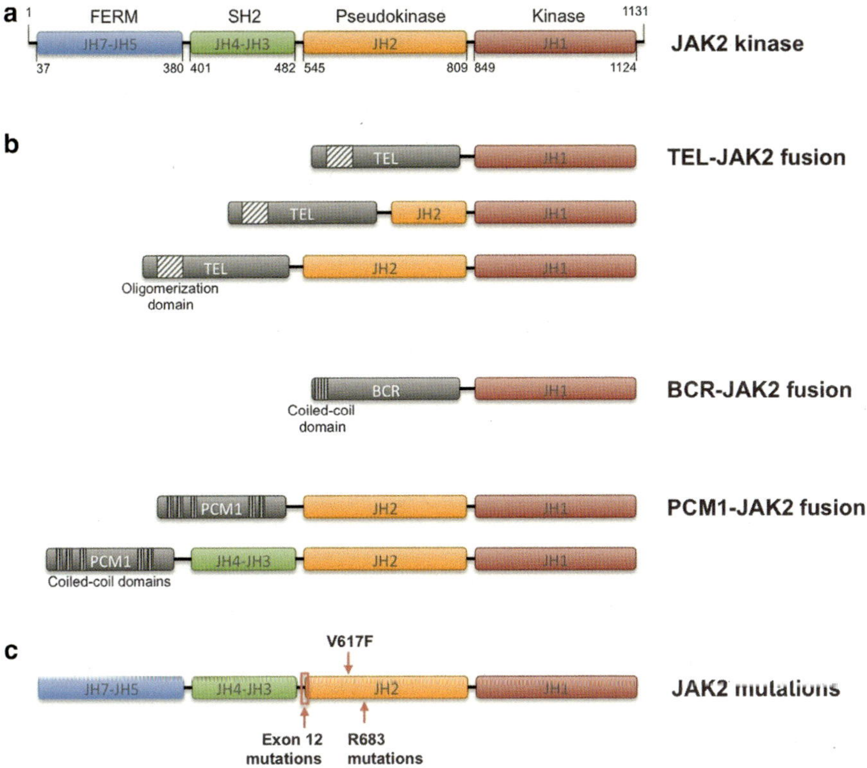

Fig. 1 Domain structures of JAK2, JAK2 fusion proteins and recurrent JAK2 mutations. (**a**) The protein structure of JAK2 comprises seven Janus homology (JH) domains including a tyrosine kinase (*JH1*) and a pseudokinase (*JH2*) domain. (**b**) JAK2 fusion proteins. All reported fusion proteins contain the entire JH1 kinase domain of JAK2. Various inclusion of the JH2 pseudokinase domain has been observed for TEL-JAK2 fusion proteins (Lacronique et al. 1997; Peeters et al. 1997), whereas all detected PCM1-JAK2 fusion products involve the entire pseudokinase domain of JAK2 (Reiter et al. 2005; Bousquet et al. 2005; Murati et al. 2005; Dargent et al. 2011). All BCR-JAK2 cases reported so far showed a fusion of the JAK2 kinase domain to the coiled-coil domain of BCR (Griesinger et al. 2005; Cirmena et al. 2008; Lane et al. 2008). (**c**) Recurrent JAK2 mutations. The V617F mutation in the pseudokinase domain of JAK2 represents the most frequent genetic alteration of JAK2 (Campbell and Green 2006). Other recurrent genetic lesions include mutations of JAK2 exon 12 that locate to the region between the pseudokinase and the JH3 domain (Pardanani et al. 2008; Scott et al. 2007) and mutations of arginine residue 683 (*R683*) that occur frequently in children with Down syndrome and acute lymphoblastic leukemia (Bercovich et al. 2008; Gaikwad et al. 2009; Kearney et al. 2009)

that the disease phenotype in TEL-JAK2 positive leukemias depends on constitutive STAT5 activation and underline the essential role of STAT5 in TEL-JAK2 induced myelo- and lymphoproliferation (Schwaller et al. 2000). Interestingly, STAT5 has been recently reported as indispensable for the maintenance of the leukemic phenotype in other leukemic disorders such as BCR-ABL positive leukemias (Hoelbl et al. 2010).

In a different mouse model, transgenic mice with the TEL-JAK2 complementary DNA under the control of a lymphoid cell-specific promotor developed a fatal T-cell leukemia at 4–22 weeks of age (Carron et al. 2000). However the correlation between STAT5 activation and leukemic phenotype was not addressed in this study, leukemic cells of TEL-JAK2 transgenic mice showed increased activation of STAT5 and STAT1. In order to investigate the oncogenic activity of TEL-JAK2 in the B-cell lineage, dos Santos et al. generated transgenic TEL-JAK2 mice that were deficient for CD3ε, an essential component of the T-cell receptor. These mice showed impaired T-cell development and acquired either T- or B-cell lineage malignancies. Notably, T-cell disorders manifested around 23 weeks of age, whereas B-cell lymphomas developed after a longer time period (50 weeks) (dos Santos and Ghysdael 2006). Further investigations of the TEL-JAK2 transgenic mouse model demonstrated that the differentiation of $CD8^+$ T-cells was altered by the presence of the fusion protein (dos Santos et al. 2007). On the contrary, transplantation of human cord blood cells transduced with a vector encoding TEL-JAK2 into immunodeficient mice induced pathophysiological features of myeloproliferative disorders including myelofibrosis, but no lymphoproliferation (Kennedy et al. 2006).

Since the identification of the t(9;22)(p34;q11.2) translocation that results in a fusion of the abelson murine leukemia viral oncogene homolog 1 (*ABL1*) gene on chromosome 9p34 to the breakpoint cluster region (*BCR*) on chromosome 22, chronic myeloid leukemia (CML) is typically characterized by the presence of the BCR-ABL1 fusion protein. Notably, cases with chromosomal rearrangements involving a region on chromosome 9 that is telomeric to *ABL1* can yield a fusion between the *BCR* region on chromosome 22 and the *JAK2* gene. The first case of BCR-JAK2 was identified in a patient with apparently clinically typical chronic myeloid leukemia, who did not respond to imatinib therapy (Griesinger et al. 2005). The BCR-JAK2 fusion protein was shown to contain the coiled-coil dimerization domain of BCR and the tyrosine kinase domain of JAK2, suggesting that it establishes constitutive activation of the JAK2 kinase and its downstream signaling targets as previously described for TEL-JAK2. Although further cases of BCR-JAK2 were reported in acute myeloid leukemia and leukemia cutis in a patient with atypical CML, t(9;22) translocations with BCR-JAK2 fusion products seem to be overall rare events among the myeloid neoplasms (Fig. 1) (Cirmena et al. 2008; Lane et al. 2008).

More frequent than the t(9;22)(p34;q11.2) translocation are rearrangements that involve chromosomes 8 and 9. In contrast to TEL-JAK2 and BCR-JAK2, the t(8;9)(p22;p24) translocation was detected in various hematologic malignancies such as atypical CML, secondary acute myeloid leukemia, chronic eosinophilic leukemia, myelodysplastic/myeloproliferative disorders, pre-B-cell acute lymphoblastic leukemia and T-cell lymphoma (Reiter et al. 2005; Bousquet et al. 2005; Murati et al. 2005; Adelaide et al. 2006; Dargent et al. 2011). Although varying breakpoints of the t(8;9) translocation were detected, it was shown to consistently fuse the entire kinase domain of JAK2 to several coiled-coil domains of the human autoantigen pericentriolar material (*PCM1*) gene on chromosome 8p22 (Fig. 1). Presumably, the coiled-coil motifs of PCM1 serve for dimerization of the fusion protein and induce constitutive activation of the JAK2 kinase domain in a similar way as it was described for TEL-JAK2. PCM1-JAK2 does not target myeloid or lymphoid cells exclusively, but seems to play a significant role in neoplastic disorders of both hematopoietic lineages. In addition, some investigators have observed that the incidence of acquired JAK2 rearrangements is significantly higher in male patients compared to female subjects (Bacher et al. 2006; Tirado et al. 2010). However, further investigations are needed to confirm a possible gender-specificity of these chromosomal defects.

Other reported rearrangements involving the *JAK2* gene include JAK2-NFE2 and JAK2-AML1, both identified in patients with myelodysplastic syndrome (MDS) (Najfeld et al. 2007). High expression levels of the transcription factor NF-E2 have previously been reported in myeloid progenitor cells of polycythemia vera (PV) patients and the *AML1* gene (also known as *RUNX1*) has been described as a translocation partner in various hematologic malignancies (Goerttler et al. 2005; De Braekeleer et al. 2009). Moreover, recent data indicate that *NFE2* is a target gene of the transcription factor AML1 and that aberrant expression of AML1 mediates increased NFE2 expression, as observed in patients with myeloproliferative neoplasms (MPN) (Wang et al. 2010). Taken together, these findings suggest a role for NFE2 and AML1 as well as their fusion products with JAK2 in MPN and MDS disease pathogenesis. Furthermore, unique cases of JAK2 rearrangements with the ribophorin I gene (RPN1-JAK2), the single-stranded DNA binding protein 2 gene (SSBP2-JAK2), the transcription factor PAX5 (PAX5-JAK2) and, most recently, the SEC13A gene (SEC13A-JAK2) have been detected in different hematopoietic neoplasms (Mark et al. 2006; Poitras et al. 2008; Nebral et al. 2009; Van Roosbroeck et al. 2011).

So far, only chromosomal translocations and fusion products involving the *JAK2* gene, but not the other Janus kinase family members have been identified in patients with hematopoietic neoplasms. In order to investigate potential functional differences between the catalytic domains of the JAK kinase family members, Lacronique et al. generated chimeric proteins in which the JH1 domains of JAK1, JAK3 and TYK2 were fused to the oligomerization domain of TEL (Lacronique et al. 2000). When expressed in the hematopoietic cell line Ba/F3, all TEL-JAK chimeras exhibited constitutive tyrosine kinase activity, induced IL3-independent cell growth and promoted constitutive activation of STATs and their downstream targets. Although cell line experiments demonstrated their transforming potential, none of these TEL-JAK fusion proteins (other than TEL-JAK2) have been detected in patients to the present. Furthermore, these results indicate that constitutive tyrosine kinase activity of JAK2 fusion proteins requires oligomerization, which is mediated by the respective domains of TEL, BCR and PCM1.

While all of the identified oncogenic fusion products contain the entire JH1 kinase domain of JAK2, some of them also comprise varying parts of the JAK2 pseudokinase (JH2) domain. Interestingly, JAK2 fusion proteins including the JH2 domain have mostly been found in patients with chronic hematopoietic malignancies such as MPN, CML and T-cell lymphoma, whereas fusion proteins that contain only the kinase domain but not or only parts of the pseudokinase domain were identified in more severe and acute disorders such as ALL. Hence, it has been postulated that various inclusion of the pseudokinase domain in oncogenic JAK2 fusion proteins might influence the disease phenotype in hematopoietic neoplasms with JAK2 rearrangements (Murati et al. 2005). However, further investigations are necessary in order to elucidate the functional impact of the JH2 domain on the activity of JAK2 fusion proteins.

Activating JAK Mutations

The finding that the *JAK2* gene was rearranged in a number of myelo- and lymphoproliferative disorders drew attention to the JAK family of tyrosine kinases and suggested that genetic alterations of these genes might in general contribute to tumorigenesis of the hematopoietic system. Among other approaches, this hypothesis paved the way for the identification of the JAK2-V617F mutation that was shown to be a major pathogenetic factor in the disease evolution of the myeloproliferative neoplasms. Following the detection of the JAK2-V617F mutation, further efford was taken in order to reveal the frequency and relevance of other mutations of the JAK kinases in hematopoietic malignancies.

DNA sequencing studies in patients with acute leukemia revealed recurrent somatic mutations of the *JAK1* gene (Flex et al. 2008; Jeong et al. 2008; Xiang et al. 2008; Mullighan et al. 2009). However, the exact frequency of *JAK1* mutations in hematopoietic malignancies remains unclear, as recent studies served conflicting results on their incidence (Asnafi et al. 2010). Another unexpected finding was that certain identified mutations of *JAK1* were demonstrated as devoid of transforming activity, although some induced a slight increase in cytokine hypersensitivity (Flex et al. 2008; Xiang et al. 2008). Thus, the exact pathogenetic function of *JAK1* mutations in leukemic disorders remains unknown. The observation that patients with mutations of the *JAK1* gene have a poor prognostic outcome suggests that *JAK1* alterations are unfavorable genetic defects and associated with disease progression (Flex et al. 2008; Mullighan et al. 2009). With the aim of gaining further insights into the functional characteristics of *JAK1* mutations, a screening of randomly mutated *JAK1* cDNAs and an analysis of their transforming abilities was performed recently (Gordon et al. 2010). The results of this study indicated that transforming mutations of the *JAK1* gene primary localize to the kinase domain, that distinct *JAK1* mutations activate different downstream signaling pathways, that – unlike mutations of the *JAK2* gene – not all *JAK1* mutations require an intact FERM domain in order to induce transformation, and that *JAK1* mutations result in increased STAT1 phosphorylation. Nevertheless, further studies are needed in order to define the exact frequency and role of *JAK1* mutations in the disease evolution of the hematopoietic neoplasms.

In contrast to JAK1 and JAK2, early studies of JAK3 already suggested that this kinase was involved in cancer pathogenesis. First results of gene expression analysis revealed that JAK3 was predominantly expressed in B-cell malignancies and cell lines, raising the hypothesis that JAK3 contributed to leukemic transformation in hematopoietic cells (Tortolani et al. 1995). Furthermore, splice variants of JAK3 were identified in cancer cells of hematopoietic and epithelial origin and signaling through JAK3 was implicated in the activation of proto-oncogenes such as c-fos and c-myc (Kawahara et al. 1995; Lai et al. 1995). Peripheral blood T-cells transformed with the human T-cell lymphotropic virus type 1 (HTLV-1), already known to be a major cause of adult T-cell leukemia, were reported to exhibit constitutive phosphorylation of JAK3 and IL-2-independent cell growth (Migone et al. 1995; Xu et al. 1995). One more line of evidence for a role of JAK3 in cancer pathogenesis was served by the finding that a TEL-JAK3 fusion protein containing the oligomerization domain of TEL and the kinase domain of JAK3 – although not found in patients so far – constitutively activated JAK3 and induced factor-independent growth of Ba/F3 cells (Carron et al. 2000). Furthermore, deletions of the JH2 domain of JAK3 have been shown to result in constitutive activation of the JH1 domain, thus suggesting an autoinhibitory function of the pseudokinase domain on the catalytic kinase activity of JAK3, similarly to what has been suggested for JAK2 (Saharinen and Silvennoinen 2002). Residues V617 to E621 of JAK2 have been predicted to serve autoinhibition of the kinase domain and V617 is conserved among JAK1, JAK2, and TYK2, but interestingly not in JAK3 (Lindauer et al. 2001). Moreover, the substitution of the M592 residue of JAK3, that is homologous to V617 of JAK2, does not result in constitutive kinase activity (James et al. 2005). Thus, it seems unlikely that mutations of this region could establish an aberrant phenotype in hematopoietic cells.

The TYK2 tyrosine kinase is predominantly associated to cytokine receptors that share the gp130 subunit (e.g. IL-6 receptor) or type II cytokine receptors such as the interferon or the IL-10 receptor family. TYK2 has been shown to play a role in immune cell signaling and rare alterations of TYK2 have been implicated in primary immunodeficiencies (Minegishi et al. 2006; Ghoreschi et al. 2009). In contrast to activating gain-of-function mutations in other members of the Janus kinase family, the TYK2 mutations identified so far were loss-of-function mutations. JAK3 represents the only exception to this observation, as both, gain and loss-of-function mutations have been reported for JAK3. However, JAK3 deficiency does not result in malignant disorders of the hematopoietic system but manifests in human severe combined immunodeficiencies (Ghoreschi et al. 2009).

The following paragraphs are aimed to serve a detailed description of activating mutations of the JAK kinases that were identified in patients with hematopoietic neoplasms to the present.

Activating JAK Mutations in Acute Leukemia

Somatically acquired mutations in the *JAK1* gene have been most prominently implicated in the pathogenesis of acute lymphoblastic leukemia. A recent study showed that various JAK1 mutations account for more than 18% of adult T-cell precursor ALL (T-ALL) cases, whereas only 3% of B-cell precursor ALL (B-ALL) were affected (Flex et al. 2008). Moreover, patients with JAK1 mutations had higher age at diagnosis, poor response to therapy as well as poor overall prognosis compared to ALL patients without alterations of JAK1. In contrast to pathogenetic JAK2 mutations, which exclusively alter the pseudokinase domain, the somatic JAK1 mutations identified in this study were spread among all domains of the JAK1 kinase. Furthermore, a screening of JAK1 alterations in pediatric ALL cases revealed no mutations in 85 patients with B-ALL and only 1 pediatric case of 49 with T-ALL (JAK1-L653F) (Flex et al. 2008). In a larger screening approach including 187 BCR-ABL1 negative high-risk pediatric ALL cases, several mutations in JAK1 ($n = 3$) and JAK3 ($n = 1$), but most prominently in JAK2 ($n = 16$) were identified. Thus, mutations of the JAK kinases are present in all together 20% of high-risk pediatric ALL patients and are associated with a poor outcome compared to cases without JAK mutations (Mullighan et al. 2009).

Investigations in patients with acute myeloid leukemia (AML) revealed the presence of JAK1 mutations in about 2% of cases. The identified somatic mutations, JAK1-T478S and JAK1-V623A, involved highly conserved residues of the JAK1 gene (Xiang et al. 2008). Experiments in cell lines and primary murine hematopoietic progenitor cells failed to demonstrate a difference in proliferative advantage or transformation to leukemia between JAK1 wild-type and the two mutant forms. However, cells expressing JAK1-T478S or JAK1-V623A showed increased phosphorylation of STAT1 and its downstream effectors after interferon stimulation. Taken together, these data suggest that mutant forms of the JAK1 kinase do not directly induce leukemic transformation but alter downstream signaling pathways in response to other growth signals. In a screening study for mutations of the JAK1, JAK3 and TYK2 genes in 494 samples from various human cancers, several mutations of JAK1 and JAK3, but none in TYK2 were identified (Jeong et al. 2008). Three identical JAK1-V658F mutations were detected in two patients with T-ALL and one AML case with a t(15;17)(q22;q12) rearrangement. Interestingly, the V658F mutation of JAK1 represents the homologous mutation to the V617F mutation of the JAK2 kinase. A third case with T-ALL harbored a JAK1-L783F mutation, whereas two patients with solid tumors, being lung adenocarcinoma and invasive ductal breast carcinoma, carried a JAK1-T782M and a JAK1-K647Y mutation, respectively. Mutations of the JAK3 gene were also identified in this study, however only in 2% of cases with breast or gastric carcinomas. Previous investigations also report low frequencies of JAK1 and JAK3 mutations in solid tumors (less than 5%) and non-recurrence of the identified mutant forms (Bardelli et al. 2003; Greenman et al. 2007). Furthermore, mutations of JAK1 and JAK3 have not been found in adult T-cell leukemia/lymphoma (Kameda et al. 2010). These data indicate that mutations in JAK kinases might represent rare passenger mutations in solid tumors and might be rather specific to rare hematologic malignancies such as the JAK2-V617F mutation in polycythemia vera and JAK1 mutations in T-ALL.

Recently, a mutation of the JAK2 gene was detected in a patient with childhood acute megakaryoblastic leukemia (AMKL) (Malinge et al. 2008). In contrast to known JAK2 mutations such as the V617F mutation, JAK2-M535I was shown neither to induce cytokine-independent growth in a transduced cell line nor to affect downstream signaling effectors. Thus, it is unlikely that this mutation plays a pathogenetic role in childhood AMKL. In a recent study, molecular analysis of the JAK2 gene was performed in 286 children with ALL. Although no V617F mutations of the JAK2 gene were identified, a c.1832T>C transition resulting in a leucine to serine substitution in the JH2 domain of JAK2 was detected in a 3-year old girl with precursor-B-ALL (Kratz et al. 2006). As the mutations was present at the time of diagnosis, but was absent in a subsequent bone marrow sample at remission, the authors concluded that the JAK2-L611S mutation was carried by the pathogenetic clone that was eliminated during therapy. Similarly, in a different screening study including 558 samples from common human cancers such as colon, breast, lung and acute adulthood leukemias, three mutations in the JAK2 gene were identified. Of the 113 cases of acute myelogenous leukemia included into the study, two patients carried JAK2-V617F and one had a newly identified K607N mutation in the pseudokinase domain of JAK2 (Lee et al. 2006). Interestingly, no mutations of JAK2 were found in other cancer types, indicating that the pathogenetic impact of mutations in the JAK2 kinase is restricted to the hematopoietic tissues.

Children with Down syndrome (DS) have a 10–20-fold increased risk of developing either acute lymphoblastic or myeloid leukemia (Hasle et al. 2000). Mutations in the *JAK2* gene occur frequently in DS-ALL. Recent investigations show that up to 28% of DS-ALL patients carry a mutation in the *JAK2* gene, mostly point mutations occurring at arginine residue on position 683 (R683) (Fig. 1) (Bercovich et al. 2008; Gaikwad et al. 2009; Kearney et al. 2009). Most of these mutations cause a replacement of the highly basic amino acid arginine by a neutral amino acid, suggesting that this amino acid change alters protein binding and that a selective pressure for mutations at this position exists. Furthermore, an acquired 5-amino acid deletion within the pseudokinase domain of JAK2 (JAK2DeltaIREED) was identified in a patient with DS-ALL and shown to induce growth factor-independent proliferation in a transduced hematopoietic cell line. Interestingly, the mutation was homozygous due to a loss of heterozygosity at the respective region on the short arm of chromosome 9 (Malinge et al. 2007).

Several mutations of JAK3 have been identified in children with Down syndrome and leukemia. Whereas the JAK3-A572V mutation was only found in the megakaryoblastic cell line CMK, the JAK3-V722I and P132T mutations have been reported in patients with acute megakaryoblastic leukemia (AMKL). However, all three variants transform Ba/F3 cells to cytokine-independent growth. Moreover, the JAK3-A572V mutation establishes several pathogenetic features of AMKL as well as a T-cell lymphoproliferative disorder in a murine bone marrow transplant model (Walters 2006). However, as full leukemic transformation is not achieved by sole expression of JAK3-A572V in a mouse model and DS-AMKL patients with JAK3 mutations also harbor GATA-1 mutations and trisomy 21, these genetic defects might need to cooperate in order to induce leukemogenesis in AMKL.

JAK2 Mutations in Myeloproliferative Neoplasms

First results indicating that alterations in the JH2 pseudokinase domain of JAK2 were involved in the pathogenesis of leukemic disorders were obtained from investigations in *Drosophila melanogaster*. A point mutation in a conserved region of the pseudokinase domain of *hopscotch* (*hop*), the homologue of mammalian Janus kinase genes in *Drosophila*, was identified to cause a glutamic acid to lysine substitution at amino acid residue 695 (E695K) (Luo et al. 1997). Overexpression of this mutant induced hyperphosphorylation and hyperactivation of D-Stat, the *Drosophila* STAT protein. Furthermore, expression of the mutant Jak2(E665K), which corresponds to the Drosophila E695K mutation, in a murine cell line was shown to cause hyperactivation of murine Stat5. Subsequent investigations showed that mutations in the *hopscotch* locus such as E695K induce hematopoietic neoplasias resembling human leukemia (Luo et al. 1997; Harrison et al. 1995).

In 2005, an acquired mutation in the Janus kinase 2 (*JAK2*) gene was described in patients with myeloproliferative neoplasms (MPN) (Fig. 1) (James et al. 2005; Kralovics et al. 2005; Baxter et al. 2005; Levine et al. 2005). The transversion of a guanine to a thymidine in exon 14 of the *JAK2* gene was identified to result in a valine for phenylalanine substitution at codon 617 (JAK2-V617F) in the catalytically inactive pseudokinase domain (JH2) of JAK2. As the JH2 domain is believed to serve autoinhibitory function, the V617F substitution, accordingly, induces constitutive activation of the JAK2 tyrosine kinase, which results in enhanced activation of multiple downstream signaling pathways such as the STATs, the mitogen activated protein kinase (MAPK) and the phosphoinositol 3-kinase (PI3K)-Akt pathway. These signaling tracks, in turn, modulate the expression of genes that regulate cell proliferation and survival, resulting in a growth and selective advantage of the cell that acquired the JAK2-V617F mutation.

The reported frequencies of JAK2-V617F in MPN are up to 98% in patients with polycythemia vera and about 50% in essential thrombocythemia and primary myelofibrosis, respectively (Campbell and Green 2006). However, the JAK2-V617F mutation does not exclusively occur in MPN, but has also been detected in patients with chronic myelomonocytic leukemia, myelodypslastic syndrome or acute myeloid leukemia (Jelinek et al. 2005; Levine et al. 2006; Steensma et al. 2005; Scott et al. 2005). In fact, activation of the JAK-STAT signaling pathway is common to a variety of human neoplasms (Verma et al. 2003). Conversely, the JAK2-V617F mutation itself seems to be restricted to myeloid malignancies, as it has never been detected in neoplasms of the lymphoid lineage or in solid tumors. Initial studies using in-vitro cultures of hematopoietic progenitor cells showed that progenitors from MPN patients exhibit hypersensitivity to cytokines and growth factors such as EPO, IGF-1, IL-3 and GM-CSF (Dai et al. 1991, 1992; Correa et al. 1994). Furthermore, spontaneous growth of megakaryocytic and erythroid progenitor cells in complete absence of cytokines were observed, a phenomenon that was restricted to patients with primary forms of polycythemia (Prchal and Axelrad 1974). With the identification and functional characterization of the JAK2-V617F mutation, it became clear that JAK2-V617F underlies these MPN-specific features as it induces cytokine-independent growth of hematopoietic cells (James et al. 2005; Kralovics et al. 2005; Baxter et al. 2005; Levine et al. 2005).

Murine bone marrow transplant models served further data on the pathogenetic effect of the JAK2-V617F mutation. First in-vivo evidences that JAK2-V617F influences the peripheral blood count were provided by results showing that mice transplanted with JAK2-V617F transduced bone marrow cells developed severe erythrocytosis (James et al. 2005). In a different experimental setting, the transplantation of murine primary bone marrow cells expressing either wild-type JAK2 or JAK2-V617F induced phenotypical features in recipient mice that were similar to those in human PV and myelofibrosis (Wernig et al. 2006). The JAK2-V617F transduced animals showed marked elevation of hematocrit as well as leukocytosis, megakaryocyte hyperplasia, reticulin fibrosis of the bone marrow, splenomegaly and extramedullary hematopoiesis. Additional studies served further evidences that the JAK2-V617F mutation was sufficient to induce a PV phenotype and to promote progression to myelofibrosis (Lacout et al. 2006). Nevertheless, the JAK2-V617F mutation is a comparably weak gain-of-function mutation, as expression of wild-type JAK2 negatively influences constitutive activation of STAT5 by JAK2-V617F (James et al. 2005). Furthermore, the high incidence of acquired uniparental disomy on chromosome 9p that causes homozygosity for the V617F mutation indicates that the loss of the wild-type JAK2 allele confers a stronger proliferative advantage to the aberrant cell. Furthermore, investigations have shown that the FERM domain of JAK2 is crucial for binding to cytokine receptors and constitutive signaling induced by JAK2-V617F. Mutations in the FERM domain disrupt the binding of the mutant JAK2 kinase to cytokine receptors and abolish its constitutive activity (Royer et al. 2005; Wernig et al. 2008).

Several reports suggest that the JAK2-V617F mutation occurs in a multipotent lympho-myeloid progenitor cell (Jamieson et al. 2006; Delhommeau et al. 2007; Li et al. 2007). JAK2-V617F was detected in hematopoietic stem cells (defined as $CD34^+$, $CD38^-$, $CD90^+$ and lineage negative) and their progeny and shown to direct hematopoiesis towards increased erythroid differentiation (Jamieson et al. 2006). Subsequent studies demonstrated that the JAK2-V617F mutation was present in B cells and natural killer cells in some patients with PV and about half of cases with primary myelofibrosis (PMF), being even present in T cells in a small proportion of patients (Delhommeau et al. 2007). Experiments in mouse models have shown that $CD34^+$ cells from JAK2-V617F positive PV and PMF patients were capable of repopulating immunodeficient mice. Interestingly, the V617F mutant allele burden was lower after engraftment than in the initial $CD34^+$ graft sample from the patients, indicating that JAK2-V617F negative stem cells persisted and contributed to bone marrow repopulation (Li et al. 2007). Recent investigations provided data on the involvement of endothelial cells in the disease clone of MPN, suggesting that the JAK2-V617F mutation might be acquired in a common hematopoietic-endothelial progenitor cell (Sozer et al. 2009; Teofili et al. 2011).

Although many insights have been gained into MPN pathogenesis since the identification and characterization of the JAK2-V617F mutation, it still remains unclear of how a single mutation can promote several distinct disease phenotypes.

Several lines of evidence suggested that the MPN phenotype might depend on the JAK2-V617F allele burden. Firstly, homozygosity for JAK2-V617F with high mutant allele burden as a result of mitotic recombination is mostly associated with a PV phenotype, whereas ET patients usually carry monoallelic mutations of JAK2 and have a lower JAK2-V617F allele burden. Furthermore, results of a JAK2-V617F transgenic mouse model indicated that the ratio of expression between mutant and wild-type JAK2 determine the effective phenotype (Tiedt et al. 2008). On the other hand, bone marrow transplantations of primary cells transduced with Jak2-V617F into lethally irradiated mice were shown to induce a PV-like disease with strain-specific differences in the phenotype (Wernig et al. 2006). This observation suggested that the genetic background might influence the manifest phenotype that is induced by JAK2-V617F. Further data supporting this hypothesis were provided by a subsequent genotype-phenotype association study in MPN patients. In search for host genetic factors that influence the disease phenotype in MPN, the investigators screened for variants of single nucleotide polymorphisms (SNPs) that associate with a certain entity of the MPN. Indeed, certain SNP variants inside the *JAK2* gene locus were revealed to significantly correlate with PV, but not with PMF or essential thrombocythemia (ET) (Pardanani et al. 2008). In 2009, three independent groups reported the presence of a distinct haplotype that contains the *JAK2* gene and predisposes to the acquisition of the JAK2-V617F mutation (Olcaydu et al. 2009a; Jones et al. 2009; Kilpivaara et al. 2009). These findings made evident that inherited genetic factors influence somatic mutability and modify the disease course and phenotype in MPN. The mechanism of how certain genetic variation might predispose to somatic mutagenesis remains to be elucidated.

Although JAK2-V617F represents the most frequent oncogenic mutation in MPN, further sequencing efforts resulted in the detection of other mutations in the *JAK2* gene. Most importantly, mutations in exon 12 of *JAK2* were identified in about 20% of JAK2-V617F negative PV patients (Fig. 1) (Pardanani et al. 2008; Scott et al. 2007). MPN patients with *JAK2* exon 12 mutations exhibit a different disease phenotype than patients with JAK2-V617F, as they mostly present with isolated erythrocytosis without leukocytosis or thrombocytosis. Thus, it is believed that exon 12 mutations introduce a different alteration of kinase function than the V617F mutation. This might also explain why *JAK2* exon 12 mutations exclusively occur in PV patients and have not been observed in cases with PMF or ET. Furthermore, a recent study showed that the *JAK2* haplotype that predisposes to the acquisition of JAK2-V617F also preferentially acquires exon 12 mutations of *JAK2* (Olcaydu et al. 2009b). Other reported rare mutations of the JAK2 pseudokinase domain in hematopoietic neoplasms include isolated D620E and E627E mutations or double mutants such as JAK2-V617FD620E, JAK2-V617FC616Y and JAK2-V617FC618R (Bacher et al. 2006; Grunebach et al. 2006; Zhang et al. 2007; Karow et al. 2008). Furthermore, amplifications of the *JAK2* gene of up to 20 copies have been detected in patients with PV, non-Hodgkin lymphoma, multiple myeloma and MDS (Najfeld et al. 2007). A study of approximately 20,000 DNA samples of patients with apparent MPN revealed various rare

JAK2 mutations in exons 12–15 that were mostly located in the JH2 domain (Ma et al. 2009). Preforming reverse transcription PCR with direct sequencing of *JAK2*, a recent study could demonstrate that about one third of patients with confirmed or suspected MPN carry a deletion of *JAK2* exon 14 (JAK2 Deltaexon 14), which was shown to result in expression of a truncated JAK2 protein (Ma et al. 2010). Further investigations are needed in order to determine the relevance of these JAK2 alterations in MPN disease pathogenesis.

After its identification, the JAK2-V617F mutation was soon approved as a diagnostically relevant molecular marker in MPN and was implicated in the revised classification and diagnostic criteria of the myeloid neoplasms (Tefferi and Vardiman 2007). Being present in the majority of patients with MPN, the JAK2-V617F mutation serves as an important diagnostic marker and especially aids the differentiation between secondary forms of cytosis and primary MPN. Apart from its significance in diagnostic approaches, the JAK2 mutation might also represent a potential therapeutic target. The same is true for other genetic defects that induce constitutive JAK-STAT signaling such as *JAK2* exon 12 mutations or JAK2 fusion proteins. Several trials of selective and non-selective JAK2 inhibitors have been launched during the last years. First results indicate that JAK2 inhibitors are beneficial for patients in alleviating disease-related and overall constitutional symptoms, especially in MPN patients with splenomegaly. However, a complete molecular remission and eradication of the JAK2-V617F positive clone is not achieved with the compounds studied so far (Chen and Prchal 2010).

Although early observations indicated a role for alterations of the JAKs in human neoplasias, it was the identification of the JAK2-V617F mutation that moved this family of protein tyrosine kinases into the center of attention. A great number of succeeding studies revealed the importance of functional JAK-STAT signaling in normal hematopoiesis and uncovered various alterations of this pathway that play a causative role in hematologic malignancies. Accordingly, currently ongoing trials with compounds that are targeting aberrant tyrosine kinases such as the JAK2 inhibitors are expected to serve novel therapeutic options in the treatment of patients with aberrations of the Janus kinases. Still, many questions regarding genetic and functional aspects of the JAKs remain to be resolved. Further studies utilizing novel technologies will aid the detailed characterization of genetic aberrations of the JAK-STAT pathway and decipher their pathogenetic impact on disease evolution in hematopoietic neoplasms.

Concluding Remarks

A surprisingly large number of acquired myeloid malignancies are caused by JAK2 kinase mutations. These gain-of-function JAK2 mutants or fusion genes exhibit variable levels of kinase activity that either cause a relatively benign chronic disease or, in some cases, a severe malignant disease. Detection of these JAK2 oncogenes has diagnostic value and allows genetic stratification of patients that may influence the choice of therapeutic intervention. As the hyperactivation of the

JAK-STAT signaling pathway by various oncogenic mutations proved to be driving a number of chronic and acute hematological phenotypes, small molecule inhibitors targeting the JAK family kinases may offer cure for a number of disease entities in the near future.

References

Adelaide J et al (2006) A t(8;9) translocation with PCM1-JAK2 fusion in a patient with T-cell lymphoma. Leukemia 20:536–537, England

Asnafi V et al (2010) JAK1 mutations are not frequent events in adult T-ALL: a GRAALL study. Br J Haematol 148(1):178–179

Bacher U et al (2006) A combination of cytomorphology, cytogenetic analysis, fluorescence in situ hybridization and reverse transcriptase polymerase chain reaction for establishing clonality in cases of persisting hypereosinophilia. Haematologica 91(6):817–820

Bardelli A et al (2003) Mutational analysis of the tyrosine kinome in colorectal cancers. Science 300(5621):949

Baxter EJ et al (2005) Acquired mutation of the tyrosine kinase JAK2 in human myeloproliferative disorders. Lancet 365(9464):1054–1061

Bercovich D et al (2008) Mutations of JAK2 in acute lymphoblastic leukaemias associated with Down's syndrome. Lancet 372:1484–1492, England

Bousquet M et al (2005) The t(8;9)(p22;p24) translocation in atypical chronic myeloid leukaemia yields a new PCM1-JAK2 fusion gene. Oncogene 24:7248–7252. doi:10.1038/sj.onc.1208850, published online 8 August 2005.: England

Campbell PJ, Green AR (2006) The myeloproliferative disorders. N Engl J Med 355(23):2452–2466

Carron C et al (2000) TEL-JAK2 transgenic mice develop T-cell leukemia. Blood 95(12):3891–3899

Chen AT, Prchal JT (2010) JAK2 kinase inhibitors and myeloproliferative disorders. Curr Opin Hematol 17(2):110–116

Cirmena G et al (2008) A BCR-JAK2 fusion gene as the result of a t(9;22)(p24;q11) in a patient with acute myeloid leukemia. Cancer Genet Cytogenet 183:105–108, 2008 Elsevier Inc.: United States

Correa PN, Eskinazi D, Axelrad AA (1994) Circulating erythroid progenitors in polycythemia vera are hypersensitive to insulin-like growth factor-1 in vitro: studies in an improved serum-free medium. Blood 83(1):99–112

Dai CH et al (1991) Polycythemia vera blood burst-forming units-erythroid are hypersensitive to interleukin-3. J Clin Invest 87(2):391–396

Dai CH et al (1992) Polycythemia vera. II. Hypersensitivity of bone marrow erythroid, granulocyte-macrophage, and megakaryocyte progenitor cells to interleukin-3 and granulocyte-macrophage colony-stimulating factor. Blood 80(4):891–899

Dargent JL et al (2011) Pathology of the bone marrow and spleen in a case of myelodysplastic/myeloproliferative neoplasm associated with t(8;9)(p22;p24) involving PCM1 and JAK2 genes. Eur J Haematol 86(1):87–90

De Braekeleer E, Ferec C, De Braekeleer M (2009) RUNX1 translocations in malignant hemopathies. Anticancer Res 29(4):1031–1037

Delhommeau F et al (2007) Evidence that the JAK2 G1849T (V617F) mutation occurs in a lymphomyeloid progenitor in polycythemia vera and idiopathic myelofibrosis. Blood 109(1):71–77

dos Santos NR, Ghysdael J (2006) A transgenic mouse model for TEL-JAK2-induced B-cell lymphoma/leukemia. Leukemia 20:182–185, England

dos Santos NR et al (2007) Pre-TCR expression cooperates with TEL-JAK2 to transform immature thymocytes and induce T-cell leukemia. Blood 109(9):3972–3981

Flex E et al (2008) Somatically acquired JAK1 mutations in adult acute lymphoblastic leukemia. J Exp Med 205(4):751–758

Gaikwad A et al (2009) Prevalence and clinical correlates of JAK2 mutations in Down syndrome acute lymphoblastic leukaemia. Br J Haematol 144:930–932, England
Ghoreschi K, Laurence A, O'Shea JJ (2009) Janus kinases in immune cell signaling. Immunol Rev 228(1):273–287
Goerttler P et al (2005) The Jak2V617F mutation, PRV-1 overexpression, and EEC formation define a similar cohort of MPD patients. Blood 106(8):2862–2864
Gordon GM et al (2010) Transforming JAK1 mutations exhibit differential signalling, FERM domain requirements and growth responses to interferon-gamma. Biochem J 432(2):255–265
Greenman C et al (2007) Patterns of somatic mutation in human cancer genomes. Nature 446(7132):153–158
Griesinger F et al (2005) A BCR-JAK2 fusion gene as the result of a t(9;22)(p24;q11.2) translocation in a patient with a clinically typical chronic myeloid leukemia. Genes Chromosomes Cancer 44(3):329–333
Grunebach F et al (2006) Detection of a new JAK2 D620E mutation in addition to V617F in a patient with polycythemia vera. Leukemia 20(12):2210–2211
Harrison DA et al (1995) Activation of a Drosophila Janus kinase (JAK) causes hematopoietic neoplasia and developmental defects. EMBO J 14(12):2857–2865
Hasle H, Clemmensen IH, Mikkelsen M (2000) Risks of leukaemia and solid tumours in individuals with Down's syndrome. Lancet 355:165–169, England
Ho JM et al (1999) Fusion of the ets transcription factor TEL to Jak2 results in constitutive Jak-Stat signaling. Blood 93(12):4354–4364
Hoelbl A et al (2010) Stat5 is indispensable for the maintenance of bcr/abl-positive leukaemia. EMBO Mol Med 2(3):98–110
James C et al (2005) A unique clonal JAK2 mutation leading to constitutive signalling causes polycythaemia vera. Nature 434(7037):1144–1148
Jamieson CH et al (2006) The JAK2 V617F mutation occurs in hematopoietic stem cells in polycythemia vera and predisposes toward erythroid differentiation. Proc Natl Acad Sci USA 103(16):6224–6229
Jelinek J et al (2005) JAK2 mutation 1849G>T is rare in acute leukemias but can be found in CMML, Philadelphia chromosome-negative CML, and megakaryocytic leukemia. Blood 106(10):3370–3373
Jeong EG et al (2008) Somatic mutations of JAK1 and JAK3 in acute leukemias and solid cancers. Clin Cancer Res 14(12):3716–3721
Jones AV et al (2009) JAK2 haplotype is a major risk factor for the development of myeloproliferative neoplasms. Nat Genet 41:446–449
Kameda T et al (2010) Absence of gain of function JAK1 and JAK3 mutations in adult T cell leukemia/lymphoma. Int J Hematol 92(2):320–325
Karow A et al (2008) JAK2 mutations other than V617F: a novel mutation and mini review. Leuk Res 32(2):365–366
Kawahara A et al (1995) Critical role of the interleukin 2 (IL-2) receptor gamma-chain-associated Jak3 in the IL-2-induced c-fos and c-myc, but not bcl-2, gene induction. Proc Natl Acad Sci USA 92(19):8724–8728
Kearney L et al (2009) Specific JAK2 mutation (JAK2R683) and multiple gene deletions in Down syndrome acute lymphoblastic leukemia. Blood 113:646–648
Kennedy JA et al (2006) Expression of TEL-JAK2 in primary human hematopoietic cells drives erythropoietin-independent erythropoiesis and induces myelofibrosis in vivo. Proc Natl Acad Sci USA 103(45):16930–16935
Kilpivaara O et al (2009) A germline JAK2 SNP is associated with predisposition to the development of JAK2(V617F)-positive myeloproliferative neoplasms. Nat Genet 41:455–459
Kralovics R et al (2005) A gain-of-function mutation of JAK2 in myeloproliferative disorders. N Engl J Med 352(17):1779–1790
Kratz CP et al (2006) Mutational screen reveals a novel JAK2 mutation, L611S, in a child with acute lymphoblastic leukemia. Leukemia 20:381–383, England

Lacout C et al (2006) JAK2V617F expression in murine hematopoietic cells leads to MPD mimicking human PV with secondary myelofibrosis. Blood 108(5):1652–1660

Lacronique V et al (1997) A TEL-JAK2 fusion protein with constitutive kinase activity in human leukemia. Science 278(5341):1309–1312

Lai KS et al (1995) A kinase-deficient splice variant of the human JAK3 is expressed in hematopoietic and epithelial cancer cells. J Biol Chem 270(42):25028–25036

Lane SW et al (2008) Leukaemia cutis in atypical chronic myeloid leukaemia with a t(9;22) (p24; q11.2) leading to BCR-JAK2 fusion. Br J Haematol 142:503, England

Lee JW et al (2006) The JAK2 V617F mutation in de novo acute myelogenous leukemias. Oncogene 25(9):1434–1436

Levine RL et al (2005) Activating mutation in the tyrosine kinase JAK2 in polycythemia vera, essential thrombocythemia, and myeloid metaplasia with myelofibrosis. Cancer Cell 7 (4):387–397

Levine RL et al (2006) X-inactivation-based clonality analysis and quantitative JAK2V617F assessment reveal a strong association between clonality and JAK2V617F in PV but not ET/MMM, and identifies a subset of JAK2V617F-negative ET and MMM patients with clonal hematopoiesis. Blood 107(10):4139–4141

Li Z et al (2007) Erlotinib effectively inhibits JAK2V617F activity and polycythemia vera cell growth. J Biol Chem 282(6):3428–3432

Lindauer K et al (2001) Prediction of the structure of human Janus kinase 2 (JAK2) comprising the two carboxy-terminal domains reveals a mechanism for autoregulation. Protein Eng 14 (1):27–37

Luo H et al (1997) Mutation in the Jak kinase JH2 domain hyperactivates *Drosophila* and mammalian Jak-Stat pathways. Mol Cell Biol 17(3):1562–1571

Ma W et al (2009) Mutation profile of JAK2 transcripts in patients with chronic myeloproliferative neoplasias. J Mol Diagn 11(1):49–53

Ma W et al (2010) JAK2 exon 14 deletion in patients with chronic myeloproliferative neoplasms. PLoS One 5(8):e12165

Malinge S et al (2006) Activation of the NF-kappaB pathway by the leukemogenic TEL-Jak2 and TEL-Abl fusion proteins leads to the accumulation of antiapoptotic IAP proteins and involves IKKalpha. Oncogene 25(25):3589–3597

Malinge S et al (2007) Novel activating JAK2 mutation in a patient with Down syndrome and B-cell precursor acute lymphoblastic leukemia. Blood 109(5):2202–2204

Malinge S et al (2008) Activating mutations in human acute megakaryoblastic leukemia. Blood 112(10):4220–4226

Mark HF et al (2006) Chronic idiopathic myelofibrosis (CIMF) resulting from a unique 3;9 translocation disrupting the janus kinase 2 (JAK2) gene. Exp Mol Pathol 81(3):217–223

Migone TS et al (1995) Constitutively activated Jak-STAT pathway in T cells transformed with HTLV-I. Science 269(5220):79–81

Minegishi Y et al (2006) Human tyrosine kinase 2 deficiency reveals its requisite roles in multiple cytokine signals involved in innate and acquired immunity. Immunity 25(5):745–755

Mullighan CG et al (2009) JAK mutations in high-risk childhood acute lymphoblastic leukemia. Proc Natl Acad Sci USA 106(23):9414–9418

Murati A et al (2005) PCM1-JAK2 fusion in myeloproliferative disorders and acute erythroid leukemia with t(8;9) translocation. Leukemia 19:1692–1696, England

Najfeld V et al (2007) Numerical gain and structural rearrangements of JAK2, identified by FISH, characterize both JAK2617V>F-positive and -negative patients with Ph-negative MPD, myelodysplasia, and B-lymphoid neoplasms. Exp Hematol 35(11):1668–1676

Nebral K et al (2009) Incidence and diversity of PAX5 fusion genes in childhood acute lymphoblastic leukemia. Leukemia 23(1):134–143

Nguyen MH et al (2001) TEL-JAK2 mediates constitutive activation of the phosphatidylinositol 3′-kinase/protein kinase B signaling pathway. J Biol Chem 276(35):32704–32713

Olcaydu D et al (2009a) A common JAK2 haplotype confers susceptibility to myeloproliferative neoplasms. Nat Genet 41:450–454

Olcaydu D et al (2009b) The 'GGCC' haplotype of JAK2 confers susceptibility to JAK2 exon 12 mutation-positive polycythemia vera. Leukemia 23:1924–1926

Pardanani A et al (2008) Host genetic variation contributes to phenotypic diversity in myeloproliferative disorders. Blood 111(5):2785–2789

Peeters P et al (1997) Fusion of TEL, the ETS-variant gene 6 (ETV6), to the receptor-associated kinase JAK2 as a result of t(9;12) in a lymphoid and t(9;15;12) in a myeloid leukemia. Blood 90(7):2535–2540

Poitras JL et al (2008) Novel SSBP2-JAK2 fusion gene resulting from a t(5;9)(q14.1;p24.1) in pre-B acute lymphocytic leukemia. Genes Chromosomes Cancer 47(10):884–889

Prchal JF, Axelrad AA (1974) Letter: bone-marrow responses in polycythemia vera. N Engl J Med 290(24):1382

Reiter A et al (2005) The t(8;9)(p22;p24) is a recurrent abnormality in chronic and acute leukemia that fuses PCM1 to JAK2. Cancer Res 65:2662–2667, United States

Royer Y et al (2005) Janus kinases affect thrombopoietin receptor cell surface localization and stability. J Biol Chem 280(29):27251–27261

Saharinen P, Silvennoinen O (2002) The pseudokinase domain is required for suppression of basal activity of Jak2 and Jak3 tyrosine kinases and for cytokine-inducible activation of signal transduction. J Biol Chem 277(49):47954–47963

Santos SC et al (2001) Involvement of the NF-kappaB pathway in the transforming properties of the TEL-Jak2 leukemogenic fusion protein. FEBS Lett 497:148–152, Netherlands

Schwaller J et al (1998) Transformation of hematopoietic cell lines to growth-factor independence and induction of a fatal myelo- and lymphoproliferative disease in mice by retrovirally transduced TEL/JAK2 fusion genes. EMBO J 17(18):5321–5333

Schwaller J et al (2000) Stat5 is essential for the myelo- and lymphoproliferative disease induced by TEL/JAK2. Mol Cell 6:693–704, United States

Scott LM et al (2005) The V617F JAK2 mutation is uncommon in cancers and in myeloid malignancies other than the classic myeloproliferative disorders. Blood 106(8):2920–2921

Scott LM et al (2007) JAK2 exon 12 mutations in polycythemia vera and idiopathic erythrocytosis. N Engl J Med 356(5):459–468

Sozer S et al (2009) Human CD34+ cells are capable of generating normal and JAK2V617F positive endothelial like cells in vivo. Blood Cells Mol Dis 43(3):304–312

Steensma DP et al (2005) The JAK2 V617F activating tyrosine kinase mutation is an infrequent event in both. Blood 106(4):1207–1209

Tefferi A, Vardiman J (2007) Classification and diagnosis of myeloproliferative neoplasms: the 2008 World Health Organization criteria and point-of-care diagnostic algorithms. Leukemia 22:14–22

Teofili L et al (2011) Endothelial progenitor cells are clonal and exhibit the JAK2(V617F) mutation in a subset of thrombotic patients with Ph-negative myeloproliferative neoplasms. Blood 117(9):2700–2707

Tiedt R et al (2008) Ratio of mutant JAK2-V617F to wild-type Jak2 determines the MPD phenotypes in transgenic mice. Blood 111(8):3931–3940

Tirado CA et al (2010) Novel JAK2 rearrangement resulting from a t(9;22)(p24;q11.2) in B-acute lymphoblastic leukemia. Leuk Res 34(12):1674–1676

Tortolani PJ et al (1995) Regulation of JAK3 expression and activation in human B cells and B cell malignancies. J Immunol 155(11):5220–5226

Van Roosbroeck K et al (2011) JAK2 rearrangements, including the novel SEC31A-JAK2 fusion, are recurrent in classical Hodgkin lymphoma. Blood 117(15):4056–4064

Vardiman JW (2010) The World Health Organization (WHO) classification of tumors of the hematopoietic and lymphoid tissues: an overview with emphasis on the myeloid neoplasms. Chem Biol Interact 184(1–2):16–20

Verma A et al (2003) Jak family of kinases in cancer. Cancer Metastasis Rev 22(4):423–434

Walters D (2006) Activating alleles of JAK3 in acute megakaryoblast.... Cancer Cell 10:65–75, PubMed result
Wang W et al (2010) AML1 is overexpressed in patients with myeloproliferative neoplasms and mediates JAK2V617F-independent overexpression of NF-E2. Blood 116(2):254–266
Wernig G et al (2006) Expression of Jak2V617F causes a polycythemia vera-like disease with associated myelofibrosis in a murine bone marrow transplant model. Blood 107(11):4274–4281
Wernig G et al (2008) The Jak2V617F oncogene associated with myeloproliferative diseases requires a functional FERM domain for transformation and for expression of the Myc and Pim proto-oncogenes. Blood 111(7):3751–3759
Xiang Z et al (2008) Identification of somatic JAK1 mutations in patients with acute myeloid leukemia. Blood 111(9):4809–4812
Xu X et al (1995) Constitutive activation of different Jak tyrosine kinases in human T cell leukemia virus type 1 (HTLV-1) tax protein or virus-transformed cells. J Clin Invest 96(3):1548–1555
Zhang SJ et al (2007) The investigation of JAK2 mutation in Chinese myeloproliferative diseases-identification of a novel C616Y point mutation in a PV patient. Int J Lab Hematol 29(1):71–72

Activation of Janus Kinases During Tumorigenesis

Jeffrey W. Schmidt and Kay-Uwe Wagner

Abstract

Janus tyrosine kinases (JAKs) are important for the growth and homeostasis of a variety of normal tissues. Specifically, JAK1 and JAK2 are essential for mammalian development, and conventional knockout models in mice show that the absence of just one of these two kinases causes prenatal and postnatal lethality. Recent studies using JAK2 conditional knockout mice show that this tyrosine kinase plays key roles in mammary gland development, fertility, pancreatic β cell homeostasis, and the suppression of fatty liver disease in adult animals. Somatically acquired point mutations or structural abnormalities in the *JAK2* gene contribute to various hematopoietic malignancies. In contrast, a sustained activation of JAK1 and JAK2 in solid human cancers, such as those of the breast, prostate, lung, head and neck, skin, and gastrointestinal tract, is caused mainly by alternative mechanisms. These include the epigenetic silencing of negative regulators of JAKs as well as an aberrant autocrine stimulation of growth factors such as PRL, EPO, and IL-6. In addition to the canonical pathway through Signal Transducers and Activators of Transcription (STATs), JAKs are an integral part of a crosstalk with receptor tyrosine kinases and their substrates that promote the progression of solid cancers. The biological significance of JAKs within wider signaling networks, however, depends on the cell type and the stage of neoplastic

J.W. Schmidt
Eppley Institute for Research in Cancer and Allied Diseases, University of Nebraska Medical Center, 985950 Nebraska Medical Center, DRCII, Rm. 5012, Omaha, NE 68198-5950, USA

K.-U. Wagner (✉)
Eppley Institute for Research in Cancer and Allied Diseases, University of Nebraska Medical Center, 985950 Nebraska Medical Center, DRCII, Rm. 5012, Omaha, NE 68198-5950, USA

Department of Pathology and Microbiology, University of Nebraska Medical Center, Omaha, NE, USA
e-mail: kuwagner@unmc.edu

Th. Decker and M. Müller (eds.), *Jak-Stat Signaling: From Basics to Disease*,
DOI 10.1007/978-3-7091-0891-8_15,

progression. For example, recent studies in breast cancer models that are conditionally deficient in JAK2 show that the importance of this kinase changes during disease initiation and progression, which may have significant implications for targeting this Janus kinase in a chemopreventive or therapeutic setting.

Preface

The four members of the Janus kinase family (JAK1, JAK2, TYK2, and JAK3) meditate signaling from multiple hormone and cytokine receptors that are crucial for normal development as well as the initiation and progression of hematopoietic malignancies and solid cancers. This chapter briefly summarizes the main biologically relevant functions of JAKs in normal tissue homeostasis and the mechanisms that mediate an aberrant activation of Janus kinases in human cancers. Also highlighted in this chapter will be the importance of JAKs as components of broader signaling networks, in particular their association to receptor tyrosine kinases and downstream effectors that are known to have pivotal roles in the genesis of solid cancers. In the main section of this chapter, there will be an overview about changes in the activation of Janus kinases in specific solid (i.e. non-hematopoietic) human cancers and their suggested effects on the proliferation, survival and invasive properties of cancer cells. Finally, there will be a discussion of important issues related to the targeted inhibition of JAKs for the prevention and treatment of human cancers. This book chapter will not review the protein structure of JAKs and the significance of specific posttranslational protein modifications that regulate the functionality of JAKs. The reader should refer to recent reviews to gain further insight into these specific molecular events (Ghoreschi et al. 2009; Schindler et al. 2007; Schindler and Plumlee 2008).

Janus Kinases and Normal Tissue Homeostasis

Janus tyrosine kinases (JAKs) are expressed in most tissues and mediate the downstream signaling of more than 50 cytokines and peptide hormones. Upon ligand binding to their corresponding receptors, the receptor-associated Janus kinases autophosphorylate themselves and become activated. Active JAKs phosphorylate specific tyrosine residues on the receptors thereby creating docking sites for signal transducers and activators of transcription (STATs). Following their recruitment to the receptors, STATs are subsequently activated by the JAKs through phosphorylation of critical tyrosine residues that serve as binding sites for the SH2 domains of other STAT proteins, mediating their homo- or heterodimerization. Active STATs undergo nuclear translocation and function as latent transcription factors by binding to consensus recognition sites. In mammals, the JAK family is composed of four members, JAK1, JAK2, JAK3, and TYK2,

Table 1 Phenotypes of Janus kinase knockouts

Gene	Knockout method	Phenotype	Reference(s)
Jak1	Conventional	Early postnatal lethality; neurological defects, SCID; cytokine insensitivity	Rodig et al. (1998)
Jak2	Conventional	Embryonic lethality; defective erythropoiesis	Krempler et al. (2004), Neubauer et al. (1998), Parganas et al. (1998)
	Conditional (mammary)	Impaired alveolar development and maintenance	Wagner et al. (2004)
	Conditional (neuroendocrine)	Impaired fertility and reproductive development	Wu et al. (2011)
	Conditional (pancreatic β-cells)	Impaired β cell homeostasis	Choi et al. (2011)
	Conditional (hepatocyte)	Fatty liver phenotype	Sos et al. (2011)
Jak3	Conventional	Viable and fertile; SCID; defective lymphoid development	Nosaka et al. (1995), Park et al. (1995), Thomis et al. (1995)
Tyk2	Conventional	Viable and fertile; impaired response to LPS and IL-12 signaling; defective cytokine signaling	Karaghiosoff et al. (2000, 2003), Shimoda et al. (2000)

SCID severe combined immunodeficiency, *IL* interleukin, *LPS* lipopolysaccharide

each capable of activating its own set of STATs. [For references regarding the coupling of specific JAKs to STATs and the ligands that activate them, see reviews by Schindler and Strehlow (2000), Kisseleva et al. (2002), and Rane and Reddy (2000).]

Since each Janus kinase transduces signals from multiple cytokines and receptors, gene knockout models have been generated in an attempt to better understand physiological functions of individual JAKs in vivo (Table 1). JAK1 knockout mice die perinatally due to neurological defects and also show insensitivity to cytokine signaling such as IL-2, IL-6, IFN and IL-10 (Rodig et al. 1998). JAK2 knockout mice are embryonic lethal and die at gestation day 12.5 due to defective erythropoiesis (Neubauer et al. 1998; Parganas et al. 1998; Krempler et al. 2004). The expression of JAK3 is largely limited to lymphoid tissues. Consequently, the knockout phenotype for the gene encoding this kinase is less severe than that of JAK1 or JAK2, and JAK3 deficient mice are both viable and fertile; however, they exhibit a severe combined immunodeficiency-like phenotype with defects in lymphoid development (Nosaka et al. 1995; Park et al. 1995; Thomis et al. 1995). The fourth member of the JAK family, the Tyrosine kinase 2 (TYK2), has been implicated in IFN-α, IL-6, IL-10 and IL-12 signaling. TYK2 knockout mice have an impaired response to LPS and IL-12 signaling (Karaghiosoff et al. 2000; Shimoda et al. 2000) and are unable to integrate signaling from multiple cytokine receptors (Karaghiosoff et al. 2003).

Since JAK2 deficiency results in early embryonic lethality, the role of this kinase in tissue homeostasis of postnatal animals is difficult to examine. The approach of transplanting cells and tissue fragments from JAK2 knockout embryos into adult mice was used by Shillingford and colleagues (2002) to study essential functions of JAK2 during mammary gland development. This study showed that JAK2 is required for epithelial cell proliferation, and the ablation of this kinase results in impaired formation of secretory alveoli. In order to better study the role of JAK2 in differentiated tissues of adult mice, Krempler et al. (2004) generated JAK2 conditional knockout mice by placing *loxP* sites around the first coding exon of the *Jak2* locus. In general, a Cre-mediated deletion of a gene such as *Jak2* provides a unique opportunity to assess important biological functions of a gene of interest beyond the initial block in development that often results from a conventional knockout. The final section of this chapter will describe how such a model can also be applied to investigate the role of Janus kinases in tumor initiation versus progression. To selectively ablate JAK2 in mammary epithelial cells in a spatially and temporally controlled manner at particular stages of mammary gland development, transgenic mice were used that express Cre recombinase in different epithelial subtypes in virgin, pregnant, and lactating females (Wagner et al. 1997, 2001). Specifically, the mouse mammary tumor virus (MMTV)-Cre-mediated deletion of *Jak2* from ductal progenitors led to a loss of activation of STAT5 in response to prolactin signaling, but deficiency in JAK2 had no effect on ductal elongation and branching morphogenesis (Wagner et al. 2004). Essential functions of JAK2 during postnatal mammary gland development in this model were restricted to alveolar cells in virgin females. In order to ablate JAK2 from differentiating alveolar cells, a whey acidic protein (WAP)-Cre-based JAK2 conditional knockout model was generated that exhibited a strong negative selection of JAK2-deficient secretory alveolar cells during late pregnancy and lactation. Collectively, both conditional knockout models demonstrated that JAK2 is required for the proliferation of alveolar progenitors and the maintenance of functionally differentiated alveolar cells during pregnancy and lactation. On a mechanistic level, these models provided clear evidence that JAK2 is an essential link between prolactin signaling and STAT5 activation in the normal mammary gland, which, as will be discussed later in this chapter, has important implications for the prevention of mammary cancer.

In addition to studying the role of JAK2 in mammary gland development, conditional knockout mice have been pivotal for the assessment of essential functions of this kinase in neuroendocrine cells, pancreatic β cells, and hepatocytes. The gonadotropin releasing hormone (GnRH) is a major regulator of the reproductive and sexual behavior of mammals. Loss of JAK2 in GnRH-producing neurons causes a number of abnormalities implicating this kinase to normal reproductive development and fertility in female mice (Wu et al. 2011). These conditional knockout mice exhibit a reduction in GnRH and luteinizing hormones, which causes a significant delay in puberty, first estrus, and irregular estrous cyclicity. Female mice showed impaired fertility as characterized by the prolonged time to produce their first litter, fewer pregnancies, and significantly smaller litter sizes. Another very recent study by Choi and colleagues (2011) using the JAK2 conditional knockout showed that erythropoietin (EPO) signaling through this particular kinase protects against the development of type 1 and type 2 diabetes. This study uncovered a key signaling pathway important for β-cell homeostasis with relevance for the treatment and prevention of diabetes. Finally, work by Sos and colleagues (2011) demonstrated that the Alb-Cre-mediated deletion of *Jak2 in* hepatocytes results in a profound fatty liver phenotype. This appears to be the result of a complex mechanism that starts with a reduction in serum insulin-like growth factor-1 (IGF-1), which, in turn, leads to an increase in serum growth hormone (GH) secretion due to a lack of feedback inhibition in the hypothalamus. While the liver-specific loss of JAK2 impairs hepatocellular GH signaling, this pathway is retained in adipocytes, and the increased level of serum GH causes enhanced lipolysis thereby releasing excess free fatty acids. These free fatty acids are then taken up by the GH-resistant hepatocytes at an enhanced rate due to increased expression of the free fatty acid transporter CD36. In support of this proposed mechanism, the fatty liver syndrome caused by the hepatocyte-specific deletion of *Jak2* could be completely reversed through abrogation of GH secretion.

Collectively, conventional and conditional knockout models for Janus kinases have been important to evaluate biologically relevant functions in signal transduction and tissue homeostasis. However there continues to be a need for these and similar kind of models to study cytokine and hormone signaling in normal versus neoplastic cell types. While JAK3 and TYK2 conventional null mutants are probably sufficient to examine the role of these kinases in hematopoietic cells, where these kinases are predominantly expressed, the availability of JAK2 conditional knockout mice now provides unique opportunities to study the role of this tyrosine kinase in a wide variety of tissues and in the context of neoplastic transformation. Similar to JAK2, JAK1 is activated in many normal tissues of adult mice, and, as discussed later, this kinase may be aberrantly activated by peptide hormones in cancer cells as part of a shift in signaling networks due to abnormal expression and activation of cytokine receptors and receptor tyrosine kinases. The necessity to genetically decipher these biological phenomena on a molecular level clearly underlines the need for the long overdue generation of a JAK1 conditional knockout mouse.

Mechanisms that Mediate an Aberrant Activation of JAKs in Cancer

Transgenic models expressing hyperactive JAK2 and STAT5 provide direct experimental evidence that Janus kinases can play a role as proto-oncogenes in the genesis of solid tumors. Particularly, a mammary-specific expression of the kinase domain of JAK2 linked to STAT5A and the transactivation domain of STAT6 prolongs cell survival and suppresses apoptosis that, in turn, induces the formation of sporadic adenocarcinomas (Iavnilovitch et al. 2002). The ability of hyperactive JAK2 to cause neoplastic transformation within the mammary epithelium depends largely on the activation of STAT5 as the main downstream effector of this Janus kinase. This notion is supported by the fact that mice overexpressing wildtype or a hyperactive mutant of STAT5 develop mammary cancer following a similar latency period (Iavnilovitch et al. 2002; Vafaizadeh et al. 2010). Although these models are able to assess the mechanisms by which JAK2 and STAT5 contribute to the initiation of solid tumors, the precise genetic and epigenetic alterations that lead to the activation of these signaling mediators in cancer cells are not accurately recapitulated. In humans, missense mutations within Janus kinases and JAK fusion gene products are linked to the initiation and progression of myeloproliferative disorders and hematopoietic malignancies (Brisken et al. 2002; Brockman and Schuler 2005; Ruchatz et al. 2003; Slupianek et al. 2002). Although TEL-JAK2, BCR-JAK2, and PCM1-JAK2 fusion proteins are observed in various leukemia subtypes [for references see reviews by Valentino and Pierre (2006) and Ghoreschi et al. (2009)], chromosomal translocations that lead to the formation of hyper-activated Jak2 are not frequently detected in solid human tumors, and these cancers acquire active JAKs through alternative mechanisms.

Missense Mutations Within JAKs

Following the discovery and characterization of constitutively activating mutations of *JAK2* (e.g. $JAK2^{V617F}$) in myeloproliferative disorders (Baxter et al. 2005; James et al. 2005; Kralovics et al. 2005; Levine et al. 2005; Vainchenker and Constantinescu 2005), considerable effort has been placed in determining whether hyperactive JAK/STAT signaling observed in several human cancers is due, at least in part, to activating mutations in Janus kinases. Several recent studies, however, suggest that the occurrence of the $JAK2^{V617F}$ aberration and of homologous mutations in other JAKs remain a rare event in solid cancers. Thus far, these sequencing efforts identified mostly silent mutations and polymorphisms (Lee et al. 2006a; Motte et al. 2007). In prostate cancer, activation of STAT5 is associated with cancer cell survival and a high histological grade of primary tumors. Nonetheless, as a recent report by Gu and colleagues (2010c) showed a gain-of-function of JAK2 through the prominent V617F mutation is not the underlying cause for the increase in STAT5 phosphorylation in this type of cancer. Similarly, this mutation does not contribute to the activation of STAT3 in pancreatic cancers (Kocher et al. 2007). Sequencing efforts to identify mutations in *JAK1*, *JAK3*, and *TYK2* revealed a rare presence of somatic *JAK1* and *JAK3* missense mutations in breast, lung, and hepatocellular carcinomas (Table 2) (Jeong et al. 2008; Xie et al. 2009). The biological importance and functionality of these JAK mutations in carcinogenesis, however, remain unknown.

Epigenetic Silencing of Suppressors of JAK/STAT Signaling

A disruption of negative feedback loops occurs very frequently in human cancer, which is exemplified by the common loss of tumor suppressors that are antagonistic to proto-oncogenic pathways. Negative regulators of the JAK/STAT pathway

Table 2 Somatic JAK missense mutations observed in solid cancers

Gene	Solid cancer	Cancer subtype	Predicted amino acid change	Domain effected	References
JAK1	Lung	Non-small cell lung cancer	T782M	JH2	Jeong et al. (2008)
JAK1	Breast	Invasive ductal carcinoma	H647Y	JH2	Jeong et al. (2008)
JAK1	Liver	Hepatocellular carcinoma	Q646H	JH2	Xie et al. (2009)
JAK1	Liver	Hepatocellular carcinoma	H647F	JH2	Xie et al. (2009)
JAK3	Breast	Invasive ductal carcinoma	V715I	JH2	Jeong et al. (2008)

JH2 Janus kinase homology 2 (pseudokinase) domain

include the protein tyrosine phosphatases (PTPs), suppressors of cytokine signaling (SOCS), and protein inhibitors of activated STATs (PIAS) (Greenhalgh and Hilton 2001). JAK phosphorylation is a reversible process, and PTPs are a major regulator of JAK inactivation through catalyzing its dephosphorylation. SOCS proteins, on the other hand, suppress cytokine signaling through at least three distinct mechanisms. These include direct interaction with activated JAKs (i.e. SOCS1), association with phosphorylated residues on the receptor to block the binding of SH2 and PTB-domain containing proteins such as STATs (i.e. CIS), a combination of the two (i.e. SOCS3), or enhancing the proteosome-dependent degradation of Janus kinases (Greenhalgh and Hilton 2001; Kamizono et al. 2001; Ram and Waxman 1999). Unlike SOCS proteins that are upregulated in response to cytokine stimulation to silence the activation of JAKs, the PIAS family of proteins is ubiquitously expressed. They function by binding STATs directly and alter their localization, DNA binding, transcriptional activation, and additional STAT activities (O'Shea and Watford 2004). Among the three families of negative regulators of JAK/STAT signaling, PTPs and SOCS are epigenetically silenced through DNA methylation in a variety of human cancers. For example, the *SOCS1* gene is found to be aberrantly methylated in 60%–65% of hepatocellular carcinomas (Okochi et al. 2003; Yoshikawa et al. 2001), 50% of pancreatic tumors (Komazaki et al. 2004), and in a subset of colorectal cancers (Fujitake et al. 2004; Xu et al. 2009). Additional JAK/STAT inhibitors that are silenced through promoter methylation in hepatocellular carcinoma and colon cancer include the cytokine-inducible SH2-containing proteins (*CIS*), *SOCS2*, *SOCS3*, and *SHP-1* (Calvisi et al. 2006; Xu et al. 2009). Other reports have shown that *SOCS1* and *SOCS2* are hypermethylated in 14–24% of primary ovarian cancers, and silencing of the *SOCS1* gene also occurs in 9% of primary breast cancer cases (Sutherland et al. 2004). Recently, Sasi and colleagues (2010) examined the expression levels of SOCS1–7 during breast cancer progression. This study showed that a higher expression of *SOCS* genes was correlated with factors such as earlier tumor stage, disease-free survival, lack of disease recurrence, and an overall better clinical outcome. In light of these observations, the authors suggested that utilizing DNA methyltransferase inhibitors might provide an additive effect in a targeted therapy against hyperactivated JAKs and STATs. Finally, the SOCS-related protein, caveolin-1, which is a potent suppressor of JAK2/STAT5 signaling, has been demonstrated to be epigenetically silenced through promoter methylation in a variety of human cancers (Chen et al. 2004; Cui et al. 2001; Park et al. 2002; Wiechen et al. 2001). Similar to transgenic models with a gain-of-function of JAK2 and STAT5, a knockout of caveolin-1 in mice accelerates the formation of multi-focal dysplastic lesions and mammary tumors (Park et al. 2002; Williams et al. 2003). The development of mammary neoplasms in this breast cancer model is closely associated with an increase in the activation of JAK2 and STAT5, which promotes mammary epithelial proliferation and premature differentiation in response to prolactin and other lactogenic hormones during pregnancy.

Autocrine Signaling

The hyper-activation of autocrine signaling networks is common in many human cancers. Rather than relying upon a constant supply of hormones and locally produced growth factors, cancer cells initiate the production of cytokines that bind to their corresponding receptors, which, in turn, activate growth and survival pathways. For example, autocrine signaling mediated by IL-6 is implicated in lung, colon, prostate and breast tumorigenesis (Giri et al. 2001; Grivennikov and Karin 2008; Sansone et al. 2007; Shirota et al. 1990). Similarly, the peptide hormone prolactin (PRL), which signals through JAK2 and STAT5, plays an important role in the etiology of breast cancer. High circulating levels of PRL are associated with an increased risk of developing breast cancer in humans (Tworoger and Hankinson 2006), and a sustained elevation of PRL has been shown to cause mammary cancers in transgenic mice (Tornell et al. 1991; Wennbo et al. 1997). In addition to the PRL that is released from the pituitary gland, breast cancer cells gain the ability to locally synthesize this hormone and enhance the expression of the PRL receptor (Clevenger et al. 1995; Ginsburg and Vonderhaar 1995). The importance of a PRL autocrine loop in the initiation of neoplastic transformation was further verified in a transgenic model (NRL-PRL) that expresses this hormone specifically in the mammary epithelium, and these transgenic mice develop both estrogen-receptor positive and negative lesions (Rose-Hellekant et al. 2003). Using JAK2 conditional knockout mice, its has been recently demonstrated that the mammary-specific ablation of this kinase completely prevented the onset of PRL-induced mammary tumors (Sakamoto et al. 2010). Therefore, the initiation of mammary tumors in the PRL overexpression model requires the activation of JAK2 and STAT5.

Collectively, the various studies that have assessed mutations within the coding regions of JAKs and STATs as well as the expression and activation of this pathway suggest that, unlike in hematopoietic malignancies, genomic alterations in these signal transducers are rare events in solid cancers. The activation of JAKs and their downstream mediators appears to be primarily regulated by alternative mechanisms that include the epigenetic silencing of negative regulators of JAK/STAT signaling in addition to an enhanced activation of Janus kinases through elevated autocrine stimulation of growth factor receptors. Unfortunately, there are only a limited number of genetically engineered in vivo model systems available today that can be employed to systematically address the importance of negative regulators of JAK/STAT signaling in disease initiation and progression.

JAKs as Components of Broader Signaling Networks in Cancer

Within specific cell types, individual JAKs and STATs are in the lines of fire of diverse cytokine and hormone receptors such as receptor tyrosine kinases (RTKs) and their downstream mediators that are part of broader signaling networks. JAKs and STATs are an integral component of receptor crosstalk in normal cells, and it is known that their biological significance can change within signaling networks following malignant transformation. Therefore, the extent and type of receptor crosstalk that utilize JAKs and STATs not only depends on the cell type but also the stage of neoplastic progression. Examples for an extensive association of JAK/STAT signaling with RTKs and their downstream effectors that play pivotal roles in the genesis of solid cancers are members of the ERBB family and the PI3K/AKT pathway.

ERBB Family

Signaling through the ERBB family of receptor tyrosine kinases (EGFR; ERBB2-4) is frequently altered in human cancers through activating mutations, gene amplifications, or overexpression of individual receptors. Mutations within the human epidermal growth factor receptor (EGFR) are a common feature in adenocarcinomas of the lung. Previous studies have identified a close relationship between the extent of EGFR expression and a phosphorylation of JAK2 and STAT3 as well as a selective activation of JAK/STAT signaling by mutant EGFR in lung cancer cells (Lo et al. 2008; Sordella et al. 2004). Interestingly, the combined inhibition of the JAK/STAT pathway and the EGFR have been shown to inhibit tumor growth and cell survival more effectively than either agent alone (Dowlati et al. 2004; Lo et al. 2008). This suggests that inhibiting JAK/STAT signaling could serve as a synergistic approach to a targeted therapy against the EGFR. Such a therapeutic strategy would also affect the ability of the mutant EGFR to induce a JAK-dependent activation of STAT3 via upregulation of IL-6 production as recently suggested by Gao and coworkers (2007).

ERBB2 (HER2, neu) is amplified in a significant subset of breast cancer cases, and overexpression of this receptor tyrosine kinase is also frequently observed in lung cancer, ovarian cancer, and, at a lesser frequency, in colon cancer (Arteaga 2003; Brabender et al. 2001; Hellstrom et al. 2001; Hirsch and Langer 2004; Ochs et al. 2004). Following the binding of prolactin (PRL) or growth hormone (GH) to their corresponding receptors, JAK2 becomes activated and phosphorylates the cytoplasmic domains of the EGFR and ERBB2 (Yamauchi et al. 1997, 2000). The downstream activation of MAP kinases by GH and PRL has been shown to depend on the phosphorylation of these two receptors. Both hormones, GH and PRL, also activate the cytoplasmic tyrosine kinase SRC, which phosphorylates various residues on the EGFR that leads to increased receptor signaling (Biscardi et al. 1999). On the other hand, the SRC family of tyrosine kinases is suggested to possess non-catalytic functions, and it was reported recently that SRC serves as a scaffold for the PRL-induced activation of JAK2 (Garcia-Martinez et al. 2010). JAK2 and STAT5 are important for the proliferation of epithelial subtypes in the mammary gland that are highly susceptible to ERBB2-induced neoplastic transformation (Henry et al. 2004). It has been demonstrated recently that the conditional deletion of the *Jak2* gene in these epithelial subtypes completely prevents the formation of mammary tumors in response to increased ERBB2 expression (Sakamoto et al. 2009). This study provides experimental evidence for the importance of receptor crosstalk between JAK/STAT signaling and RTK activation during the process of neoplastic transformation.

In adenocarcinomas of the breast and other organs, the ERBB2 receptor forms stable heterodimers with ERBB3, and both are suggested to function as an oncogenic unit (Holbro et al. 2003; Kim et al. 2005). Signaling through ERBB2/ERBB3 receptor complexes has also been shown to rapidly activate TYK2 and JAK3 and subsequently STAT3 and STAT5 in the lung epithelium (Liu and Kern 2002). Another example for important receptor crosstalk between ERBBs and JAK/STAT signaling is the proposed function of ERBB4 as a nuclear chaperone of active STAT5A in the mammary gland (Long et al. 2003; Williams et al. 2004). Clark and colleagues (2005) also demonstrated that ERBB4 modulates the activity of STAT5 by regulating the phosphorylation of additional serine residues besides the known JAK2-mediated tyrosine phosphorylation of this signal transducer. Collectively, these studies show that, in addition to their classical role as RTKs, ERBBs also possess important scaffold functions that can significantly modulate the activity of JAKs and STATs in normal and neoplastic cell types.

PI3K/AKT

The functional interactions between JAKs and ERBBs are prime examples for the engagement of different types of receptor tyrosine kinases involved in receptor crosstalk, but there are other RTKs such as the insulin receptor and the insulin-like growth factor-1(IGF-1) receptor that specifically interact with JAK1 and JAK2 and that play a role in a variety of solid cancers (Gual et al. 1998; Himpe and Kooijman 2009). In fact, many of these RTKs have overlapping biological activities in cancer cells. For example, ERBB2-overxpressing breast cancer cells are able to evade the antiproliferative action of a targeted therapy with trastuzumab through up-regulation of IGF-1 receptor expression or loss of PTEN function (Hynes and Lane 2005). These shifts in the signaling network are facilitated by downstream mediators such as phosphoinositide-3 kinase (PI3K) and AKT/PKB that are synchronously activated by various RTKs. The importance of PI3K signaling in cancer is highlighted by the fact that this is one of the most frequently deregulated pathways in human cancers

(Yuan and Cantley 2008). JAK/STAT signaling and the PI3K/AKT cascade share a number of similarities. Both promote survival, proliferation, and metabolism in a variety of cell types, and it is therefore reasonable to propose that these pathways converge within a signaling network. For example, it has been shown that active STAT3 and STAT5 can associate directly with the p85 regulatory subunit of the PI3 kinase in hematopoietic cells to initiate an activation of the PI3K/AKT pathway (Pfeffer et al. 1997; Santos et al. 2001). The functional association of p85 and STAT5 is suggested to play a role in myeloid leukemia (Harir et al. 2007; Nyga et al. 2005; Rosa Santos et al. 2000; Santos et al. 2001), but this interaction is not restricted to normal or neoplastic hematopoietic cells. Phosphorylated STAT5 was also shown to bind to p85 in mammary epithelial cells in vitro and in vivo following stimulation with PRL (Sakamoto et al. 2007). Prolactin signaling and activated JAK2 have previously been shown to promote PI3K activity (Tessier et al. 2001; Yamauchi et al. 1998), and it was proposed that active STAT5 may directly stimulate the activity of the PI3 kinase in the mammary epithelium. This notion is supported by the observation that the conditional deletion of the *Jak2* gene, which causes lack of STAT5 activation, leads to a synchronous reduction in the expression and activation of AKT1 (Sakamoto et al. 2007). This was not a consequence of a functional inhibition of SRC or MAP kinases since these signal transducers were still activated by PRL in the absence of JAK2. Deficiency in JAK2 leads to a dramatic reduction in the total levels of AKT1, and it was recently demonstrated that STAT5 controls the transcriptional expression of the *Akt1* gene in mammary epithelial cells. Nuclear STAT5 binds directly to consensus sites within the *Akt1* locus in a growth factor dependant manner and initiates transcription from a novel, mammary-specific promoter (Creamer et al. 2010). This proposed mechanism of a direct modulation of AKT1 expression and activation through STAT5 was verified in transgenic mice that overexpress a hyperactive mutant of this transcription factor in the mammary epithelium. The gain-of-function of STAT5 in vivo caused a sustained transcriptional upregulation of *Akt1*. The phenotypic consequence of this molecular association was a prolonged survival of functionally differentiated mammary epithelial cells despite activation of pro-apoptotic signaling pathways (Creamer et al. 2010). This phenotype is virtually identical to transgenic mice that overexpress wildtype or hyperactive AKT1 under the regulation of the MMTV LTR (Ackler et al. 2002; Hutchinson et al. 2001; Schwertfeger et al. 2001). Collectively, the results of these studies clearly show that JAK2/STAT5 signaling and the PI3K/AKT pathway can converge at various levels in particular cell types to execute similar biological functions.

Activation of JAKs in Specific Human Cancer Types

Breast Cancer

Elevated levels of prolactin (PRL) have been implicated in the occurrence of human breast cancer (Hankinson et al. 1999; Tworoger and Hankinson 2006), and this peptide hormone is suggested to play an important role in the establishment of an aberrant autocrine loop that fuels the multiplication of breast cancer cells (Clevenger et al. 1995; Ginsburg and Vonderhaar 1995). Since PRL signals mainly through its receptor and the JAK2/STAT5 pathway in luminal breast epithelial cells, it is evident that these signal transducers are key for the genesis of human breast cancer subtypes that originate from this epithelial compartment (Wagner and Rui 2008). Unlike in normal mammary epithelial cells, PRL is also capable of activating STAT3 in human breast cancer cell lines (Cataldo et al. 2000), suggesting that signaling networks undergo a substantial rewiring process during neoplastic progression. Nelson and coworkers (2007) have recently shown that PRL activates JAK1 in a JAK2-dependent manner, and this may provide an underlying mechanism by which this hormone activates STAT3 and MAP kinases that play a pivotal role in breast cancer progression. The activation of JAK1 may correlate with particular breast cancer subtypes that result in a poor prognosis, and it has been reported that the inhibition of estrogen receptor expression in MCF-7 cells leads to an increase in the activation of JAK1 (Yeh et al. 2007).

Although PRL is a major growth factor for the multiplication of normal and neoplastic mammary epithelial cells, the activation of JAK/STAT signaling cascades in breast cancer is not restricted to this peptide hormone. Another growth factor that activates JAK2 and is known to play a significant role in human breast cancer is erythropoietin (EPO). It was reported recently that the receptor for erythropoietin (EpoR) is expressed in a significant subset of human breast tumor specimens and breast cancer cell lines (Larsson et al. 2009; Liang et al. 2010). Major side effects in patients treated with erythropoiesis-stimulating agents prompted the US Food and Drug Administration to issue a black-box warning for both epoetin alfa and darbepoetin alfa in 2008. It had been found that when these agents were given to patients with advanced breast cancer to achieve a target hemoglobin concentration, it shortened their overall survival and increased disease progression (Crouch and DeSantis 2009). Recently it has been shown that recombinant human EPO is also capable of counteracting the treatment of ERBB2-positive breast cancer cells with trastuzumab (Liang et al. 2010). The EPO-mediated activation of JAK2 and SRC as well as the inactivation of PTEN were identified as underlying mechanisms for this biological phenomenon. In addition to PRL and EPO, interleukin-6 (IL-6) is upregulated in primary human breast cancer specimens, and elevated expression of this cytokine is a poor prognostic indicator for breast cancer patients (Berishaj et al. 2007; Knupfer and Preiss 2007). IL-6 plays a key role in the activation of glycoprotein 130 (GP130) receptor-associated JAKs that are known mediators of STAT3 phosphorylation. Treatment of breast cancer cells with a pan-Jak inhibitor, blockade of the GP130 receptor, or sequestration of the IL-6 ligand each led to a decrease of active STAT3 in breast cancer cells (Berishaj et al. 2007). This may suggest that an inhibition of the IL-6/GP130-induced activation of JAK1 and JAK2 might be an effective strategy to target STAT3 in breast cancer. Besides the cytokines and their receptors that are known to directly activate JAK/STAT signaling such as PRL, EPO and IL-6, a recent report highlighted the role of a chemokine-like extracellular matrix protein, osteopontin (OPN), for the activation of STAT3 in a JAK2-dependent manner (Behera et al. 2010). The study suggests that OPN promotes enhanced tumor growth and that increased expression of OPN and pSTAT3 correlates with breast cancer progression in clinical specimens.

It is evident that JAK1 and JAK2 are critical for the activation of STATs in breast cancer cells. Most clinical studies, however, focus solely on the examination of phosphorylated STATs (in particular STAT1, STAT3, and STAT5A/B) as biological readouts for the activation of JAK/STAT signaling cascades. The initial examination of the expression and activation of STAT5A in breast cancer showed that this signal transducer is nuclear localized and tyrosine phosphorylated in approximately 76% of human breast tumors, and its activation was positively correlated with tumor differentiation (Cotarla et al. 2004). Results from a larger study where more than 1,100 breast cancer specimens were analyzed revealed that active STAT5 is consistently present in healthy breast tissue. Its activity, however, is gradually lost during malignant progression, and less than 20% of metastases expressed active STAT5 (Nevalainen et al. 2004). Collectively, this study showed that STAT5 is as an independent prognostic factor for overall patient survival, but the molecular mechanism responsible for this phenomenon remained unknown. Recently, Johnson and colleagues (2010) provided evidence that an upregulation of the protein tyrosine phosphatase 1B (PTP1B) may account for the reduction in activate STAT5 in metastatic breast cancer cells. PTP1B functions as an inhibitor for active JAK2 by catalyzing the dephosphorylation of the Janus kinase, which consequently suppresses the activation of STAT5. The suggested biological role of active STAT5 as a proposed suppressor of metastasis is supported by the observation that STAT5 promotes differentiation as observed by the homotypic clustering of breast cancer cells, a reduction in invasive characteristics, and an increase in the cell surface levels of the adhesion molecule E-cadherin (Sultan et al. 2005). As the expression and activation of STAT5 declines, the level of phosphorylated STAT3 increases significantly during malignant progression. In fact, approximately 50–60% of primary breast cancers exhibit a constitutive activation of STAT3, and it has been shown in various independent studies that this member of the STAT family plays a key role in breast cancer cell growth, survival, and metastatic progression (Barbieri et al. 2010; Berishaj et al. 2007; Burke et al. 2001; Kunigal et al. 2009; Proietti et al. 2009; Ranger et al. 2009). As discussed earlier, STAT3 is synchronously phosphorylated in breast cancer cells by JAK1 and JAK2 in response to the aberrant activation of the receptors for PRL, EPO, and IL-6 as well as the crosstalk with cytoplasmic and receptor tyrosine kinases. It can therefore be expected that the targeted inhibition of STAT3, either directly or indirectly through inhibition of JAKs, will be of therapeutic value to treat advanced breast cancers.

Prostate Cancer

Recent evidence suggests a significant association between JAK/STAT activity and the development of androgen-refractory prostate cancer. Specifically, an increased expression of the IL-6 receptor and active cytoplasmic STAT3 have been linked to early relapse and reduced patient survival (Tam et al. 2007). This is in agreement with previous work by Drachenberg and colleagues (1999) who observed a correlative increase in serum IL-6 in patients with hormone-refractory prostate cancer. In addition, cell culture studies have demonstrated that treatment with IL-6 or expression of constitutively active STAT3 are sufficient to promote androgen-independent growth (Lee et al. 2004). In support of this notion, the inhibition of STAT3 was reported to induce apoptosis in IL-6-dependent prostate cancer cells (Barton et al. 2004). In contrast to breast cancer, where an inverse relationship between STAT3 and STAT5 has been observed, both STATs are active in advanced prostate cancers where they promote disease progression. While STAT5 is required for cell viability and growth of the primary tumor, STAT3 is suggested to be an important driver for metastasis (Gu et al. 2010a, b). In contrast to STAT3, which can be activated by a variety of tyrosine kinases (e.g. JAK1, JAK2, JAK3, EGFR/HER family), the activation of STAT5 in the prostate epithelium is largely mediated by JAK2 in response to systemic and autocrine PRL signaling (Dagvadorj et al. 2007; Li et al. 2004; Nevalainen et al. 1997). The PRL/JAK2/STAT5 cascade has been implicated as a critical pathway for the growth, survival, and malignant progression of prostate cancer, and it is therefore a valid target for clinical therapy (Liao et al. 2010). This is supported by a number of studies demonstrating the efficacy of antagonizing PRL, JAK2, or STAT5 for the treatment and/or sensitization of prostate cancer cells by inhibiting their growth and viability (Dagvadorj et al. 2007, 2008; Li et al. 2004; Wu et al. 2007). STAT5, which is found to be active in 95% of clinical hormone-refractory prostate cancers, interacts with the ligand-bound androgen-receptor (AR) to synergistically promote the transcriptional activation of both AR and STAT5 (Tan et al. 2008). Since it has been shown that active STAT5 is necessary for the survival of androgen-sensitive as well as androgen-independent human prostate cancer cells, the therapeutic value for targeting JAK2 or STAT5 in hormone-refractory prostate cancer is of high clinical importance (Ahonen et al. 2003).

Lung Cancer

Atypical growth factor signaling is especially common in human lung cancers and frequently occurs through mutations in the epidermal growth factor receptor (EGFR). These mutant receptors transduce anti-apoptotic signaling selectively through the Phosphatidylinositol-3-Kinase (PI3K)/AKT and STAT signaling pathways. Similar to blocking EGFR signaling with gefinitinib, inhibiting the PI3K or JAKs was reported to cause extensive apoptosis in non-small-cell lung cancers (NSCLC) that are resistant to conventional chemotherapy (Sordella et al. 2004). The matrix metalloproteinase-10 (MMP-10) is a major contributor of lung tumor development and expansion through degradation of the extracellular matrix. Interestingly, protein levels of MMP-10 are significantly elevated in NSCLC compared to normal lung tissue, and this increase was reported to be JAK2-dependent through activation of the IL-6/JAK2/STAT3 signaling cascade (Zhang et al. 2009). Despite a clear relationship between JAKs and advanced lung cancer, additional studies are needed to better decipher their role in disease initiation and their potential for preventative and/or therapeutic intervention.

Head and Neck Cancer

STAT5 and STAT3 are found to be highly expressed and activated in a number of squamous cell carcinomas of the head and neck (SCCHN), where these signal transducers contribute to cancer cell survival and proliferation (Lai and Johnson 2010). The constitutive activation of STAT3 is frequently observed in SCCHN and is suggested to be an early event in carcinogenesis (Grandis et al. 2000; Nagpal et al. 2002). STAT5, on the other hand, is thought to be activated during disease progression in response to EPO signaling. The abundance of EPO and its receptor in tumor biopsies correlates significantly with disease progression, and their highest expression is observed in the most malignant and invasive specimens. In support of these observations, Mohyeldin and coworkers (2005) showed that inhibition of JAK2 was sufficient to reduce both basal and EPO-induced invasiveness. In a study by Xi et al. (2003), active STAT5 was consistently elevated in head and neck tumors compared to normal epithelium, and there was a close correlation between phosphorylation of STAT5 and malignant progression. Particularly, the targeted inhibition of STAT5B resulted in a reduced proliferation of SCCHN cancer cells in vitro and tumor growth in vivo (Leong et al. 2002). Collectively, both studies suggest that targeting JAK2 and STAT5B could be clinically relevant for the treatment of advanced head and neck cancer.

Melanoma

As with cancers of the prostate and head and neck, STAT3 and STAT5 display significant levels of activity in melanocytes as they progress from a normal into a malignant stage, and both STATs have been shown to be needed for the survival and growth of melanoma cells (Hassel et al. 2008; Kortylewski et al. 2005; Mirmohammadsadegh et al. 2006; Niu et al. 2002). A previous study by Niu et al. (2002) suggested that STAT3 and STAT5 are predominantly activated by the SRC kinase, but a new report by Huang et al. (2008) showed that the increase in STAT3 activation was accompanied by an upregulation of JAK2 and a decrease in SOCS-1. The utilization of interferons (i.e. IFN-α) in the treatment of melanoma to provoke an anti-tumor response requires JAK/STAT signaling. Specifically, STAT1, which is activated by JAK1 and TYK2, is critical for the anti-proliferative effect of IFN-α and INF-γ (Kortylewski et al. 2004; Tassiulas et al. 2004). Samples from melanoma patients clinically resistant to IFN-α therapy frequently exhibited dysfunctional JAK/STAT signaling, including a reduction in STAT1 activity (Pansky et al. 2000). In addition to inducing an anti-tumor response via STAT1, it was also found that IFN-α causes phosphorylation of STAT5, which leads to resistance to cytokine-mediated antiproliferative therapy (Wellbrock et al. 2005). Therefore, overcoming interferon resistance in melanoma may lie in the ability to discriminate between the activation of particular STAT family members.

Gastrointestinal Cancers

The IL-6 and GH-mediated activation of STAT3 and STAT5 via JAK1 and JAK2 plays a key role in hepatocellular carcinoma (HCC) (Chow et al. 1996; Fuke et al. 2007; Lee et al. 2006b; Tan et al. 2010). In particular, expression and activation of STAT5B, which is the main target of JAK2 in response to GH signaling, is associated with a young age at tumor onset, metastatic progression, and overall poor patient survival (Fuke et al. 2007). Paradoxically, the loss of STAT5A/B in hepatocytes results in liver fibrosis and enhances chemically-induced tumor formation, presumably through increased activation of TGF-β and STAT3 (Hosui et al. 2009). Active STAT3 is found in approximately 50% of HCC specimens and 75% of metastatic lesions, whereas little or no activity of this signal transducer is observed in adjacent normal tissue. Inhibition of Janus kinases with AG490 has been shown to lower the activation of STAT3, which results in reduced cell proliferation and viability and enhances TNF-related apoptosis-inducing ligand (TRAIL)-induced apoptosis (Fuke et al. 2007). This suggests that inhibition of Janus kinases might be a suitable strategy for sensitizing HCC cells for TRAIL agonists currently under development.

The importance of JAK/STAT signaling in pancreatic cancer was recently highlighted in a study by Thoennissen and colleagues (2009) who tested the efficacy of the experimental anti-cancer drug Cucurbitacin B. This agent causes significant cell cycle arrest and apoptosis of pancreatic cancer cells associated with inhibition of JAK2, STAT3, and STAT5. Another study by Lee et al. (2007) suggested that enhanced JAK2 activity, which is observed in pancreatic cancers, might be a result of elevated levels of reactive oxygen species (ROS). According to this report, growth factor signaling promotes the formation of ROS, which, in turn, prolongs JAK2 phosphorylation and cell survival by inhibiting the low molecular weight-protein tyrosine phosphatase (LMW-PTP) responsible for JAK2 inactivation.

Sustained activation of Janus kinases 1 and 2 is commonly observed in human colon cancers, and the JAK1/2-mediated activation of STAT3 is suggested to control key events during early colonic tumorigenesis and its progression into malignant adenocarcinoma. According to a study by Xiong et al. (2008), the expression of active STAT3 increased from 26.7% in the normal epithelium to virtually 100% in adenocarcinomas as determined by immunostaining. This was accompanied by an increase in pJAK2 staining from 46.7% to 81.6%, respectively. The authors further demonstrated that JAK1, JAK2, and STAT3 are involved in controlling the growth, survival, and metastatic capabilities of colorectal cancer cells. Beside these changes, a functional loss of p53 is observed in the majority of colon cancers and has been shown to confer resistance to irinotecan, a topoisomerase 1 inhibitor. On the other hand, irinotecan is able to enhance TRAIL-induced apoptosis in a p53-independent manner, and this cellular phenomenon is reported to be a consequence of an inhibition of JAK2/STAT3/5 signaling which leads to reduced colon cancer metastases (Ravi et al. 2004). Finally, JAK2 is also implicated in mediating the growth-promoting anti-apoptotic effects of the potent growth factor glycine-extended gastrin (G-Gly). Overexpression of G-Gly in transgenic mice promotes colonic proliferation, and colon cancers are known to upregulate G-Gly as part of an autocrine signaling mechanism (Koh et al. 1999; Stepan et al. 1999; Watson et al. 1999). According to a study by Beales and Ogunwobi (2006), JAK2 is activated in response to G-Gly and promotes the subsequent activation of AKT and STAT3. The authors suggested that targeting G-Gly directly or through key signaling nodes (i.e. JAKs) might be a useful approach for sensitizing colon cancers to chemotherapeutic agents. In a follow-up study, Ogunwobi and Beales (2007) also found that JAK2, in conjunction with AKT and STAT3, is required for the anti-apoptotic effects of leptin in colon cancers. Collectively, signaling through JAK1 and JAK2 and possibly even JAK3 (Lin et al. 2005; Mori et al. 2005) regulates important biological events during colon cancer initiation and progression, but it needs to be experimentally verified that JAKs are genuine targets for colon cancer prevention and therapy.

Additional Human Cancer Types

There is emerging evidence from a number of recent reports that JAK/STAT signaling plays a key role in the genesis of an even wider variety of malignant tumor subtypes including cancers of the brain, cervix and ovary. For example, JAK2 and STAT5 were suggested to be constitutively active in most brain tumors compared to normal brain tissue, which has a significantly lower activity of JAK2 and STAT5. This abnormal stimulation of JAK2 and STAT5 was reported to be a consequence of both ligand-dependent and ligand-independent mechanisms (Kondyli et al. 2010). Similar to breast cancer, enhanced signaling through the EPO and IL-6 receptors play a key role in cervical and ovarian cancers. Specifically, the EPO ligand and receptor are expressed in 88% and 92% of cervical tumor samples, respectively (Leo et al. 2006). In a new study, Sobti and colleagues (2010) observed a significant increase in STAT5B expression in cervical tumors compared to nearly undetectable levels of this signal transducer in normal tissues. The authors also reported that the expression of STAT5B was associated with the severity of the disease. Finally, a recent study by Colomiere et al. (2009) shows that ovarian cancers exhibit a significant increase in the activity of JAK2 and STAT3 compared to normal tissues, which was reported to be the consequence of receptor crosstalk between the EGFR and the IL6-R signaling through JAK2 and STAT3 during the process of epithelial to mesenchymal transition (Colomiere et al. 2009). Collectively, all these studies suggest that aberrant activation of JAK/STAT signaling cascades seem to play important roles during particular stages of disease initiation and malignant progression.

V. Targeting JAKs for the Prevention and Treatment of Cancers

The development of small molecular inhibitors against JAKs, in particular JAK2, to treat myeloproliferative disorders (MPDs) and hematopoietic cancers invigorates the concept to utilize these new agents for a pharmacological inhibition of Janus kinases in solid cancers. There is also sufficient experimental evidence to suggest that some of the newly developed drugs might be successful in future clinical settings. For example,

JANEX-1, a small molecule inhibitor of JAK3, is reported to be effective in preventing intestinal tumor development in the APC^{min} model for spontaneous intestinal adenoma formation (Uckun and Dibirdik 2010). This finding is encouraging since JAK3 has been recognized as a poor prognostic indicator in colon cancer (Lin et al. 2005; Mori et al. 2005). However, this finding brings along with it a note of caution as systematic inhibition of JAK3 can lead to severe combined immunodeficiency. Whether this or similar compounds will be useful as a chemopreventive drugs remains to be seen as these types of agents would have to be administered over long periods of time. Currently, there are more than a dozen investigational studies underway to test putative JAK2 inhibitors to treat MPDs (Geron et al. 2008; Pardanani 2008; Pardanani et al. 2010; Wernig et al. 2008). Among these agents, INCB018424 from Incyte might be of interest for the treatment of solid cancer since this drug was reported to inhibit both JAK1 and JAK2 that are often simultaneously hyperactive in solid cancers (Verstovsek et al. 2010). Despite these advances, it still remains to be determined whether all these new agents are genuine inhibitors that specifically target their corresponding Janus kinase(s). Also, unlike MPDs and other hematopoietic malignancies that originate through point mutations or structural abnormalities in the *JAK2* gene, the vast majority of solid cancers exhibit a constitutive activation of wildtype JAKs and STATs through alternative mechanisms. Therefore, it might be unreasonable to expect that drugs specifically designed to target the mutant JAK2 will have the same efficacy in the treatment of solid cancers that exhibit an upregulation of active, wildtype Janus kinases.

Much of the work demonstrating the importance of JAK/STAT signaling in human cancers has been performed using cancer cell lines by knocking down JAKs, STAT3, and STAT5 or by utilizing various JAK inhibitors, which have been shown to alter tumorigenic properties such as growth, survival, and invasion. While these studies provide detailed insights into particular pathways that promote malignant properties of cancer cells, they do so under non-physiological conditions in primary or metastatic cells. Specifically, the activation of Janus kinases in established cancers requires a ligand-inducible stimulation of the JAK-associated hormone and cytokine receptors, but there are known inter-species-related incompatibilities between growth factors that can significantly alter the outcome of a preclinical study (Wagner et al. 2004). For example, the hormone PRL is suggested to fuel the proliferation of breast cancer cells. While some breast cancer cell lines might synthesize PRL as part of an autocrine loop, the systemic hormone produced in the mouse failed to induce biologic responses mediated by the human PRL receptor such as cell clustering, proliferation, and signal transduction through STAT5, STAT3, ERK1/2, and AKT (Utama et al. 2006). Hence, in order to adequately reflect the endocrine environment in breast cancer patients, it would be necessary to “humanize” the recipient animal model through expression of hormones and cytokines that activate their receptors and downstream JAKs and STATs in a physiological manner.

Although the efficacy and specificity of putative JAK inhibitors remain to be more thoroughly assessed in patients and animal models that permit an activation of JAKs and STATs at physiologically relevant levels, there are appropriate genetic tools available to date to examine whether particular JAKs and STATs are required for disease initiation and progression. Specifically, conditional knockout mice for JAK2,

STAT5, STAT3, and STAT1 are available (Krempler et al. 2004; Cui et al. 2004; Takeda et al. 1998; Klover et al. 2010) that can be used to delete these signal transducers in a temporally controlled manner specifically within primary or metastatic cancer cells (i.e. following neoplastic transformation). In addition, transgenic mice that express wildtype or hyperactive STAT5 under regulation of the tetracycline-controlled transactivator have been generated (Creamer et al. 2010; Yamaji et al. 2009), and those can now be used to downregulate exogenous STAT5 in solid tumors or hematopoietic malignancies. Such an experimental design would address whether STATs that contribute to neoplastic transformation are equally required for the maintenance of a neoplasm. Based on these experimental concepts, Sakamoto and colleagues recently discriminated the importance of JAK2 in mammary tumor initiation versus progression in two established breast cancer models (Sakamoto et al. 2009, 2010). Collectively, the results of these two studies clearly show that the deletion of the *Jak2* gene from the mammary epithelium prior to tumor onset completely protected female mice from developing mammary tumors in response to an overexpression of ERBB2 as well as PRL. This suggests that signaling through JAK2 in the cancer-initiating epithelial subtype is required for neoplastic transformation, and therefore targeting JAK2 might be a suitable strategy for cancer prevention. In principle, this experimental design is similar to previous animal model studies on the basis of conventional knockout mice to assess the appearance of neoplasms in the absence of a gene-of-interest, for example *Stat5a* or *Cyclin D1* (Humphreys and Hennighausen 1999; Yu et al. 2001), and this also includes models that co-express Cre recombinase and an oncogene such as ERBB2 from a bicistronic construct in a conditional knockout background (Klover et al. 2010; Ursini-Siegel et al. 2008). The lack of tumorigenesis in particular knockout models, for example Cyclin D1 null mice (Yu et al. 2001), prompted the authors to suggest that targeting these genes is therapeutically relevant to treat established cancers. These conclusions are premature since these mice neither developed cancer nor expressed the "therapeutic target" a single day in their lives (Matulka and Wagner 2005). The use of the conditional JAK2 knockout mice has demonstrated that the timing of the functional ablation of a gene is critical for its impact on tumorigenesis. While the deletion of *Jak2* prior to tumor initiation was protective against mammary cancer, the ablation of this Janus kinase from fully neoplastic cells had no impact on tumor cell survival and proliferation in vitro or in vivo (Sakamoto et al. 2009, 2010). Collectively, both studies show that JAK2 is a moving target during neoplastic progression, and the gain-of-function of other tyrosine kinases, in particular RTKs, might substitute for the loss of JAK2 in particular tumor types. The role of JAK1 alone or in combination with JAK2 in tumorigenesis still needs to be determined once a conditional knockout mouse for JAK1 becomes available.

Concluding Remarks

Janus kinases have important functions in normal tissue homeostasis, and their constitutive activation can promote neoplastic transformation and cancer progression. Specifically, a sustained activation of Janus kinases 1 and 2 is commonly observed in a variety of solid human cancers. However, unlike hematopoietic malignancies that originate through point mutations or structural abnormalities in the *JAK2* gene, solid cancers almost exclusively exhibit a hyper-activation of wildtype JAKs and STATs through alternative mechanisms. These include the epigenetic silencing of negative regulators of JAK/STAT signaling as well as an enhanced activation of Janus kinases through aberrant autocrine stimulation of growth factors such as PRL, EPO, and IL-6. Janus kinases and their associated STATs are an integral component of receptor crosstalk in normal cells, and recent studies in genetically engineered models show that their biological significance can change within signaling networks following malignant transformation. The extent and type of receptor crosstalk that utilize JAKs and STATs therefore not only depends on the cell type but also the stage of neoplastic progression. This is one reason why JAKs can become moving targets for chemoprevention and therapy. Due to the heterogeneity of solid cancers, the outcome for a successful treatment of primary and metastatic tumors with JAK inhibitors requires a stratification of cancer subtypes according to their molecular characteristics that should not be restricted to gene expression profiles but rather include the activation of particular JAK/STAT pathways. First reports on the successful development of JAK1/2 inhibitors and their use in the clinic to treat MPDs and other hematopoietic malignancies are encouraging. The efficacy and specificity of these agents remain to be thoroughly evaluated in patients with solid tumors or preclinical animal models for human cancers. In addition, genetically engineered mice in which individual JAKs and STATs can be deleted from normal and neoplastic cell types can be utilized to address whether they are required for disease initiation and progression. They may also determine whether targeting individual JAK/STAT signaling pathways are relevant for the prevention and/or treatment of specific cancer subtypes.

Acknowledgements This work was supported by the Public Health Service grants CA117930 from the National Cancer Institute to K.-U. Wagner. J.W. Schmidt received a graduate fellowship through the UNMC Cancer Research Training Program (CA009476), a Program of Excellence Graduate Assistantship from the UNMC Graduate Studies Office as well as a Breast Cancer Predoctoral Traineeship Award from the Department of Defense Congressionally Directed Medical Research Program (BC100147).

References

Ackler S, Ahmad S, Tobias C, Johnson MD, Glazer RI (2002) Delayed mammary gland involution in MMTV-AKT1 transgenic mice. Oncogene 21(2):198–206

Ahonen TJ, Xie J, LeBaron MJ, Zhu J, Nurmi M, Alanen K, Rui H, Nevalainen MT (2003) Inhibition of transcription factor Stat5 induces cell death of human prostate cancer cells. J Biol Chem 278:27287–27292

Arteaga CL (2003) Trastuzumab, an appropriate first-line single-agent therapy for HER2-overexpressing metastatic breast cancer. Breast Cancer Res 5(2):96–100

Barbieri I, Quaglino E, Maritano D, Pannellini T, Riera L, Cavallo F, Forni G, Musiani P, Chiarle R, Poli V (2010) Stat3 is required for anchorage-independent growth and metastasis but not for mammary tumor development downstream of the ErbB-2 oncogene. Mol Carcinog 49:114–120

Barton BE, Karras JG, Murphy TF, Barton A, Huang HF (2004) Signal transducer and activator of transcription 3 (STAT3) activation in prostate cancer: direct STAT3 inhibition induces apoptosis in prostate cancer lines. Mol Cancer Ther 3:11–20

Baxter EJ, Scott LM, Campbell PJ, East C, Fourouclas N, Swanton S, Vassiliou GS, Bench AJ, Boyd EM, Curtin N et al (2005) Acquired mutation of the tyrosine kinase JAK2 in human myeloproliferative disorders. Lancet 365:1054–1061

Beales IL, Ogunwobi O (2006) Glycine-extended gastrin inhibits apoptosis in colon cancer cells via separate activation of Akt and JNK pathways. Mol Cell Endocrinol 247:140–149

Behera R, Kumar V, Lohite K, Karnik S, Kundu GC (2010) Activation of JAK2/STAT3 signaling by osteopontin promotes tumor growth in human breast cancer cells. Carcinogenesis 31:192–200

Berishaj M, Gao SP, Ahmed S, Leslie K, Al-Ahmadie H, Gerald WL, Bornmann W, Bromberg JF (2007) Stat3 is tyrosine-phosphorylated through the interleukin-6/glycoprotein 130/Janus kinase pathway in breast cancer. Breast Cancer Res 9:R32

Biscardi JS, Maa MC, Tice DA, Cox ME, Leu TH, Parsons SJ (1999) c-Src-mediated phosphorylation of the epidermal growth factor receptor on Tyr845 and Tyr1101 is associated with modulation of receptor function. J Biol Chem 274(12):8335–8343

Brabender J, Danenberg KD, Metzger R, Schneider PM, Park J, Salonga D, Holscher AH, Danenberg PV (2001) Epidermal growth factor receptor and HER2-neu mRNA expression in non-small cell lung cancer is correlated with survival. Clin Cancer Res 7:1850–1855

Brisken C, Ayyannan A, Nguyen C, Heineman A, Reinhardt F, Tan J, Dey SK, Dotto GP, Weinberg RA (2002) IGF-2 is a mediator of prolactin-induced morphogenesis in the breast. Dev Cell 3:877–887

Brockman JL, Schuler LA (2005) Prolactin signals via Stat5 and Oct-1 to the proximal cyclin D1 promoter. Mol Cell Endocrinol 239:45–53

Burke WM, Jin X, Lin HJ, Huang M, Liu R, Reynolds RK, Lin J (2001) Inhibition of constitutively active Stat3 suppresses growth of human ovarian and breast cancer cells. Oncogene 20:7925–7934

Calvisi DF, Ladu S, Gorden A, Farina M, Conner EA, Lee JS, Factor VM, Thorgeirsson SS (2006) Ubiquitous activation of Ras and Jak/Stat pathways in human HCC. Gastroenterology 130:1117–1128

Cataldo L, Chen NY, Yuan Q, Li W, Ramamoorthy P, Wagner TE, Sticca RP, Chen WY (2000) Inhibition of oncogene STAT3 phosphorylation by a prolactin antagonist, hPRL-G129R, in T-47D human breast cancer cells. Int J Oncol 17(6):1179–1185

Chen ST, Lin SY, Yeh KT, Kuo SJ, Chan WL, Chu YP, Chang JG (2004) Mutational, epigenetic and expressional analyses of caveolin-1 gene in breast cancers. Int J Mol Med 14:577–582

Choi D, Schroer SA, Lu SY, Wang L, Wu X, Liu Y, Zhang Y, Gaisano HY, Wagner KU, Wu H, Retnakaran R, Woo M (2010) Erythropoietin protects against diabetes through direct effects on pancreatic beta cells. J Exp Med 207(13):2831–2842

Chow JC, Ling PR, Qu Z, Laviola L, Ciccarone A, Bistrian BR, Smith RJ (1996) Growth hormone stimulates tyrosine phosphorylation of JAK2 and STAT5, but not insulin receptor substrate-1 or SHC proteins in liver and skeletal muscle of normal rats in vivo. Endocrinology 137:2880–2886

Clark DE, Williams CC, Duplessis TT, Moring KL, Notwick AR, Long W, Lane WS, Beuvink I, Hynes NE, Jones FE (2005) ERBB4/HER4 potentiates STAT5A transcriptional activity by regulating novel STAT5A serine phosphorylation events. J Biol Chem 280(25):24175–24180

Clevenger CV, Chang WP, Ngo W, Pasha TL, Montone KT, Tomaszewski JE (1995) Expression of prolactin and prolactin receptor in human breast carcinoma. Evidence for an autocrine/paracrine loop. Am J Pathol 146:695–705

Colomiere M, Ward AC, Riley C, Trenerry MK, Cameron-Smith D, Findlay J, Ackland L, Ahmed N (2009) Cross talk of signals between EGFR and IL-6R through JAK2/STAT3 mediate epithelial-mesenchymal transition in ovarian carcinomas. Br J Cancer 100:134–144

Cotarla I, Ren S, Zhang Y, Gehan E, Singh B, Furth PA (2004) Stat5a is tyrosine phosphorylated and nuclear localized in a high proportion of human breast cancers. Int J Cancer 108:665–671

Creamer BA, Sakamoto K, Schmidt JW, Triplett AA, Moriggl R, Wagner KU (2010) Stat5 promotes survival of mammary epithelial cells through transcriptional activation of a distinct promoter in Akt1. Mol Cell Biol 30:2957–2970

Cui J, Rohr LR, Swanson G, Speights VO, Maxwell T, Brothman AR (2001) Hypermethylation of the caveolin-1 gene promoter in prostate cancer. Prostate 46:249–256

Cui Y, Riedlinger G, Miyoshi K, Tang W, Li C, Deng CX, Robinson GW, Hennighausen L (2004) Inactivation of Stat5 in mouse mammary epithelium during pregnancy reveals distinct functions in cell proliferation, survival, and differentiation. Mol Cell Biol 24(18):8037–8047

Dagvadorj A, Collins S, Jomain JB, Abdulghani J, Karras J, Zellweger T, Li H, Nurmi M, Alanen K, Mirtti T et al (2007) Autocrine prolactin promotes prostate cancer cell growth via Janus kinase-2-signal transducer and activator of transcription-5a/b signaling pathway. Endocrinology 148:3089–3101

Dagvadorj A, Kirken RA, Leiby B, Karras J, Nevalainen MT (2008) Transcription factor signal transducer and activator of transcription 5 promotes growth of human prostate cancer cells in vivo. Clin Cancer Res 14:1317–1324

Dowlati A, Nethery D, Kern JA (2004) Combined inhibition of epidermal growth factor receptor and JAK/STAT pathways results in greater growth inhibition in vitro than single agent therapy. Mol Cancer Ther 3:459–463

Drachenberg DE, Elgamal AA, Rowbotham R, Peterson M, Murphy GP (1999) Circulating levels of interleukin-6 in patients with hormone refractory prostate cancer. Prostate 41:127–133

Fujitake S, Hibi K, Okochi O, Kodera Y, Ito K, Akiyama S, Nakao A (2004) Aberrant methylation of SOCS-1 was observed in younger colorectal cancer patients. J Gastroenterol 39:120–124

Fuke H, Shiraki K, Sugimoto K, Tanaka J, Beppu T, Yoneda K, Yamamoto N, Ito K, Masuya M, Takei Y (2007) Jak inhibitor induces S phase cell-cycle arrest and augments TRAIL-induced apoptosis in human hepatocellular carcinoma cells. Biochem Biophys Res Commun 363:738–744

Gao SP, Mark KG, Leslie K, Pao W, Motoi N, Gerald WL, Travis WD, Bornmann W, Veach D, Clarkson B et al (2007) Mutations in the EGFR kinase domain mediate STAT3 activation via IL-6 production in human lung adenocarcinomas. J Clin Invest 117:3846–3856

Garcia-Martinez JM, Calcabrini A, Gonzalez L, Martin-Forero E, Agullo-Ortuno MT, Simon V, Watkin H, Anderson SM, Roche S, Martin-Perez J (2010) A non-catalytic function of the Src family tyrosine kinases controls prolactin-induced Jak2 signaling. Cell Signal 22(3):415–426

Geron I, Abrahamsson AE, Barroga CF, Kavalerchik E, Gotlib J, Hood JD, Durocher J, Mak CC, Noronha G, Soll RM, Tefferi A, Kaushansky K, Jamieson CH (2008) Selective inhibition of JAK2-driven erythroid differentiation of polycythemia vera progenitors. Cancer Cell 13(4): 321–330

Ghoreschi K, Laurence A, O'Shea JJ (2009) Janus kinases in immune cell signaling. Immunol Rev 228:273–287

Ginsburg E, Vonderhaar BK (1995) Prolactin synthesis and secretion by human breast cancer cells. Cancer Res 55:2591–2595

Giri D, Ozen M, Ittmann M (2001) Interleukin-6 is an autocrine growth factor in human prostate cancer. Am J Pathol 159:2159–2165

Grandis JR, Drenning SD, Zeng Q, Watkins SC, Melhem MF, Endo S, Johnson DE, Huang L, He Y, Kim JD (2000) Constitutive activation of Stat3 signaling abrogates apoptosis in squamous cell carcinogenesis in vivo. Proc Natl Acad Sci USA 97:4227–4232

Greenhalgh CJ, Hilton DJ (2001) Negative regulation of cytokine signaling. J Leukoc Biol 70:348–356

Grivennikov S, Karin M (2008) Autocrine IL-6 signaling: a key event in tumorigenesis? Cancer Cell 13:7–9

Gu L, Dagvadorj A, Lutz J, Leiby B, Bonuccelli G, Lisanti MP, Addya S, Fortina P, Dasgupta A, Hyslop T et al (2010a) Transcription factor Stat3 stimulates metastatic behavior of human prostate cancer cells in vivo, whereas Stat5b has a preferential role in the promotion of prostate cancer cell viability and tumor growth. Am J Pathol 176:1959–1972

Gu L, Vogiatzi P, Puhr M, Dagvadorj A, Lutz J, Ryder A, Addya S, Fortina P, Cooper C, Leiby B et al (2010b) Stat5 promotes metastatic behavior of human prostate cancer cells in vitro and in vivo. Endocr Relat Cancer 17:481–493

Gu L, Zhu XH, Visakorpi T, Alanen K, Mirtti T, Edmonston TB, Nevalainen MT (2010c) Activating mutation (V617F) in the tyrosine kinase JAK2 is absent in locally-confined or castration-resistant prostate cancer. Anal Cell Pathol (Amsterdam) 33:55–59

Gual P, Baron V, Lequoy V, Van Obberghen E (1998) Interaction of Janus kinases JAK-1 and JAK-2 with the insulin receptor and the insulin-like growth factor-1 receptor. Endocrinology 139(3):884–893

Hankinson SE, Willett WC, Michaud DS, Manson JE, Colditz GA, Longcope C, Rosner B, Speizer FE (1999) Plasma prolactin levels and subsequent risk of breast cancer in postmenopausal women. J Natl Cancer Inst 91:629–634

Harir N, Pecquet C, Kerenyi M, Sonneck K, Kovacic B, Nyga R, Brevet M, Dhennin I, Gouilleux-Gruart V, Beug H et al (2007) Constitutive activation of Stat5 promotes its cytoplasmic localization and association with PI3-kinase in myeloid leukemias. Blood 109:1678–1686

Hassel JC, Winnemoller D, Schartl M, Wellbrock C (2008) STAT5 contributes to antiapoptosis in melanoma. Melanoma Res 18:378–385

Hellstrom I, Goodman G, Pullman J, Yang Y, Hellstrom KE (2001) Overexpression of HER-2 in ovarian carcinomas. Cancer Res 61:2420–2423

Henry MD, Triplett AA, Oh KB, Smith GH, Wagner KU (2004) Parity-induced mammary epithelial cells facilitate tumorigenesis in MMTV-neu transgenic mice. Oncogene 23 (41):6980–6985

Himpe E, Kooijman R (2009) Insulin-like growth factor-I receptor signal transduction and the Janus kinase/signal transducer and activator of transcription (JAK-STAT) pathway. Biofactors 35(1):76–81

Hirsch FR, Langer CJ (2004) The role of HER2/neu expression and trastuzumab in non-small cell lung cancer. Semin Oncol 31:75–82

Holbro T, Beerli RR, Maurer F, Koziczak M, Barbas CF 3rd, Hynes NE (2003) The ErbB2/ErbB3 heterodimer functions as an oncogenic unit: ErbB2 requires ErbB3 to drive breast tumor cell proliferation. Proc Natl Acad Sci USA 100(15):8933–8938

Hosui A, Kimura A, Yamaji D, Zhu BM, Na R, Hennighausen L (2009) Loss of STAT5 causes liver fibrosis and cancer development through increased TGF-{beta} and STAT3 activation. J Exp Med 206:819–831

Huang FJ, Steeg PS, Price JE, Chiu WT, Chou PC, Xie K, Sawaya R, Huang S (2008) Molecular basis for the critical role of suppressor of cytokine signaling-1 in melanoma brain metastasis. Cancer Res 68:9634–9642

Humphreys RC, Hennighausen L (1999) Signal transducer and activator of transcription 5a influences mammary epithelial cell survival and tumorigenesis. Cell Growth Differ 10(10): 685–694

Hutchinson J, Jin J, Cardiff RD, Woodgett JR, Muller WJ (2001) Activation of Akt (protein kinase B) in mammary epithelium provides a critical cell survival signal required for tumor progression. Mol Cell Biol 21(6):2203–2212

Hynes NE, Lane HA (2005) ERBB receptors and cancer: the complexity of targeted inhibitors. Nat Rev Cancer 5(5):341–354

Iavnilovitch E, Groner B, Barash I (2002) Overexpression and forced activation of stat5 in mammary gland of transgenic mice promotes cellular proliferation, enhances differentiation, and delays postlactational apoptosis. Mol Cancer Res 1:32–47

James C, Ugo V, Le Couedic JP, Staerk J, Delhommeau F, Lacout C, Garcon L, Raslova H, Berger R, Bennaceur-Griscelli A et al (2005) A unique clonal JAK2 mutation leading to constitutive signalling causes polycythaemia vera. Nature 434:1144–1148

Jeong EG, Kim MS, Nam HK, Min CK, Lee S, Chung YJ, Yoo NJ, Lee SH (2008) Somatic mutations of JAK1 and JAK3 in acute leukemias and solid cancers. Clin Cancer Res 14:3716–3721

Johnson KJ, Peck AR, Liu C, Tran TH, Utama FE, Sjolund AB, Schaber JD, Witkiewicz AK, Rui H (2010) PTP1B suppresses prolactin activation of Stat5 in breast cancer cells. Am J Pathol 177:2971–2983

Kamizono S, Hanada T, Yasukawa H, Minoguchi S, Kato R, Minoguchi M, Hattori K, Hatakeyama S, Yada M, Morita S et al (2001) The SOCS box of SOCS-1 accelerates ubiquitin-dependent proteolysis of TEL-JAK2. J Biol Chem 276:12530–12538

Karaghiosoff M, Neubauer H, Lassnig C, Kovarik P, Schindler H, Pircher H, McCoy B, Bogdan C, Decker T, Brem G et al (2000) Partial impairment of cytokine responses in Tyk2-deficient mice. Immunity 13:549–560

Karaghiosoff M, Steinborn R, Kovarik P, Kriegshauser G, Baccarini M, Donabauer B, Reichart U, Kolbe T, Bogdan C, Leanderson T et al (2003) Central role for type I interferons and Tyk2 in lipopolysaccharide-induced endotoxin shock. Nat Immunol 4:471–477

Kim A, Liu B, Ordonez-Ercan D, Alvarez KM, Jones LD, McKimmey C, Edgerton SM, Yang X, Thor AD (2005) Functional interaction between mouse erbB3 and wild-type rat c-neu in transgenic mouse mammary tumor cells. Breast Cancer Res 7(5):R708–R718

Kisseleva T, Bhattacharya S, Braunstein J, Schindler CW (2002) Signaling through the JAK/STAT pathway, recent advances and future challenges. Gene 285(1–2):1–24

Klover PJ, Muller WJ, Robinson GW, Pfeiffer RM, Yamaji D, Hennighausen L (2010) Loss of STAT1 from mouse mammary epithelium results in an increased neu-induced tumor burden. Neoplasia 12(11):899–905

Knupfer H, Preiss R (2007) Significance of interleukin-6 (IL-6) in breast cancer (review). Breast Cancer Res Treat 102:129–135

Kocher HM, Mears L, Lea NC, Raj K, Mufti GJ (2007) JAK V617F missense mutation is absent in pancreatic cancer. Gut 56:1174–1175

Koh TJ, Dockray GJ, Varro A, Cahill RJ, Dangler CA, Fox JG, Wang TC (1999) Overexpression of glycine-extended gastrin in transgenic mice results in increased colonic proliferation. J Clin Invest 103(8):1119–1126

Komazaki T, Nagai H, Emi M, Terada Y, Yabe A, Jin E, Kawanami O, Konishi N, Moriyama Y, Naka T et al (2004) Hypermethylation-associated inactivation of the SOCS-1 gene, a JAK/STAT inhibitor, in human pancreatic cancers. Jpn J Clin Oncol 34:191–194

Kondyli M, Gatzounis G, Kyritsis A, Varakis J, Assimakopoulou M (2010) Immunohistochemical detection of phosphorylated JAK-2 and STAT-5 proteins and correlation with erythropoietin receptor (EpoR) expression status in human brain tumors. J Neurooncol 100(2):157–164

Kortylewski M, Komyod W, Kauffmann ME, Bosserhoff A, Heinrich PC, Behrmann I (2004) Interferon-gamma-mediated growth regulation of melanoma cells: involvement of STAT1-dependent and STAT1-independent signals. J Invest Dermatol 122:414–422

Kortylewski M, Jove R, Yu H (2005) Targeting STAT3 affects melanoma on multiple fronts. Cancer Metastasis Rev 24:315–327

Kralovics R, Passamonti F, Buser AS, Teo SS, Tiedt R, Passweg JR, Tichelli A, Cazzola M, Skoda RC (2005) A gain-of-function mutation of JAK2 in myeloproliferative disorders. N Engl J Med 352:1779–1790

Krempler A, Qi Y, Triplett AA, Zhu J, Rui H, Wagner KU (2004) Generation of a conditional knockout allele for the Janus kinase 2 (Jak2) gene in mice. Genesis 40:52–57

Kunigal S, Lakka SS, Sodadasu PK, Estes N, Rao JS (2009) Stat3-siRNA induces Fas-mediated apoptosis in vitro and in vivo in breast cancer. Int J Oncol 34:1209–1220

Lai SY, Johnson FM (2010) Defining the role of the JAK-STAT pathway in head and neck and thoracic malignancies: implications for future therapeutic approaches. Drug Resist Updat 13:67–78

Larsson AM, Jirstrom K, Fredlund E, Nilsson S, Ryden L, Landberg G, Pahlman S (2009) Erythropoietin receptor expression and correlation to tamoxifen response and prognosis in breast cancer. Clin Cancer Res 15(17):5552–5559

Lee SO, Lou W, Johnson CS, Trump DL, Gao AC (2004) Interleukin-6 protects LNCaP cells from apoptosis induced by androgen deprivation through the Stat3 pathway. Prostate 60:178–186

Lee JW, Soung YH, Kim SY, Nam SW, Park WS, Lee JY, Yoo NJ, Lee SH (2006a) Absence of JAK2 V617F mutation in gastric cancers. Acta Oncol 45:222–223

Lee TK, Man K, Poon RT, Lo CM, Yuen AP, Ng IO, Ng KT, Leonard W, Fan ST (2006b) Signal transducers and activators of transcription 5b activation enhances hepatocellular carcinoma aggressiveness through induction of epithelial-mesenchymal transition. Cancer Res 66:9948–9956

Lee JK, Edderkaoui M, Truong P, Ohno I, Jang KT, Berti A, Pandol SJ, Gukovskaya AS (2007) NADPH oxidase promotes pancreatic cancer cell survival via inhibiting JAK2 dephosphorylation by tyrosine phosphatases. Gastroenterology 133:1637–1648

Leo C, Horn LC, Rauscher C, Hentschel B, Liebmann A, Hildebrandt G, Hockel M (2006) Expression of erythropoietin and erythropoietin receptor in cervical cancer and relationship to survival, hypoxia, and apoptosis. Clin Cancer Res 12:6894–6900

Leong PL, Xi S, Drenning SD, Dyer KF, Wentzel AL, Lerner EC, Smithgall TE, Grandis JR (2002) Differential function of STAT5 isoforms in head and neck cancer growth control. Oncogene 21:2846–2853

Levine RL, Wadleigh M, Cools J, Ebert BL, Wernig G, Huntly BJ, Boggon TJ, Wlodarska I, Clark JJ, Moore S et al (2005) Activating mutation in the tyrosine kinase JAK2 in polycythemia vera, essential thrombocythemia, and myeloid metaplasia with myelofibrosis. Cancer Cell 7:387–397

Li H, Ahonen TJ, Alanen K, Xie J, LeBaron MJ, Pretlow TG, Ealley EL, Zhang Y, Nurmi M, Singh B et al (2004) Activation of signal transducer and activator of transcription 5 in human prostate cancer is associated with high histological grade. Cancer Res 64:4774–4782

Liang K, Esteva FJ, Albarracin C, Stemke-Hale K, Lu Y, Bianchini G, Yang CY, Li Y, Li X, Chen CT, Mills GB, Hortobagyi GN, Mendelsohn J, Hung MC, Fan Z (2010) Recombinant human erythropoietin antagonizes trastuzumab treatment of breast cancer cells via Jak2-mediated Src activation and PTEN inactivation. Cancer Cell 18(5):423–435

Liao Z, Lutz J, Nevalainen MT (2010) Transcription factor Stat5a/b as a therapeutic target protein for prostate cancer. Int J Biochem Cell Biol 42:186–192

Lin Q, Lai R, Chirieac LR, Li C, Thomazy VA, Grammatikakis I, Rassidakis GZ, Zhang W, Fujio Y, Kunisada K et al (2005) Constitutive activation of JAK3/STAT3 in colon carcinoma tumors and cell lines: inhibition of JAK3/STAT3 signaling induces apoptosis and cell cycle arrest of colon carcinoma cells. Am J Pathol 167:969–980

Liu J, Kern JA (2002) Neuregulin-1 activates the JAK-STAT pathway and regulates lung epithelial cell proliferation. Am J Respir Cell Mol Biol 27:306–313

Lo HW, Cao X, Zhu H, Ali-Osman F (2008) Constitutively activated STAT3 frequently coexpresses with epidermal growth factor receptor in high-grade gliomas and targeting STAT3 sensitizes them to Iressa and alkylators. Clin Cancer Res 14:6042–6054

Long W, Wagner KU, Lloyd KC, Binart N, Shillingford JM, Hennighausen L, Jones FE (2003) Impaired differentiation and lactational failure of Erbb4-deficient mammary glands identify ERBB4 as an obligate mediator of STAT5. Development 130(21):5257–5268

Matulka LA, Wagner KU (2005) Models of breast cancer. Drug Discov Today 2(1):1–6

Mirmohammadsadegh A, Hassan M, Bardenheuer W, Marini A, Gustrau A, Nambiar S, Tannapfel A, Bojar H, Ruzicka T, Hengge UR (2006) STAT5 phosphorylation in malignant melanoma is important for survival and is mediated through SRC and JAK1 kinases. J Invest Dermatol 126:2272–2280

Mohyeldin A, Lu H, Dalgard C, Lai SY, Cohen N, Acs G, Verma A (2005) Erythropoietin signaling promotes invasiveness of human head and neck squamous cell carcinoma. Neoplasia 7:537–543

Mori D, Nakafusa Y, Miyazaki K, Tokunaga O (2005) Differential expression of Janus kinase 3 (JAK3), matrix metalloproteinase 13 (MMP13), heat shock protein 60 (HSP60), and mouse double minute 2 (MDM2) in human colorectal cancer progression using human cancer cDNA microarrays. Pathol Res Pract 201:777–789

Motte N, Saulnier P, Le Couedic JP, Soria JC, Delaloge S, Boige V, Pautier P, Vainchenker W, Bidart JM, Villeval JL (2007) Mutations in JAK2V617F homologous domain of JAK genes are uncommon in solid tumors. Int J Cancer 121:2113–2115
Nagpal JK, Mishra R, Das BR (2002) Activation of Stat-3 as one of the early events in tobacco chewing-mediated oral carcinogenesis. Cancer 94:2393–2400
Neilson LM, Zhu J, Xie J, Malabarba MG, Sakamoto K, Wagner KU, Kirken RA, Rui H (2007) Coactivation of janus tyrosine kinase (Jak)1 positively modulates prolactin-Jak2 signaling in breast cancer: recruitment of ERK and signal transducer and activator of transcription (Stat)3 and enhancement of Akt and Stat5a/b pathways. Mol Endocrinol 21:2218–2232
Neubauer H, Cumano A, Muller M, Wu H, Huffstadt U, Pfeffer K (1998) Jak2 deficiency defines an essential developmental checkpoint in definitive hematopoiesis. Cell 93:397–409
Nevalainen MT, Valve EM, Ingleton PM, Nurmi M, Martikainen PM, Harkonen PL (1997) Prolactin and prolactin receptors are expressed and functioning in human prostate. J Clin Invest 99:618–627
Nevalainen MT, Xie J, Torhorst J, Bubendorf L, Haas P, Kononen J, Sauter G, Rui H (2004) Signal transducer and activator of transcription-5 activation and breast cancer prognosis. J Clin Oncol 22:2053–2060
Niu G, Bowman T, Huang M, Shivers S, Reintgen D, Daud A, Chang A, Kraker A, Jove R, Yu H (2002) Roles of activated Src and Stat3 signaling in melanoma tumor cell growth. Oncogene 21:7001–7010
Nosaka T, van Deursen JM, Tripp RA, Thierfelder WE, Witthuhn BA, McMickle AP, Doherty PC, Grosveld GC, Ihle JN (1995) Defective lymphoid development in mice lacking Jak3. Science 270:800–802
Nyga R, Pecquet C, Harir N, Gu H, Dhennin-Duthille I, Regnier A, Gouilleux-Gruart V, Lassoued K, Gouilleux F (2005) Activated STAT5 proteins induce activation of the PI 3-kinase/Akt and Ras/MAPK pathways via the Gab2 scaffolding adapter. Biochem J 390:359–366
O'Shea JJ, Watford W (2004) A peek at PIAS. Nat Immunol 5:875–876
Ochs AM, Wong L, Kakani V, Neerukonda S, Gorske J, Rao A, Riggs M, Ward H, Keville L (2004) Expression of vascular endothelial growth factor and HER2/neu in stage II colon cancer and correlation with survival. Clin Colorectal Cancer 4:262–267
Ogunwobi OO, Beales IL (2007) The anti-apoptotic and growth stimulatory actions of leptin in human colon cancer cells involves activation of JNK mitogen activated protein kinase, JAK2 and PI3 kinase/Akt. Int J Colorectal Dis 22:401–409
Okochi O, Hibi K, Sakai M, Inoue S, Takeda S, Kaneko T, Nakao A (2003) Methylation-mediated silencing of SOCS-1 gene in hepatocellular carcinoma derived from cirrhosis. Clin Cancer Res 9:5295–5298
Pansky A, Hildebrand P, Fasler-Kan E, Baselgia L, Ketterer S, Beglinger C, Heim MH (2000) Defective Jak-STAT signal transduction pathway in melanoma cells resistant to growth inhibition by interferon-alpha. Int J Cancer 85:720–725
Pardanani A (2008) JAK2 inhibitor therapy in myeloproliferative disorders: rationale, preclinical studies and ongoing clinical trials. Leukemia 22(1):23–30
Pardanani A, Vannucchi AM, Passamonti F, Cervantes F, Barbui T, Tefferi A (2010) JAK inhibitor therapy for myelofibrosis: critical assessment of value and limitations. Leukemia 25 (2):218–225
Parganas E, Wang D, Stravopodis D, Topham DJ, Marine JC, Teglund S, Vanin EF, Bodner S, Colamonici OR, van Deursen JM et al (1998) Jak2 is essential for signaling through a variety of cytokine receptors. Cell 93:385–395
Park SY, Saijo K, Takahashi T, Osawa M, Arase H, Hirayama N, Miyake K, Nakauchi H, Shirasawa T, Saito T (1995) Developmental defects of lymphoid cells in Jak3 kinase-deficient mice. Immunity 3:771–782
Park DS, Lee H, Frank PG, Razani B, Nguyen AV, Parlow AF, Russell RG, Hulit J, Pestell RG, Lisanti MP (2002) Caveolin-1-deficient mice show accelerated mammary gland development

during pregnancy, premature lactation, and hyperactivation of the Jak-2/STAT5a signaling cascade. Mol Biol Cell 13:3416–3430

Pfeffer LM, Mullersman JE, Pfeffer SR, Murti A, Shi W, Yang CH (1997) STAT3 as an adapter to couple phosphatidylinositol 3-kinase to the IFNAR1 chain of the type I interferon receptor. Science 276(5317):1418–1420

Proietti CJ, Rosemblit C, Beguelin W, Rivas MA, Diaz Flaque MC, Charreau EH, Schillaci R, Elizalde PV (2009) Activation of Stat3 by heregulin/ErbB-2 through the co-option of progesterone receptor signaling drives breast cancer growth. Mol Cell Biol 29:1249–1265

Ram PA, Waxman DJ (1999) SOCS/CIS protein inhibition of growth hormone-stimulated STAT5 signaling by multiple mechanisms. J Biol Chem 274:35553–35561

Rane SG, Reddy EP (2000) Janus kinases: components of multiple signaling pathways. Oncogene 19(49):5662–5679

Ranger JJ, Levy DE, Shahalizadeh S, Hallett M, Muller WJ (2009) Identification of a Stat3-dependent transcription regulatory network involved in metastatic progression. Cancer Res 69 (17):6823–6830

Ravi R, Jain AJ, Schulick RD, Pham V, Prouser TS, Allen H, Mayer EG, Yu H, Pardoll DM, Ashkenazi A et al (2004) Elimination of hepatic metastases of colon cancer cells via p53-independent cross-talk between irinotecan and Apo2 ligand/TRAIL. Cancer Res 64:9105–9114

Rodig SJ, Meraz MA, White JM, Lampe PA, Riley JK, Arthur CD, King KL, Sheehan KC, Yin L, Pennica D et al (1998) Disruption of the Jak1 gene demonstrates obligatory and nonredundant roles of the Jaks in cytokine-induced biologic responses. Cell 93:373–383

Rosa Santos SC, Dumon S, Mayeux P, Gisselbrecht S, Gouilleux F (2000) Cooperation between STAT5 and phosphatidylinositol 3-kinase in the IL-3-dependent survival of a bone marrow derived cell line. Oncogene 19:1164–1172

Rose-Hellekant TA, Arendt LM, Schroeder MD, Gilchrist K, Sandgren EP, Schuler LA (2003) Prolactin induces ERalpha-positive and ERalpha-negative mammary cancer in transgenic mice. Oncogene 22:4664–4674

Ruchatz H, Coluccia AM, Stano P, Marchesi E, Gambacorti-Passerini C (2003) Constitutive activation of Jak2 contributes to proliferation and resistance to apoptosis in NPM/ALK-transformed cells. Exp Hematol 31:309–315

Sakamoto K, Creamer BA, Triplett AA, Wagner KU (2007) The Janus kinase 2 is required for expression and nuclear accumulation of cyclin D1 in proliferating mammary epithelial cells. Mol Endocrinol 21:1877–1892

Sakamoto K, Lin WC, Triplett AA, Wagner KU (2009) Targeting janus kinase 2 in Her2/neu-expressing mammary cancer: implications for cancer prevention and therapy. Cancer Res 69:6642–6650

Sakamoto K, Triplett AA, Schuler LA, Wagner KU (2010) Janus kinase 2 is required for the initiation but not maintenance of prolactin-induced mammary cancer. Oncogene 29:5359–5369

Sansone P, Storci G, Tavolari S, Guarnieri T, Giovannini C, Taffurelli M, Ceccarelli C, Santini D, Paterini P, Marcu KB et al (2007) IL-6 triggers malignant features in mammospheres from human ductal breast carcinoma and normal mammary gland. J Clin Invest 117:3988–4002

Santos SC, Lacronique V, Bouchaert I, Monni R, Bernard O, Gisselbrecht S, Gouilleux F (2001) Constitutively active STAT5 variants induce growth and survival of hematopoietic cells through a PI 3-kinase/Akt dependent pathway. Oncogene 20:2080–2090

Sasi W, Jiang WG, Sharma A, Mokbel K (2010) Higher expression levels of SOCS 1,3,4,7 are associated with earlier tumour stage and better clinical outcome in human breast cancer. BMC Cancer 10:178

Schindler C, Plumlee C (2008) Inteferons pen the JAK-STAT pathway. Semin Cell Dev Biol 19:311–318

Schindler C, Levy DE, Decker T (2007) JAK-STAT signaling: from interferons to cytokines. J Biol Chem 282:20059–20063

Schindler C, Strehlow I (2000) Cytokines and STAT signaling. Adv Pharmacol 47:113–174
Schwertfeger KL, Richert MM, Anderson SM (2001) Mammary gland involution is delayed by activated Akt in transgenic mice. Mol Endocrinol 15(6):867–881
Shillingford JM, Miyoshi K, Robinson GW, Grimm SL, Rosen JM, Neubauer H, Pfeffer K, Hennighausen L (2002) Jak2 is an essential tyrosine kinase involved in pregnancy-mediated development of mammary secretory epithelium. Mol Endocrinol 16:563–570
Shimoda K, Kato K, Aoki K, Matsuda T, Miyamoto A, Shibamori M, Yamashita M, Numata A, Takase K, Kobayashi S et al (2000) Tyk2 plays a restricted role in IFN alpha signaling, although it is required for IL-12-mediated T cell function. Immunity 13:561–571
Shirota K, LeDuy L, Yuan SY, Jothy S (1990) Interleukin-6 and its receptor are expressed in human intestinal epithelial cells. Virchows Arch B Cell Pathol Incl Mol Pathol 58:303–308
Slupianek A, Hoser G, Majsterek I, Bronisz A, Malecki M, Blasiak J, Fishel R, Skorski T (2002) Fusion tyrosine kinases induce drug resistance by stimulation of homology-dependent recombination repair, prolongation of G(2)/M phase, and protection from apoptosis. Mol Cell Biol 22:4189–4201
Sobti RC, Singh N, Hussain S, Suri V, Bharadwaj M, Das BC (2010) Deregulation of STAT-5 isoforms in the development of HPV-mediated cervical carcinogenesis. J Recept Signal Transduct Res 30:178–188
Sordella R, Bell DW, Haber DA, Settleman J (2004) Gefitinib-sensitizing EGFR mutations in lung cancer activate anti-apoptotic pathways. Science 305:1163–1167
Sos BC, Harris C, Nordstrom SM, Tran JL, Balázs M, Caplazi P, Febbraio M, Applegate MAB, Wagner KU, Weiss EJ (2011) Profound fatty liver in mice with hepatocyte-specific deletion of JAK2 is completely rescued by abrogation of growth hormone secretion. J Clin Invest (in press)
Stepan VM, Krametter DF, Matsushima M, Todisco A, Delvalle J, Dickinson CJ (1999) Glycine-extended gastrin regulates HEK cell growth. Am J Physiol 277:R572–R581
Sultan AS, Xie J, LeBaron MJ, Ealley EL, Nevalainen MT, Rui H (2005) Stat5 promotes homotypic adhesion and inhibits invasive characteristics of human breast cancer cells. Oncogene 24:746–760
Sutherland KD, Lindeman GJ, Choong DY, Wittlin S, Brentzell L, Phillips W, Campbell IG, Visvader JE (2004) Differential hypermethylation of SOCS genes in ovarian and breast carcinomas. Oncogene 23:7726–7733
Takeda K, Kaisho T, Yoshida N, Takeda J, Kishimoto T, Akira S (1998) Stat3 activation is responsible for IL-6-dependent T cell proliferation through preventing apoptosis: generation and characterization of T cell-specific Stat3-deficient mice. J Immunol 161(9):4652–4660
Tam L, McGlynn LM, Traynor P, Mukherjee R, Bartlett JM, Edwards J (2007) Expression levels of the JAK/STAT pathway in the transition from hormone-sensitive to hormone-refractory prostate cancer. Br J Cancer 97:378–383
Tan SH, Dagvadorj A, Shen F, Gu L, Liao Z, Abdulghani J, Zhang Y, Gelmann EP, Zellweger T, Culig Z et al (2008) Transcription factor Stat5 synergizes with androgen receptor in prostate cancer cells. Cancer Res 68:236–248
Tan SM, Li F, Rajendran P, Prem Kumar A, Hui KM, Sethi G (2010) Identification of {beta}-escin as a novel inhibitor of STAT3/JAK2 signaling pathway that suppresses proliferation and induces apoptosis in human hepatocellular carcinoma cells. J Pharmacol Exp Ther 334 (1):285–293
Tassiulas I, Hu X, Ho H, Kashyap Y, Paik P, Hu Y, Lowell CA, Ivashkiv LB (2004) Amplification of IFN-alpha-induced STAT1 activation and inflammatory function by Syk and ITAM-containing adaptors. Nat Immunol 5:1181–1189
Tessier C, Prigent-Tessier A, Ferguson-Gottschall S, Gu Y, Gibori G (2001) PRL antiapoptotic effect in the rat decidua involves the PI3K/protein kinase B-mediated inhibition of caspase-3 activity. Endocrinology 142:4086–4094
Thoennissen NH, Iwanski GB, Doan NB, Okamoto R, Lin P, Abbassi S, Song JH, Yin D, Toh M, Xie WD et al (2009) Cucurbitacin B induces apoptosis by inhibition of the JAK/STAT pathway and potentiates antiproliferative effects of gemcitabine on pancreatic cancer cells. Cancer Res 69:5876–5884

Thomis DC, Gurniak CB, Tivol E, Sharpe AH, Berg LJ (1995) Defects in B lymphocyte maturation and T lymphocyte activation in mice lacking Jak3. Science 270:794–797

Tornell J, Rymo L, Isaksson OG (1991) Induction of mammary adenocarcinomas in metallothionein promoter-human growth hormone transgenic mice. Int J Cancer 49:114–117

Tworoger SS, Hankinson SE (2006) Prolactin and breast cancer risk. Cancer Lett 243:160–169

Uckun FM, Dibirdik I (2010) Chemoprevention of colorectal cancer by targeting Janus kinase 3 with a rationally designed small molecule inhibitor. Nutr Cancer 62:968–972

Ursini-Siegel J, Hardy WR, Zuo D, Lam SH, Sanguin-Gendreau V, Cardiff RD, Pawson T, Muller WJ (2008) ShcA signalling is essential for tumour progression in mouse models of human breast cancer. EMBO J 27(6):910–920

Utama FE, LeBaron MJ, Neilson LM, Sultan AS, Parlow AF, Wagner KU, Rui H (2006) Human prolactin receptors are insensitive to mouse prolactin: implications for xenotransplant modeling of human breast cancer in mice. J Endocrinol 188:589–601

Vafaizadeh V, Klemmt P, Brendel C, Weber K, Doebele C, Britt K, Grez M, Fehse B, Desrivieres S, Groner B (2010) Mammary epithelial reconstitution with gene-modified stem cells assigns roles to Stat5 in luminal alveolar cell fate decisions, differentiation, involution, and mammary tumor formation. Stem Cells 28:928–938

Vainchenker W, Constantinescu SN (2005) A unique activating mutation in JAK2 (V617F) is at the origin of polycythemia vera and allows a new classification of myeloproliferative diseases. Hematology Am Soc Hematol Educ Program 2005:195–200

Valentino L, Pierre J (2006) JAK/STAT signal transduction: regulators and implication in hematological malignancies. Biochem Pharmacol 71:713–721

Verstovsek S, Kantarjian H, Mesa RA, Pardanani AD, Cortes-Franco J, Thomas DA, Estrov Z, Fridman JS, Bradley EC, Erickson-Viitanen S, Vaddi K, Levy R, Tefferi A (2010) Safety and efficacy of INCB018424, a JAK1 and JAK2 inhibitor, in myelofibrosis. N Engl J Med 363 (12):1117–1127

Wagner KU, Rui H (2008) Jak2/Stat5 signaling in mammogenesis, breast cancer initiation and progression. J Mammary Gland Biol Neoplasia 13:93–103

Wagner KU, Wall RJ, St-Onge L, Gruss P, Wynshaw-Boris A, Garrett L, Li M, Furth PA, Hennighausen L (1997) Cre-mediated gene deletion in the mammary gland. Nucleic Acids Res 25:4323–4330

Wagner KU, McAllister K, Ward T, Davis B, Wiseman R, Hennighausen L (2001) Spatial and temporal expression of the Cre gene under the control of the MMTV-LTR in different lines of transgenic mice. Transgenic Res 10:545–553

Wagner KU, Krempler A, Triplett AA, Qi Y, George NM, Zhu J, Rui H (2004) Impaired alveologenesis and maintenance of secretory mammary epithelial cells in Jak2 conditional knockout mice. Mol Cell Biol 24:5510–5520

Watson SA, Morris TM, Varro A, Michaeli D, Smith AM (1999) A comparison of the therapeutic effectiveness of gastrin neutralisation in two human gastric cancer models: relation to endocrine and autocrine/paracrine gastrin mediated growth. Gut 45(6):812–817

Wellbrock C, Weisser C, Hassel JC, Fischer P, Becker J, Vetter CS, Behrmann I, Kortylewski M, Heinrich PC, Schartl M (2005) STAT5 contributes to interferon resistance of melanoma cells. Curr Biol 15:1629–1639

Wernig G, Kharas MG, Okabe R, Moore SA, Leeman DS, Cullen DE, Gozo M, McDowell EP, Levine RL, Doukas J, Mak CC, Noronha G, Martin M, Ko YD, Lee BH, Soll RM, Tefferi A, Hood JD, Gilliland DG (2008) Efficacy of TG101348, a selective JAK2 inhibitor, in treatment of a murine model of JAK2V617F-induced polycythemia vera. Cancer Cell 13(4):311–320

Wennbo H, Gebre-Medhin M, Gritli-Linde A, Ohlsson C, Isaksson OG, Tornell J (1997) Activation of the prolactin receptor but not the growth hormone receptor is important for induction of mammary tumors in transgenic mice. J Clin Invest 100:2744–2751

Wiechen K, Diatchenko L, Agoulnik A, Scharff KM, Schober H, Arlt K, Zhumabayeva B, Siebert PD, Dietel M, Schafer R et al (2001) Caveolin-1 is down-regulated in human ovarian carcinoma and acts as a candidate tumor suppressor gene. Am J Pathol 159:1635–1643

Williams CC, Allison JG, Vidal GA, Burow ME, Beckman BS, Marrero L, Jones FE (2004) The ERBB4/HER4 receptor tyrosine kinase regulates gene expression by functioning as a STAT5A nuclear chaperone. J Cell Biol 167(3):469–478

Williams TM, Cheung MW, Park DS, Razani B, Cohen AW, Muller WJ, Di Vizio D, Chopra NG, Pestell RG, Lisanti MP (2003) Loss of caveolin-1 gene expression accelerates the development of dysplastic mammary lesions in tumor-prone transgenic mice. Mol Biol Cell 14:1027–1042

Wu S, Divall S, Hoffman GE, Le WW, Wagner KU, Wolfe A (2011) Jak2 is necessary for neuroendocrine control of female reproduction. J Neurosci 31(1):184–192

Wu W, Zanello L, Walker AM (2007) S179D prolactin sensitizes human prostate cancer cells such that physiological concentrations of 1, 25 dihydroxy vitamin D3 result in growth inhibition and cell death. Prostate 67:1498–1506

Xi S, Zhang Q, Gooding WE, Smithgall TE, Grandis JR (2003) Constitutive activation of Stat5b contributes to carcinogenesis in vivo. Cancer Res 63:6763–6771

Xie HJ, Bae HJ, Noh JH, Eun JW, Kim JK, Jung KH, Ryu JC, Ahn YM, Kim SY, Lee SH et al (2009) Mutational analysis of JAK1 gene in human hepatocellular carcinoma. Neoplasma 56:136–140

Xiong H, Zhang ZG, Tian XQ, Sun DF, Liang QC, Zhang YJ, Lu R, Chen YX, Fang JY (2008) Inhibition of JAK1, 2/STAT3 signaling induces apoptosis, cell cycle arrest, and reduces tumor cell invasion in colorectal cancer cells. Neoplasia 10:287–297

Xu SB, Liu XH, Li BH, Zhang Y, Yuan J, Yuan Q, Li PD, Yang XZ, Li F, Zhang WJ (2009) DNA methylation regulates constitutive expression of Stat6 regulatory genes SOCS-1 and SHP-1 in colon cancer cells. J Cancer Res Clin Oncol 135:1791–1798

Yamaji D, Na R, Feuermann Y, Pechhold S, Chen W, Robinson GW, Hennighausen L (2009) Development of mammary luminal progenitor cells is controlled by the transcription factor STAT5A. Genes Dev 23(20):2382–2387

Yamauchi T, Kaburagi Y, Ueki K, Tsuji Y, Stark GR, Kerr IM, Tsushima T, Akanuma Y, Komuro I, Tobe K et al (1998) Growth hormone and prolactin stimulate tyrosine phosphorylation of insulin receptor substrate-1, -2, and -3, their association with p85 phosphatidylinositol 3-kinase (PI3-kinase), and concomitantly PI3-kinase activation via JAK2 kinase. J Biol Chem 273:15719–15726

Yamauchi T, Ueki K, Tobe K, Tamemoto H, Sekine N, Wada M, Honjo M, Takahashi M, Takahashi T, Hirai H, Tushima T, Akanuma Y, Fujita T, Komuro I, Yazaki Y, Kadowaki T (1997) Tyrosine phosphorylation of the EGF receptor by the kinase Jak2 is induced by growth hormone. Nature 390(6655):91–96

Yamauchi T, Yamauchi N, Ueki K, Sugiyama T, Waki H, Miki H, Tobe K, Matsuda S, Tsushima T, Yamamoto T, Fujita T, Taketani Y, Fukayama M, Kimura S, Yazaki Y, Nagai R, Kadowaki T (2000) Constitutive tyrosine phosphorylation of ErbB-2 via Jak2 by autocrine secretion of prolactin in human breast cancer. J Biol Chem 275(43):33937–33944

Yeh YT, Ou-Yang F, Chen IF, Yang SF, Su JH, Hou MF, Yuan SS (2007) Altered p-JAK1 expression is associated with estrogen receptor status in breast infiltrating ductal carcinoma. Oncol Rep 17:35–39

Yoshikawa H, Matsubara K, Qian GS, Jackson P, Groopman JD, Manning JE, Harris CC, Herman JG (2001) SOCS-1, a negative regulator of the JAK/STAT pathway, is silenced by methylation in human hepatocellular carcinoma and shows growth-suppression activity. Nat Genet 28:29–35

Yu Q, Geng Y, Sicinski P (2001) Specific protection against breast cancers by cyclin D1 ablation. Nature 411(6841):1017–1021

Yuan TL, Cantley LC (2008) PI3K pathway alterations in cancer: variations on a theme. Oncogene 27(41):5497–5510

Zhang X, Yin P, Di D, Luo G, Zheng L, Wei J, Zhang J, Shi Y, Xu N (2009) IL-6 regulates MMP-10 expression via JAK2/STAT3 signaling pathway in a human lung adenocarcinoma cell line. Anticancer Res 29:4497–4501

STAT1 and STAT3 Transcription Factors in Inflammation-associated Colon Cancer

Paulina Rampetsreiter and Robert Eferl

Abstract

Inflammation is a strong promoter of colorectal cancer formation. Colorectal tumor cells establish heterotypic interactions with inflammatory cells in the stroma that are important for tumor angiogenesis and invasiveness. Recent studies in genetically modified mice have identified transcription factors and signaling networks that are implicated in these heterotypic tumor-stroma interactions and modulate preconditions of tumor formation such as chronic inflammation. Here, tumor-promoting and tumor-adverse effects of cytokine activated transcription factors STAT1 and STAT3 in inflammatory cells and colorectal cancer cells are discussed.

Introduction

Colorectal Cancer (CRC) in Humans

Colorectal cancer (CRC) originates from the epithelial cells lining the colon or rectum of the gastrointestinal tract and represents the third most common form of cancer worldwide. The histopathological sequence of tumor progression has been well defined which was mainly due to the easy accessibility and the frequent appearance

P. Rampetsreiter
Ludwig Boltzmann Institute for Cancer Research (LBICR), Währinger Straße 13a, 1090 Vienna, Austria

R. Eferl (✉)
Ludwig Boltzmann Institute for Cancer Research (LBICR), Währinger Straße 13a, 1090 Vienna, Austria

Institute for Cancer Research, Medical University of Vienna (MUV), Vienna, Austria
e-mail: robert.eferl@lbicr.lbg.ac.at

Th. Decker and M. Müller (eds.), *Jak-Stat Signaling: From Basics to Disease*,
DOI 10.1007/978-3-7091-0891-8_16,

of this tumor type. The most commonly used staging system is represented by the TNM classification defined by the American Joint Committee on Cancer (AJCC). The criteria used by that classification comprise local invasiveness, presence of lymph node metastases and far distant metastasis. Cancers that invade only the colonic submucosa (TNM stage I) or the muscularis propria (TNM stage II) can be cured with surgery. Prognosis is worse when these cancers spread to regional lymph nodes (TNM stage III) but still >70% are curable by surgery and chemotherapy. However, when cancers form far distant metastases (TNM stage IV), they are usually not curable and chemotherapy can only extend survival (Markowitz and Bertagnolli 2009).

Genetic and Epigenetic Changes in CRC

The easy accessibility of CRC biopsy samples that represent different stages of tumor development has greatly facilitated the identification of cell-autonomous tumor-inducing genetic and epigenetic events. At early stages, mutations in adenomatous polyposis coli (Apc) have been identified in ~90% of cases indicating that this gene encodes a tumor suppressor whose deletion is an almost essential event for CRC formation. Apc is also mutated in the germline of patients suffering from the inherited disease FAP (familal adenomatous polyposis coli) that leads to development of thousands of adenomatous polyps in the colon thereby increasing the risk for development of CRC substantially (de Lau et al. 2007). Apc encodes a protein that is present in a complex consisting of Axin, Glycogen synthetase kinase 3ß (GSK3ß) and ß-Catenin and integrates Wnt signals from the cell surface. Loss of Apc results in ß-Catenin stabilization and its constitutive presence in the nucleus where it associates and activates transcription factors of the Lef/Tcf family (Korinek et al. 1997; Morin et al. 1997). Wnt/ß-Catenin signaling is required for maintenance of mucosal integrity and proper cell turnover in the intestine. Paneth cells at the bottom of the intestinal crypts secrete Wnts and other factors that are essential for Lgr5-positive intestinal stem cells and activate ß-Catenin signaling (Sato et al. 2011). When the stem cell progeny moves upwards the crypts, ß-Catenin activity decreases and cells differentiate into intestinal cell lineages. Therefore, loss of Apc and constitutive presence of ß-Catenin in the nucleus keeps intestinal epithelial cells in a proliferating undifferentiated state (Sansom et al. 2004). In 10% of colon tumors, mutations in Apc can be replaced by mutations in GSK3ß or activating mutations in ß-Catenin (de Lau et al. 2007; Markowitz and Bertagnolli 2009). Additional mutations in these undifferentiated cells promote further tumor progression. Most important are activating mutations in K-ras that occur relatively early during tumor progression in ~50% of CRCs. Alternative mutations in B-raf have also been observed that are mutually exclusive with K-ras mutations. During adenoma to carcinoma progression, upregulation of the cyclooxygenase 2 (COX2) gene is frequently observed. This gene encodes an enzyme that is implicated in prostaglandin E biosynthesis, a known promoter of CRC formation. At a later stage of CRC progression, loss of a chromosomal region on chromosome 18q that harbours the gene for the TGFß signaling component Smad4 have been identified. The importance of downregulation of TGFß signaling is further supported by alternative mutations in the genes for activin receptors or the TGFß receptor II. 50–70% of CRCs harbour p53 mutations that occur at late stages of tumor development. It is likely that these mutations prevent p53-induced apoptosis that is triggered by deregulated Wnt-signaling. In some cases p53 is not mutated but the p53-targed gene BAX is, indicating that BAX executes the p53-mediated apoptotic program in CRC cells (Markowitz and Bertagnolli 2009). Moreover, p53 mutations might promote genomic instability in CRC. Similarly, defects in the apparatus for DNA mismatch repair (MMR) are the cause for CRC development in patients with the inherited disease HNPCC (hereditary non-polyposis colon cancer). However, only about 20% of CRCs are inherited whereas the majority of cases develop without a familial basis (Rustgi 2007).

Relevance of Inflammation in CRC

Several risk factors for development of CRC have been identified. They include age, presence of adenomatous polyps, presence of other cancers (ovary, uterus or breast in women), heredity, smoking, diet, concentration of lithocholic acid (bile acid) in the colon, presence of sclerosing cholangitis in the liver, low levels of selenium, estrogen levels and most importantly presence of inflammatory bowel disease (IBD). Approximately 20% of human cancers are estimated to develop in close association with chronic inflammation. Similarly, CRC can develop without overt inflammation or is associated with inflammatory bowel disease (IBD). The latter subtype of CRC is called colitis-associated cancer (CAC) and develops at high frequency in IBD patients (Terzic et al. 2010). The pathological appearance and stages of tumor progression are comparable between non-inflammatory CRC and CAC and they share similar tumor cell-autonomous genetic changes. However, it has to be considered that CRCs that developed in an environment without prominent inflammation display robust inflammatory infiltration of the tumor tissue and expression of pro-inflammatory cytokines. Among innate immune cells present in CRCs are neutrophils, T-cells, dendritic cells (DCs), natural killer cells (NKs), tumor-associated macrophages (TAMs) and myeloid derived suppressor cells (MDSCs). The latter represent a population of $CD11b^{+}$ $GR1^{+}$ cells that suppress immunologic anti-tumor responses. Cells of the adaptive immune system such as B- and T-cells are also present (Terzic et al. 2010). Inflammation promotes tumorigenesis but inflammatory infiltrates can also contain cells that execute a profound anti-tumorigenic activity such as cytotoxic T-cells. Inflammation can promote CRC formation in various ways. Inflammatory cells produce reactive oxygen species (ROS) and cytokines that promote ROS production in tumor cells. ROS can directly damage DNA leading to additional mutations (Terzic et al. 2010) or can modify DNA mismatch repair enzymes at the protein level thereby increasing genomic instability (Colotta et al. 2009). Moreover, cytokines and chemokines produced by inflammatory cells can promote tumor cell proliferation, survival, tumor angiogenesis and suppress immunologic anti-tumor responses (Terzic et al. 2010). Most of these cytokines are regulated by the transcription factors NFκB and STAT3 and interfering with these factors in inflammatory cells might interfere with CAC formation which has been demonstrated for NFκB (Greten et al. 2004).

Mouse Models for Intestinal Cancer

Genetic Models

Several genetic and chemical models have been employed to study intestinal cancer. The most commonly used genetic models exploit the strong tumor-inducing potential of Apc mutations upon loss of heterozygosity (Uronis and Threadgill 2009). The Apc^{Min} mutant allele was originally identified in a mouse mutagenesis screen. Since then, many investigators have used Apc^{Min} mice as an intestinal tumor model. Unfortunately, this mouse model does not recapitulate the tumor location of humans with non-hereditary CRCs and hereditary FAP where tumors develop predominantly in the colon. Recently, the Pirc (polyposis in rat colon) rat CRC model was established that is also based on a mutation in Apc (Amos-Landgraf et al. 2007). In Pirc rats, tumors develop frequently in the colon thereby mimicking human CRC. In addition, Apc^{Min} mice die quite early due to intestinal blockage. Pirc rats live longer because their intestinal diameter is larger than that of mice which extents the time frame for investigation of tumor progression. Apart from the Apc^{Min} mutation, other mutations have been introduced into the Apc gene by homologous recombination. Mice harbouring either the $Apc^{\Delta 716}$ or Apc^{1638N} mutant alleles also develop intestinal tumors albeit at different frequency than Apc^{Min} mice (Uronis and Threadgill 2009).

Apart from Apc mutants, mice deficient for mismatch repair (MMR) genes Mhl1, Msh2 and Msh6 that represent a model for HNPCC develop intestinal cancer (Uronis and Threadgill 2009). However, these mice develop also tumors in other organs and have not been frequently used as intestinal tumor models. Similarly, mice with genetic modifications that affect TGFß signaling develop intestinal tumors but have also not been widely used as cancer models. Mice lacking TGFß1 die postnatally because of severe inflammation but can be rescued in a Rag2-deficient background where T- and B-cell functions are ablated. Combined $TGFß1^{-/-}$ $Rag2^{-/-}$ mice develop tumors in the colonic and rectal regions. Mice deficient in the TGFß signaling component SMAD3 develop also CRC that is even metastatic. However, mutations in SMAD3 have not been reported in human CRCs (where rather SMAD2 and SMAD4 are mutated). Moreover, the TGFß-induced intestinal tumor models lack constitutive activation of the Wnt-signaling pathway that is commonly observed in human CRC (Uronis and Threadgill 2009).

Chemical Models

Chemical CRC models have been developed that closely mimic location and pathologic appearance of human CRC. A plethora of different carcinogens have been employed for CRC induction (Rosenberg et al. 2009). Most frequently, repetitive injections of Dimethylhydrazine (DMH) or its metabolite Azoxymethane (AOM) are used. Importantly, these carcinogens lead to stabilizing mutations in the gene for ß-Catenin thereby constitutively activating Wnt-signaling. The AOM/DSS protocol is an extension of the AOM protocol and induces CAC. The protocol is based on a single injection of AOM followed by repeated treatment of mice with Dextransulfate (DSS) in the drinking water. DSS of the right molecular weight induces a pronounced inflammation in the colon thereby promoting CAC formation. The AOM/DSS protocol has been employed in knock-out or conditional knock-out mice to unravel important functions of transcription factors NFκB and STAT3 in colorectal cancer (Bollrath et al. 2009; Greten et al. 2004).

Signaling by the Janus Kinase/Signal Transducer and Activator of Transcription (JAK/STAT) System

STATs are cytoplasmic transcription factors that mediate signal transduction from various growth factors and cytokines to the nucleus (Murray 2007). The seven members of the STAT family (STAT1–4, STAT5a, STAT5b and STAT6) share several structural domains (Fig. 1). Cytokine binding to corresponding cytokine receptors activates associated JAKs (JAK1, JAK2, JAK3 or TYK2) that mediate receptor phosphorylation thereby creating docking sites for the SH2 domains of STATs. Recruitment of STATs is followed by STAT tyrosine and (in some cases) serine phosphorylation predominantly by JAKs but also other closely associated kinases. After phosphorylation, STATs form

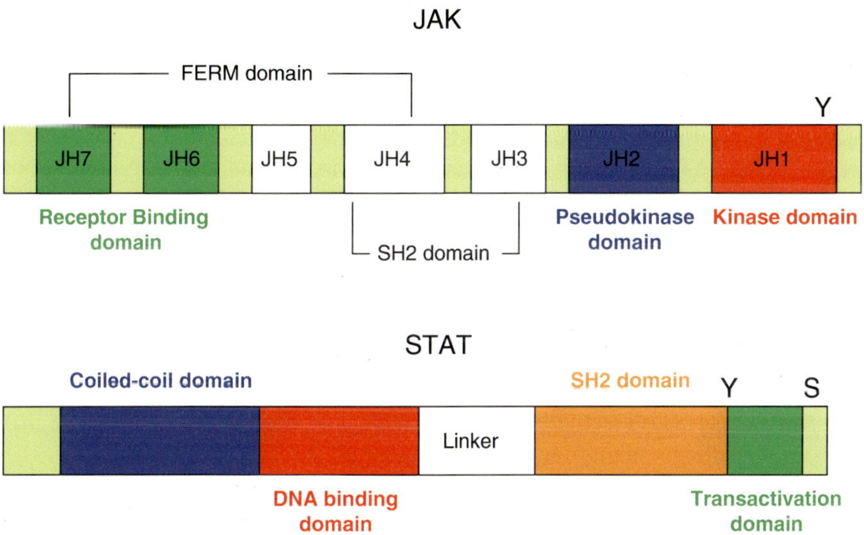

Fig. 1 *Domain structure of JAKs and STATs.* (**a**) JAK proteins (JAK1, JAK2, JAK3, TYK2) consist of 1,130–1,142 amino acid residues and contain seven JAK homology (*JH1–7*) regions that are similar in sequence. The JH1 region corresponds to the kinase domain and contains a regulatory tyrosine residue (Y). Moreover, a Pseudo-kinase domain (*JH2*), a SH2 domain (*JH3–4*) and an N-terminal FERM domain are present in a typical JAK. The FERM domain acts as a common membrane binding module that directly interacts with cytoplasmic domains of cytokine receptors. (**b**) The seven STAT proteins (STAT1–4, STAT5a, STAT5b, STAT6) consist of 748–851 amino acid residues and contain a coiled-coil domain for STAT dimerization, a DNA-binding domain, a SH2 domain for interaction with tyrosine-phosphorylated proteins and a transactivation domain. The latter contains regulatory tyrosine (Y) and serine (S) residues that can be phosphorylated by JAKs. In the case of STAT1 and STAT3, alternative splicing generates two isoforms (STAT1α/β, STAT3 α/β) that differ in the C-terminal transactivation domain

homo- or heterodimers and migrate to the nucleus where they bind to promotors of target genes. In addition to JAK/STAT activation, cytokine receptors can also activate the MAP kinase cascade (Murray 2007).

Specific Cytokines Activate STAT1 and STAT3

STAT1 is predominantly activated by type I (α, β) and type II (γ) interferons (IFN) via JAK1, JAK2 or TYK2 and induces a cohort of genes that are often termed as "IFN signature". These STAT1 target genes are either regulated by the ISGF3 transcription factor complex that contains STAT1, STAT2 and IRF9 or by STAT1 homodimers. STAT1:STAT3 heterodimers have also been described but their physiological function remains unknown (Murray 2007). STAT3 is predominantly activated by cytokines that bind a common gp130 receptor subunit that is associated with a specific subunit. For example, IL-6-mediated STAT3 activation is mediated through IL-6R/gp130 receptors that activate JAK1 or JAK2. Other STAT3 activating cytokines are IL-10, IL-11, IL-22, Oncostatin M, LIF or CNTF. Moreover, STAT3 is activated by receptor tyrosine kinases such as epidermal growth factor receptor (EGFR) and non-receptor tyrosine kinases such as Src (Murray 2007). The activity of STATs is regulated by phosphorylation, methylation, acetylation, ubiquitinylation and sumoylation. Tyrosine phosphorylation is the best understood modification which mediates dimerization and localization to the nucleus. Crucial tyrosine residues for STAT1 and STAT3 activation are located at positions 701 and 705, respectively, but additional serine residues are also phosphorylated (Regis et al. 2008). However, non-canonical functions of STAT1 and STAT3 have been described that do not depend on tyrosine phosphorylation (Cheon et al. 2011). Negative regulation of STAT activity is accomplished by various inhibitory factors and mechanisms. Most important are proteins of the suppressor of cytokine signalling (SOCS) family that consists of eight members (CIS and SOCS1–7). SOCS proteins are induced by cytokines thereby establishing a negative feedback loop that keeps STAT activity under control. SOCS members inhibit STAT activity by various molecular mechanisms and ensure that cytokine signalling remains transient (Murray 2007).

STAT1 and STAT3 in Tumorigenesis

Expression and activity of STAT1 and STAT3 is perturbed in a wide range of malignancies (O'Sullivan et al. 2007). Both transcription factors fulfil cell-autonomous functions in tumor cells and non-cell-autonomous functions in inflammatory cells of the tumor stroma. Concerning tumorigenesis, cell proliferation and cell survival, STAT1 and STAT3 are supposed to play opposing roles (Regis et al. 2008). Although STAT1 can support development of leukemia (Kovacic et al. 2006), it is generally considered to be a tumor suppressor since cancer incidence is increased in STAT1-deficient mice upon loss of p53 or challenge with the carcinogen Methylcholanthrene (MCA) (Shankaran et al. 2001). In contrast, STAT3 signaling is considered to promote tumorigenesis. It has been demonstrated that constitutive active STAT3 can transform 3T3 fibroblasts and interference with STAT3 activity in transplanted tumors reduced tumor load (Bromberg et al. 1999). STAT3 is a promoter of tumor cell proliferation and survival which is reflected by the spectrum of target genes that are either activated or repressed by STAT3. Consistent with their tumor-suppressive and tumor-promoting functions, STAT1 is frequently downregulated whereas STAT3 is often constitutively activated in many tumors (Yu et al. 2007).

The dual function of STAT1 and STAT3 in tumor cells and inflammatory stromal cells is linked by paracrine tumor-stroma interactions. For example, many cytokines, chemokines and growth factors such as IL-6 or IL-1β are produced by tumor cells in a STAT3-dependent manner (Yu et al. 2009). These factors in turn activate STAT3 in stromal cells that also start to produce STAT3-activating cytokines thereby establishing a positive feedback loop between tumor cells and cells in the tumor microenvironment (Yu et al. 2007).

STAT1 and STAT3 in Intestinal Cancer

STAT1 in Intestinal Cancer

STAT1 is considered to be a tumor suppressor that could interfere with CRC formation by at least two distinct mechanisms: (1) STAT1 promotes anti-tumorigenic activities of inflammatory stromal cells. (2) STAT1 integrates anti-proliferative and pro-apoptotic IFN signals in a cell-autonomous manner in tumor cells (Fig. 2). The function of STAT1 in immunological tumor defence could already counteract early stages of colorectal tumors in IBD patients. IBD is characterised by the increased expression of pro-inflammatory cytokines that include TNF-α, IL-6 and IFN-γ, a potent inducer of STAT1 phosphorylation. Consistently, STAT1 was found upregulated in mucosal samples of patients with ulcerative colitis and Crohn's disease and IHC analysis demonstrated that pY-STAT1 (STAT1 phosphorylated at tyrosine 701) was present predominantly in monocytic and neutrophilic cells of the inflamed mucosa (Schreiber et al. 2002). However, upregulation of STAT1 activity in inflammatory cells does not only promote immune-mediated anti-tumorigenic effects but also aggravates the severity of inflammation. The latter is due to the pro-apoptotic activity of STAT1 that leads to tissue damage and by regulating genes that are important for immune cell recruitment and activation such as ICAM1 and CXCL10 (Regis et al. 2008). This might suggest that STAT1 is not exclusively tumor-suppressive but could also promote formation of CRCs by its pro-inflammatory activity. Similarly, STAT1 can regulate gene expression in immune cells that support tumor formation by paracrine mechanisms. For example, TAMs produce IL-1ß in a STAT1-dependent manner that acts on colon cancer cells and promotes nuclear localization of ß-Catenin (Kaler et al. 2009). The pro-inflammatory role of STAT1 is supported by the observation that DSS-induced colitis is reduced in $STAT1^{-/-}$ mice (Bandyopadhyay et al. 2008). Moreover, reduction of the gene dosage of SOCS1, a potent inhibitor of STAT1 activity, in $SOCS1^{+/-}$ mice resulted in more severe DSS-induced colitis than in $SOCS1^{+/+}$ mice

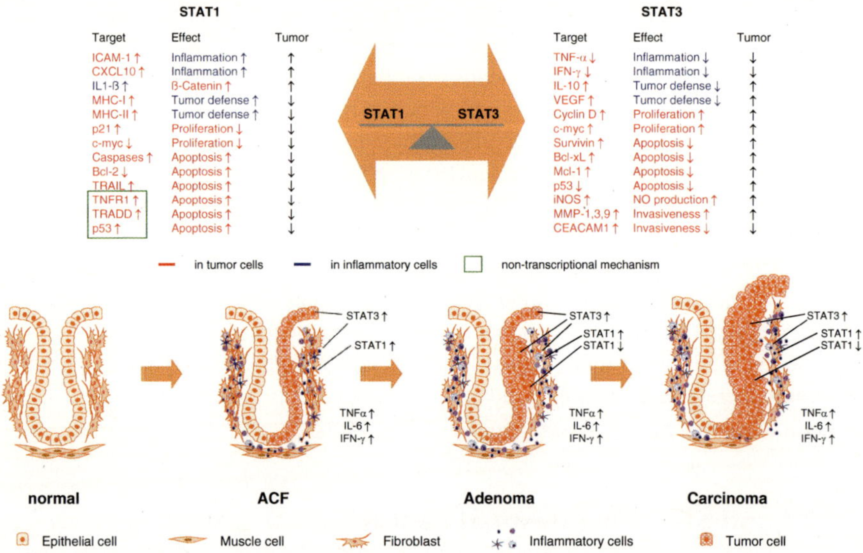

Fig. 2 *Functions of STAT1 and STAT3 during colorecal tumor progression.* A normal colonic crypt, a colonic crypt with inflammation and an altered cellular focus (ACF), an adenoma and a carcinoma is shown. Up- or down regulation of STAT1 and STAT3 in the colonic mucosa, microenvironment and tumor cells are indicated. Cytokines that activate STAT1 and STAT3 are produced during inflammatory conditions. However, tumor cells employ different mechanisms to shut down IFN/STAT1 signals. Important STAT target genes or protein-protein interactions and their effects on colorectal tumor formation are indicated in the tables above

(Horino et al. 2008). Conditional inactivation of SOCS1 in T-cells phenocopied this effect indicating that STAT1 might promote the severity of DSS-induced colitis via T-cell activation. Consistently, STAT1 is required for development and function of Th1 lymphocytes (Agnello et al. 2003) suggesting that this transcription factor is a central mediator of inflammatory responses. Recent evidence has suggested an alternative non-canonical signaling mechanism for STAT1 that does not depend on STAT1 tyrosine phosphorylation (unphosphorylated STAT1 or U-STAT1) (Cheon et al. 2011). It remains to be determined which immune-regulatory functions depend on canonical or non-canonical STAT1 signaling. STAT1 can also induce tumor cell autonomous defence mechanisms. Tumors that are insensitive to IFN-γ can more potently evade defence mechanisms than IFN-γ-sensitive tumors (Klampfer 2008). Activation of STAT1 by IFN induces expression of MHC class I and class II molecules thereby promoting recognition of tumor cells by the immune system (Regis et al. 2008). Consistently, a correlation between nuclear localization of STAT1, MHC class I expression and T-cell infiltration was established in human colorectal tumors that can be used as a biomarker for prognosis of patient survival (Simpson et al. 2010).

The anti-proliferative and pro-apoptotic activities of IFNs in tumor cells are integrated by activated STAT1. Among STAT1 target genes that mediate these effects are cyclins, p21, p27, c-myc, IRF-1, caspases, Fas, FasL, TRAIL, KILLER/DR5, mdm2, iNos, bcl-xL and bcl-2 (Klampfer 2008; Regis et al. 2008). These genes are either activated or repressed by STAT1. Two important STAT1 target genes that repress or promote tumor cell proliferation are p21 and c-myc, respectively. The p21 gene harbours multiple STAT1 binding sites in its promoter. Consistently, ablation of STAT1 impedes p21 induction in response to various agents that inhibit cell cycle progression. The importance of p21 for intestinal tumorigenesis has been demonstrated in the Apc^{1638N} and AOM/DSS-induced mouse models for intestinal cancer (Poole et al. 2004; Yang et al. 2001) which displayed enhanced tumor formation in a p21-negative genetic background. In contrast to p21, c-myc expression is negatively regulated by STAT1. Interestingly, IFN-γ signalling inhibits c-myc expression in the presence of STAT1 but activates c-myc and promotes cell proliferation in the absence of STAT1 (Klampfer 2008). This indicates that STAT1 represents a molecular switch that modulates the outcome of IFN signals on c-myc expression.

In addition to its anti-proliferative effect, STAT1 is a mediator of IFN-induced apoptosis. Similar to the molecular switch function during regulation of c-myc expression, the outcome of IFN-γ-induced apoptosis depends on the presence or absence of STAT1. Colon cancer cells with intact STAT1 respond to IFN-γ with apoptosis whereas the same signal promotes proliferation in the absence of STAT1 (Klampfer 2008). Activation of caspase and TRAIL expression as well as repression of bcl-xL and bcl-2 expression may at least in part underlie the pro-apoptotic activity of STAT1. Moreover, STAT1 can regulate apoptosis by non-transcriptional mechanisms via protein-protein interactions with TNFR1 and TRADD or interaction with p53 thereby modulating DNA damage-induced apoptosis (Regis et al. 2008). However, under certain circumstances STAT1 can also protect from apoptosis which results in higher resistance of tumor cells towards chemotherapy. A protective function of STAT1 has been described in radiation-induced cell death, cisplatin-induced DNA damage and cell death induced by HDAC inhibitors (Klampfer 2008).

Consistent with the anti-proliferative and pro-apoptic activities of IFN/STAT1 signals, tumor cells try to escape this pathway and acquire resistance by different means. Some tumor cells downregulate IFN-γ receptors whereas others have deleted the genomic IFN-γ locus. Alternatively, STAT1 activity or the activity of other factors implicated in IFN signalling can be downregulated by DNA methylation in tumors making them refractory to IFN signals (Klampfer 2008). Sometimes, IFN/STAT1 target genes are silenced by methylation. For example, IRF8 expression is silenced by DNA methylation in human colon carcinoma cells (McGough et al. 2008). Importantly, downregulation of IRF8 correlated with the metastatic phenotype of human CRCs. The expression of IRF8 is induced by IFN-γ and STAT1. However, binding of STAT1 to the IRF8 promoter is not prevented by methylation. Instead, the PIAS1 protein is recruited to the methylated IRF8 promoter which is a potent inhibitor of pY-STAT1 (McGough et al. 2008). Methylation as a mechanism to downregulate STAT1 expression has also been demonstrated in HT29 colon carcinoma cells. Reactivation of STAT1 expression by the demethylating agent 5-Aza-2-deoxycytidine restored sensitivity of HT29 cells to IFN-α (Klampfer 2008). Recently, mir-145 microRNA-mediated downregulation of STAT1 expression has been reported in colon cancer suggesting alternative mechanisms to DNA methylation-induced silencing of STAT1 (Gregersen et al. 2010). Interestingly, another mechanism that interferes with STAT1 activity has been linked to oncogenic K-ras. In colon cancer cell lines that harbour mutant K-ras, expression of STAT1 and corresponding target genes was reduced when compared to colon cancer cell lines without K-ras mutations. This was due to K-ras-mediated interference with STAT1 transcription. Consistently, colon cancer cell lines harbouring oncogenic K-ras mutants are less sensitive to IFN-γ than cell lines without K-ras mutations. Among the various downstream effector pathways that are activated by oncogenic K-ras, the PI3K and MAPK pathways have been shown to interfere with JAK/STAT signalling and are therefore possible candidates for K-ras-mediated downregulation of STAT1 (Klampfer 2008).

Despite the predicted function of STAT1 as a tumor suppressor in CRC, experiments in genetically modified mouse models have given contradictory results. Tumor formation was not altered in STAT1-deficient Apc^{Min} mice when compared to STAT1-proficient Apc^{Min} controls (Liddle and Frank 2008). Moreover, spontaneous formation of CRC has been observed in $SOCS1^{-/-}$ mice that were rescued from perinatal lethality by transgenic reconstitution of SOCS1 expression in T- and B-cells (SOCS1-deficient mice die at birth because of hyperactive lymphocytes). Genetic experiments demonstrated that tumor formation was IFN-γ-dependent and coincided with strong activation of STAT1 in the colonic epithelium (Hanada et al. 2006). Therefore, additional studies in mice with conditional deletion of STAT1 in the intestinal epithelium and individual inflammatory cell types of the tumor stroma are required to define its function in intestinal tumorigenesis.

STAT3 in Intestinal Cancer

STAT3 is strongly activated in the inflamed mucosa and inflammatory cells of human IBD patients and DSS-treated mice (Klampfer 2008). However, activation of STAT3 does not necessarily support inflammation but might represent a negative feedback mechanism that limits inflammation (Fig. 2). This has been demonstrated by several experiments in genetically-modified mice. Conditional ablation of STAT3 in macrophages or the whole hematopoietic system using LysMCre or MxCre mice, respectively, led to severe immune-mediated colitis (Alonzi et al. 2004; Takeda et al. 1999). STAT3 is a downstream component of anti-inflammatory IL-10 signaling in myeloid cells thereby limiting inflammation. Moreover, activation of IL-10/STAT3 can directly inhibit IFN-induced gene expression through downregulation of STAT1 activity which represents an antagonistic link between STAT3 and STAT1 in inflammation (Regis et al. 2008).

Similar to inflammation, STAT3 and STAT1 might have opposite activities in tumor defence. Cancer cells exploit the ability of STAT3 to induce expression of several anti-inflammatory cytokines including VEGF and IL-10 that interfere with the functional maturation of dendritic cells. In addition, STAT3 represses the expression of pro-inflammatory cytokines and chemokines such as IFN-γ, TNF-α, IL-6 and IP-10 in tumor cells that can activate the innate immune system and tumor-specific T-cell responses (Klampfer 2008; Regis et al. 2008). Moreover, tumor transplantation experiments have demonstrated that ablation of STAT3 in the hematopoietic compartment triggers an intrinsic immune defence system that inhibits tumor growth and metastasis. Tumor-bearing mice with STAT3-deficient hematopoietic cells displayed enhanced anti-tumorigenic functions of DCs, T-cells, NKs and neutrophils. Blocking STAT3 with a small-molecule inhibitor induced T-cell- and NK-dependent growth inhibition of established tumors that were otherwise resistant to direct killing by the drug (Kortylewski et al. 2005). These data indicate that STAT3 interferes with immunological tumor defence by tumor cell-autonomous and non-cell-autonomous mechanisms.

In contrast to STAT1, STAT3 is considered to act as an oncogene in intestinal tumor cells. This has been demonstrated in colon cancer cell lines such as SW480 and HT29 that underwent apoptosis and cell cycle arrest upon inhibition of STAT3 which was associated with downregulation of Bcl-2 family members and upregulation of p21. Similarly, colon cancer xenografts displayed reduced proliferation and tumor growth upon inhibition of STAT3. In primary human CRCs, STAT3 has been shown to be activated at the protein level by tyrosine phosphorylation or at the DNA level by increased gene transcription (Klampfer 2008). Further analyses are needed to investigate if STAT3 can promote CRC formation by non-canonical mechanisms that depend on unphosphorylated rather than tyrosine phosphorylated STAT3 (U-STAT3) or on STAT3 that is located in the mitochondria (Gough et al. 2009; Yang et al. 2007).

Direct evidence for the requirement of IL-6 and STAT3 in colon cancer came from tumor studies in IL-6$^{-/-}$ mice and mice with conditional deletion of STAT3 in the intestinal epithelium (STAT3$^{\Delta IEC}$). Both types of mice displayed reduced tumorigenicity in response to AOM/DSS when compared to corresponding controls. These studies have also demonstrated that STAT3 activity is implicated in tumor initiation of colitis-associated cancer since it promotes mucosal healing after inflammatory tissue damage (Bollrath et al. 2009; Grivennikov et al. 2009). Cells in the tumor microenvironment such as myeloid cells or T-cells are a major source for the potent STAT3 activator IL-6. T-cells seem to produce IL-6 in a TGFß dependent manner (TGFß being a negative regulator of IL 6 production) thereby increasing the levels of pY STAT3 in tumor cells and promoting tumorigenesis by a paracrine mechanism (Becker et al. 2004). Similarly, conditional inactivation of SOCS3, a potent negative regulator of STAT3 activity, in the intestinal epithelium promoted AOM/DSS-induced colon cancer formation that was associated with increased STAT3 activity (Rigby et al. 2007).

The oncogenic potential of STAT3 in colon cancer can be explained by regulation of target genes that modulate intestinal tumor formation but also by post-transcriptional mechanisms. Inhibition of STAT3 activity has been shown to induce export of ß-Catenin from the nucleus to the cytoplasm by an unknown mechanism thereby interfering with Lef/Tcf-regulated gene expression (Kawada et al. 2006). However, a more recent report has shown that Apc^{Min}-induced intestinal tumors lacking STAT3 displayed a more pronounced nuclear localization of ß-Catenin than STAT3-proficient tumors that was associated with downregulation of the cell adhesion molecule CEACAM1 (Musteanu et al. 2010). These data indicate that STAT3 can positively or negatively modulate nuclear localization of ß-Catenin by signals that have not yet been identified.

Besides these post-transcriptional mechanisms, STAT3 can directly regulate many genes implicated in cell survival and cell proliferation. Constitutive STAT3 activity leads to upregulation of Bcl-xL, survivin, Mcl1, c-myc or D-type cyclins, thereby inhibiting apoptosis and promoting cell proliferation. STAT3 can also modulate p53 signals. STAT3 can directly bind to the p53 promoter, thereby repressing p53 expression and keeping the pro-apoptotic activity of p53 under control. Moreover, IL-22-induced induction of iNOS expression depends on STAT3 suggesting that STAT3 regulates NO production in CRCs (Klampfer 2008; Regis et al. 2008).

Immunohistochemistry identified activated pY-STAT3 (STAT3 phosphorylated at tyrosine 705) in 12.5% of human colonic adenomas and 72.7% of carcinomas suggesting that STAT3 is not only implicated in growth but also in progression of colorectal tumors. Moreover, the intensity of pY-STAT3 signals correlated positively with tumor invasiveness. pY-STAT3 was not only found in colonic tumor cells but also in infiltrating lymphocytes present in the tumor stroma (Huang 2007; Klampfer 2008). The stromal compartment of intestinal tumors might be the source of cytokines that activate STAT3 in tumor cells. This is suggested by the observation that isolated colonic cancer cells cultured in vitro without tumor stroma frequently lost persistent STAT3 activity (Corvinus et al. 2005). Several molecular mechanisms have been described that might contribute to STAT3-mediated progression of colorectal tumors. STAT3 regulates expression of matrix metalloproteinases MMP-1, MMP-3 and MMP-9 thereby increasing the invasive potential of colon cancer cells (Tsareva et al. 2007). Moreover, STAT3 is a downstream signaling component of VEGF and regulates expression of trefoil factor 3 (TFF3) which is involved in proliferation, survival and invasiveness of colon cancer cells (Rivat et al. 2005). However, the positive implication of STAT3 in colorectal cancer progression and invasiveness is not unambiguous. A recent report has described a significantly improved survival of CRC patients with strong pY-STAT3 levels in tumors (Monnien et al. 2010). Moreover, deletion of STAT3 in the intestinal epithelium of Apc^{Min} mice resulted in a higher percentage of carcinomas that displayed a more aggressive invasive behaviour when compared to STAT3-proficient control mice. Strikingly, the few carcinomas that developed in STAT3-proficient controls showed downregulation of pY-STAT3 at the invasive front into the muscularis mucosa indicating that STAT3 is a negative regulator of intestinal tumor invasion (Musteanu et al. 2010). Therefore, additional studies are needed to better define oncogenic and anti-oncogenic activities of STAT3 in CRC.

STAT1/STAT3 Functional Interaction and Interference

The antagonistic activities of STAT1 and STAT3 in inflammation, tumor defence, apoptosis and proliferation are not considered as independent entities but rather as mutually interfering interactions. This is not only due to the possible formation of STAT1:STAT3 heterodimers but also due to shared signaling pathways for STAT activation. The latter are of particular importance when STAT1 or STAT3 activities are selectively ablated. For example, deletion of STAT1 can switch the pro-apoptotic and anti-proliferative activities of IFNs to survival and proliferation signals in various cell types such as mouse embryonic fibroblasts (MEFs), bone marrow-derived macrophages or T-cells. This functional switch is at least in part due to aberrant activation of STAT3 by IFN signals that only occurs in the absence of STAT1. In contrast, STAT1 can be activated by gp130-inducing signals when STAT3 is ablated. For example, STAT3-deficient fibroblasts respond to IL-6 with substantial STAT1 activation that results in unusual IL-6-mediated activation of IFN responsive genes. This mutual interplay explains the differential response of cells with selective ablation of STAT1 or STAT3 to corresponding activating signals. Therefore it is not surprising that several pathological conditions are associated with unbalanced expression or activation of STAT1 and STAT3 (Regis et al. 2008). The molecular mechanisms that underlie the reciprocal influence of STAT1 and STAT3 activation is still a matter of debate but competition for common cytokine receptors, the implication of SOCS proteins and the mutual inactivation of STATs by sequestering in STAT1:STAT3 heterodimer complexes have been suggested (Murray 2007). Therefore, it has to be considered that application of a drug specific for STAT1 might have a profound effect on STAT3 activity and vice versa.

Concluding Remarks

Inflammatory conditions and ensuing accumulation of gp130 cytokines and interferons activate STAT transcription factors in colorectal cancer cells and stromal cells. Some of these events promote CRC formation, others (e.g. STAT1 activation in tumor cells) might interfere with tumor growth. However, the overall beneficial effects of inflammation and STAT activation encourages cancer cells to arrange with and to counterselect against adverse effects by various genetic and epigenetic mechanisms. Moreover, tumor-promoting and tumor-adverse functions of STATs might change with the progression stage of colorectal cancers. The dual activity of STATs in stromal and tumor cells, potential adaptive mechanisms against adverse STAT effects and progression-specific requirements for STATs should be taken into account before therapeutic application of JAK/STAT inhibitors.

Acknowledgements This work was supported by the Austrian Science Fund FWF grant SFB F28 (RE), GENAU "Austromouse" (RE) and Doktoratskolleg plus IAI (Inflammation and Immunity; RE).

References

Agnello D, Lankford CS, Bream J, Morinobu A, Gadina M, O'Shea JJ, Frucht DM (2003) Cytokines and transcription factors that regulate T helper cell differentiation: new players and new insights. J Clin Immunol 23:147–161

Alonzi T, Newton IP, Bryce PJ, Di Carlo E, Lattanzio G, Tripodi M, Musiani P, Poli V (2004) Induced somatic inactivation of STAT3 in mice triggers the development of a fulminant form of enterocolitis. Cytokine 26:45–56

Amos-Landgraf JM, Kwong LN, Kendziorski CM, Reichelderfer M, Torrealba J, Weichert J, Haag JD, Chen KS, Waller JL, Gould MN et al (2007) A target-selected Apc-mutant rat kindred enhances the modeling of familial human colon cancer. Proc Natl Acad Sci USA 104:4036–4041

Bandyopadhyay SK, de la Motte CA, Kessler SP, Hascall VC, Hill DR, Strong SA (2008) Hyaluronan-mediated leukocyte adhesion and dextran sulfate sodium-induced colitis are attenuated in the absence of signal transducer and activator of transcription 1. Am J Pathol 173:1361–1368

Becker C, Fantini MC, Schramm C, Lehr HA, Wirtz S, Nikolaev A, Burg J, Strand S, Kiesslich R, Huber S et al (2004) TGF-beta suppresses tumor progression in colon cancer by inhibition of IL-6 trans-signaling. Immunity 21:491–501

Bollrath J, Phesse TJ, von Burstin VA, Putoczki T, Bennecke M, Bateman T, Nebelsiek T, Lundgren-May T, Canli O, Schwitalla S et al (2009) gp130-mediated Stat3 activation in enterocytes regulates cell survival and cell-cycle progression during colitis-associated tumorigenesis. Cancer Cell 15:91–102

Bromberg JF, Wrzeszczynska MH, Devgan G, Zhao Y, Pestell RG, Albanese C, Darnell JE Jr (1999) Stat3 as an oncogene. Cell 98:295–303

Cheon H, Yang J, Stark GR (2011) The functions of signal transducers and activators of transcriptions 1 and 3 as cytokine-inducible proteins. J Interferon Cytokine Res 31:33–40

Colotta F, Allavena P, Sica A, Garlanda C, Mantovani A (2009) Cancer-related inflammation, the seventh hallmark of cancer: links to genetic instability. Carcinogenesis 30:1073–1081

Corvinus FM, Orth C, Moriggl R, Tsareva SA, Wagner S, Pfitzner EB, Baus D, Kaufmann R, Huber LA, Zatloukal K et al (2005) Persistent STAT3 activation in colon cancer is associated with enhanced cell proliferation and tumor growth. Neoplasia 7:545–555

de Lau W, Barker N, Clevers H (2007) WNT signaling in the normal intestine and colorectal cancer. Front Biosci 12:471–491

Gough DJ, Corlett A, Schlessinger K, Wegrzyn J, Larner AC, Levy DE (2009) Mitochondrial STAT3 supports Ras-dependent oncogenic transformation. Science (New York) 324:1713–1716

Gregersen LH, Jacobsen AB, Frankel LB, Wen J, Krogh A, Lund AH (2010) MicroRNA-145 targets YES and STAT1 in colon cancer cells. PLoS One 5:e8836

Greten FR, Eckmann L, Greten TF, Park JM, Li ZW, Egan LJ, Kagnoff MF, Karin M (2004) IKKbeta links inflammation and tumorigenesis in a mouse model of colitis-associated cancer. Cell 118:285–296

Grivennikov S, Karin E, Terzic J, Mucida D, Yu GY, Vallabhapurapu S, Scheller J, Rose-John S, Cheroutre H, Eckmann L et al (2009) IL-6 and Stat3 are required for survival of intestinal epithelial cells and development of colitis-associated cancer. Cancer Cell 15:103–113

Hanada T, Kobayashi T, Chinen T, Saeki K, Takaki H, Koga K, Minoda Y, Sanada T, Yoshioka T, Mimata H et al (2006) IFNgamma-dependent, spontaneous development of colorectal carcinomas in SOCS1-deficient mice. J Exp Med 203:1391–1397

Horino J, Fujimoto M, Terabe F, Serada S, Takahashi T, Soma Y, Tanaka K, Chinen T, Yoshimura A, Nomura S et al (2008) Suppressor of cytokine signaling-1 ameliorates dextran sulfate sodium-induced colitis in mice. Int Immunol 20:753–762

Huang S (2007) Regulation of metastases by signal transducer and activator of transcription 3 signaling pathway: clinical implications. Clin Cancer Res 13:1362–1366

Kaler P, Augenlicht L, Klampfer L (2009) Macrophage-derived IL-1beta stimulates Wnt signaling and growth of colon cancer cells: a crosstalk interrupted by vitamin D3. Oncogene 28:3892–3902

Kawada M, Seno H, Uenoyama Y, Sawabu T, Kanda N, Fukui H, Shimahara Y, Chiba T (2006) Signal transducers and activators of transcription 3 activation is involved in nuclear accumulation of beta-catenin in colorectal cancer. Cancer Res 66:2913–2917

Klampfer L (2008) The role of signal transducers and activators of transcription in colon cancer. Front Biosci 13:2888–2899

Korinek V, Barker N, Morin PJ, van Wichen D, de Weger R, Kinzler KW, Vogelstein B, Clevers H (1997) Constitutive transcriptional activation by a beta-catenin-Tcf complex in APC-/- colon carcinoma. Science (New York) 275:1784–1787

Kortylewski M, Kujawski M, Wang T, Wei S, Zhang S, Pilon-Thomas S, Niu G, Kay H, Mule J, Kerr WG et al (2005) Inhibiting Stat3 signaling in the hematopoietic system elicits multicomponent antitumor immunity. Nat Med 11:1314–1321

Kovacic B, Stoiber D, Moriggl R, Weisz E, Ott RG, Kreibich R, Levy DE, Beug H, Freissmuth M, Sexl V (2006) STAT1 acts as a tumor promoter for leukemia development. Cancer Cell 10:77–87

Liddle FJ, Frank DA (2008) STAT1 expression is not required for polyp formation in Min mice. Mol Carcinog 47:75–79

Markowitz SD, Bertagnolli MM (2009) Molecular origins of cancer: molecular basis of colorectal cancer. N Engl J Med 361:2449–2460

McGough JM, Yang D, Huang S, Georgi D, Hewitt SM, Rocken C, Tanzer M, Ebert MP, Liu K (2008) DNA methylation represses IFN-gamma-induced and signal transducer and activator of transcription 1-mediated IFN regulatory factor 8 activation in colon carcinoma cells. Mol Cancer Res 6:1841–1851

Monnien F, Zaki H, Borg C, Mougin C, Bosset JF, Mercier M, Arbez-Gindre F, Kantelip B (2010) Prognostic value of phosphorylated STAT3 in advanced rectal cancer: a study from 104 French patients included in the EORTC 22921 trial. J Clin Pathol 63:873–878

Morin PJ, Sparks AB, Korinek V, Barker N, Clevers H, Vogelstein B, Kinzler KW (1997) Activation of beta-catenin-Tcf signaling in colon cancer by mutations in beta-catenin or APC. Science (New York) 275:1787–1790

Murray PJ (2007) The JAK-STAT signaling pathway: input and output integration. J Immunol 178:2623–2629

Musteanu M, Blaas L, Mair M, Schlederer M, Bilban M, Tauber S, Esterbauer H, Mueller M, Casanova E, Kenner L et al (2010) Stat3 is a negative regulator of intestinal tumor progression in Apc(Min) mice. Gastroenterology 138:1003–1011, e1001–1005

O'Sullivan LA, Liongue C, Lewis RS, Stephenson SE, Ward AC (2007) Cytokine receptor signaling through the Jak-Stat-Socs pathway in disease. Mol Immunol 44:2497–2506

Poole AJ, Heap D, Carroll RE, Tyner AL (2004) Tumor suppressor functions for the Cdk inhibitor p21 in the mouse colon. Oncogene 23:8128–8134

Regis G, Pensa S, Boselli D, Novelli F, Poli V (2008) Ups and downs: the STAT1:STAT3 seesaw of Interferon and gp130 receptor signalling. Semin Cell Dev Biol 19:351–359

Rigby RJ, Simmons JG, Greenhalgh CJ, Alexander WS, Lund PK (2007) Suppressor of cytokine signaling 3 (SOCS3) limits damage-induced crypt hyper proliferation and inflammation-associated tumorigenesis in the colon. Oncogene 26:4833–4841

Rivat C, Rodrigues S, Bruyneel E, Pietu G, Robert A, Redeuilh G, Bracke M, Gespach C, Attoub S (2005) Implication of STAT3 signaling in human colonic cancer cells during intestinal trefoil factor 3 (TFF3) – and vascular endothelial growth factor-mediated cellular invasion and tumor growth. Cancer Res 65:195–202

Rosenberg DW, Giardina C, Tanaka T (2009) Mouse models for the study of colon carcinogenesis. Carcinogenesis 30:183–196

Rustgi AK (2007) The genetics of hereditary colon cancer. Genes Dev 21:2525–2538

Sansom OJ, Reed KR, Hayes AJ, Ireland H, Brinkmann H, Newton IP, Batlle E, Simon-Assmann P, Clevers H, Nathke IS et al (2004) Loss of Apc in vivo immediately perturbs Wnt signaling, differentiation, and migration. Genes Dev 18:1385–1390

Sato T, van Es JH, Snippert HJ, Stange DE, Vries RG, van den Born M, Barker N, Shroyer NF, van de Wetering M, Clevers H (2011) Paneth cells constitute the niche for Lgr5 stem cells in intestinal crypts. Nature 469:415–418

Schreiber S, Rosenstiel P, Hampe J, Nikolaus S, Groessner B, Schottelius A, Kuhbacher T, Hamling J, Folsch UR, Seegert D (2002) Activation of signal transducer and activator of transcription (STAT) 1 in human chronic inflammatory bowel disease. Gut 51:379–385

Shankaran V, Ikeda H, Bruce AT, White JM, Swanson PE, Old LJ, Schreiber RD (2001) IFNgamma and lymphocytes prevent primary tumour development and shape tumour immunogenicity. Nature 410:1107–1111

Simpson JA, Al-Attar A, Watson NF, Scholefield JH, Ilyas M, Durrant LG (2010) Intratumoral T cell infiltration, MHC class I and STAT1 as biomarkers of good prognosis in colorectal cancer. Gut 59:926–933

Takeda K, Clausen BE, Kaisho T, Tsujimura T, Terada N, Forster I, Akira S (1999) Enhanced Th1 activity and development of chronic enterocolitis in mice devoid of Stat3 in macrophages and neutrophils. Immunity 10:39–49

Terzic J, Grivennikov S, Karin E, Karin M (2010) Inflammation and colon cancer. Gastroenterology 138:2101–2114, e2105

Tsareva SA, Moriggl R, Corvinus FM, Wiederanders B, Schutz A, Kovacic B, Friedrich K (2007) Signal transducer and activator of transcription 3 activation promotes invasive growth of colon carcinomas through matrix metalloproteinase induction. Neoplasia 9:279–291

Uronis JM, Threadgill DW (2009) Murine models of colorectal cancer. Mamm Genome 20:261–268

Yang W, Velcich A, Mariadason J, Nicholas C, Corner G, Houston M, Edelmann W, Kucherlapati R, Holt PR, Augenlicht LH (2001) p21(WAF1/cip1) is an important determinant of intestinal cell response to sulindac in vitro and in vivo. Cancer Res 61:6297–6302

Yang J, Liao X, Agarwal MK, Barnes L, Auron PE, Stark GR (2007) Unphosphorylated STAT3 accumulates in response to IL-6 and activates transcription by binding to NFkappaB. Genes Dev 21:1396–1408

Yu H, Kortylewski M, Pardoll D (2007) Crosstalk between cancer and immune cells: role of STAT3 in the tumour microenvironment. Nat Rev 7:41–51

Yu H, Pardoll D, Jove R (2009) STATs in cancer inflammation and immunity: a leading role for STAT3. Nat Rev Cancer 9:798–809

Universal and Specific Functions of STAT3 in Solid Tumours

Lidia Avalle*, Gabriella Regis*, and Valeria Poli

Abstract

STAT3 is constitutively activated in a high percentage of tumours and tumour-derived cells of both liquid and solid origin, often correlating with aggressive disease and bad prognosis. Persistent STAT3 activity, to which tumours often become addicted, is mostly due to the aberrant activation of pro-oncogenic/pro-inflammatory signals that can trigger its phosphorylation, such as oncogenes, growth factor receptors and cytokines. Among STAT3-mediated functions are increased survival and proliferation, enhanced angiogenesis, motility and invasion, and down-modulation of anti-tumour immune responses. Moreover, STAT3 was recently shown to play unexpected roles in regulating cell metabolism and mitochondrial activity via both transcriptional and non-transcriptional mechanisms. Here, we review the main knowledge about the role of STAT3 in solid tumours, with a particular focus on breast cancer and our recent work with mouse models.

Introduction

Signal Transducers and Activators of Transcription (STAT) factors mediate the signalling downstream of cytokine and growth factor receptors, and often their activity is deregulated in cancer (Turkson and Jove 2000; Siddiquee et al. 2007). Once activated by tyrosine-phosphorylation via receptor-associated JAK kinases,

*These two authors equally contributed to this work.

L. Avalle • G. Regis • V. Poli (✉)
Molecular Biotechnology Center (MBC) and Department of Genetics, Biology and Biochemistry, University of Turin, Via Nizza 52, Turin 10126, Italy
e-mail: valeria.poli@unito.it

Th. Decker and M. Müller (eds.), *Jak-Stat Signaling: From Basics to Disease*,
DOI 10.1007/978-3-7091-0891-8_17, © Springer-Verlag Wien 2012

STATs form parallel dimers that concentrate into the nucleus regulating the expression of target genes (Schindler et al. 2007). The family member STAT3 can be activated by a wide variety of cytokines and growth factors (e.g. IL-6 family, leptin, IL-12, IL-17, IL-10, Interferons, G-CSF, EGF, PDGF) and by a number of oncogenes such as Src, Abl, Sis, Fps, Ros, Met and ErbB2 (Turkson and Jove 2000). Accordingly, STAT3 is found to be constitutively tyrosine-phosphorylated in a high percentage of tumours and tumour-derived cell lines of both liquid and solid origin, which often become addicted to its activity for continuous survival and growth (Yu et al. 1995; Turkson and Jove 2000; Kortylewski et al. 2005; Siddiquee et al. 2007), and is considered a good target for anti-cancer therapy. Indeed, STAT3 tyrosine phosphorylation and consequent transcriptional activation was shown to be required for cell transformation downstream of several oncogenes, the prototype being v-Src (Yu et al. 1995; Bromberg et al. 1998; Silva 2004). Although a unique core activity determining addiction to STAT3 by a wide spectrum of biologically distinct tumors has still not been identified, STAT3-mediated gene expression signature is mostly consistent with tumour cell survival and proliferation (Pensa 2008; Yu et al. 2009). In addition, STAT3 constitutive activity in tumour cells can down-modulate anti-tumour immune responses (Yu et al. 2009) as well as promote tumour angiogenesis (Niu et al. 2002a). STAT3 can also regulate cell movement, contributing to cytoskeleton reorganization and controlling cell adhesion properties, and is thought to play a role in tumour invasion and metastasis by inducing the expression of matrix metalloproteinases (MMP) and promoting the epithelial to mesenchymal transition (EMT) (Pensa 2008; Yu et al. 2009). Finally, recent work has shown that STAT3 acts as an important regulator of cell metabolism, promoting aerobic glycolysis and downregulating mitochondrial activity via its canonical, nuclear functions (Demaria et al. 2010) while preserving mitochondrial respiratory activity via mitochondrial localization of its serine-phosphorylated form (Gough et al. 2009; Wegrzyn et al. 2009). Both activities contribute to tumor transformation downstream of distinct signals, which promote STAT3 phosphorylation on either tyrosine or serine. The pro-oncogenic role of STAT3 was first directly demonstrated in vitro by the finding that overexpression of the constitutively active mutant form STAT3C can transform fibroblasts and epithelial cells (Bromberg et al. 1999; Dechow et al. 2004) followed by in vivo experiments in transgenic or knock-in mice demonstrating oncogenic potential in the lung, skin and breast (Li et al. 2007; Chan et al. 2008; Barbieri et al. 2010a).

Many aspects of STAT3 biology in tumours have been extensively reviewed. Here, we chose to summarize the main knowledge about the role of STAT3 in solid tumours, with a particular focus on breast cancer and our recent work with mouse models (Barbieri et al. 2010a, b; Demaria et al. 2010).

The solid tumours where STAT3 has been found to be constitutively active, as well as its main functions and target genes are sketched in Table 1 and detailed below.

Table 1 How activated STAT3 affects different aspects of tumorigenesis. The table highlights the experimental findings concerning the role of activated STAT3 in prognosis, proliferation/survival, metastasis/angiogenesis, drug resistance, immune response, cancer stem cells in different solid tumours, indicating in which model system the data were obtained and the STAT3 target genes identified in each system

Tumor	Role in tumorigenesis[a]	Model[b]	Target genes	References
Epidermal non-melanoma tumours	Proliferation/survival	Mouse	Ciclin D1, Bcl-xL, c-Myc	Chan et al. (2004a, b, 2008)
	Metastasis/angiogenesis	Cell, clinic.	bFGF	Jee et al. (2004), Suiqing et al. (2005)
Melanoma	Correlation with prognosis not evaluated (90%)	Clinic.		Niu et al. (2002a)
	Proliferation/survival	Cell	Bcl-xL, Mcl1	Niu et al. (2002a)
	Metastasis/angiogenesis	Cell	MMP2, VEGF, bFGF	Niu et al. (2002a), Xie et al. (2004, 2006)
	Immune response	Cell		Wang et al. (2004)
Head and neck cancer	Poor prognosis/highest expression in early stages (40–80%)	Clinic.		Masuda et al. (2002b), Nagpal et al. (2002)
	Proliferation/survival	Cell	Cyclin D1, Bcl-xL, Bcl2	Masuda et al. (2002a, b, 2007)
	Metastasis/angiogenesis	Cell	VEGF	Masuda et al. (2002a, 2007), Lui et al. (2009)
	Drug resistance	Cell		Masuda et al. (2002c), Gu et al. (2010)
	Immune response	Cell		Albesiano et al. (2010)
	Cancer stem cells	Clinic.		Chen et al. (2008b)
Colorectal carcinoma	Poor prognosis, metastasis (50%)	Clinic.		Kusaba et al. (2005, 2006)
	Proliferation/survival	Cell, clinic.	Bcl-xL, survivin, miR-21, miR-181b1	Turkson et al. (2004), Lassmann et al. (2007), Iliopoulos et al. (2010)
	Metastasis	Cell	MMP1, MMP3	Tsareva et al. (2007)
	Immune response	Cell		Nefedova et al. (2004), Wang et al. (2004)
Hepatocellular carcinoma	Poor prognosis (60–75%)	Clinic.		Feng et al. (2001), Zhang et al. (2010)
	Proliferation/survival	Mouse	Bcl-xL, cyclin D1	Yoshida et al. (2002)
	Metastasis/angiogenesis	Cell	VEGF, MMP2	Li et al. (2006)

(continued)

Table 1 (continued)

Tumor	Role in tumorigenesis[a]	Model[b]	Target genes	References
Pancreatic cancer	Proliferation/survival	Cell		Wei et al. (2003)
	Metastasis/angiogenesis	Cell	VEGF	Wei et al. (2003)
	Immune response	Cell		Bharadwaj et al. (2007)
Gastric carcinoma	Poor prognosis (30%)	Clinic.		Lee et al. (2009)
	Proliferation/survival	Cell, clinic.	Survivin, Bcl2	Kanda et al. (2004), Choi et al. (2006)
	Angiogenesis	Clinic.	VEGF	Choi et al. (2006)
Ovarian cancer	Poor prognosis (90%)	Clinic.		Rosen et al. (2006)
	Proliferation/survival	Cell	Cyclin D1, Bcl-xL	Huang et al. (2000)
	Metastasis/angiogenesis	Cell		Silver et al. (2004), Nilsson et al. (2005)
	Drug resistance	Cell	Bcl-xL	Burke et al. (2001)
Prostate cancer	Highest malignancy (80%)	Clinic.		Mora et al. (2002)
	Proliferation/survival	Cell	Bcl2, cyclin D1, c-Myc, Bcl-xL, Mcl1	Lou et al. (2000), Gao et al. (2001, 2005), Flowers et al. (2005), Hellsten et al. (2008)
	Metastasis/angiogenesis	Cell	HIF1, VEGF, E-cadherin	Xu et al. (2005), Azare et al. (2007), Abdulghani et al. (2008)
	Drug resistance	Cell	Bcl-xL (partially)	Pu et al. (2004)
	Cancer stem cells	Cell		Mathews et al. (2010)
Lung carcinoma	Poor prognosis but not significant (70%)	Clinic.		van Cruijsen et al. (2009)
	Proliferation/survival	Cell	Bcl2, Bcl-xL, Mcl1, survivin, cyclin D1, c-Myc	Weerasinghe et al. (2007)
	Metastasis/angiogenesis	Cell	VEGF	Weerasinghe et al. (2007), Pfeiffer et al. (2009)
	Immune response	Mouse	Ccl5, IL-6, VEGF	Li et al. (2007)
Glioblastoma	Correlation with prognosis not evaluated (95%)	Clinic.		Rahaman et al. (2002)
	Proliferation/survival	Cell, clinic.	Bcl-xL, Mcl1, Bcl2	Rahaman et al. (2002), Sherry et al. (2009)
	Metastasis/angiogenesis	Cell	MMP9; VEGF	Loeffler et al. (2005), Liu et al. (2010)
	Immune response	Cell		Hussain et al. (2007)
	Cancer stem cells	Cell		Sherry et al. (2009), Villalva et al. (2010)

Rabdo-myosarcoma	Proliferation/survival	Mouse, cell	Bcl-xL	Lee et al. (2006), Chen et al. (2007b)
	Immune response	Cell		Nabarro et al. (2005)
Renal carcinoma	Poor prognosis, metastasis (100%)	Clinic.		Horiguchi et al. (2002b)
	Proliferation/survival	Cell	Survivin, Bcl-xL, Bcl2, cyclin E; VEGF	Horiguchi et al. (2002a), Xin et al. (2009)
	Angiogenesis	Cell	VEGF	Jung et al. (2007), Xin et al. (2009)
	Immune response	Cell		Xin et al. (2009)
Cervical cancer	Poor prognosis (60%)	Clinic.		Takemoto et al. (2009)
	Proliferation/survival	Clinic.	Bcl-xL, survivin, Mcl1	Chen et al. (2007a)
	Angiogenesis	Clinic.	VEGF	Wei et al. (2003)
Bladder cancer	Proliferation/survival	Cell	Bcl2, Bcl-xL, survivin, cyclin D1	Chen et al. (2008a)
	Metastasis	Cell	MMP1	Itoh et al. (2006)
Breast cancer	Better prognosis of node-negative breast cancer; metastasis in regional lymph node; poor response to therapy (30–60%)	Clinic.		Dolled-Filhart et al. (2003), Dechow et al. (2004), Hsieh et al. (2005), Diaz et al. (2006)
	Proliferation/survival	Cell	TIMP3	Garcia et al. (2001), Selander et al. (2004)
	Metastasis/angiogenesis	Cell	VEGF	Niu et al. (2002a), Selander et al. (2004)
	Drug resistance	Cell	Bcl2	Real et al. (2002)
	Immune response	Cell	SOCS3	Sun et al. (2006)
	Cancer stem cells	Cell		Zhou et al. (2007)

[a]The percentage refers to the clinical samples where activated STAT3 was detected in the cited reference/s
[b]Model: *mouse* mouse model, *cell* cell lines, *clinic.* clinical samples

Epidermal Non-Melanoma Tumours

STAT3 constitutive activation was observed in several types of human epidermal non melanoma cutaneous tumours, correlating with poor differentiation, tumour invasion and metastasis in clinical samples of cutaneous squamous cell carcinoma (SCC) (Suiqing et al. 2005). STAT3 involvement with metastatic potential was also confirmed in xenograft experiments of basal cell carcinoma cells (BCC) over-expressing IL-6, where IL-6-mediated angiogenesis supported tumour development in part via STAT3 activation (Jee et al. 2004).

The two-stage chemical carcinogenesis model is considered a good model of epithelial carcinogenesis, recapitulating the different phases from tumor initiation to progression (Chan et al. 2004a). The first experimental evidence that STAT3 activation is required for epithelial tumorigenesis in vivo was obtained using this model, where STAT3 ablation in keratinocytes completely abrogated skin tumour development (Chan et al. 2004b). STAT3-deficient keratinocytes were more sensitive to DMBA-induced apoptosis and STAT3 inhibition with an oligonucleotide decoy injected into primary skin papillomas led to significant reduction of tumour volume. STAT3 was also implicated in ultraviolet B (UVB)-induced skin carcinogenesis by the observation that UVB irradiation promoted proliferation and survival of keratinocytes via STAT3 activation (Sano et al. 2005). Conversely, transgenic mice overexpressing the constitutively active form STAT3C in keratinocytes developed skin tumours with a shorter latency and in greater number compared to non-transgenic mice (Kim et al. 2007; Chan et al. 2008). STAT3C acted in both tumour initiation and promotion by upregulating genes involved in survival, proliferation, angiogenesis and metastasis.

One key step in STAT3-mediated epithelial tumorigenesis may be the induction of Bcl-xL, which plays a fundamental role in early skin carcinogenesis by enhancing survival of keratinocyte stem cells in the bulge region of the hair follicle, where mutations are believed to arise during the initiation stage and whose clonal expansion occurs during tumour promotion, as shown by studies in Bcl-xL deficient mice (Kim et al. 2009).

Melanoma

STAT3 is constitutively active in the vast majority of melanoma tumours and cell lines (Niu et al. 2002a), most often downstream of activated c-Src (Kortylewski et al. 2005), where it favours proliferation, escape from apoptosis and angiogenesis via induction of Bcl-xL, Mcl1 and VEGF (Niu et al. 2002a; Xie et al. 2006). STAT3 activation is apparently crucial to promote the metastatic process in melanoma, through direct upregulation of MMP2, VEGF and bFGF expression (Xie et al. 2004, 2006). STAT3 activation was shown to promote an immunosuppressive environment leading to impaired dendritic cell (DC) maturation and tumour-specific T cell response in melanoma B16 cells (Wang et al. 2004). Accordingly WP1066, a STAT3 inhibitor that blocks melanoma cells growth, was found to interfere with melanoma brain metastasis by inhibiting the production of immunosuppressive cytokines such as TGF-b, MCP1, RANTES and VEGF by tumour cells, thus enhancing cytotoxic T lymphocyte responses and inhibiting T regulatory (Treg) cells differentiation (Kong et al. 2008).

Head and Neck Cancer

STAT3 constitutive activation has been observed in 40–80% of human head and neck squamous cell cancer (HNSCC), correlating with poor prognosis and with proliferation and apoptosis resistance via induction of cyclin D1, Bcl2 and Bcl-xL expression (Masuda et al. 2002b; Nagpal et al. 2002). Apoptosis could be reinstated and tumour growth blocked by inhibiting STAT3 activity (Leong et al. 2003; Jing et al. 2006). Therapeutic blockade of STAT3 activity also resulted in impaired angiogenesis due to direct STAT3-mediated regulation of VEGF expression (Masuda et al. 2007). STAT3 activation occurring upon EBV infection or deregulated EGFR signalling in nasopharingeal cancer promotes anchorage-independent growth and invasion (Lui et al. 2009; Wheeler et al. 2010). High STAT3 phosphorylation levels in cells from HNSCC patients are associated with the expression of CD44 and aldehyde dehydrogenase 1 (ALDH1) stem cell markers and with typical features of Cancer Stem Cells (CSCs) such as high tumorigenicity, radioresistance, expression of the stemness markers Bmi, Oct4 and Nanog and of the EMT genes Snail and Twist (Chen et al. 2008b). STAT3 inhibition in these cells reinstated responsiveness to chemotherapy, favoured differentiation and impaired tumorigenesis and metastasis formation (Chen et al. 2008b). Moreover, anti-tumour immune responses were affected, with enhanced production of proinflammatory cytokines and chemokines which, in turn, triggered DC activation and lymphocytes migration and prompted anti-tumour immune response (Albesiano et al. 2010).

Colorectal Carcinoma

STAT3 constitutive activity was observed in colorectal carcinoma cells in about 50% of clinical samples (Kusaba et al. 2005, 2006), correlating with proliferation and tumour growth rate (Becker et al. 2005; Corvinus et al. 2005), and with tumour invasion, lymph node metastasis and poor prognosis (Kusaba et al. 2005, 2006). STAT3-related enhanced invasiveness correlates with strong expression of the MMP1, -3, -7, and -9. MMP1 and MMP3 are direct STAT3 targets, and activated STAT3 colocalizes with MMP1 in tumour specimens (Tsareva et al. 2007). Importantly, both NF-kB and STAT3 activation have been shown to be crucial for inflammation-driven colon carcinogenesis (Greten et al. 2004; Bollrath et al. 2009; Grivennikov et al. 2009). NF-kB activation in myeloid cells drives IL-6 expression, whose levels are indeed increased in patients serum, and in turn IL-6 is responsible for STAT3 constitutive activation in colon tumour cells, a paradigm that is thought to hold true for several other inflammation-related tumours (Grivennikov and Karin 2010).

Mice lacking STAT3 in intestinal epithelial cells (IECs) showed an almost complete protection from the development of the AOM/DSS model of colitis-associated cancer (CAC), correlating with decreased epithelial proliferation and enhanced sensitivity to treatment-induced apoptosis (Bollrath et al. 2009; Grivennikov et al. 2009). However, the role of STAT3 in colon tumorigenesis appears to be context-dependent, as we have recently shown in Apc^{Min} mice that this factor promotes early tumorigenesis steps but impairs tumour progression at later stages, via regulation of the adhesion molecule CEACAM1 (Musteanu et al. 2010).

STAT3-mediated immune suppression was shown to play an important role in colon cancer cells (Nefedova et al. 2004; Wang et al. 2004), where tumour-derived factors, inducing STAT3 activation in infiltrating immature myeloid cells, prevented their differentiation into mature dendritic cells (Nefedova et al. 2004).

Other Tumours of the Gastrointestinal Tract

STAT3 constitutive activation plays a role in several other tumours that develop in the gastrointestinal tract. In hepatocellular carcinoma (HCC) its activity is often induced by HCV infection or IL-6 and other inflammatory cytokines and drives tumorigenesis by promoting proliferation, survival (via Bcl-xL and cyclin D1 induction) and anchorage-independent growth (Yoshida et al. 2002; He et al. 2010). STAT3 inhibition in HCC cells impaired growth, angiogenesis and metastasis while enhancing apoptosis and sensitivity to chemotherapy (Li et al. 2006; Choudhari et al. 2007; Sun et al. 2008).

Constitutively activated STAT3 is widely observed also in pancreatic cancer, often downstream of IL-6 or G-CSF, promoting tumour cell growth, metastasis and angiogenesis (Wei et al. 2003) and impairing dendritic cells differentiation and activation (Bharadwaj et al. 2007).

STAT3 phosphorylation is relatively infrequent in gastric carcinoma, but when present it correlates with tumour cell proliferation, survival and angiogenesis via induction of Bcl2, VEGF and survivin expression (Kanda et al. 2004; Choi et al. 2006). The analysis of human gastric cancer specimens identified correlations between STAT3 activation and lymph node metastasis (Deng et al. 2010). Moreover, STAT3 uncontrolled activity can lead to gastric cancer, as shown by the spontaneous development of gastric carcinomas following disruption of the integrity of mucosal epithelium in gp130 knock-in mutant mice ($gp130^{757F}$) that are unable to respond to SOCS3-mediated negative feedback (Tebbutt et al. 2002).

Ovarian Cancer

Constitutive activation of STAT3 was detected in a high percentage of ovarian cancer cell lines and human tumour specimens (94%) with respect to normal ovary epithelium, correlating with aggressive clinical behavior and tumour progression, and it was shown to enhance proliferation and inhibition of apoptosis through induction of cyclin D1 and Bcl-xL expression, respectively (Huang et al. 2000; Rosen et al. 2006). Moreover, studies in cell lines have shown that enhanced

STAT3 activity and expression contribute to resistance to apoptosis in response to chemotherapeutic drugs (Burke et al. 2001; Duan et al. 2006). STAT3 inhibition leads to decreased Bcl-xL expression, sensitizing tumour cells to chemotherapy-induced apoptosis (Burke et al. 2001). One of the main signals responsible for STAT3 constitutive activation in ovarian cancer is its canonical activator IL-6, which promotes angiogenesis, leading to tumour proliferation and dissemination of malignant cells (Nilsson et al. 2005).

Prostate Cancer

Phosphorylated STAT3 was detected in the majority of human prostate cancers, correlating with the degree of malignancy (Mora et al. 2002) and with JAK2/IL-6 signalling, which enhances proliferation and survival (Lou et al. 2000; Flowers et al. 2005). STAT3 inhibition lead to apoptosis of tumour cell lines, both in vitro and in vivo, through downregulation of Bcl2, cyclin D1, c-Myc, Bcl-xL and Mcl1 expression (Lou et al. 2000; Jing et al. 2004; Turkson et al. 2004; Gao et al. 2005; Hellsten et al. 2008), and inhibited angiogenesis and tumour growth via downregulation of both HIF1a and VEGF expression (Xu et al. 2005). The expression of the constitutively active form STAT3C in immortalized prostate epithelial cells caused tumour transformation and enhanced cell motility by decreasing E-cadherin level and increasing the number of lamellipodia and stress fibers (Azare et al. 2007), suggesting a role in EMT and metastasis that was subsequently confirmed by the observation of constitutive STAT3 activation in clinical samples of prostate cancer metastasis, where it promoted cell motility by reorganizing the actin and microtubule network (Abdulghani et al. 2008). IL-6 dependent STAT3 activation was shown to contribute to resistance of human prostate cancer cells to chemotherapy (Pu et al. 2004). There appear to be correlations between STAT3 activation, tumour invasion and CSCs. In particular, invasive prostate cancer cells were shown to display promoter methylation patterns reminiscent of those observed in CSCs, with many differentially methylated genes belonging to the IL-6/STAT3 pathway (Mathews et al. 2010). Additionally, STAT3 was shown to interact with the CSC marker SOX1, whose silencing decreased STAT3 activation and in vitro invasiveness (Mathews et al. 2010). Accordingly, IL-6 was recently shown to induce the conversion of prostate non stem cancer cells (NSCCs) into sphere-forming CSCs, similar to what observed in breast cancer cells (Iliopoulos et al. 2011a).

Lung Carcinoma

About 50–70% of human non-small cell lung carcinomas (NSCLC) and cell lines were shown to display constitutive STAT3 activation, correlating with enhanced proliferation and survival (Song et al. 2003; Haura et al. 2005; van Cruijsen et al. 2009). STAT3 inhibition in these cells lead to decreased expression of a number of known STAT3 targets (e.g. Bcl2, Bcl-xL, Mcl1, survivin, VEGF, cyclin D1 and c-Myc), thereby promoting apoptosis, impairing proliferation and reducing angiogenesis (Weerasinghe et al. 2007). Interestingly, mutant EGFR forms in primary human lung adenocarcinomas lead to STAT3 activation via IL-6 upregulation (Gao et al. 2007). In contrast with these data, Pfeiffer and colleagues demonstrated that STAT3 constitutive activation is characteristic of primary tumour samples from patients with small cell lung cancer (SCLC) but not from NSCLC, and that blocking STAT3 activation impaired anchorage-independent tumour cell growth, suggesting the implication of STAT3 in the rapid metastasizing phenotype of SCLC (Pfeiffer et al. 2009).

The lung was the first tissue where over-expressed constitutively active STAT3 was shown to play an autonomous pro-oncogenic role, since transgenic expression of the STAT3C mutant form in alveolar type II epithelial cells induced lung bronchoalveolar adenocarcinomas preceded by remarkable infiltration of inflammatory cells (Li et al. 2007). Tumour development correlated with enhanced secretion of pro-inflammatory molecules and with reactivation of genes critical for epithelial cell growth during embryonic lung development, similar to what observed in human bronchoalveolar adenocarcinomas (Li et al. 2007). Accordingly, STAT3 downstream genes were proposed to serve as biomarkers in human lung adenocarcinoma and chronic obstructive pulmonary disease, which are both induced by chronic inflammation of the lung (Qu et al. 2009).

Glioblastoma

High levels of STAT3 activation are also detected in about 95% of glioblastoma cell lines and tumour samples, inducing proliferation and apoptosis resistance through upregulation of Bcl-xL, Mcl1 and Bcl2 expression (Rahaman et al. 2002), and promoting angiogenesis, invasion and metastasis via upregulation of VEGF and MMP9 expression (Loeffler et al. 2005; Liu et al. 2010). Hypoxia resistance is a common feature of both stem cells and CSCs, which are thought to act as tumour initiating cells (TICs) in different types of tumours, including glioblastoma (Hemmati et al. 2003; Zhou and Zhang 2008). Resistance of these cells to chemotherapy is often responsible for relapses and/or metastasis (Villalva et al. 2010). The highly hypoxic glioblastoma microenvironment triggers STAT3-mediated induction of VEGF, HIF1, MMP2 and Twist1, which in turn promote angiogenesis and tumour invasion (Kang et al. 2010). Interestingly, STAT3 activation was shown to be essential for glioblastoma stem cells proliferation and ability to form neurospheres, and inhibition of its activity triggered the downregulation of genes associated with the stem cell phenotype (Sherry et al. 2009) and sensitization to chemotherapeutic treatment, suggesting that combined chemotherapy and STAT3 inhibition may allow more efficient killing of CSCs. STAT3 activation in glioblastoma is often supported by the constitutive expression of IL-6 in tumour cells, and indeed $IL6^{-/-}$ mice were protected from glioblastoma development (Brantley and Benveniste 2008). Abnormal activation of the FGFR and EGFR pathways also correlated with STAT3 phosphorylation (Brantley and Benveniste 2008). Interestingly, however, while STAT3 could cooperate with the oncogenic mutant form EGFRvIII to mediate cell transformation, it accelerated disease progression in glioblastomas induced by PTEN-loss. Thus, depending on the genetic background, STAT3 activity in glioblastoma can be either tumour-suppressive or tumour-promoting (de la Iglesia et al. 2008).

Other Solid Tumours

A low percentage of rhabdomyosarcomas showed STAT3 activation that is linked to enhanced proliferation and resistance to apoptosis (Chen et al. 2007b) and correlating with the overexpression of the stem cell marker Piwil2, recently found associated to different tumours. Piwil2 can activate STAT3, which in turn enhances tumour cell survival through Bcl-xL induction (Lee et al. 2006). Moreover, STAT3 can interact with PAX3-FKHR, an oncogenic fusion protein specifically associated with an aggressive rhabdomyosarcoma metastatic subtype. This association leads to a reduction in tumour MHC expression and to an altered cytokine microenvironment that inhibits inflammatory cells action and hampers immune detection of tumour (Nabarro et al. 2005).

STAT3 activation was observed in 100% of renal carcinomas, correlating with poor prognosis and metastatic disease and promoting proliferation and survival (Horiguchi et al. 2002a, b). Pharmacological inhibition of STAT3 not only favoured the apoptotic action of chemotherapeutic agents on tumour cells, but also downmodulated their angiogenic and metastatic potential while improving antitumour immune response by reducing myeloid suppressor and Tregs cells (Xin et al. 2009).

About 60% of cervical cancers display STAT3 phosphorylation, correlating with poor prognosis (Takemoto et al. 2009) and linked to increased proliferation and apoptosis resistance via induced expression of Bcl2, survivin, Mcl1 (Chen et al. 2007a) and enhanced angiogenesis mediated by VEGF (Wei et al. 2003).

Finally STAT3 activation in bladder cancer cells, although limited, was implicated in tumour cells proliferation and invasion (Itoh et al. 2006; Chen et al. 2008a).

Breast Cancer

Persistently phosphorylated STAT3 is detected in 30–60% of primary breast carcinomas (Garcia et al. 2001) correlating with poor response to therapy (Diaz et al. 2006) and with regional lymph node metastasis (Hsieh et al. 2005), although a correlation with a good prognosis of node-negative cancers was suggested (Dolled-Filhart et al. 2003). High STAT3 phosphorylation levels are detected in several human breast cancer cell lines, where its inactivation leads to growth arrest and cell death (Garcia et al. 1997, 2001). Similar to most other solid tumours, STAT3 activity in breast cancer has been linked to enhanced proliferation and survival, to resistance to apoptosis and to cell movement, invasion and metastasis.

Pathways Leading to Persistent STAT3 Activation in Breast Cancer

Despite the wide range of tumours where STAT3 is constitutively active, so far no activating genetic mutations have been described, suggesting that abnormal STAT3 activity in neoplastic cells must be triggered by deregulated upstream signalling. In breast cancer, STAT3 activation shows positive correlation with EGF and ErbB2 receptors overexpression and with Src activation (Berclaz et al. 2001; Diaz et al. 2006; Leslie et al. 2006), all of which have been shown to lead to STAT3 phosphorylation, albeit not directly (Berishaj et al. 2007). V-Src was the first oncogene whose transforming activities were shown to require STAT3 (Bromberg et al. 1998). Additionally, STAT3 was reported to be a substrate of the breast tumour kinase (Brk), distantly related to the Src family (Liu et al. 2006).

An impressive body of data points towards IL-6 as the main trigger for STAT3 aberrant activation in solid tumours, which at hindsight is perhaps not surprising since IL-6 and its family of related cytokines are among the most prominent inducers of STAT3 activity. In breast cancer patients, serum IL-6 levels are elevated (Jiang et al. 2000; Kozłowski et al. 2003), and correlate with advanced tumour stage (Kozłowski et al. 2003), increased number of metastatic sites (Salgado et al. 2003) and overall poor prognosis (Zhang and Adachi 1999) (Bachelot et al. 2003; Salgado et al. 2003). High local IL-6 production is also detected, correlating with tumour grade (Chavey et al. 2007). Indeed, inflammation-induced IL-6 produced either systemically or locally by tumour infiltrating inflammatory cells is believed to start a positive loop by activating STAT3 in cancer cells (Grivennikov and Karin 2010). This in turn induces the secretion of soluble factors promoting STAT3 activation and anergy in the antigen presenting cells, finally leading to enhanced tumour cell survival and growth both via cell autonomous and immune-mediated mechanisms (Yu et al. 2009). An oncogene-driven inflammatory loop was also implicated in the initial stages of tumour transformation. Indeed, transient Src activation generates an inflammatory signal which triggers an epigenetic switch to cancer cells via a positive feedback loop involving NF-kB, Lin28, let-7 and IL-6 (Iliopoulos et al. 2009). IL-6-activated STAT3 is essential for this switch via direct induction of miR-21 and miR-181b-1, which target the PTEN and CYLD tumour suppressor genes, respectively. Their downregulation in turn leads to NF-kB activation, required to maintain the transformed state (Iliopoulos et al. 2010). The importance of this circuit was first demonstrated in transformation of mammary epithelial cells and subsequently confirmed in prostate, colon, lung and hepatocellular carcinoma cells.

IL-6-induced STAT3 activation is normally transient, due to tight negative feedback control such as that mediated by SOCS3 (Yoshimura 2005). What are the mechanisms helping to maintain persistent STAT3 phosphorylation in tumours?

Loss of negative feedback via silencing of SOCS factors has been shown to occur in several systems (Baltayiannis et al. 2008). Recently, it was shown that the low but constitutive activation of STAT3 in different tumours, including the breast, is at least partly mediated by the elevated expression of S1PR1, the receptor for the lysophospholipid sphingosine-1-phospate. S1PR1 is a STAT3 transcriptional target which in turn upregulates IL-6 expression and enhances STAT3 activation, establishing a positive feedback loop resulting in STAT3 persistent activation in both the tumour cells and the tumour microenvironment, accelerated tumour growth and malignant progression (Lee et al. 2010).

Other cytokines belonging to the IL-6 family, such as LIF (Quaglino et al. 2007) and leptin (Park et al. 2010), are also elevated in breast tumours, driving STAT3 activation. In particular, adipocyte-derived leptin is present at high concentrations within the mammary gland of obese individuals, is considered as a risk factor in several types of cancers and is proposed to correlate with breast cancer progression (Garofalo et al. 2004). Estrogen receptor alpha was shown to enhance leptin-mediated STAT3 activation (Binai et al. 2010), and inactivation of the peripheral leptin receptor attenuates tumour progression and metastasis in an MMTV-PyMT model of breast cancer, via inactivation of the ERK1/2 and Jak2/STAT3 pathways (Park et al. 2010).

STAT3-Mediated Features: Proliferation and Survival

Most cell lines displaying persistent STAT3 phosphorylation are addicted to its activity for proliferation and survival, both in vitro and in vivo (Garcia et al. 2001; Hsieh et al. 2005; Diaz et al. 2006), at least partly correlating with the induction of the anti-apoptotic genes survivin/BIRC5 and Bcl-xL and of cyclin D1 (Siddiquee et al. 2007). Indeed, high levels of activated STAT3 correlate positively with elevated cyclin D1 mRNA and protein expression in breast tumours and cell lines (Leslie et al. 2006) and STAT3 can directly bind to the promoter of the human cyclin D1 gene (Leslie et al. 2006; Saxena et al. 2007). Moreover, cyclin D1 appears to be required for mouse fibroblasts anchorage-independent growth downstream of constitutively active STAT3C or v-Src (Leslie et al. 2006). Interestingly, the progesteron receptor was shown to act as STAT3 coactivator by inducing ErbB2 nuclear translocation and the assembly of a transcriptional complex on the cyclin D1 promoter (Béguelin et al. 2010).

Immunohistochemical analyses of invasive breast carcinomas also showed a positive correlation between activated Src, phosphorylated STAT3 and the expression of survivin, a member of the inhibitor of apoptosis protein family (Diaz et al. 2006). Like cyclin D1, also survivin is a direct STAT3 transcriptional target, and STAT3 silencing leads to survivin downregulation and apoptotic death in a human breast cancer cell line (Gritsko et al. 2006). In addition to downregulating survivin and Bcl-xL expression, STAT3 silencing in human breast cancer cells was recently shown to lead to Fas-mediated intrinsic apoptotic pathway via the activation of caspases-8, -9, -3 and PARP1 cleavage (Kunigal et al. 2009).

The pro-survival role of STAT3 might be exploited for therapeutic purposes in combined treatments. For example, the inhibition of STAT3 in metastatic breast cancer cells enhanced the proapoptotic effects of doxorubicin, at least in part interfering with survivin and Bcl-xL expression (Gariboldi et al. 2007). Recently, ErbB2-activated STAT3 was shown to directly upregulate the p21(Cip1) gene in breast cancer cells, resulting in increased Taxol resistance and suggesting that Src and STAT3 inhibitors may be used in Taxol sensitization of ErbB2-overexpressing breast cancers (Hawthorne et al. 2009).

STAT3-Mediated Features: Migration, Invasion and Metastasis

Activated STAT3 levels have been reported to correlate with invasiveness and metastasis in breast cancer (Hsieh et al. 2005), and indeed a leading role for STAT3 in driving migration, invasion and metastatic disease of breast cancer cells has emerged in the past years, and thoroughly explored in mouse models of ErbB2-driven tumorigenesis genetically modified for STAT3 (see next section). Both transcriptional and non-transcriptional mechanisms have been proposed to drive STAT3-induced migration.

Intriguingly, activated STAT3 was shown for the first time in ovarian cancer cells to localize not only to the nucleus but also to the focal adhesions, interacting with activated paxillin and focal adhesion kinase, implying local regulation of focal adhesions and integrin-mediated cell movement (Silver et al. 2004). We also have observed STAT3 localization to focal adhesions in mouse breast cancer cell lines derived from MMTV-Her2 transgenic tumours, which was enhanced in cells derived from mice expressing constitutively active STAT3C and displaying more aggressive and invasive tumour phenotype (Barbieri et al. 2010a). Cytoplasmic, non-phosphorylated STAT3 was reported to induce cell migration by interacting with, and inhibiting, the microtubules destabilizer stathmin, thus enhancing microtubules polymerization in murine embryonic fibroblasts (Ng et al. 2006). Conversely, several microtubule-based drugs were shown to modulate STAT3 activity by reducing its phosphorylation in breast tumour cell lines, possibly explaining part of their therapeutic mechanism (Walker et al. 2010).

STAT3-mediated invasion has been linked to the ability to directly upregulate the transcription of MMP9, whose expression levels correlated with those of phosphorylated STAT3 in primary breast cancers. MMP9 was required for mammary epithelial cells transformation mediated by constitutively active STAT3 (Dechow et al. 2004), and its downregulation by the trimeric resveratrol derivative LYR71 correlates with suppression of STAT3 activation, tumour migration and invasion in mouse breast cancer cells (Kim et al. 2008). Additionally, upregulation of the Fra-1 oncogene in response to tumour associated macrophages lead to a malignant switch in breast tumour cells, via activation of the IL-6/JAK/STAT3 loop and increased release of MMP9, VEGF and TGF-b (Luo et al. 2010). Interestingly, STAT3 apparently regulates different subsets of MMPs in different kinds of cancer including MMP2 in melanoma (Xie et al. 2004), MMP1 in the bladder (Itoh et al. 2006) and MMP1, -3, -7, and -9 in the colon (Tsareva et al. 2007).

STAT3-driven metastasis formation was also linked to its ability to induce anchorage-independent growth, EMT and angiogenesis. Impaired in vivo metastasis due to reduced angiogenesis was reported to occur as a consequence of inhibiting STAT3 activation by expressing a dominant negative form of gp130 in a human breast cancer cell line (Selander et al. 2004), correlating with increased expression of the tissue inhibitor of metallo-proteinase 3 (TIMP-3). On the other hand, VEGF is a direct STAT3 transcriptional target, and its upregulated production by STAT3 is believed to induce angiogenesis in different cancer types including the breast (Niu et al. 2002b).

Metastatic breast cancer cells display increased expression of the EMT transcription factor Twist1, which is required for EMT and breast cancer metastasis. A strong positive correlation between active STAT3 and Twist1 levels was detected in late stage breast cancer tissues and in subpopulations of human breast cancer cell lines displaying enhanced invasiveness (Lo et al. 2007; Cheng et al. 2008c). In these cells, STAT3 inhibition lead to Twist1 downregulation correlating with impaired migration, invasion and colony formation, all of which could be rescued by Twist1 re-expression. Interestingly, it was recently shown that IL-6, the canonical STAT3 activator, induces EMT in the ERα-positive human breast cancer cell line MCF-7, including impaired E-cadherin expression and induction of vimentin, N-cadherin, Snail and Twist1. Conversely, constitutive expression of Twist1 triggered aberrant IL-6 production and STAT3 activation, suggesting a positive loop promoting autocrine IL-6 production (Sullivan et al. 2009). Finally, STAT3 was required for EGF-induced Twist1 upregulation in human breast carcinoma cells by directly binding to its promoter (Lo et al. 2007). Correlations between STAT3 and Twist1 were observed also in mouse cells. The silencing of STAT3 in the metastatic mouse breast cancer 4T1 cell line is sufficient to impair tumour formation in vivo and invasion ability in vitro, correlating with reduction of c-Myc, activated Src and Twist1 (Ling and Arlinghaus 2005). However, no putative STAT3 binding site was detected in the murine Twist1 promoter, suggesting different modes of STAT3-dependent activation in the human and the mouse (Lo et al. 2007).

Finally, IL-6 paracrine/autocrine production and STAT3 activation were recently shown to take part in the cross-talk between cancer cells and tumour microenvironment to regulate motility, aggressiveness, angiogenesis and metastasis. Mesenchymal stem cells (MSCs), which reside in the bone marrow, are likely to come in contact with extravasated, metastasis-initiating breast cancer cells. These cells were shown to enhance tumour aggressivity and growth rates in ER-alpha-negative breast cancer cell lines via IL-6 secretion and STAT3 activation (Sasser et al. 2007). MSCs have also been shown to selectively migrate to hypoxic breast tumours, where they are thought to play a tumour-promoting role. Tumour-produced IL-6 acts as an attractant for MSCs, leading to their cytoskeletal reorganization via STAT3 activation (Rattigan et al. 2010). Once within strict contact, a positive loop is likely to get started, whereby infiltrating MSCs in turn produce IL-6 and enhance STAT3 activation in the cancer cells. Due to their specific ability to migrate to and engraft into primary breast tumours, genetically modified MSCs over-expressing Interferon-b (IFN-b) have been explored as potential therapeutic means. IFN-b producing MSCs suppressed breast cancer cells growth and pulmonary and hepatic metastases mainly via inhibition of STAT3 signalling (Ling et al. 2010).

Growth and spread of cancer is thought to be mainly driven by a small subpopulation of CSCs, the only cells capable of long-term self renewal and of generating phenotypically diverse tumour cell populations. These slowly-replicating, self-maintaining cells are resistant to most chemotherapeutics, thus driving relapse. STAT3 is prominently involved in maintaining the undifferentiated status of mouse embryonic stem cells (Burdon et al. 2002), and was shown to be critical for the viability and maintenance of the stem-like side population in the MCF-7 breast cancer cell line (Zhou et al. 2007). Additionally, experimental evidence implied IL-6 signalling in driving formation and malignancy of breast cancer stem cells. Sansone and co-authors reported that mammospheres from node invasive basal-like breast carcinoma tissues, an aggressive breast carcinoma variant showing stem cell features, produce high levels of IL-6, and that autocrine IL-6 signalling sustain the aggressive features of hypoxia-selected MCF-7 cells (Sansone et al. 2007). Recently, IL-6 was shown to drive the conversion of nonstem cancer cells in CSCs in human breast tumours and cell lines (Iliopoulos et al. 2011a). The intimate relationship of STAT3 with the IL-6 pathway leads to postulate its involvement in these systems, even though its activation was not specifically explored.

Role of STAT3 in ErbB2-Driven Mammary Tumorigenesis: Lessons from Mouse Models

Overexpression of the rat oncogenic form of the human EGF receptor ErbB2 (Neu) in the mammary gland under the MMTV promoter triggers the onset of invasive multifocal breast carcinomas at high multiplicity and is widely used as a model for human breast cancer (Guy et al. 1996). The role of STAT3 in Neu-mediated tumorigenesis has been studied by several groups including ours, suggesting a pivotal role of STAT3 in driving tumour progression and metastasis that is in agreement with the clinical and experimental observations reported above. All studies suggest that, although not required for Neu-driven breast tumours onset and growth, STAT3 is heavily implicated in the formation of lung metastasis by a variety of mechanisms. Analyzing the role of b4 integrin in ErbB2 signalling by deleting the b4 signalling domain in the context of MMTV-Neu transgenic mice, Guo and colleagues have shown that b4 integrin forms a complex with ErbB2, enhancing the activation of the transcription factors STAT3 and c-Jun. While c-Jun is required for hyperproliferation, STAT3 contributes to disruption of epithelial adhesion and polarity, and is required for cell invasion and experimental metastasis (Guo et al. 2006). In agreement with this finding, Cre-mediated STAT3 loss of function in MMTV-Neu transgenic mice has shown that STAT3 is not required for the onset and growth of breast tumours, but its deletion results in a dramatic reduction of lung metastasis by both primary and xenografted tumours (Ranger et al. 2009; Barbieri et al. 2010b). The reduced malignancy of STAT3-deficient tumours was partly due to an inhibition of both inflammatory and angiogenic responses, normally activated in a STAT3-dependent transcriptional cascade involving C/EBPδ (Ranger et al. 2009). Additionally, STAT3 is required in a cell autonomous fashion to warrant anchorage-independent growth and the ability to produce lung metastasis in immuno-depressed mice (Barbieri et al. 2010b).

In an effort to reproduce the relatively low but persistent activation of STAT3 observed in most tumours, we have generated knock-in mice expressing at physiological levels the constitutively active mutant form STAT3C. In agreement with the results obtained with the loss of function mutants, MMTV-Neu transgenic mice carrying the STAT3C allele developed earlier onset, more aggressive tumours with lower levels of spontaneous apoptosis but similar proliferation rates (Barbieri et al. 2010a). Tumour-derived STAT3C/Neu cell lines displayed enhanced migration and invasion in vitro and increased tumorigenic and metastatic potential in vivo, correlating with a profoundly modified organization of cell-cell contacts showing altered, irregular distribution of both adherent and tight junctions components such as E-cadherin, b-catenin and Zo-1. Cytoskeletal organization was also perturbed, with actin redistributing from a cortical localization typical of well differentiated epithelial cells to form abundant actin stress fibres, typical of highly migratory cells (Pellegrin and Mellor 2007). Several genes consistently expressed at higher levels in all three STAT3C/Neu cell lines are known players in regulating cell migration and/or tumour metastasis, including the STAT3 transcriptional target Twist1, involved in tumour invasiveness and EMT (Lo et al. 2007; Cheng et al. 2008c).

In addition, we identified the atypical tensin family member Cten as a novel STAT3 target. Cten was recently shown to mediate EGF-induced migration (Katz et al. 2007), to promote colon cancer tumorigenicity and cell motility (Albasri et al. 2009; Liao et al. 2009), and to correlate positively with tumour stage in thymomas, lung tumours and gastric tumours (Sasaki et al. 2003a, b; Sakashita et al. 2008), all displaying constitutive STAT3 activity. It is the most consistently upregulated gene in both STAT3C-expressing cell lines and tumours, and is involved in both their increased migration and disruption of cell junctions organization (Fig. 1, adapted from Barbieri et al. [2010a]). Moreover, we could show that Cten is induced by IL-6 in MCF10 mammary epithelial cells. IL-6-mediated induction is STAT3-dependent, suggesting that indeed Cten may represent an important functional mediator in the inflammation-STAT3-migration-metastasis loop. Indeed, CTEN expression is particularly elevated in the extremely aggressive and invasive inflammatory breast cancers, correlating with high EGFR and HER2 levels, loss of oestrogen receptor, high tumour grade and node metastasis (Katz et al. 2007). Thus, CTEN may represent an important point of functional convergence between inflammation-driven STAT3 activity, altered EGFR/ErbB2-mediated signalling and invasion of the surrounding tissues.

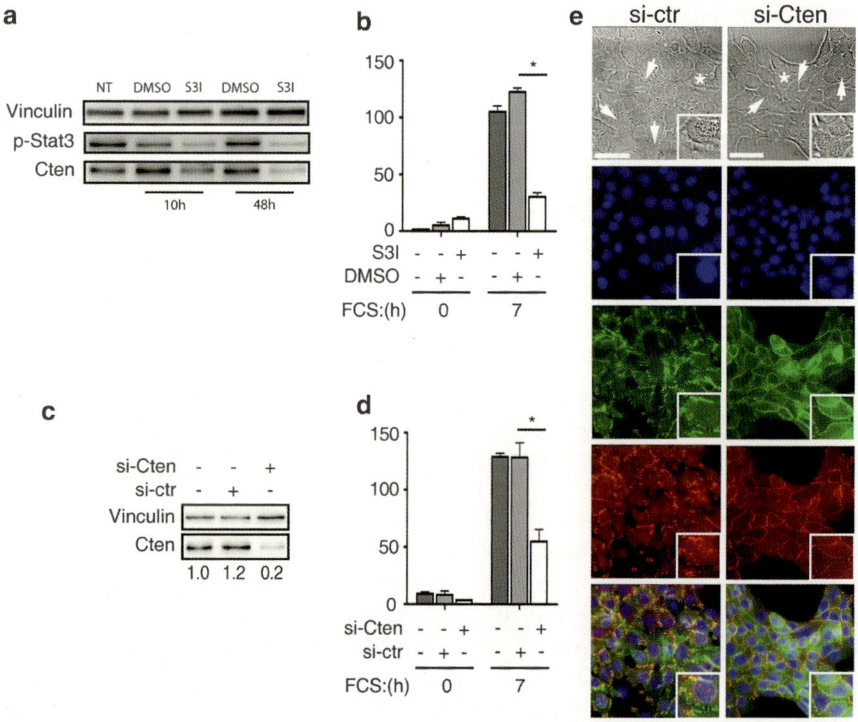

Fig. 1 Both Stat3 inhibition and Cten silencing partially revert the aggressive phenotype of Stat3C expressing cells. Adapted from REF. STAT3C/Neu cells were either treated with the S3I inhibitor for the indicated lengths of time (**a**, **b**) or transfected with an siRNA against Cten (**c**–**e**). S3I treatment downregulates STAT3 phosphorylation (p-Stat3) and Cten expression, as shown by Western blot (**a**). Both treatments significantly impaired FCS-stimulated Transwell migration (**b**, **d**). Values are shown as mean numbers ± SEM of migrated cells per microscopic field (20×) of triplicates in one representative experiment out of two independently performed ($p < 0.05$). (**e**) phase contrast and immunofluorescence images of Cten-silenced cells. *Arrows* indicate evident discontinuous (si-ctrl) versus tight (si-Cten) cell-cell contacts. *Blue*, nuclei; *green*, β-catenin; *red*, Zo-1. The insects (4× magnification) correspond to the areas indicated by an *asterisk*. Scale bar, 20 μm

Concluding Remarks

STAT3 has come a long way since its discovery in the 90s as STAT1's little brother. Initially thought to be almost an IL-6-family-dedicated factor, it has soon emerged as one of the most pleiotropic STATs from many points of view, all contributing to its widespread role in tumours. First, its ever-growing number of upstream activating pathways including many that are aberrantly active in tumours, as initially hinted by the lethal phenotype of STAT3 null embryos. Second, the tissue-dependent variety of target genes, reflected in its variegated functions. Third, its novel non-canonical roles, which apparently do not involve its transcriptional activities. Importantly, its improperly prolonged activity is pro-oncogenic both in tumour and stromal cells, and indeed STAT3 is emerging as a key factor in mediating the cross talk between microenvironment and tumour cells and a main player in inflammation-driven tumorigenesis. Despite the intense research for STAT3 inhibitors, transcription factors are certainly not easily druggable targets. The understanding of STAT3 biology therefore, including which upstream events drive its activation and which are its main effectors in specific tumours, is still highly relevant on the agenda.

References

Abdulghani J, Gu L, Dagvadorj A, Lutz J, Leiby B, Bonuccelli G, Lisanti MP, Zellweger T, Alanen K, Mirtti T, Visakorpi T, Bubendorf L, Nevalainen MT (2008) STAT3 promotes metastatic progression of prostate cancer. Am J Pathol 172:1717–1728

Albasri A, Seth R, Jackson D, Benhasouna A, Crook S, Nateri AS, Chapman R, Ilyas M (2009) C-terminal tensin-like (CTEN) is an oncogene which alters cell motility possibly through repression of E-cadherin in colorectal cancer. J Pathol 218:57–65

Albesiano E, Davis M, See AP, Han JE, Lim M, Pardoll DM, Kim Y (2010) Immunologic consequences of signal transducers and activators of transcription 3 activation in human squamous cell carcinoma. Cancer Res 70:6467–6476

Azare J, Leslie K, Al-Ahmadie H, Gerald W, Weinreb PH, Violette SM, Bromberg J (2007) Constitutively activated STAT3 induces tumorigenesis and enhances cell motility of prostate epithelial cells through integrin beta 6. Mol Cell Biol 27:4444–4453

Bachelot T, Ray-Coquard I, Menetrier-Caux C, Rastkha M, Duc A, Blay J-Y (2003) Prognostic value of serum levels of interleukin 6 and of serum and plasma levels of vascular endothelial growth factor in hormone-refractory metastatic breast cancer patients. Br J Cancer 88:1721–1726

Baltayiannis G, Baltayiannis N, Tsianos EV (2008) Suppressors of cytokine signaling as tumor repressors. Silencing of SOCS3 facilitates tumor formation and growth in lung and liver. J BUON 13:263–265

Barbieri I, Pensa S, Pannellini T, Quaglino E, Maritano D, Demaria M, Voster A, Turkson J, Cavallo F, Watson CJ, Provero P, Musiani P, Poli V (2010a) Constitutively active STAT3 enhances neu-mediated migration and metastasis in mammary tumors via upregulation of Cten. Cancer Res 70:2558–2567

Barbieri I, Quaglino E, Maritano D, Pannellini T, Riera L, Cavallo F, Forni G, Musiani P, Chiarle R, Poli V (2010b) STAT3 is required for anchorage-independent growth and metastasis but not for mammary tumor development downstream of the ErbB-2 oncogene. Mol Carcinog 49:114–120

Becker C, Fantini MC, Wirtz S, Nikolaev A, Lehr HA, Galle PR, Rose-John S, Neurath MF (2005) IL-6 signaling promotes tumor growth in colorectal cancer. Cell Cycle 4:217–220

Béguelin W, Díaz Flaqué MC, Proietti CJ, Cayrol F, Rivas MA, Tkach M, Rosemblit C, Tocci JM, Charreau EH, Schillaci R, Elizalde PV (2010) Progesterone receptor induces ErbB-2 nuclear translocation to promote breast cancer growth via a novel transcriptional effect: ErbB-2 function as a coactivator of STAT3. Mol Cell Biol 30:5456–5472

Berclaz G, Altermatt HJ, Rohrbach V, Siragusa A, Dreher E, Smith PD (2001) EGFR dependent expression of STAT3 (but not STAT1) in breast cancer. Int J Oncol 19:1155–1160

Berishaj M, Gao SP, Ahmed S, Leslie K, Al-Ahmadie H, Gerald WL, Bornmann W, Bromberg JF (2007) STAT3 is tyrosine-phosphorylated through the interleukin-6/glycoprotein 130/Janus kinase pathway in breast cancer. Breast Cancer Res 9:R32

Bharadwaj U, Li M, Zhang R, Chen C, Yao Q (2007) Elevated interleukin-6 and G-CSF in human pancreatic cancer cell conditioned medium suppress dendritic cell differentiation and activation. Cancer Res 67:5479–5488

Binai NA, Damert A, Carra G, Steckelbroeck S, Löwer J, Löwer R, Wessler S (2010) Expression of estrogen receptor alpha increases leptin-induced STAT3 activity in breast cancer cells. Int J Cancer 127:55–66

Bollrath J, Phesse TJ, von Burstin VA, Putoczki T, Bennecke M, Bateman T, Nebelsiek T, Lundgren-May T, Canli O, Schwitalla S, Matthews V, Schmid RM, Kirchner T, Arkan MC, Ernst M, Greten FR (2009) gp130-mediated STAT3 activation in enterocytes regulates cell survival and cell-cycle progression during colitis-associated tumorigenesis. Cancer Cell 15:91–102

Brantley EC, Benveniste EN (2008) Signal transducer and activator of transcription-3: a molecular hub for signaling pathways in gliomas. Mol Cancer Res 6:675–684

Bromberg JF, Horvath CM, Besser D, Lathem WW, Darnell JE (1998) STAT3 activation is required for cellular transformation by v-src. Mol Cell Biol 18:2553–2558

Bromberg JF, Wrzeszczynska MH, Devgan G, Zhao Y, Pestell RG, Albanese C, Darnell JE (1999) STAT3 as an oncogene. Cell 98:295–303

Burdon T, Smith A, Savatier P (2002) Signalling, cell cycle and pluripotency in embryonic stem cells. Trends Cell Biol 12:432–438

Burke WM, Jin X, Lin HJ, Huang M, Liu R, Reynolds RK, Lin J (2001) Inhibition of constitutively active STAT3 suppresses growth of human ovarian and breast cancer cells. Oncogene 20:7925–7934

Chan KS, Carbajal S, Kiguchi K, Clifford J, Sano S, DiGiovanni J (2004a) Epidermal growth factor receptor-mediated activation of STAT3 during multistage skin carcinogenesis. Cancer Res 64:2382–2389

Chan KS, Sano S, Kiguchi K, Anders J, Komazawa N, Takeda J, DiGiovanni J (2004b) Disruption of STAT3 reveals a critical role in both the initiation and the promotion stages of epithelial carcinogenesis. J Clin Invest 114:720–728

Chan KS, Sano S, Kataoka K, Abel E, Carbajal S, Beltran L, Clifford J, Peavey M, Shen J, Digiovanni J (2008) Forced expression of a constitutively active form of STAT3 in mouse epidermis enhances malignant progression of skin tumors induced by two-stage carcinogenesis. Oncogene 27:1087–1094

Chavey C, Bibeau F, Gourgou-Bourgade S, Burlinchon S, Boissière F, Laune D, Roques S, Lazennec G (2007) Oestrogen receptor negative breast cancers exhibit high cytokine content. Breast Cancer Res 9:R15

Chen CL, Hsieh FC, Lieblein JC, Brown J, Chan C, Wallace JA, Cheng G, Hall BM, Lin J (2007a) STAT3 activation in human endometrial and cervical cancers. Br J Cancer 96:591–599

Chen CL, Loy A, Cen L, Chan C, Hsieh FC, Cheng G, Wu B, Qualman SJ, Kunisada K, Yamauchi-Takihara K, Lin J (2007b) Signal transducer and activator of transcription 3 is involved in cell growth and survival of human rhabdomyosarcoma and osteosarcoma cells. BMC Cancer 7:111

Chen CL, Cen L, Kohout J, Hutzen B, Chan C, Hsieh FC, Loy A, Huang V, Cheng G, Lin J (2008a) Signal transducer and activator of transcription 3 activation is associated with bladder cancer cell growth and survival. Mol Cancer 7:78

Chen YW, Chen KH, Huang PI, Chen YC, Chiou GY, Lo WL, Tseng LM, Hsu HS, Chang KW, Chiou SH (2008b) Cucurbitacin I suppressed stem-like property and enhanced radiation-induced apoptosis in head and neck squamous carcinoma-derived CD44(+)ALDH1(+) cells. Mol Cancer Ther 9:2879–2892

Cheng GZ, Zhang WZ, Sun M, Wang Q, Coppola D, Mansour M, Xu LM, Costanzo C, Cheng JQ, Wang L-H (2008) Twist is transcriptionally induced by activation of STAT3 and mediates STAT3 oncogenic function. J Biol Chem 283:14665–14673

Choi JH, Ahn MJ, Park CK, Han HX, Kwon SJ, Lee YY, Kim IS (2006) Phospho-STAT3 expression and correlation with VEGF, p53, and Bcl-2 in gastric carcinoma using tissue microarray. APMIS 114:619–625

Choudhari SR, Khan MA, Harris G, Picker D, Jacob GS, Block T, Shailubhai K (2007) Deactivation of Akt and STAT3 signaling promotes apoptosis, inhibits proliferation, and enhances the sensitivity of hepatocellular carcinoma cells to an anticancer agent. Atiprimod Mol Cancer Ther 6:112–121

Corvinus FM, Orth C, Moriggl R, Tsareva SA, Wagner S, Pfitzner EB, Baus D, Kaufmann R, Huber LA, Zatloukal K, Beug H, Ohlschlager P, Schutz A, Halbhuber KJ, Friedrich K (2005) Persistent STAT3 activation in colon cancer is associated with enhanced cell proliferation and tumor growth. Neoplasia 7:545–555

de la Iglesia N, Konopka G, Puram SV, Chan JA, Bachoo RM, You MJ, Levy DE, Depinho RA, Bonni A (2008) Identification of a PTEN-regulated STAT3 brain tumor suppressor pathway. Genes Dev 22:449–462

Dechow TN, Pedranzini L, Leitch A, Leslie K, Gerald WL, Linkov I, Bromberg JF (2004) Requirement of matrix metalloproteinase-9 for the transformation of human mammary epithelial cells by STAT3-C. Proc Natl Acad Sci USA 101:10602–10607

Demaria M, Giorgi C, Lebiedzinska M, Esposito G, D'Angeli L, Bartoli A, Gough DJ, Turkson J, Levy DE, Watson CJ, Wieckowski MR, Provero P, Pinton P, Poli V (2010) A STAT3-mediated metabolic switch is involved in tumour transformation and STAT3 addiction. Aging (Albany NY) 2:823–842

Deng JY, Sun D, Liu XY, Pan Y, Liang H (2010) STAT-3 correlates with lymph node metastasis and cell survival in gastric cancer. World J Gastroenterol 16:5380–5387

Diaz N, Minton S, Cox C, Bowman T, Gritsko T, Garcia R, Eweis I, Wloch M, Livingston S, Seijo E, Cantor AB, Lee J-H, Beam CA, Sullivan D, Jove R, Muro-Cacho CA (2006) Activation of STAT3 in primary tumors from high-risk breast cancer patients is associated with elevated levels of activated SRC and survivin expression. Clin Cancer Res 12:20–28

Dolled-Filhart M, Camp RL, Kowalski DP, Smith BL, Rimm DL (2003) Tissue microarray analysis of signal transducers and activators of transcription 3 (STAT3) and phospho-STAT3 (Tyr705) in node-negative breast cancer shows nuclear localization is associated with a better prognosis. Clin Cancer Res 9:594–600

Duan Z, Foster R, Bell DA, Mahoney J, Wolak K, Vaidya A, Hampel C, Lee H, Seiden MV (2006) Signal transducers and activators of transcription 3 pathway activation in drug-resistant ovarian cancer. Clin Cancer Res 12:5055–5063

Feng DY, Zheng H, Tan Y, Cheng RX (2001) Effect of phosphorylation of MAPK and STAT3 and expression of c-fos and c-jun proteins on hepatocarcinogenesis and their clinical significance. World J Gastroenterol 7:33–36

Flowers LO, Subramaniam PS, Johnson HM (2005) A SOCS-1 peptide mimetic inhibits both constitutive and IL-6 induced activation of STAT3 in prostate cancer cells. Oncogene 24:2114–2120

Gao B, Shen X, Kunos G, Meng Q, Goldberg ID, Rosen EM, Fan S (2001) Constitutive activation of JAK-STAT3 signaling by BRCA1 in human prostate cancer cells. FEBS Lett 488:179–184

Gao L, Zhang L, Hu J, Li F, Shao Y, Zhao D, Kalvakolanu DV, Kopecko DJ, Zhao X, Xu DQ (2005) Down-regulation of signal transducer and activator of transcription 3 expression using vector-based small interfering RNAs suppresses growth of human prostate tumor in vivo. Clin Cancer Res 11:6333–6341

Gao SP, Mark KG, Leslie K, Pao W, Motoi N, Gerald WL, Travis WD, Bornmann W, Veach D, Clarkson B, Bromberg JF (2007) Mutations in the EGFR kinase domain mediate STAT3 activation via IL-6 production in human lung adenocarcinomas. J Clin Invest 117:3846–3856

Garcia R, Yu CL, Hudnall A, Catlett R, Nelson KL, Smithgall T, Fujita DJ, Ethier SP, Jove R (1997) Constitutive activation of STAT3 in fibroblasts transformed by diverse oncoproteins and in breast carcinoma cells. Cell Growth Differ 8:1267–1276

Garcia R, Bowman TL, Niu G, Yu H, Minton S, Muro-Cacho CA, Cox CE, Falcone R, Fairclough R, Parsons S, Laudano A, Gazit A, Levitzki A, Kraker A, Jove R (2001) Constitutive activation of STAT3 by the Src and JAK tyrosine kinases participates in growth regulation of human breast carcinoma cells. Oncogene 20:2499–2513

Gariboldi MB, Ravizza R, Molteni R, Osella D, Gabano E, Monti E (2007) Inhibition of STAT3 increases doxorubicin sensitivity in a human metastatic breast cancer cell line. Cancer Lett 258:181–188

Garofalo C, Sisci D, Surmacz E (2004) Leptin interferes with the effects of the antiestrogen ICI 182,780 in MCF-7 breast cancer cells. Clin Cancer Res 10:6466–6475

Gough DJ, Corlett A, Schlessinger K, Wegrzyn J, Larner AC, Levy DE (2009) Mitochondrial STAT3 supports Ras-dependent oncogenic transformation. Science 324:1713–1716

Greten FR, Eckmann L, Greten TF, Park JM, Li ZW, Egan LJ, Kagnoff MF, Karin M (2004) IKKbeta links inflammation and tumorigenesis in a mouse model of colitis-associated cancer. Cell 118:285–296

Gritsko T, Williams A, Turkson J, Kaneko S, Bowman T, Huang M, Nam S, Eweis I, Diaz N, Sullivan D, Yoder S, Enkemann S, Eschrich S, Lee J-H, Beam CA, Cheng J, Minton S, Muro-Cacho CA, Jove R (2006) Persistent activation of STAT3 signaling induces survivin gene expression and confers resistance to apoptosis in human breast cancer cells. Clin Cancer Res 12:11–19

Grivennikov SI, Karin M (2010) Dangerous liaisons: STAT3 and NF-kappaB collaboration and crosstalk in cancer. Cytokine Growth Factor Rev 21:11–19

Grivennikov S, Karin E, Terzic J, Mucida D, Yu GY, Vallabhapurapu S, Scheller J, Rose-John S, Cheroutre H, Eckmann L, Karin M (2009) IL-6 and STAT3 are required for survival of intestinal epithelial cells and development of colitis-associated cancer. Cancer Cell 15:103–113

Gu F, Ma Y, Zhang Z, Zhao J, Kobayashi H, Zhang L, Fu L (2010) Expression of STAT3 and Notch1 is associated with cisplatin resistance in head and neck squamous cell carcinoma. Oncol Rep 23:671–676

Guo W, Pylayeva Y, Pepe A, Yoshioka T, Muller WJ, Inghirami G, Giancotti FG (2006) Beta 4 integrin amplifies ErbB2 signaling to promote mammary tumorigenesis. Cell 126:489–502

Guy CT, Cardiff RD, Muller WJ (1996) Activated neu induces rapid tumor progression. J Biol Chem 271:7673–7678

Haura EB, Zheng Z, Song L, Cantor A, Bepler G (2005) Activated epidermal growth factor receptor-STAT-3 signaling promotes tumor survival in vivo in non-small cell lung cancer. Clin Cancer Res 11:8288–8294

Hawthorne VS, Huang W-C, Neal CL, Tseng L-M, Hung M-C, Yu D (2009) ErbB2-mediated Src and signal transducer and activator of transcription 3 activation leads to transcriptional up-regulation of p21Cip1 and chemoresistance in breast cancer cells. Mol Cancer Res 7:592–600

He G, Yu GY, Temkin V, Ogata H, Kuntzen C, Sakurai T, Sieghart W, Peck-Radosavljevic M, Leffert HL, Karin M (2010) Hepatocyte IKKbeta/NF-kappaB inhibits tumor promotion and progression by preventing oxidative stress-driven STAT3 activation. Cancer Cell 17:286–297

Hellsten R, Johansson M, Dahlman A, Dizeyi N, Sterner O, Bjartell A (2008) Galiellalactone is a novel therapeutic candidate against hormone-refractory prostate cancer expressing activated Stat3. Prostate 68:269–280

Hemmati HD, Nakano I, Lazareff JA, Masterman-Smith M, Geschwind DH, Bronner-Fraser M, Kornblum HI (2003) Cancerous stem cells can arise from pediatric brain tumors. Proc Natl Acad Sci USA 100:15178–15183

Horiguchi A, Oya M, Marumo K, Murai M (2002a) STAT3, but not ERKs, mediates the IL-6-induced proliferation of renal cancer cells, ACHN and 769P. Kidney Int 61:926–938

Horiguchi A, Oya M, Shimada T, Uchida A, Marumo K, Murai M (2002b) Activation of signal transducer and activator of transcription 3 in renal cell carcinoma: a study of incidence and its association with pathological features and clinical outcome. J Urol 168:762–765

Hsieh F-C, Cheng G, Lin J (2005) Evaluation of potential STAT3-regulated genes in human breast cancer. Biochem Biophys Res Commun 335:292–299

Huang M, Page C, Reynolds RK, Lin J (2000) Constitutive activation of STAT 3 oncogene product in human ovarian carcinoma cells. Gynecol Oncol 79:67–73

Hussain SF, Kong LY, Jordan J, Conrad C, Madden T, Fokt I, Priebe W, Heimberger AB (2007) A novel small molecule inhibitor of signal transducers and activators of transcription 3 reverses immune tolerance in malignant glioma patients. Cancer Res 67:9630–9636

Iliopoulos D, Hirsch HA, Struhl K (2009) An epigenetic switch involving NF-kappaB, Lin28, Let-7 MicroRNA, and IL6 links inflammation to cell transformation. Cell 139:693–706

Iliopoulos D, Jaeger SA, Hirsch HA, Bulyk ML, Struhl K (2010) STAT3 activation of miR-21 and miR-181b-1 via PTEN and CYLD are part of the epigenetic switch linking inflammation to cancer. Mol Cell 39:493–506

Iliopoulos D, Hirsch HA, Wang G, Struhl K (2011) Inducible formation of breast cancer stem cells and their dynamic equilibrium with non-stem cancer cells via IL6 secretion. Proc Natl Acad Sci USA 108:1397–1402

Itoh M, Murata T, Suzuki T, Shindoh M, Nakajima K, Imai K, Yoshida K (2006) Requirement of STAT3 activation for maximal collagenase-1 (MMP-1) induction by epidermal growth factor and malignant characteristics in T24 bladder cancer cells. Oncogene 25:1195–1204

Jee SH, Chu CY, Chiu HC, Huang YL, Tsai WL, Liao YH, Kuo ML (2004) Interleukin-6 induced basic fibroblast growth factor-dependent angiogenesis in basal cell carcinoma cell line via JAK/STAT3 and PI3-kinase/Akt pathways. J Invest Dermatol 123:1169–1175

Jiang XP, Yang DC, Elliott RL, Head JF (2000) Reduction in serum IL-6 after vacination of breast cancer patients with tumour-associated antigens is related to estrogen receptor status. Cytokine 12:458–465

Jing N, Li Y, Xiong W, Sha W, Jing L, Tweardy DJ (2004) G-quartet oligonucleotides: a new class of signal transducer and activator of transcription 3 inhibitors that suppresses growth of prostate and breast tumors through induction of apoptosis. Cancer Res 64:6603–6609

Jing N, Zhu Q, Yuan P, Li Y, Mao L, Tweardy DJ (2006) Targeting signal transducer and activator of transcription 3 with G-quartet oligonucleotides: a potential novel therapy for head and neck cancer. Mol Cancer Ther 5:279–286

Jung JE, Kim HS, Lee CS, Park DH, Kim YN, Lee MJ, Lee JW, Park JW, Kim MS, Ye SK, Chung MH (2007) Caffeic acid and its synthetic derivative CADPE suppress tumor angiogenesis by blocking STAT3-mediated VEGF expression in human renal carcinoma cells. Carcinogenesis 28:1780–1787

Kanda N, Seno H, Konda Y, Marusawa H, Kanai M, Nakajima T, Kawashima T, Nanakin A, Sawabu T, Uenoyama Y, Sekikawa A, Kawada M, Suzuki K, Kayahara T, Fukui H, Sawada M, Chiba T (2004) STAT3 is constitutively activated and supports cell survival in association with survivin expression in gastric cancer cells. Oncogene 23:4921–4929

Kang SH, Yu MO, Park KJ, Chi SG, Park DH, Chung YG (2010) Activated STAT3 regulates hypoxia-induced angiogenesis and cell migration in human glioblastoma. Neurosurgery 67:1386–1395, discussion 1395

Katz M, Amit I, Citri A, Shay T, Carvalho S, Lavi S, Milanezi F, Lyass L, Amariglio N, Jacob-Hirsch J, Ben-Chetrit N, Tarcic G, Lindzen M, Avraham R, Liao Y-C, Trusk P, Lyass A, Rechavi G, Spector NL, Lo SH, Schmitt F, Bacus SS, Yarden Y (2007) A reciprocal tensin-3-Cten switch mediates EGF-driven mammary cell migration. Nat Cell Biol 9:961–969

Kim DJ, Chan KS, Sano S, Digiovanni J (2007) Signal transducer and activator of transcription 3 (STAT3) in epithelial carcinogenesis. Mol Carcinog 46:725–731

Kim JE, Kim HS, Shin YJ, Lee CS, Won C, Lee SA, Lee JW, Kim Y, Kang JS, Ye SK, Chung MH (2008) LYR71, a derivative of trimeric resveratrol, inhibits tumorigenesis by blocking STAT3-mediated matrix metalloproteinase 9 expression. Exp Mol Med 40:514–522

Kim DJ, Kataoka K, Sano S, Connolly K, Kiguchi K, DiGiovanni J (2009) Targeted disruption of Bcl-xL in mouse keratinocytes inhibits both UVB- and chemically induced skin carcinogenesis. Mol Carcinog 48:873–885

Kong LY, Abou-Ghazal MK, Wei J, Chakraborty A, Sun W, Qiao W, Fuller GN, Fokt I, Grimm EA, Schmittling RJ, Archer GE Jr, Sampson JH, Priebe W, Heimberger AB (2008) A novel inhibitor of signal transducers and activators of transcription 3 activation is efficacious against established central nervous system melanoma and inhibits regulatory T cells. Clin Cancer Res 14:5759–5768

Kortylewski M, Jove R, Yu H (2005) Targeting STAT3 affects melanoma on multiple fronts. Cancer Metastasis Rev 24:315–327

Kozłowski L, Zakrzewska I, Tokajuk P, Wojtukiewicz MZ (2003) Concentration of interleukin-6 (IL-6), interleukin-8 (IL-8) and interleukin-10 (IL-10) in blood serum of breast cancer patients. Rocz Akad Med Bialymst 48:82–84

Kunigal S, Lakka SS, Sodadasu PK, Estes N, Rao JS (2009) STAT3-siRNA induces Fas-mediated apoptosis in vitro and in vivo in breast cancer. Int J Oncol 34:1209–1220

Kusaba T, Nakayama T, Yamazumi K, Yakata Y, Yoshizaki A, Nagayasu T, Sekine I (2005) Expression of p-STAT3 in human colorectal adenocarcinoma and adenoma; correlation with clinicopathological factors. J Clin Pathol 58:833–838

Kusaba T, Nakayama T, Yamazumi K, Yakata Y, Yoshizaki A, Inoue K, Nagayasu T, Sekine I (2006) Activation of STAT3 is a marker of poor prognosis in human colorectal cancer. Oncol Rep 15:1445–1451

Lassmann S, Schuster I, Walch A, Gobel H, Jutting U, Makowiec F, Hopt U, Werner M (2007) STAT3 mRNA and protein expression in colorectal cancer: effects on STAT3-inducible targets linked to cell survival and proliferation. J Clin Pathol 60:173–179

Lee JH, Schütte D, Wulf G, Füzesi L, Radzun H-J, Schweyer S, Engel W, Nayernia K (2006) Stem-cell protein Piwil2 is widely expressed in tumors and inhibits apoptosis through activation of STAT3/Bcl-XL pathway. Hum Mol Genet 15:201–211

Lee J, Kang WK, Park JO, Park SH, Park YS, Lim HY, Kim J, Kong J, Choi MG, Sohn TS, Noh JH, Bae JM, Kim S, Lim do H, Kim KM, Park CK (2009) Expression of activated signal transducer and activator of transcription 3 predicts poor clinical outcome in gastric adenocarcinoma. APMIS 117:598–606

Lee H, Deng J, Kujawski M, Yang C, Liu Y, Herrmann A, Kortylewski M, Horne D, Somlo G, Forman S, Jove R, Yu H (2010) STAT3-induced S1PR1 expression is crucial for persistent STAT3 activation in tumors. Nat Med 16:1421–1428

Leong PL, Andrews GA, Johnson DE, Dyer KF, Xi S, Mai JC, Robbins PD, Gadiparthi S, Burke NA, Watkins SF, Grandis JR (2003) Targeted inhibition of STAT3 with a decoy oligonucleotide abrogates head and neck cancer cell growth. Proc Natl Acad Sci USA 100:4138–4143

Leslie K, Lang C, Devgan G, Azare J, Berishaj M, Gerald W, Kim YB, Paz K, Darnell JE, Albanese C, Sakamaki T, Pestell R, Bromberg JF (2006) Cyclin D1 is transcriptionally regulated by and required for transformation by activated signal transducer and activator of transcription 3. Cancer Res 66:2544–2552

Li WC, Ye SL, Sun RX, Liu YK, Tang ZY, Kim Y, Karras JG, Zhang H (2006) Inhibition of growth and metastasis of human hepatocellular carcinoma by antisense oligonucleotide targeting signal transducer and activator of transcription 3. Clin Cancer Res 12:7140–7148

Li Y, Du H, Qin Y, Roberts J, Cummings OW, Yan C (2007) Activation of the signal transducers and activators of the transcription 3 pathway in alveolar epithelial cells induces inflammation and adenocarcinomas in mouse lung. Cancer Res 67:8494–8503

Liao Y-C, Chen N-T, Shih Y-P, Dong Y, Lo SH (2009) Up-regulation of C-terminal tensin-like molecule promotes the tumorigenicity of colon cancer through beta-catenin. Cancer Res 69:4563–4566

Ling X, Arlinghaus RB (2005) Knockdown of STAT3 expression by RNA interference inhibits the induction of breast tumors in immunocompetent mice. Cancer Res 65:2532–2536

Ling X, Marini F, Konopleva M, Schober W, Shi Y, Burks J, Clise-Dwyer K, Wang R-Y, Zhang W, Yuan X, Lu H, Caldwell L, Andreeff M (2010) Mesenchymal stem cells overexpressing IFN-β inhibit breast cancer growth and metastases through STAT3 signaling in a syngeneic tumor model. Cancer Microenviron 3:83–95

Liu L, Gao Y, Qiu H, Miller WT, Poli V, Reich NC (2006) Identification of STAT3 as a specific substrate of breast tumor kinase. Oncogene 25:4904–4912

Liu Q, Li G, Li R, Shen J, He Q, Deng L, Zhang C, Zhang J (2010) IL-6 promotion of glioblastoma cell invasion and angiogenesis in U251 and T98G cell lines. J Neurooncol 100:165–176

Lo H-W, Hsu S-C, Xia W, Cao X, Shih J-Y, Wei Y, Abbruzzese JL, Hortobagyi GN, Hung M-C (2007) Epidermal growth factor receptor cooperates with signal transducer and activator of transcription 3 to induce epithelial-mesenchymal transition in cancer cells via up-regulation of TWIST gene expression. Cancer Res 67:9066–9076

Loeffler S, Fayard B, Weis J, Weissenberger J (2005) Interleukin-6 induces transcriptional activation of vascular endothelial growth factor (VEGF) in astrocytes in vivo and regulates VEGF promoter activity in glioblastoma cells via direct interaction between STAT3 and Sp1. Int J Cancer 115:202–213

Lou W, Ni Z, Dyer K, Tweardy DJ, Gao AC (2000) Interleukin-6 induces prostate cancer cell growth accompanied by activation of STAT3 signaling pathway. Prostate 42:239–242

Lui VW, Wong EY, Ho Y, Hong B, Wong SC, Tao Q, Choi GC, Au TC, Ho K, Yau DM, Ma BB, Hui EP, Chan AS, Tsang CM, Tsao SW, Grandis JR, Chan AT (2009) STAT3 activation contributes directly to Epstein-Barr virus-mediated invasiveness of nasopharyngeal cancer cells in vitro. Int J Cancer 125:1884–1893

Luo YP, Zhou H, Krueger J, Kaplan C, Liao D, Markowitz D, Liu C, Chen T, Chuang T-H, Xiang R, Reisfeld RA (2010) The role of proto-oncogene Fra-1 in remodeling the tumor microenvironment in support of breast tumor cell invasion and progression. Oncogene 29:662–673

Masuda M, Suzui M, Lim JT, Deguchi A, Soh JW, Weinstein IB (2002a) Epigallocatechin-3-gallate decreases VEGF production in head and neck and breast carcinoma cells by inhibiting EGFR-related pathways of signal transduction. J Exp Ther Oncol 2:350–359

Masuda M, Suzui M, Yasumatu R, Nakashima T, Kuratomi Y, Azuma K, Tomita K, Komiyama S, Weinstein IB (2002b) Constitutive activation of signal transducers and activators of transcription 3 correlates with cyclin D1 overexpression and may provide a novel prognostic marker in head and neck squamous cell carcinoma. Cancer Res 62:3351–3355

Masuda M, Toh S, Koike K, Kuratomi Y, Suzui M, Deguchi A, Komiyama S, Weinstein IB (2002c) The roles of JNK1 and STAT3 in the response of head and neck cancer cell lines to combined treatment with all-trans-retinoic acid and 5-fluorouracil. Jpn J Cancer Res 93:329–339

Masuda M, Ruan HY, Ito A, Nakashima T, Toh S, Wakasaki T, Yasumatsu R, Kutratomi Y, Komune S, Weinstein IB (2007) Signal transducers and activators of transcription 3 up-regulates vascular endothelial growth factor production and tumor angiogenesis in head and neck squamous cell carcinoma. Oral Oncol 43:785–790

Mathews LA, Hurt EM, Zhang X, Farrar WL (2010) Epigenetic regulation of CpG promoter methylation in invasive prostate cancer cells. Mol Cancer 9:267

Mora LB, Buettner R, Seigne J, Diaz J, Ahmad N, Garcia R, Bowman T, Falcone R, Fairclough R, Cantor A, Muro Cacho C, Livingston S, Karras J, Pow-Sang J, Jove R (2002) Constitutive activation of STAT3 in human prostate tumors and cell lines: direct inhibition of STAT3 signaling induces apoptosis of prostate cancer cells. Cancer Res 62:6659–6666

Musteanu M, Blaas L, Mair M, Schlederer M, Bilban M, Tauber S, Esterbauer H, Mueller M, Casanova E, Kenner L, Poli V, Eferl R (2010) STAT3 is a negative regulator of intestinal tumor progression in Apc(Min) mice. Gastroenterology 138(1003–1011):e1001–1005

Nabarro S, Himoudi N, Papanastasiou A, Gilmour K, Gibson S, Sebire N, Thrasher A, Blundell MP, Hubank M, Canderan G, Anderson J (2005) Coordinated oncogenic transformation and inhibition of host immune responses by the PAX3-FKHR fusion oncoprotein. J Exp Med 202:1399–1410

Nagpal JK, Mishra R, Das BR (2002) Activation of STAT-3 as one of the early events in tobacco chewing-mediated oral carcinogenesis. Cancer 94:2393–2400

Nefedova Y, Huang M, Kusmartsev S, Bhattacharya R, Cheng P, Salup R, Jove R, Gabrilovich D (2004) Hyperactivation of STAT3 is involved in abnormal differentiation of dendritic cells in cancer. J Immunol 172:464–474

Ng DCH, Lin BH, Lim CP, Huang G, Zhang T, Poli V, Cao X (2006) STAT3 regulates microtubules by antagonizing the depolymerization activity of stathmin. J Cell Biol 172:245–257

Nilsson MB, Langley RR, Fidler IJ (2005) Interleukin-6, secreted by human ovarian carcinoma cells, is a potent proangiogenic cytokine. Cancer Res 65:10794–10800

Niu G, Bowman T, Huang M, Shivers S, Reintgen D, Daud A, Chang A, Kraker A, Jove R, Yu H (2002a) Roles of activated Src and STAT3 signaling in melanoma tumor cell growth. Oncogene 21:7001–7010

Niu G, Wright KL, Huang M, Song L, Haura E, Turkson J, Zhang S, Wang T, Sinibaldi D, Coppola D, Heller R, Ellis LM, Karras J, Bromberg JF, Pardoll DM, Jove R, Yu H (2002b) Constitutive STAT3 activity up-regulates VEGF expression and tumor angiogenesis. Oncogene 21:2000–2008

Park J, Kusminski CM, Chua SC, Scherer PE (2010) Leptin receptor signaling supports cancer cell metabolism through suppression of mitochondrial respiration in vivo. Am J Pathol 177:3133–3144

Pellegrin S, Mellor H (2007) Actin stress fibres. J Cell Sci 120:3491–3499

Pensa S, Regis G, Boselli D, Novelli F, Poli V (2008) STAT1 and STAT3 in tumorigenesis: two sides of the same coin? JAK-STAT Pathway in Disease A. Stephanou, Landes Bioscience, Austin, 100–121

Pfeiffer M, Hartmann TN, Leick M, Catusse J, Schmitt-Graeff A, Burger M (2009) Alternative implication of CXCR4 in JAK2/STAT3 activation in small cell lung cancer. Br J Cancer 100:1949–1956

Pu YS, Hour TC, Chuang SE, Cheng AL, Lai MK, Kuo ML (2004) Interleukin-6 is responsible for drug resistance and anti-apoptotic effects in prostatic cancer cells. Prostate 60:120–129

Qu P, Roberts J, Li Y, Albrecht M, Cummings OW, Eble JN, Du H, Yan C (2009) STAT3 downstream genes serve as biomarkers in human lung carcinomas and chronic obstructive pulmonary disease. Lung Cancer 63:341–347

Quaglino A, Schere-Levy C, Romorini L, Meiss RP, Kordon EC (2007) Mouse mammary tumors display Stat3 activation dependent on leukemia inhibitory factor signaling. Breast Cancer Res 9:R69

Rahaman SO, Harbor PC, Chernova O, Barnett GH, Vogelbaum MA, Haque SJ (2002) Inhibition of constitutively active STAT3 suppresses proliferation and induces apoptosis in glioblastoma multiforme cells. Oncogene 21:8404–8413

Ranger JJ, Levy DE, Shahalizadeh S, Hallett M, Muller WJ (2009) Identification of a STAT3-dependent transcription regulatory network involved in metastatic progression. Cancer Res 69:6823–6830

Rattigan Y, Hsu J-M, Mishra PJ, Glod J, Banerjee D (2010) Interleukin 6 mediated recruitment of mesenchymal stem cells to the hypoxic tumor milieu. Exp Cell Res 316:3417–3424

Real PJ, Sierra A, De Juan A, Segovia JC, Lopez-Vega JM, Fernandez-Luna JL (2002) Resistance to chemotherapy via STAT3-dependent overexpression of Bcl-2 in metastatic breast cancer cells. Oncogene 21:7611–7618

Rosen DG, Mercado-Uribe I, Yang G, Bast RC Jr, Amin HM, Lai R, Liu J (2006) The role of constitutively active signal transducer and activator of transcription 3 in ovarian tumorigenesis and prognosis. Cancer 107:2730–2740

Sakashita K, Mimori K, Tanaka F, Kamohara Y, Inoue H, Sawada T, Hirakawa K, Mori M (2008) Prognostic relevance of tensin4 expression in human gastric cancer. Ann Surg Oncol 15:2606–2613

Salgado R, Junius S, Benoy I, Van Dam P, Vermeulen P, Van Marck E, Huget P, Dirix LY (2003) Circulating interleukin-6 predicts survival in patients with metastatic breast cancer. Int J Cancer J103:642–646

Sano S, Chan KS, Kira M, Kataoka K, Takagi S, Tarutani M, Itami S, Kiguchi K, Yokoi M, Sugasawa K, Mori T, Hanaoka F, Takeda J, DiGiovanni J (2005) Signal transducer and activator of transcription 3 is a key regulator of keratinocyte survival and proliferation following UV irradiation. Cancer Res 65:5720–5729

Sansone P, Storci G, Tavolari S, Guarnieri T, Giovannini C, Taffurelli M, Ceccarelli C, Santini D, Paterini P, Marcu KB, Chieco P, Bonafè M (2007) IL-6 triggers malignant features in mammospheres from human ductal breast carcinoma and normal mammary gland. J Clin Invest 117:3988–4002

Sasaki H, Moriyama S, Mizuno K, Yukiue H, Konishi A, Yano M, Kaji M, Fukai I, Kiriyama M, Yamakawa Y, Fujii Y (2003a) Cten mRNA expression was correlated with tumor progression in lung cancers. Lung Cancer 40:151–155

Sasaki H, Yukiue H, Kobayashi Y, Fukai I, Fujii Y (2003b) Cten mRNA expression is correlated with tumor progression in thymoma. Tumour Biol 24:271–274

Sasser AK, Sullivan NJ, Studebaker AW, Hendey LF, Axel AE, Hall BM (2007) Interleukin-6 is a potent growth factor for ER-alpha-positive human breast cancer. FASEB J 21:3763–3770

Saxena NK, Vertino PM, Anania FA, Sharma D (2007) Leptin-induced growth stimulation of breast cancer cells involves recruitment of histone acetyltransferases and mediator complex to CYCLIN D1 promoter via activation of STAT3. J Biol Chem 282:13316–13325

Schindler C, Levy DE, Decker T (2007) JAK-STAT signaling: from interferons to cytokines. J Biol Chem 282:20059–20063

Selander KS, Li L, Watson L, Merrell M, Dahmen H, Heinrich PC, Muller-Newen G, Harris KW (2004) Inhibition of gp130 signaling in breast cancer blocks constitutive activation of STAT3 and inhibits in vivo malignancy. Cancer Res 64:6924–6933

Sherry MM, Reeves A, Wu JK, Cochran BH (2009) STAT3 is required for proliferation and maintenance of multipotency in glioblastoma stem cells. Stem Cells 27:2383–2392

Siddiquee K, Zhang S, Guida WC, Blaskovich MA, Greedy B, Lawrence HR, Yip MLR, Jove R, McLaughlin MM, Lawrence NJ, Sebti SM, Turkson J (2007) Selective chemical probe inhibitor of STAT3, identified through structure-based virtual screening, induces antitumor activity. Proc Natl Acad Sci USA 104:7391–7396

Silva CM (2004) Role of STATs as downstream signal transducers in Src family kinase-mediated tumorigenesis. Oncogene 23:8017–8023

Silver DL, Naora H, Liu J, Cheng W, Montell DJ (2004) Activated signal transducer and activator of transcription (STAT) 3: localization in focal adhesions and function in ovarian cancer cell motility. Cancer Res 64:3550–3558

Song L, Turkson J, Karras JG, Jove R, Haura EB (2003) Activation of STAT3 by receptor tyrosine kinases and cytokines regulates survival in human non-small cell carcinoma cells. Oncogene 22:4150–4165

Suiqing C, Min Z, Lirong C (2005) Overexpression of phosphorylated-STAT3 correlated with the invasion and metastasis of cutaneous squamous cell carcinoma. J Dermatol 32:354–360

Sullivan NJ, Sasser AK, Axel AE, Vesuna F, Raman V, Ramirez N, Oberyszyn TM, Hall BM (2009) Interleukin-6 induces an epithelial-mesenchymal transition phenotype in human breast cancer cells. Oncogene 28:2940–2947

Sun Z, Yao Z, Liu S, Tang H, Yan X (2006) An oligonucleotide decoy for STAT3 activates the immune response of macrophages to breast cancer. Immunobiology 211:199–209

Sun X, Zhang J, Wang L, Tian Z (2008) Growth inhibition of human hepatocellular carcinoma cells by blocking STAT3 activation with decoy-ODN. Cancer Lett 262:201–213

Takemoto S, Ushijima K, Kawano K, Yamaguchi T, Terada A, Fujiyoshi N, Nishio S, Tsuda N, Ijichi M, Kakuma T, Kage M, Hori D, Kamura T (2009) Expression of activated signal transducer and activator of transcription-3 predicts poor prognosis in cervical squamous-cell carcinoma. Br J Cancer 101:967–972

Tebbutt NC, Giraud AS, Inglese M, Jenkins B, Waring P, Clay FJ, Malki S, Alderman BM, Grail D, Hollande F, Heath JK, Ernst M (2002) Reciprocal regulation of gastrointestinal homeostasis by SHP2 and STAT-mediated trefoil gene activation in gp130 mutant mice. Nat Med 8:1089–1097

Tsareva SA, Moriggl R, Corvinus FM, Wiederanders B, Schütz A, Kovacic B, Friedrich K (2007) Signal transducer and activator of transcription 3 activation promotes invasive growth of colon carcinomas through matrix metalloproteinase induction. Neoplasia 9:279–291

Turkson J, Jove R (2000) STAT proteins: novel molecular targets for cancer drug discovery. Oncogene 19:6613–6626

Turkson J, Zhang S, Palmer J, Kay H, Stanko J, Mora LB, Sebti S, Yu H, Jove R (2004) Inhibition of constitutive signal transducer and activator of transcription 3 activation by novel platinum complexes with potent antitumor activity. Mol Cancer Ther 3:1533–1542

van Cruijsen H, Ruiz MG, van der Valk P, de Gruijl TD, Giaccone G (2009) Tissue micro array analysis of ganglioside N-glycolyl GM3 expression and signal transducer and activator of transcription (STAT)-3 activation in relation to dendritic cell infiltration and microvessel density in non-small cell lung cancer. BMC Cancer 9:180

Villalva C, Martin-Lanneree S, Cortes U, Dkhissi F, Wager M, Le Corf A, Tourani JM, Dusanter-Fourt I, Turhan AG, Karayan-Tapon L (2010) STAT3 is essential for the maintenance of neurosphere-initiating tumor cells in patients with glioblastomas: a potential for targeted therapy? Int J Cancer 128:826–838

Walker SR, Chaudhury M, Nelson EA, Frank DA (2010) Microtubule-targeted chemotherapeutic agents inhibit signal transducer and activator of transcription 3 (STAT3) signaling. Mol Pharmacol 78:903–908

Wang T, Niu G, Kortylewski M, Burdelya L, Shain K, Zhang S, Bhattacharya R, Gabrilovich D, Heller R, Coppola D, Dalton W, Jove R, Pardoll DM, Yu H (2004) Regulation of the innate and adaptive immune responses by STAT-3 signaling in tumor cells. Nat Med 10:48–54

Weerasinghe P, Garcia GE, Zhu Q, Yuan P, Feng L, Mao L, Jing N (2007) Inhibition of STAT3 activation and tumor growth suppression of non-small cell lung cancer by G-quartet oligonucleotides. Int J Oncol 31:129–136

Wegrzyn J, Potla R, Chwae YJ, Sepuri NB, Zhang Q, Koeck T, Derecka M, Szczepanek K, Szelag M, Gornicka A, Moh A, Moghaddas S, Chen Q, Bobbili S, Cichy J, Dulak J, Baker DP, Wolfman A, Stuehr D, Hassan MO, Fu XY, Avadhani N, Drake JI, Fawcett P, Lesnefsky EJ, Larner AC (2009) Function of mitochondrial STAT3 in cellular respiration. Science 323:793–797

Wei D, Le X, Zheng L, Wang L, Frey JA, Gao AC, Peng Z, Huang S, Xiong HQ, Abbruzzese JL, Xie K (2003) STAT3 activation regulates the expression of vascular endothelial growth factor and human pancreatic cancer angiogenesis and metastasis. Oncogene 22:319–329

Wheeler SE, Suzuki S, Thomas SM, Sen M, Leeman-Neill RJ, Chiosea SI, Kuan CT, Bigner DD, Gooding WE, Lai SY, Grandis JR (2010) Epidermal growth factor receptor variant III mediates head and neck cancer cell invasion via STAT3 activation. Oncogene 29:5135–5145

Xie T-X, Wei D, Liu M, Gao AC, Ali-Osman F, Sawaya R, Huang S (2004) STAT3 activation regulates the expression of matrix metalloproteinase-2 and tumor invasion and metastasis. Oncogene 23:3550–3560

Xie TX, Huang FJ, Aldape KD, Kang SH, Liu M, Gershenwald JE, Xie K, Sawaya R, Huang S (2006) Activation of STAT3 in human melanoma promotes brain metastasis. Cancer Res 66:3188–3196

Xin H, Zhang C, Herrmann A, Du Y, Figlin R, Yu H (2009) Sunitinib inhibition of STAT3 induces renal cell carcinoma tumor cell apoptosis and reduces immunosuppressive cells. Cancer Res 69:2506–2513

Xu Q, Briggs J, Park S, Niu G, Kortylewski M, Zhang S, Gritsko T, Turkson J, Kay H, Semenza GL, Cheng JQ, Jove R, Yu H (2005) Targeting STAT3 blocks both HIF-1 and VEGF expression induced by multiple oncogenic growth signaling pathways. Oncogene 24:5552–5560

Yoshida T, Hanada T, Tokuhisa T, Kosai K, Sata M, Kohara M, Yoshimura A (2002) Activation of STAT3 by the hepatitis C virus core protein leads to cellular transformation. J Exp Med 196:641–653

Yoshimura A (2005) Negative regulation of cytokine signaling. Clin Rev Allergy Immunol 28:205–220

Yu CL, Meyer DJ, Campbell GS, Larner AC, Carter-Su C, Schwartz J, Jove R (1995) Enhanced DNA-binding activity of a STAT3-related protein in cells transformed by the Src oncoprotein. Science 269:81–83

Yu H, Pardoll D, Jove R (2009) STATs in cancer inflammation and immunity: a leading role for STAT3. Nat Rev Cancer 9:798–809

Zhang GJ, Adachi I (1999) Serum interleukin-6 levels correlate to tumor progression and prognosis in metastatic breast carcinoma. Anticancer Res 19:1427–1432

Zhang CH, Xu GL, Jia WD, Li JS, Ma JL, Ren WH, Ge YS, Yu JH, Liu WB, Wang W (2010) Activation of STAT3 signal pathway correlates with twist and E-cadherin expression in hepatocellular carcinoma and their clinical significance. J Surg Res (in press)

Zhou J, Zhang Y (2008) Cancer stem cells: models, mechanisms and implications for improved treatment. Cell Cycle 7:1360–1370

Zhou J, Wulfkuhle J, Zhang H, Gu P, Yang Y, Deng J, Margolick JB, Liotta LA, Petricoin E, Zhang Y (2007) Activation of the PTEN/mTOR/STAT3 pathway in breast cancer stem-like cells is required for viability and maintenance. Proc Natl Acad Sci USA 104:16158–16163

Impact of STAT5 on Normal Tissue Development and Cancer

Maria M. Caffarel and Christine J. Watson

Abstract
STAT5A and STAT5B are two closely related transcription factors that transduce signals from cytokines and growth factors. Two tissues that rely upon STAT5 for lineage commitment and differentiation are the mammary gland and the haematopoietic system. During pregnancy, mammary epithelial cells undergo extensive proliferation and differentiation to generate milk-producing alveolar structures. Alveologenesis is abrogated in the absence of both STAT5A and B and fat pad transplantation experiments demonstrated that while the ability of stem cells to generate ductal outgrowths was not affected by loss of STAT5A/B, the alveolar compartment failed to develop in pregnant mice. In contrast, in the haematopoietic system, STAT5A and B control both stem and progenitor cell fate and are essential for the development of immune cells of the T, B and NK lineages. Aberrant STAT5 activity has major consequences and can induce tumours of both the blood and breast, in addition to other tissues. The mechanisms of constitutive STAT5 activity are manifold and include mutation of the upstream kinase JAK2.

Introduction

STAT5 was first described in the lactating mammary glands of sheep and rats where it was called milk protein binding factor (MPBF) (Watson et al. 1991) or mammary gland factor (MGF) (Wakao et al. 1992) respectively and when the gene encoding STAT5 was subsequently cloned, it became apparent that it was the fifth member of the newly described family of STAT proteins (Schindler et al. 1992; Wakao et al. 1992).

M.M. Caffarel • C.J. Watson (✉)
Department of Pathology, University of Cambridge, Tennis Court Road, Cambridge CB2 1QP, UK
e-mail: cjw53@cam.ac.uk

Th. Decker and M. Müller (eds.), *Jak-Stat Signaling: From Basics to Disease*, DOI 10.1007/978-3-7091-0891-8_18, © Springer-Verlag Wien 2012

STAT5 is widely expressed and is encoded by two genes, STAT5A and STAT5B that probably arose by a gene duplication event and the respective proteins are 96% identical at the amino acid level, differing primarily in their carboxy termini (Liu et al. 1995). STAT5A and STAT5B transduce different signals and can be activated downstream of a variety of cytokines including the IL-3 family (IL-3, IL-5, and GM-CSF), the single chain family (GH, PRL, Tpo, and Epo), and the γ_C family (IL-2, IL-7, IL-9, IL-15, and IL-21). Interestingly, STAT5A preferentially forms tetramers when binding to DNA while STAT5B favours dimers (Verdier et al. 1998). STAT5A and B are not functionally redundant, although each can substitute for the other in many circumstances. In this review we will focus on the essential functions of STAT5 in the haematopoietic system and in the adult mammary gland during a reproductive cycle and discus the consequences of aberrant STAT5 activity for tumourigenesis.

STAT5 in Physiological Conditions: Adult Mammary Gland

Following the description of STAT5 in lactating mammary gland, its functional role as a regulator of milk protein gene expression was demonstrated in vivo. The promoter of the β-lactoglobulin (BLG) protein, a component of milk in ruminants and marsupials, was shown to have three STAT5 binding sites in the proximal 406 bp region (Burdon et al. 1994a) and mutation of pairs of these motifs abrogated expression of BLG. This was the first demonstration of a role for STAT5 as a transcriptional regulator in vivo (Burdon et al. 1994b) and was followed by studies showing that STAT5 regulates also the expression of other milk proteins such as whey acidic protein (WAP) in mammary gland (Li and Rosen 1995) and β-casein in HC11 mammary epithelial cells (Happ and Groner 1993). Thus, STAT5 is an important regulator of milk protein gene expression and its expression profile, being induced during pregnancy and peaking at lactation, suggested that it could have a role in the development of the milk-producing alveolar cells that arise during pregnancy.

Initial genetic studies focussed on deleting *Stat5a* and *Stat5b* individually. Disruption of *Stat5a* resulted in reduced lobuloalveolar development during pregnancy and failed lactation (Liu et al. 1997) while in contrast, *Stat5b* deficient mice had no mammary gland abnormalities but exhibited growth defects (Udy et al. 1997). However, STAT5B can partially rescue the failed lactation in *Stat5a* knockout mice in subsequent pregnancies demonstrating some functional redundancy. Combined deletion of both genes demonstrated that they mediate virtually all growth hormone (STAT5B) and prolactin (STAT5A) functions, although infertility precluded investigation of lobuloalveolar development (Teglund et al. 1998). Subsequently, conditional gene targeting revealed that STAT5 is required for lobuloalveolar development during pregnancy and that deletion of STAT5 from differentiated alveolar cells results in their rapid cell death (Cui et al. 2004), suggesting roles as a differentiation and survival factor (Fig. 1a).

Deletion of the upstream regulators of STAT5A, prolactin receptor (PRLR) and the receptor-associated kinase JAK2, essentially recapitulated the phenotype of STAT5A deficient mice (Ormandy et al. 1997; Wagner et al. 2004). Thus, STAT5 can be considered the nexus of signalling that controls lobuloalveolar development in mammary gland.

Since alveologenesis is abrogated in the absence of both STAT5A and B, it seems likely that these transcription factors could be important for specifying the alveolar lineage. In order to investigate this, mammary stem and progenitor cells were investigated by FACS analysis and fat pad transplantation using conditional deletion of *Stat5a/b* in all mammary epithelial lineages with the MMTV-Cre transgene (Yamaji et al. 2009). While loss of STAT5A and B did not affect the ability of stem cells to generate ductal outgrowths, it did affect the ability of *Stat5a/5b*$^{fl/fl}$; *Cre* stem cells to generate the alveolar compartment in pregnant mice. Expression of transgenic STAT5A rescued alveologenesis thus demonstrating a

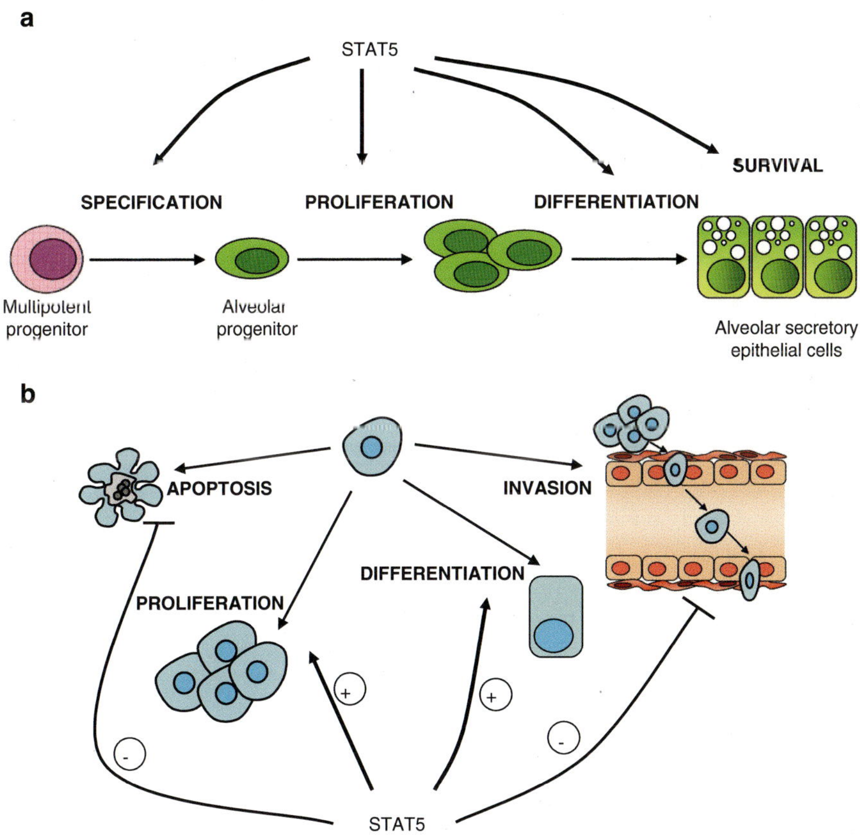

Fig. 1 Schematic representation of the role of STAT5 in lineage commitment and maintenance in normal mammary gland development (**a**) and in tumourigenesis in the breast (**b**)

requirement for STAT5A in the commitment of progenitors to the alveolar lineage (Fig. 1a).

Since ductal morphogenesis is unaffected in the absence of STAT5A and B, this distinguishes STAT5 from GATA-3, which is required in both ductal and alveolar lineages. Notably, whereas GATA-3 expression is unperturbed in *Stat5a/b* null cells, the expression of Elf5 is severely diminished (Yamaji et al. 2009). These data suggest that there are at least two alveolar lineages: a Prl-Stat5/Elf5 controlled lineage and a Stat6-GATA-3/oestrogen receptor (ERα) lineage that are dependent upon each other since ablation of either STAT5 or GATA-3 results in the death of differentiated epithelial cells suggesting an interaction between these lineages.

Much remains to be understood in terms of specificity of signalling and selection of downstream target genes. The simple signalling paradigm for the Jak/STAT pathway is clearly only part of the story. Interaction with a number of other pathways has been noted in addition to co-repressors and co-activators. Signalling through the ERBB4 receptor in mammary gland can induce STAT5 phosphorylation and notably, deletion of ERBB4 in differentiated mammary epithelial cells resulted in inhibition of cell proliferation and functional differentiation (Long et al. 2003). ERBB4 can also induce phosphorylation of serine residues in STAT5A and Schütz and colleagues (Engblom et al. 2007) demonstrated that the physical interaction of STAT5 with the glucocorticoid receptor (GR) is required for many of the functions exerted by either of these factors although deletion of GR in mammary epithelium did not affect lactation (Wintermantel et al. 2005), suggesting that this physical interaction is not essential in lactating mammary gland although it does appear to be important in hepatocytes (Engblom et al. 2007).

STAT5 in Physiological Conditions: The Haematopoietic System

STAT5A and STAT5B control stem and progenitor cell fate in the haematopoietic system (Wang et al. 2009) and are essential for the development of immune cells, playing redundant and non-redundant roles. In the absence of both STAT5A and B, mice have severely reduced development of B cells and αβ T cells, and an absence of γδ T-lymphocytes. Natural killer (NK) cells also fail to develop (Yao et al. 2006). In *STAT5*$^{-/-}$ mice, B cells do not progress beyond the uncommitted pre-pro-B cell stage in the bone marrow and these mice have severe combined immunodeficiency. This essentially recapitulates the phenotype of IL-7R, gammac, and Jak3 deficient mice. Binding of STAT5 to GAS sites in the promoter of the *Foxp3* gene (Yao et al. 2007) demonstrates that STAT5 directly regulates the transcription of the transcription factor FOXP3, a master regulator of Regulatory T (Treg) cells. Tregs suppress T-cell proliferation and function and in this way restrict immune responses against self- and nonself-antigens. Deletion of *Stat5* specifically at the onset of lymphopoiesis using conditional mutagenesis in adult mice demonstrated that the most important role of Stat5 during pro-B cell development is to maintain cell survival by activating Mcl1, a member of the Bcl2 family of anti-apoptosis regulators and to prevent premature *Igk* gene rearrangements by binding to an intronic enhancer of iEκ.

STAT5 in Pathological Situations: Cancer

Role of STAT5 in Cancers of the Blood

STAT5 has been implicated in the pathogenesis of different haematopoietic malignancies and various epithelial cancers (Ferbeyre and Moriggl 2011). The involvement of STAT5 in tumourigenesis of blood cells was demonstrated soon after its discovery in 1994. STAT5 was shown to be persistently activated in various haematopoietic malignancies: acute myelogenous leukemia (AML) and chronic myelogenous leukemia (CML), HTLV-1 and TEL-JAK2 dependent leukaemias (Migone et al. 1995; Schwaller et al. 2000) and myeloproliferative diseases like polycythemia vera, thrombocytopenia and mastocytosis (Ferbeyre and Moriggl 2011). Most frequently, STAT5 activation is induced by hyperactive tyrosine kinases such as JAK2. In fact, the JAK2 V617F mutation in the pseudokinase domain, which induces the constitutive activation of JAK2 and its downstream target STAT5, was found to be the cause of some of these myeloproliferative disorders (Baxter et al. 2005; Levine et al. 2005). In CML, it was demonstrated that the oncogenic BCR-ABL kinase is responsible for the persistent activation of STAT5 in the malignant cells (Huang et al. 2002) while in AML, constitutive activation is driven by the FLT3-receptor tyrosine kinase (Levis et al. 2002). In addition, naturally truncated forms of STAT5 have been implicated in blood-cell cancer, and they seem to play a role in the relapse of the disease (Xia et al. 2001). Recently, it was shown that phosphorylation of STAT5 on serines 725 and 779 is a prerequisite for haematopoietic transformation (Friedbichler et al. 2010) and formation of STAT5 tetramers in preference to dimers is associated with leukemogenesis (Moriggl et al. 2005).

Some of the STAT5 downstream targets that play a role in tumourigenesis were identified first in these haematopoietic malignancies. STAT5 can prevent apoptosis in cancer cells by increasing the expression of the anti-apoptotic proteins Bcl-x_L (Gesbert and Griffin 2000; Horita et al. 2000) and Mcl1 (Huang et al. 2002). Additionally, c-Myc is a well known regulator of apoptosis and cell proliferation (Dang 1999) and STAT5 has been shown to regulate its expression (Huang et al. 2002). Furthermore, many growth-factor signalling pathways are known to regulate cell proliferation by enhancing the levels of cyclins, which activate cyclin-dependent kinases. STAT5 activity can promote the expression of cyclin D1 and D2 (Martino et al. 2001; Hennighausen and Robinson 2005). Other genes that have been involved in the pro-survival and anti-apoptotic role of STAT5 are Bcl-2, Pim-1, A1, serine protease inhibitors Spi2.1 and Spi2.2 and Mcl-1 (Ferbeyre and Moriggl 2011). Interestingly, the pro-apoptotic micro RNAs, miR15/16, which inactivate Bcl-2 and Bcl- x_L can be suppressed by STAT5 (Li et al. 2010).

Another aspect of tumourigenesis that warrants consideration is the role of STAT proteins in the regulation on the immune system, particularly with regard to anti-tumour immune responses. STAT3 is well established as a modulator of the tumour microenvironment (Pensa et al. 2009). STAT5 also promotes tumourigenesis indirectly through a role in the expansion of T regulatory cells, which promote tumour progression by inhibiting anti-tumour immune responses (Yu et al. 2009).

Role of STAT5 in Breast Cancer

As described in the previous sections, STAT5 plays a key role in mouse mammary gland development and has been shown be active in lactating human breast (Watson and Miller 1995). It is critically important for the specification and proliferation of the alveolar progenitors and for the survival of terminally differentiated secretory epithelial cells. These cells are located at the ends of the ductal tree, a region known as terminal duct lobular units (TDLU) in humans. It is generally thought that the TDLUs are the site in the human breast where neoplastic lesions arise, as they contain hormone-responsive epithelial cells that are highly susceptible to transformation (Wellings et al. 1975). There is increasing evidence that STAT5 deregulation plays a role in breast tumourigenesis. However, data obtained in human breast cancer samples suggests that STAT5 activation can be a good prognostic factor and that STAT5 expression is lost during breast cancer progression (Wagner and Rui 2008). In the following sections, we will summarise the apparently conflicting data about the role of STAT5 in breast tumourigenesis using different animal models of mammary cancer, breast cancer cell lines and human breast cancer samples.

STAT5 in Breast Cancer Cell Lines and in vivo Models

Several studies have demonstrated that STAT5 promotes tumourigenesis of rodent mammary gland (Tan and Nevalainen 2008). Mouse mammary gland tumour models have been widely used, but there is also some evidence that STAT5 is important in rat mammary gland tumourigenesis (Shan et al. 2004). Using a chemical carcinogen induced rat mammary gland tumour model, this study showed nuclear immunostaining of STAT5A in 65% of carcinomas, while STAT5A was cytosolic in control mammary gland tissue. They also demonstrated that STAT5A nuclear localization was associated with high-grade carcinomas and correlated with the degree of proliferation in the tumours, determined by PCNA nuclear staining (Shan et al. 2004).

The role of STAT5A in mouse mammary cancer models has been studied in more depth. The deletion of STAT5A delayed mammary tumour development in a WAP driven transforming growth factor-α-induced cancer model (Humphreys and Hennighausen 1999). In another study, hemizygous loss of the STAT5A allele was sufficient to delay tumour formation and to reduce the number and the size of the tumours in a WAP induced SV40T antigen mammary tumour model (Ren et al. 2002). In addition, the over-expression of wild-type (wt) or constitutively active

STAT5A promoted the occurrence of sporadic and well differentiated mammary tumours after latency periods of 8–12 months (Iavnilovitch et al. 2004). The fact that only a small proportion of these mice developed tumours after a long latency period suggests that STAT5A cannot be considered to be a potent oncogene for the transformation of mammary cells in vivo. Interestingly, in the same study the over-expression of a carboxyl-terminally truncated dominant negative STAT5 induced also the transformation of mammary epithelial cells. In contrast to the tumours generated by over-expression of STAT5 wt, these tumours were poorly differentiated (Iavnilovitch et al. 2004). It is possible that this different phenotype is due to the inhibition of the endogenous wt STAT5. It is also reasonable to think that STAT5 lacking the C-terminal transactivation domain maintains some biological functions necessary for neoplastic transformation. In a later study by the same group, it was shown that the tumours originated by over-expression of wt STAT5A are highly dependent on parity and that they only occur in aged and postestropausal females (Eilon et al. 2007). It is possible that deregulation of STAT5A during the reproductive cycle, probably during the highly proliferative stage of pregnancy, initiates a sequence of events leading later to tumourigenesis (Barash 2006). Non-differentiated mammary epithelial cells that do not undergo cell death during involution could retain this deregulated STAT5 which could cooperate later with other oncogenic events thereby inducing mammary tumours (Wagner et al. 2002). Using lentiviral infection to introduce a constitutively active STAT5 (cS5-F), Vafaizadeh and colleagues showed that expression of this variant of STAT5 caused epithelial hyperproliferation, thickening of the ducts and precocious formation of alveoli in non-pregnant mice (Vafaizadeh et al. 2010). Interestingly, cS5-F expression resulted in the formation of ER^+PR^+ adenocarcinomas.

STAT5 has also a pro-survival role in human breast cancer cells. It has been described that STAT5B is constitutively active in human breast cancer cell lines and that over-expression of a dominant negative variant of STAT5 suppresses ERα transcriptional activity and induces apoptosis in luminal human breast cancer cells (Yamashita et al. 2003).

Different molecular targets of STAT5 could be involved in its pro-survival role in breast cancer cells. STAT5 deregulation may modulate cell cycle progression by enhancing Cyclin D1 transcription through its binding site in the cyclin D1 promoter (Matsumura et al. 1999; Brockman et al. 2002). Over-expression of cyclin D1 in transgenic mouse models leads to hyperproliferation of mammary cells and to mammary carcinomas (Wang et al. 1994). Akt1 could also be a target of STAT5 in its pro-survival role since it has been described that STAT5 can promote survival of mammary epithelial cells through transcriptional activation of a mammary gland specific promoter of Akt1 (Creamer et al. 2010). In addition, it could inhibit apoptosis through activation of the NF-κB pathway via RANK ligand, which is another STAT5 target (Hennighausen and Robinson 2005). STAT5 has been shown to interact with different proteins that have a role in breast cancer such as BRCA-1 and -2, ERK, ErbB2 and ERα (Olayioye et al. 1999; Barash 2006), and this could also explain its role in initiating mammary tumourigenesis.

The idea that over-activation of STAT5 plays a role in mammary transformation is supported by studies in caveolin-1 (Cav-1) knockout (ko) mice (Williams et al. 2003; Sotgia et al. 2006). Cav-1 has been shown to repress JAK2/STAT5 signalling in mammary epithelial cells (Park et al. 2002) and to inhibit breast cancer growth and metastasis (Sloan et al. 2004). The mammary glands of Cav-1 ko mice harbour a hyperactive JAK2/STAT5 pathway and developed hyperplastic and well differentiated mammary tumours (Sotgia et al. 2006). This hyper-activation has been associated with ERα and cyclin D1 expression (Li et al. 2006; Williams et al. 2006). Recent work from our laboratory has shown that expression of the constitutively active JAK2 mutant, V617F, in MCF-7 breast cancer cells promotes a more invasive phenotype in xenografts (Caffarel et al. 2012). The classical mechanism of STAT5 activation in mammary epithelial cells is through prolactin (PRL) receptor-JAK2 signalling. The role of PRL and PRL receptor in breast cancer has been extensively investigated. Over-expression of PRL in the mammary gland induced proliferation of mammary epithelial cells and ERα positive and negative mammary tumours (Rose-Hellekant et al. 2003). Likewise, loss of PRL receptor delayed mammary tumour formation and reduced mammary cancer cell proliferation in a model where transplants of mouse mammary epithelium expressing the SV40 Tantigen oncogene were implanted in wt or PRL receptor ko recipients (Oakes et al. 2007). It has yet to be determined whether this effect is due to inhibition of JAK2/STAT5 signalling or the result of the modulation of other PRL receptor targets as Src, Akt or Erk. Moreover, a variant of the PRL receptor exhibiting constitutive activity has been recently identified in breast cancer patients (Fernandez et al. 2010), supporting a role for the PRL signalling pathway in breast tumourigenesis.

Although there is sufficient evidence to support the notion that STAT5 signalling can induce tumourigenesis and promote cancer cell proliferation in in vivo mouse models of breast cancer and in some breast cancer cell lines, it appears that STAT5 has a tumour suppressive role in later stages of tumour development and metastasis. STAT5 activation by PRL has an invasion-suppressive role in human luminal breast cancer cell lines, as determined by an increase in the surface levels of E-cadherin and an inhibition of metalloproteinase activity, invasiveness and migration of these cells (Sultan et al. 2005). Inhibition of PRL-induced AP-1 signalling and Bcl6 expression by STAT5 in breast cancer cell lines could explain, at least in part, this invasion-suppressive role (Gutzman et al. 2007; Tran et al. 2010). AP-1 proteins are critical transcriptional regulators of the invasive phenotype of cancer cells (Ozanne et al. 2007). Reduction of STAT5 in breast cancer cells increased PRL-induced AP-1 signalling, leading to increased MMP-2 expression and associated invasive behaviour (Gutzman et al. 2007). On the other hand, Bcl6 expression has been correlated with high-grade and metastatic breast cancer and STAT5A has been shown to suppress Bcl6 expression in a wide array of breast cancer cell lines (Tran et al. 2010).

STAT5 in Human Tumour Samples

Although genomic alterations (i.e. gene amplifications or activating mutations) in STAT5 or JAK2 have not been reported in breast tumours, STAT5 seems to be active in a subset of these. The first evidence of STAT activation in human breast cancer was reported more than 15 years ago in a small samples of breast biopsies where it was shown that carcinoma samples had significantly higher STAT3 binding activity than benign lesions and normal breast tissues (Watson and Miller 1995). Although the activation of STAT5 was not detected in this study, it was shown later by immunohistochemical techniques in several studies and STAT5 activation has been shown to positively correlate with the differentiation status of the tumour (Cotarla et al. 2004; Nevalainen et al. 2004; Yamashita et al. 2006). In a series of 83 primary breast adenocarcinomas, STAT5A was activated in 76% of the samples and its activation was positively associated with the degree of differentiation of these tumours. No correlation was found with other markers of prognosis such as tumour size or lymph node mestastases (Cotarla et al. 2004). In a larger study, the analysis of more than 1,000 primary breast cancer specimens and 19 samples of healthy breast tissue confirmed a correlation between STAT5 activation and the differentiation status of the tumours (Nevalainen et al. 2004). Expression of phosphorylated STAT5 in the nucleus was associated with favourable prognosis and was gradually lost during breast cancer progression: STAT5 activation was shown in all the non transformed tissues and in less than 20% of node-positive breast cancers and mestastases (Nevalainen et al. 2004). Other studies confirmed the inverse correlation between STAT5 activation and the histological grade of the tumours and the positive correlation between STAT5 and good prognosis (Yamashita et al. 2006). In addition, they showed for the first time that STAT5 activation predicted a better response to endocrine therapy (Yamashita et al. 2006). Interestingly, it has been described that STAT5A activation is particularly increased in secretory carcinomas compared to the more common in situ or invasive ductal carcinomas (Strauss et al. 2006). Moreover, other specialised histological types of breast cancer such as apocrine metaplasia or mucinous carcinoma did not exhibit STAT5 activity (Strauss et al. 2006). This may reflect the cell of origin of breast cancer.

Further studies need to be carried out to determine whether STAT5 inactivation in the later stages of breast tumourigenesis and mestastasis is causally involved in the promotion of mestastasis, or whether it is just a correlative marker of epithelial-mesenchymal transition, being lost as cells lose their epithelial phenotype. The molecular mechanism underlying STAT5 inactivation in breast cancer progression is unknown. Possible mechanisms could be genetic or epigenetic changes that diminish STAT5 expression or modulate the regulators of STAT5. Supporting the latter idea, it has been described that the levels of SHP-1, a phosphatase that can inactivate STATs by de-phosphorylation (Shuai and Liu 2003), are increased in breast cancers compared to normal breast epithelial cells (Yip et al. 2000). The tyrosine phosphatase PTPB1 was able to suppress PRL mediated activation of STAT5 in breast cancer cell lines and a negative correlation between levels of active STAT5 and PTPB1 expression has been demonstrated (Johnson et al. 2010). In addition, the levels of suppressors of cytokine signalling (SOCS), well known negative regulators of STATs (Croker et al. 2008), seem to be altered in breast cancer. High levels of SOCS-1 and 3 were found in breast carcinomas and breast cancer cell lines compared to non-transformed breast tissue and cell lines (Raccurt et al. 2003). Conversely, SOCS-2 expression has been inversely correlated with the histological grade of breast cancer and correlated with higher survival rates (Haffner et al. 2007). SOCS-2 is primarily a negative regulator of STAT5 while SOCS1 and SOCS3 can also regulated STATs 1 and 3.

Clearly, STAT5 has a complex role in breast cancer (Fig. 1b). While it promotes tumourigenesis in mouse models, it is a marker of good prognosis in human breast tumours and inhibits the metastatic behaviour of human breast cancer cell lines. It seems, therefore, that STAT5 has a dual role in this particular type of cancer. STAT5 may promote the earlier steps of tumour progression, but maintain the differentiation status of established breast cancers, inhibiting their ability to metastasise. Another possible explanation is that the roles of STAT5A/B may be different in human and mouse mammary tumourigenesis. Further studies that address these questions need to be undertaken, as well as others that clarify the possibly different roles of STAT5A and B. In breast cancer cell lines, we know so far that constitutively active STAT5A seems to be more potent than constitutively active STAT5B in inducing survival and anchorage-independent growth and in inhibiting cell migration (Tang et al. 2010). Moreover, STAT5A and B seem to regulate differentially ERα and ERβ transcription (Frasor et al. 2001).

In addition, new mouse genetic models need to be found to study the role of STAT5 in breast tumourigenesis. The conventional models used so far: knock-out or over-expressing models (e.g. STAT5A or PRL receptor deficient mice) in combination with oncogene-expressing transgenes, are useful but have their limitations. New models where STAT5 could be deleted or over-expressed in the tumour cells once the tumours are formed would be more therapeutically relevant and would allow us to study the role of STAT5 in later stages of mammary tumourigenesis. The current models only give clues about the role of STAT5 in tumour initiation and, therefore, are useful only as models for cancer prevention (Wagner and Rui 2008). Furthermore, it is not clear to what extent the cell of origin or the initiating oncogene can dictate the type of tumour that arises. It seems likely, from the breast cancer studies discussed above, that STAT5 will be important in a subset of breast cell types.

Role of STAT5 in Other Solid Tumours

STAT5 is expressed and activated in other solid tumours such as prostate cancer (Liao et al. 2010), head and neck squamous cell carcinoma (HNSCC) (Lai and Johnson 2010) and melanoma (Mirmohammadsadegh et al. 2006), where it contributes to cell survival and proliferation. In addition, it has been found in non-small cell lung cancer tumours (Sanchez-Ceja et al. 2006) and in nasopharyngeal carcinomas (Hsiao et al. 2003). In these two types of cancer, STAT5 activation did not correlate with the progression of the disease. Conversely, in nasopharyngeal carcinomas it was associated with good prognosis (Hsiao et al. 2003). The role of STAT5 in liver cancer is controversial. STAT5B activation was correlated with tumour progression in hepatocellular carcinoma (HCC) samples and it enhanced HCC cells aggressiveness through induction of epithelial-mesenchymal transition (Lee et al. 2006). On the other hand, STAT5 deletion in transgenic mice led to liver fibrosis and cancer development through TGF-β and STAT3 activation (Hosui et al. 2009). In agreement with this tumour suppressor role of STAT5 in liver cancer, it has been published recently that STAT5 deletion in liver tissue led to enhanced cell cycle progression through the inactivation of the CDK inhibitors p15 and p21 (Yu et al. 2010).

Prostate Cancer

STAT5 is a critical survival factor for the growth of prostate cancer cells in vitro and in vivo. This has been demonstrated with experiments in prostate cancer cell lines and in the TRAMP (transgenic adenocarcinoma of mouse prostate) mouse prostate cancer model (Tan and Nevalainen 2008). Inhibition of STAT5A/B in STAT5-positive prostate cancer cells induces apoptosis and reduces their growth as xenografts in nude mice (Ahonen et al. 2003; Dagvadorj et al. 2008). Cyclin D1 and BCL-X_L were identified as important targets of STAT5 in prostate cancer cells (Dagvadorj et al. 2008). It has also been described that the inhibition of STAT5, by expression of a truncated STAT5B mutant, decreased the growth of mouse prostate tumour cells derived from TRAMP mice in soft agar and reduced tumour formation by these cells in nude mice (Kazansky et al. 2003). STAT5 can also promote metastatic behaviour of prostate cancer cells (Gu et al. 2010).

Studies in human prostate tumour samples have shown that STAT5 is constitutively activated in prostate tumours but not in the epithelium of adjacent normal prostate tissue (Ahonen et al. 2003). Moreover, its activation correlates with the histological grade of the tumours and is a prognostic marker of early disease recurrence (Li et al. 2004, 2005). STAT5 seems to be particularly important in hormone-refractory prostate cancer, where androgen receptor (AR) signalling remains active despite low levels of circulatory androgens (Isaacs and Isaacs 2004). It has been described that STAT5 is activated in 95% of these tumours and that it transcriptionally enhances AR expression (Tan and Nevalainen 2008). On the other hand, ligand-bound AR increases the transcriptional activity of STAT5 and it has been demonstrated that AR and STAT5 physically interact (Tan and Nevalainen 2008). The molecular mechanisms underlying constitutive activation of STAT5 in primary and hormone-refractory prostate tumours need to be elucidated. It has been proposed that they may involve autocrine PRL signalling in prostate cancer cells, as PRL and PRL receptor expression are associated with high histological grades of human prostate cancer (Li et al. 2004). Another possible mechanism is the amplification of STAT5A/B genes, as these genes are located in chromosome 17 (Clark et al. 2003) which is frequently altered in both incidental and hereditary prostate cancer (Gillanders et al. 2004) however further studies are required.

Head and Neck Squamous Cell Carcinoma (HNSCC)

STAT5 and STAT3 play a role in HNSCC disease progression. STAT5 activation correlated with progression to a malignant phenotype in HNSCC samples (Xi et al. 2003). Moreover, constitutive activation of STAT5 in HNSCC cells led to increased cell growth, migration, invasion, anchorage-independent growth and tumour volumes in xenografts in addition to epithelial-mesenchymal transition and resistance to cisplatin and erlotinib (an epidermal growth factor receptor tyrosine kinase inhibitor) (Koppikar et al. 2008). Interestingly, STAT5B has a more important role than STAT5A in HNSCC cell growth (Leong et al. 2002). In a xenograft model, blocking of STAT5B, but not STAT5A, resulted in tumour growth inhibition and abrogation of STAT5 targets (Xi et al. 2003). STAT5A could have a dominant role in regulating cell invasion. It has been shown that erythropoietin activates the JAK2-STAT5 pathway, resulting in HNSCC invasion, mainly through the STAT5A isoform (Lai et al. 2005).

Concluding Remarks

STAT5 has a clear cut and essential role in normal development of the mammary gland and is essential for controlling cell fate in the haematopoietic system. STAT5A is the more important isoform in mammary gland where it is the mediator of PRL signalling via JAK2 while in haematopoietic cells, both STAT5A and STAT5B are required although they have both redundant and non-redundant functions. Constitutive activation of STAT5 results in mammary cancers and other solid tumours and in leukemias and myeloproliferative disorders. An interesting question for future research is whether the role of STAT5 in establishing and maintaining particular lineages is reflected in the types of tumours that arise and the cell of origin of the cancer.

References

Ahonen TJ, Xie J et al (2003) Inhibition of transcription factor Stat5 induces cell death of human prostate cancer cells. J Biol Chem 278(29):27287–27292

Barash I (2006) Stat5 in the mammary gland: controlling normal development and cancer. J Cell Physiol 209(2):305–313

Baxter EJ, Scott LM et al (2005) Acquired mutation of the tyrosine kinase JAK2 in human myeloproliferative disorders. Lancet 365(9464):1054–1061

Brockman JL, Schroeder MD et al (2002) PRL activates the cyclin D1 promoter via the Jak2/Stat pathway. Mol Endocrinol 16(4):774–784

Burdon TG, Demmer J et al (1994a) The mammary factor MPBF is a prolactin-induced transcriptional regulator which binds to STAT factor recognition sites. FEBS Lett 350(2–3):177–182

Burdon TG, Maitland KA et al (1994b) Regulation of the sheep beta-lactoglobulin gene by lactogenic hormones is mediated by a transcription factor that binds an interferon-gamma activation site-related element. Mol Endocrinol 8(11):1528–1536

Caffarel MM, Zaragoza R et al (2012) Constitutive activation of JAK2 in mammary epithelium elevates Stat5 signalling, promotes alveologenesis and resistance to cell death, and contributes to tumourigenesis. Cell Death Differ 19(3):511–522

Clark J, Edwards S et al (2003) Genome-wide screening for complete genetic loss in prostate cancer by comparative hybridization onto cDNA microarrays. Oncogene 22(8):1247–1252

Cotarla I, Ren S et al (2004) Stat5a is tyrosine phosphorylated and nuclear localized in a high proportion of human breast cancers. Int J Cancer 108(5):665–671

Creamer BA, Sakamoto K et al (2010) Stat5 promotes survival of mammary epithelial cells through transcriptional activation of a distinct promoter in Akt1. Mol Cell Biol 30(12):2957–2970

Croker BA, Kiu H et al (2008) SOCS regulation of the JAK/STAT signalling pathway. Semin Cell Dev Biol 19(4):414–422

Cui Y, Riedlinger G et al (2004) Inactivation of Stat5 in mouse mammary epithelium during pregnancy reveals distinct functions in cell proliferation, survival, and differentiation. Mol Cell Biol 24(18):8037–8047

Dagvadorj A, Kirken RA et al (2008) Transcription factor signal transducer and activator of transcription 5 promotes growth of human prostate cancer cells in vivo. Clin Cancer Res 14(5):1317–1324

Dang CV (1999) c-Myc target genes involved in cell growth, apoptosis, and metabolism. Mol Cell Biol 19(1):1–11

Eilon T, Groner B et al (2007) Tumors caused by overexpression and forced activation of Stat5 in mammary epithelial cells of transgenic mice are parity-dependent and developed in aged, postestropausal females. Int J Cancer 121(9):1892–1902

Engblom D, Kornfeld JW et al (2007) Direct glucocorticoid receptor-Stat5 interaction in hepatocytes controls body size and maturation-related gene expression. Genes Dev 21(10):1157–1162

Ferbeyre G, Moriggl R (2011) The role of Stat5 transcription factors as tumor suppressors or oncogenes. Biochim Biophys Acta 1815:104–114

Fernandez I, Touraine P et al (2010) Prolactin and human tumourogenesis. J Neuroendocrinol 22(7):771–777

Frasor J, Park K et al (2001) Differential roles for signal transducers and activators of transcription 5a and 5b in PRL stimulation of ERalpha and ERbeta transcription. Mol Endocrinol 15 (12):2172–2181

Friedbichler K, Kerenyi MA et al (2010) Stat5a serine 725 and 779 phosphorylation is a prerequisite for hematopoietic transformation. Blood 116(9):1548–1558

Gesbert F, Griffin JD (2000) Bcr/Abl activates transcription of the Bcl-X gene through STAT5. Blood 96(6):2269–2276

Gillanders EM, Xu J et al (2004) Combined genome-wide scan for prostate cancer susceptibility genes. J Natl Cancer Inst 96(16):1240–1247

Gu L, Vogiatzi P et al (2010) Stat5 promotes metastatic behavior of human prostate cancer cells in vitro and in vivo. Endocr Relat Cancer 17(2):481–493

Gutzman JH, Rugowski DE et al (2007) Stat5 activation inhibits prolactin-induced AP-1 activity: distinct prolactin-initiated signals in tumorigenesis dependent on cell context. Oncogene 26(43):6341–6348

Haffner MC, Petridou B et al (2007) Favorable prognostic value of SOCS2 and IGF-I in breast cancer. BMC Cancer 7:136

Happ B, Groner B (1993) The activated mammary gland specific nuclear factor (MGF) enhances in vitro transcription of the beta-casein gene promoter. J Steroid Biochem Mol Biol 47(1–6):21–30

Hennighausen L, Robinson GW (2005) Information networks in the mammary gland. Nat Rev Mol Cell Biol 6(9):715–725

Horita M, Andreu EJ et al (2000) Blockade of the Bcr-Abl kinase activity induces apoptosis of chronic myelogenous leukemia cells by suppressing signal transducer and activator of transcription 5-dependent expression of Bcl-xL. J Exp Med 191(6):977–984

Hosui A, Kimura A et al (2009) Loss of STAT5 causes liver fibrosis and cancer development through increased TGF-{beta} and STAT3 activation. J Exp Med 206(4):819–831
Hsiao JR, Jin YT et al (2003) Constitutive activation of STAT3 and STAT5 is present in the majority of nasopharyngeal carcinoma and correlates with better prognosis. Br J Cancer 89(2):344–349
Huang M, Dorsey JF et al (2002) Inhibition of Bcr-Abl kinase activity by PD180970 blocks constitutive activation of Stat5 and growth of CML cells. Oncogene 21(57):8804–8816
Humphreys RC, Hennighausen L (1999) Signal transducer and activator of transcription 5a influences mammary epithelial cell survival and tumorigenesis. Cell Growth Differ 10(10):685–694
Iavnilovitch E, Cardiff RD et al (2004) Deregulation of Stat5 expression and activation causes mammary tumors in transgenic mice. Int J Cancer 112(4):607–619
Isaacs JT, Isaacs WB (2004) Androgen receptor outwits prostate cancer drugs. Nat Med 10(1):26–27
Johnson KJ, Peck AR et al (2010) PTP1B suppresses prolactin activation of Stat5 in breast cancer cells. Am J Pathol 177(6):2971–2983
Kazansky AV, Spencer DM et al (2003) Activation of signal transducer and activator of transcription 5 is required for progression of autochthonous prostate cancer: evidence from the transgenic adenocarcinoma of the mouse prostate system. Cancer Res 63(24):8757–8762
Koppikar P, Lui VW et al (2008) Constitutive activation of signal transducer and activator of transcription 5 contributes to tumor growth, epithelial-mesenchymal transition, and resistance to epidermal growth factor receptor targeting. Clin Cancer Res 14(23):7682–7690
Lai SY, Johnson FM (2010) Defining the role of the JAK-STAT pathway in head and neck and thoracic malignancies: implications for future therapeutic approaches. Drug Resist Updat 13(3):67–78
Lai SY, Childs EE et al (2005) Erythropoietin-mediated activation of JAK-STAT signaling contributes to cellular invasion in head and neck squamous cell carcinoma. Oncogene 24(27):4442–4449
Lee TK, Man K et al (2006) Signal transducers and activators of transcription 5b activation enhances hepatocellular carcinoma aggressiveness through induction of epithelial-mesenchymal transition. Cancer Res 66(20):9948–9956
Leong PL, Xi S et al (2002) Differential function of STAT5 isoforms in head and neck cancer growth control. Oncogene 21(18):2846–2853
Levine RL, Wadleigh M et al (2005) Activating mutation in the tyrosine kinase JAK2 in polycythemia vera, essential thrombocythemia, and myeloid metaplasia with myelofibrosis. Cancer Cell 7(4):387–397
Levis M, Allebach J et al (2002) A FLT3-targeted tyrosine kinase inhibitor is cytotoxic to leukemia cells in vitro and in vivo. Blood 99(11):3885–3891
Li S, Rosen JM (1995) Nuclear factor I and mammary gland factor (STAT5) play a critical role in regulating rat whey acidic protein gene expression in transgenic mice. Mol Cell Biol 15(4):2063–2070
Li H, Ahonen TJ et al (2004) Activation of signal transducer and activator of transcription 5 in human prostate cancer is associated with high histological grade. Cancer Res 64(14):4774–4782
Li H, Zhang Y et al (2005) Activation of signal transducer and activator of transcription-5 in prostate cancer predicts early recurrence. Clin Cancer Res 11(16):5863–5868
Li T, Sotgia F et al (2006) Caveolin-1 mutations in human breast cancer: functional association with estrogen receptor alpha-positive status. Am J Pathol 168(6):1998–2013
Li G, Miskimen KL et al (2010) STAT5 requires the N-domain for suppression of miR15/16, induction of bcl-2, and survival signaling in myeloproliferative disease. Blood 115(7):1416–1424
Liao Z, Lutz J et al (2010) Transcription factor Stat5a/b as a therapeutic target protein for prostate cancer. Int J Biochem Cell Biol 42(2):186–192

Liu X, Robinson GW et al (1995) Cloning and expression of Stat5 and an additional homologue (Stat5b) involved in prolactin signal transduction in mouse mammary tissue. Proc Natl Acad Sci USA 92(19):8831–8835
Liu X, Robinson GW et al (1997) Stat5a is mandatory for adult mammary gland development and lactogenesis. Genes Dev 11(2):179–186
Long W, Wagner KU et al (2003) Impaired differentiation and lactational failure of Erbb4-deficient mammary glands identify ERBB4 as an obligate mediator of STAT5. Development 130(21):5257–5268
Martino A, Holmes JH IV et al (2001) Stat5 and Sp1 regulate transcription of the cyclin D2 gene in response to IL-2. J Immunol 166(3):1723–1729
Matsumura I, Kitamura T et al (1999) Transcriptional regulation of the cyclin D1 promoter by STAT5: its involvement in cytokine-dependent growth of hematopoietic cells. EMBO J 18(5):1367–1377
Migone TS, Lin JX et al (1995) Constitutively activated Jak-STAT pathway in T cells transformed with HTLV-I. Science 269(5220):79–81
Mirmohammadsadegh A, Hassan M et al (2006) STAT5 phosphorylation in malignant melanoma is important for survival and is mediated through SRC and JAK1 kinases. J Invest Dermatol 126(10):2272–2280
Moriggl R, Sexl V et al (2005) Stat5 tetramer formation is associated with leukemogenesis. Cancer Cell 7(1):87–99
Nevalainen MT, Xie J et al (2004) Signal transducer and activator of transcription-5 activation and breast cancer prognosis. J Clin Oncol 22(11):2053–2060
Oakes SR, Robertson FG et al (2007) Loss of mammary epithelial prolactin receptor delays tumor formation by reducing cell proliferation in low-grade preinvasive lesions. Oncogene 26(4):543–553
Olayioye MA, Beuvink I et al (1999) ErbB receptor-induced activation of stat transcription factors is mediated by Src tyrosine kinases. J Biol Chem 274(24):17209–17218
Ormandy CJ, Binart N et al (1997) Mammary gland development in prolactin receptor knockout mice. J Mammary Gland Biol Neoplasia 2(4):355–364
Ozanne BW, Spence HJ et al (2007) Transcription factors control invasion: AP-1 the first among equals. Oncogene 26(1):1–10
Park DS, Lee H et al (2002) Caveolin-1-deficient mice show accelerated mammary gland development during pregnancy, premature lactation, and hyperactivation of the Jak-2/STAT5a signaling cascade. Mol Biol Cell 13(10):3416–3430
Pensa S, Watson CJ et al (2009) Stat3 and the inflammation/acute phase response in involution and breast cancer. J Mammary Gland Biol Neoplasia 14(2):121–129
Raccurt M, Tam SP et al (2003) Suppressor of cytokine signalling gene expression is elevated in breast carcinoma. Br J Cancer 89(3):524–532
Ren S, Cai HR et al (2002) Loss of Stat5a delays mammary cancer progression in a mouse model. Oncogene 21(27):4335–4339
Rose-Hellekant TA, Arendt LM et al (2003) Prolactin induces ERalpha-positive and ERalpha-negative mammary cancer in transgenic mice. Oncogene 22(30):4664–4674
Sanchez-Ceja SG, Reyes-Maldonado E et al (2006) Differential expression of STAT5 and Bcl-xL, and high expression of Neu and STAT3 in non-small-cell lung carcinoma. Lung Cancer 54(2):163–168
Schindler C, Shuai K et al (1992) Interferon-dependent tyrosine phosphorylation of a latent cytoplasmic transcription factor. Science 257(5071):809–813
Schwaller J, Parganas E et al (2000) Stat5 is essential for the myelo- and lymphoproliferative disease induced by TEL/JAK2. Mol Cell 6(3):693–704
Shan L, Yu M et al (2004) Possible role of Stat5a in rat mammary gland carcinogenesis. Breast Cancer Res Treat 88(3):263–272
Shuai K, Liu B (2003) Regulation of JAK-STAT signalling in the immune system. Nat Rev Immunol 3(11):900–911

Sloan EK, Stanley KL et al (2004) Caveolin-1 inhibits breast cancer growth and metastasis. Oncogene 23(47):7893–7897

Sotgia F, Williams TM et al (2006) Caveolin-1 deficiency (−/−) conveys premalignant alterations in mammary epithelia, with abnormal lumen formation, growth factor independence, and cell invasiveness. Am J Pathol 168(1):292–309

Strauss BL, Bratthauer GL et al (2006) STAT 5a expression in the breast is maintained in secretory carcinoma, in contrast to other histologic types. Hum Pathol 37(5):586–592

Sultan AS, Xie J et al (2005) Stat5 promotes homotypic adhesion and inhibits invasive characteristics of human breast cancer cells. Oncogene 24(5):746–760

Tan SH, Nevalainen MT (2008) Signal transducer and activator of transcription 5A/B in prostate and breast cancers. Endocr Relat Cancer 15(2):367–390

Tang JZ, Zuo ZH et al (2010) Signal transducer and activator of transcription (STAT)-5A and STAT5B differentially regulate human mammary carcinoma cell behavior. Endocrinology 151(1):43–55

Teglund S, McKay C et al (1998) Stat5a and Stat5b proteins have essential and nonessential, or redundant, roles in cytokine responses. Cell 93(5):841–850

Tran TH, Utama FE et al (2010) Prolactin inhibits BCL6 expression in breast cancer through a Stat5a-dependent mechanism. Cancer Res 70(4):1711–1721

Udy GB, Towers RP et al (1997) Requirement of STAT5b for sexual dimorphism of body growth rates and liver gene expression. Proc Natl Acad Sci USA 94(14):7239–7244

Vafaizadeh V, Klemmt P et al (2010) Mammary epithelial reconstitution with gene-modified stem cells assigns roles to Stat5 in luminal alveolar cell fate decisions, differentiation, involution, and mammary tumor formation. Stem Cells 28(5):928–938

Verdier F, Rabionet R et al (1998) A sequence of the CIS gene promoter interacts preferentially with two associated STAT5A dimers: a distinct biochemical difference between STAT5A and STAT5B. Mol Cell Biol 18(10):5852–5860

Wagner KU, Rui H (2008) Jak2/Stat5 signaling in mammogenesis, breast cancer initiation and progression. J Mammary Gland Biol Neoplasia 13(1):93–103

Wagner KU, Boulanger CA et al (2002) An adjunct mammary epithelial cell population in parous females: its role in functional adaptation and tissue renewal. Development 129(6):1377–1386

Wagner KU, Krempler A et al (2004) Impaired alveologenesis and maintenance of secretory mammary epithelial cells in Jak2 conditional knockout mice. Mol Cell Biol 24(12):5510–5520

Wakao H, Schmitt-Ney M et al (1992) Mammary gland-specific nuclear factor is present in lactating rodent and bovine mammary tissue and composed of a single polypeptide of 89 kDa. J Biol Chem 267(23):16365–16370

Wang TC, Cardiff RD et al (1994) Mammary hyperplasia and carcinoma in MMTV-cyclin D1 transgenic mice. Nature 369(6482):669–671

Wang Z, Li G et al (2009) Conditional deletion of STAT5 in adult mouse hematopoietic stem cells causes loss of quiescence and permits efficient nonablative stem cell replacement. Blood 113(20):4856–4865

Watson CJ, Miller WR (1995) Elevated levels of members of the STAT family of transcription factors in breast carcinoma nuclear extracts. Br J Cancer 71(4):840–844

Watson CJ, Gordon KE et al (1991) Interaction of DNA-binding proteins with a milk protein gene promoter in vitro: identification of a mammary gland-specific factor. Nucleic Acids Res 19(23):6603–6610

Wellings SR, Jensen HM et al (1975) An atlas of subgross pathology of the human breast with special reference to possible precancerous lesions. J Natl Cancer Inst 55(2):231–273

Williams TM, Cheung MW et al (2003) Loss of caveolin-1 gene expression accelerates the development of dysplastic mammary lesions in tumor-prone transgenic mice. Mol Biol Cell 14(3):1027–1042

Williams TM, Sotgia F et al (2006) Stromal and epithelial caveolin-1 both confer a protective effect against mammary hyperplasia and tumorigenesis: caveolin-1 antagonizes cyclin D1 function in mammary epithelial cells. Am J Pathol 169(5):1784–1801

Wintermantel TM, Bock D et al (2005) The epithelial glucocorticoid receptor is required for the normal timing of cell proliferation during mammary lobuloalveolar development but is dispensable for milk production. Mol Endocrinol 19(2):340–349

Xi S, Zhang Q et al (2003) Constitutive activation of Stat5b contributes to carcinogenesis in vivo. Cancer Res 63(20):6763–6771

Xia Z, Sait SN et al (2001) Truncated STAT proteins are prevalent at relapse of acute myeloid leukemia. Leuk Res 25(6):473–482

Yamaji D, Na R et al (2009) Development of mammary luminal progenitor cells is controlled by the transcription factor STAT5A. Genes Dev 23(20):2382–2387

Yamashita H, Iwase H et al (2003) Naturally occurring dominant-negative Stat5 suppresses transcriptional activity of estrogen receptors and induces apoptosis in T47D breast cancer cells. Oncogene 22(11):1638–1652

Yamashita H, Nishio M et al (2006) Stat5 expression predicts response to endocrine therapy and improves survival in estrogen receptor-positive breast cancer. Endocr Relat Cancer 13(3):885–893

Yao Z, Cui Y et al (2006) Stat5a/b are essential for normal lymphoid development and differentiation. Proc Natl Acad Sci USA 103(4):1000–1005

Yao Z, Kanno Y et al (2007) Nonredundant roles for Stat5a/b in directly regulating Foxp3. Blood 109(10):4368–4375

Yip SS, Crew AJ et al (2000) Up-regulation of the protein tyrosine phosphatase SHP-1 in human breast cancer and correlation with GRB2 expression. Int J Cancer 88(3):363–368

Yu H, Pardoll D et al (2009) STATs in cancer inflammation and immunity: a leading role for STAT3. Nat Rev Cancer 9(11):798–809

Yu JH, Zhu BM et al (2010) The transcription factors signal transducer and activator of transcription 5A (STAT5A) and STAT5B negatively regulate cell proliferation through the activation of cyclin-dependent kinase inhibitor 2b (Cdkn2b) and Cdkn1a expression. Hepatology 52(5):1808–1818

Crosstalk with the Jak-STAT Pathway in Inflammation

Lionel B. Ivashkiv

Abstract
The Jak-STAT signaling pathway is activated by multiple immune cytokines and plays a key role in mediating inflammatory responses. The functional outcomes of Jak-STAT signaling are modulated by signaling crosstalk with heterologous signaling pathways. Conversely, Jak-STAT signaling regulates cell responses to multiple cytokines and inflammatory factors. Emerging evidence suggests that on balance Jak-STAT signaling is pathogenic in chronic inflammatory disorders, as Jak inhibitors have demonstrated efficacy in preclinical disease models and early clinical trials in rheumatoid arthritis (RA). This review describes Jak-STAT signaling crosstalk with pathways activated by inflammatory cytokines such as TNF-α, pattern recognition receptors such as Toll-like receptors, and ITAM-associated receptors, including crosstalk at the level of chromatin modification and gene expression. The Jak-STAT pathway is placed within the context of a signaling network that determines functional responses and outcomes during inflammation and in chronic inflammatory diseases such as RA.

Overview

The discovery and initial characterization of the Jak-STAT pathway suggested a linear signal transduction pathway that could specifically activate gene expression (Darnell et al. 1994). The key event in STAT activation is Jak-mediated

L.B. Ivashkiv (✉)
Arthritis and Tissue Degeneration Program, Hospital for Special Surgery, 535 East 70th Street, New York, NY 10021, USA

Graduate Program in Immunology and Microbial Pathogenesis, Weill Cornell Graduate School of Medical Sciences, New York, NY, USA
e-mail: ivashkivl@hss.edu

Th. Decker and M. Müller (eds.), *Jak-Stat Signaling: From Basics to Disease*,
DOI 10.1007/978-3-7091-0891-8_19,

phosphorylation of a conserved tyrosine residue that allows dimerization and nuclear translocation of STATs, with subsequent gene activation events determined by specific interactions of STAT DNA binding domains with cognate DNA sequences present in the regulatory regions of STAT target genes. It rapidly became apparent that STAT transcriptional activity was upregulated by phosphorylation of conserved carboxy-terminal serine residues that can be phosphorylated by various kinases, including MAPKs and PKC. This dual activation by distinct JAK tyrosine kinases and serine kinases mediated positive crosstalk between cytokine receptors that induce tyrosine phosphorylation and various receptors that induce STAT serine phosphorylation, including antigen receptors in lymphocytes, inflammatory cytokines such as TNF-α and IL-1, and sensors of microbial products, such as Toll-like receptors (TLRs) (Varinou et al. 2003; Wen et al. 1995). Subsequently it has become clear that Jak-STAT signaling is modulated in both positive and negative directions by crosstalk with multiple heterologous signaling pathways. Jak-STAT signaling, in turn, modulates signaling by various receptor systems important for immune and inflammatory responses, including pattern recognition receptors (PRRs) that sense microbial products, inflammatory cytokines such as TNF-α and IL-1, and ITAM-associated receptors that signal via immune tyrosine-based activation motifs (ITAMs). Indeed, the Jak-STAT pathway is enmeshed in a complex signaling network, with bidirectional regulation of signaling that determines functional outcomes. It is also becoming clear that Jak-STAT signaling crosstalk occurs at the level of chromatin modification and epigenetic regulation. In this review, we will focus on more recent developments in Jak-STAT signaling crosstalk, with an emphasis on the regulation of innate immunity and inflammation. In addition, we will discuss how dysregulation of signaling crosstalk can contribute to the pathogenesis of autoimmune and inflammatory diseases, such as rheumatoid arthritis (RA) and systemic lupus erythematosus (SLE).

Crosstalk That Enhances Jak-STAT Signaling

One of the first examples of enhanced Jak-STAT signaling was the finding that pretreatment of cells with low concentrations of type I or type II IFNs (IFNα/β and IFN-γ, respectively) results in increased Jak-STAT signaling on subsequent challenge with cytokines, including IFNs themselves and IL-6 (Taniguchi and Takaoka 2001). This phenomenon, termed priming, can be effectively achieved by low concentrations of IFNs that do not actually activate cells. The function of priming is to enhance rapid and effective host defense by inducing a state of 'readiness' in the innate immune system, and by allowing low concentrations of cytokines to induce effective signaling responses. Priming also appears to occur during the course of autoimmune and inflammatory diseases (Hu et al. 2002; Ivashkiv 2003; Kono et al. 2003; Taniguchi and Takaoka 2001; van der Pouw Kraan et al. 2003), in which case enhanced inflammatory responses to cytokines and other activating factors could contribute to pathogenesis.

Enhancement of IFN-γ Jak-STAT Signaling

One example of priming involves 'self-priming' whereby low concentrations of IFN-γ sensitize macrophages for a robust STAT1-mediated response to a subsequent challenge with IFN-γ. During macrophage priming by low doses of IFN-γ, IFN-γ signaling is sensitized by a mechanism that involves increased STAT1 expression (Hu et al. 2002). Low priming concentrations of IFN-γ do not actually activate macrophages, but instead transiently induce expression of a small subset of IFN-γ-inducible genes, including STAT1. STAT1 protein accumulates in primed cells because of ongoing gene expression and the stability of STAT1 protein, which exhibits a half life of greater than 24 h. Primed macrophages, then, appear to be quiescent, but strongly activate STAT1 upon rechallenge with even very small amounts of IFN-γ, with concomitant activation of downstream STAT1-dependent genes and inflammatory functions. IFN-γ is a major activator of macrophages, and sensitization of IFN-γ signaling may be particularly important to achieve full macrophage activation early in immune responses when IFN-γ levels are low.

Several lines of evidence support a role for increased STAT1 expression in sensitization of IFN-γ signaling. First, sensitization is not accompanied by any changes in expression of IFN-γ receptors, or in the level of activation of Jak1, Jak2, or STAT3 by IFN-γ. These results indicate that IFN-γ delivers a comparable proximal signal to both non-primed and primed macrophages. Second, the rate of de-activation of STAT1 is comparable in non-primed and primed cells, indicating that priming does not inactivate a STAT1 phosphatase or suppress degradation by proteasomes. Third, sensitization of signaling is specific for STAT1 relative to STAT3 when either IFN-γ or IFN-α are used to stimulate primed cells, consistent with the relative expression levels of these STATs, and increased STAT1 activation is recapitulated by forced expression of STAT1. These data argue for a model where an increased intracellular STAT1 concentration leads to more efficient docking onto the activated IFN-γ receptor complex.

An important component of this model is that low priming doses of IFN-γ capable of activating sustained STAT1 protein expression only transiently and weakly activate expression of feedback inhibitory molecules such as SOCS1. Thus, STAT1 activation proceeds in primed macrophages unopposed by feedback mechanisms that restrain Jak-STAT signaling. In contrast, high activating doses of IFN-γ induce sustained expression of SOCS1, and thus engage feedback inhibition that counterbalances STAT1 activation. Thus, the amplitude of signaling is regulated by relative strength of induction of activating signals versus feedback inhibitory mechanisms.

STAT1 expression is dynamically regulated during the course of immune and inflammatory responses, and STAT1 expression levels thus can regulate the pattern of STAT activation by a cytokine over time. For example, elevation of STAT1 expression during the course of a viral infection results in increased STAT1 activation, and diminished STAT4 activation, in response to type I IFNs (Nguyen et al. 2002). Interestingly, elevated STAT1 mRNA and/or protein levels have been detected in several autoimmune/inflammatory conditions, including systemic lupus erythematosus, RA, and hepatitis (Baechler et al. 2003; Bennett et al. 2003;

Ehrt et al. 2001; Hong et al. 2002; Hu et al. 2002; Kuroiwa et al. 2003). Thus, increased STAT1 expression in these conditions will alter cellular responses to cytokines such as IFN-γ. The role for IFN-γ in RA has been controversial, and IFN-γ protein has been difficult to detect in RA joints (Firestein and Zvaifler 2002; Ivashkiv 1996). Recent work demonstrating that RA synovial cells express high levels of STAT1 and also express IFN-γ-inducible genes (Antoniv and Ivashkiv 2006; Hu et al. 2002; Ivashkiv and Hu 2003; van der Pouw Kraan et al. 2003) supports the notion that low levels of IFN-γ may activate gene expression through the above described autosensitization mechanism in RA synovium.

Inflammatory Shift in Type I IFN Signaling

Type I IFNs are pleiotropic cytokines that can either activate or suppress immune responses. Type I IFNs signal via the IFNAR receptor comprised of IFNAR1 and IFNAR2 subunits to activate various STATs, including STAT1:STAT2 heterodimers (which associate with IRF9 to form the ISGF3 complex that activates antiviral genes) and STAT1:STAT1 homodimers that activate inflammatory genes, similar to IFN-γ. Priming of macrophages with low concentrations of IFN-γ that induce high STAT1 levels shifts the balance of IFN-*a*-induced STAT activation towards STAT1:STAT1 homodimers, with increased expression of inflammatory STAT1 target genes such as *CXCL9* and *CXCL10* (Tassiulas et al. 2004). Preferential activation of STAT1 and subsequent stronger pro-inflammatory macrophage responses to IFN-α might have a role in the pathogenesis of IFN-mediated diseases such as SLE (Wang et al. 2008).

Preferential and increased IFN-α-induced activation of STAT1 in IFN-γ-primed macrophages requires not only increased STAT1 expression, but also an additional signaling input from ITAM-coupled receptors (Tassiulas et al. 2004). Indeed, low level basal signaling by the ITAM-containing adaptors DAP12 and FcRγ via the Syk protein tyrosine kinase enhances IFN-α-induced activation of Jaks and STAT1 even in unprimed cells (Wang et al. 2008). Basal (also termed tonic) ITAM signaling is induced by ongoing ligation of ITAM-associated receptors by as yet unknown ligands that are proposed to be constitutively expressed on myeloid cells. Calcium-dependent signaling pathways link ITAM-coupled adaptors and IFN receptors by a pathway consisting of DAP12/FcRγ-Syk-calcium-CaMK-Pyk2; Pyk2 interacts with Jaks and can amplify their activity (Wang et al. 2008). In tightly adherent murine macrophages, integrins, recently identified to be coupled to DAP12, are major contributors to tonic ITAM signaling, indicating a role for macrophage-ECM interactions in regulation of IFN responses. The calcium-dependent Pyk2 tyrosine kinase associates with Syk and with Jaks and can relay signals between these two different receptor systems. Syk can also phosphorylate STAT1 directly and thus contributes to preferential activation of STAT1 in primed macrophages that express high STAT1 concentrations (Tassiulas et al. 2004). Preferential activation of STAT1 results in stronger pro-inflammatory macrophage responses to IFN-α, and ITAM-enhanced IFN signaling may play a role in the pathogenesis of IFN-mediated diseases such as SLE. IFN-γ priming can increase the positive effects of ITAMs on IFN-α signaling by increasing STAT1 expression and modulating the expression of multiple ITAM-associated receptors. Thus, communication between ITAM-coupled receptors and IFNs is bidirectional and allows coupling and fine-tuning of responses to both sets of receptors.

TNF-α and Jak-STAT Signaling

Receptors for inflammatory cytokines in the TNF and IL-1 families do not associate with Jaks and thus do not directly activate Jak-STAT signaling. Recently we found that TNF-α activates IRF1 to induce production of very low levels (typically <20 pg/mL) of IFNβ (Yarilina et al. 2008). Production of IFNβ was sustained for up to 24 h and was sufficient to weakly activate Jak-STAT signaling via IFNAR, as detected by low levels of STAT1 and STAT2 tyrosine phosphorylation. However, the combined action of canonical and direct TNF-α signaling via MAPKs and NF-κB and low level Jak-STAT signaling was sufficient to strongly induce expression of genes, such as *CXCL9*, *CXCL10* and *CCL5* that encode inflammatory chemokines and are synergistically activated by both pathways. This combination of direct and autocrine TNF-α signaling strongly induced STAT1 expression and accumulation of STAT1 protein, and subsequent emergence of a classic IFN response characterized by expression of IRF7 and antiviral genes such as Mx1 (Yarilina et al. 2008). This delayed IFN response is likely mediated by the priming mechanisms discussed above and by high STAT1 expression. Thus, TNF-α induces a delayed IFN response mediated by Jak-STAT signaling. Interestingly, Jak inhibitors have emerged as potentially effective therapies for rheumatoid arthritis (Garber 2011), in which TNF-α is a key pathogenic factor. The target of Jak inhibitors that results in efficacy in RA therapy in not known, but is widely assumed to be a cytokine(s) that directly activates Jak-STAT signaling, such as IL-6. The results showing indirect activation of Jak-STAT signaling by TNF-α suggest that Jak inhibitors may work at least in part by targeting autocrine signaling by TNF-α, an established pathogenic cytokine in RA.

Crosstalk That Suppresses Jak-STAT Signaling

Cytokine-induced Jak-STAT signaling is quantitatively modulated and fine-tuned by a variety of inhibitory mechanisms (reviewed in Shuai and Liu 2003). Of these mechanisms, we will focus on inducible mechanisms that mediate cross-inhibition of Jak-STAT signaling by heterologous receptors.

Induction of Inhibitors/SOCS

The best known inhibitors of Jak-STAT signaling are the suppressors of cytokine signaling (SOCS). SOCS inhibit Jak-STAT signaling by several mechanisms, including competition with STATs for receptor docking sites and inhibition of the catalytic activity of receptor-associated Jaks. SOCS proteins are typically expressed at undetectable or low levels and are induced by a variety of stimuli. Initial reports showed that SOCS are induced by Jak-STAT signaling and therefore much attention has focused on these molecules as feedback inhibitors (Shuai and Liu 2003). However, as SOCS1 and SOCS3 can interact with and inhibit several receptors, it has become clear that these proteins also mediate cross-inhibition by heterologous receptors. In addition to Jak-STAT signaling, SOCS are also induced by inflammatory factors such as TLRs and TNF and can mediate communication between these receptors and type I and type II cytokines that directly activate Jak-STAT signaling. SOCS3 expression is induced by multiple factors including TLR ligands and TNF-α and preferentially inhibits signaling by the gp130 subunit of IL-6-related receptors. Thus, SOCS3 mediates fine tuning of IL-6 signaling by multiple inflammatory stimuli. As SOCS proteins are not expressed at baseline and their expression needs to be induced, there is typically a delay of several hours between stimulation and the emergence of SOCS-mediated inhibition.

Direct and Rapid Inhibitory Pathways

Our laboratory and others have described rapidly acting, direct, SOCS-independent pathways by which TLRs, IL-1, TNF-α and ITAM-coupled receptors inhibit Jak-STAT signaling (Ivashkiv and Hu 2004). In one example, inhibition occurs by a p38-dependent mechanism that leads to increased SHP-2 association with gp130 (Bode et al. 2003); p38 signals can also induce phosphorylation and internalization of receptors, likely mediated by MK2 (Radtke et al. 2010). We have recently shown that a similar p38-mediated inhibitory mechanism also blocks signaling by IL-27 in human monocytes (Kalliolias and Ivashkiv 2008). TLRs, IL-1 and TNF inhibit gp130 signaling via p38 but do not substantially inhibit signaling by IFN-α and IL-10, which requires additional calcium signaling and activation of PKCs (Ji et al. 2003; Wang et al. 2010). Thus, IL-10 and IFN signaling is more effectively inhibited by ITAM-associated receptors that activate these calcium signaling pathways. One mechanism of inhibition of IFNAR signaling involves phosphorylation of intracellular receptor sequences that target the IFNAR for internalization and destruction, and this mechanism likely mediates inhibition of IFNAR signaling by many kinases, including p38 (Bhattacharya et al. 2011).

In contrast to low level tonic ITAM signaling that enhances IFNAR signaling, as described above, high intensity activation of ITAM-associated receptors can inhibit cytokine Jak-STAT signaling. TCR crosslinking inhibits signaling by IL-2, IL-4, IL-6 and IFN-α in T cells by a mechanism dependent on PKC and downstream ERK activation, and also requires calcium signaling for full inhibition to occur (Lee et al. 1999; Zhu et al. 2000). Similarly, high avidity ligation of FcγRs or Dectin-1 inhibits IL-10 signaling in macrophages by a PKCβ/δ-dependent mechanism that induces internalization of the IL-10 receptor (Ji et al. 2003). FcγR signaling inhibits IFN-α signaling by a PKCβ- (and to a lesser extent PKCδ-) dependent mechanism that induces recruitment of SHP-2 to the IFN-α receptor and increased SHP-2 catalytic activity (Du et al. 2005). These inhibitory mechanisms do not require new protein synthesis, are independent of SOCS, and are operative rapidly (within 15 min) of ligation of the inhibitory receptor.

Glucocorticoids

Glucocortocoids are potent anti-inflammatory factors that work by activating the glucocorticoid receptor (GCR) that is a member of the nuclear receptor family. Although GCRs can interact with STATs in other systems to potentiate STAT function, in the immune system direct modulation and inhibition of STATs by GCRs has not been described. Instead, GCs suppress inflammation and inflammatory gene expression by a well established 'tethering' mechanism whereby GCRs interact with DNA-bound NF-κB and AP-1 proteins to inhibit expression of NF-κB and AP-1 target genes (Rogatsky and Ivashkiv 2006). Although GC-mediated inhibition of Jak-STAT signaling has been described (Bianchi et al. 2000; Flammer et al. 2010), in most cases inhibition is indirect, whereby GCs inhibit NF-κB-mediated expression of Jak-associating receptors, with secondary decreases in Jak-STAT responses. One exception is direct inhibition of IFN-α responses by GCs (Flammer et al. 2010). Consistent with a lack of a direct effect on Jak-STAT signaling, GCs do not suppress IFN-α-induced tyrosine phosphorylation of STATs or subsequent translocation into the nucleus. Instead, GCs directly suppress type I IFN responses by inhibiting IFN-α-induced gene transcription. The mechanism is GC-mediated suppression of stable assembly and transcriptional function of the STAT1-STAT2-IRF9 (ISGF3) complex on target gene promoters, with consequent downregulation of target gene expression (Flammer et al. 2010). Interestingly, ISGF3 uses the GCR cofactor GRIP1 as a coactivator, and the mechanism of GC-mediated inhibition appears to be competition for GRIP1 and diminished occupancy of GRIP1 at certain ISGF3-driven promoters. Thus, suppression of the nuclear function of ISGF3 contributes to the anti-inflammatory properties of GCs.

Enhancement of Inflammatory Responses by Jak-STAT Signaling

Synergy with NF-κB

Concurrent signaling via NF-κB and STAT1 synergistically activates expression of many inflammatory genes, and strong expression of some of these genes, such as *Nos2* (encoding iNOS) or *Il12b* (encoding the p40 cytokine subunit shared by IL-12 and IL-23) is dependent on both stimuli. Understanding of the mechanisms of STAT1-NF-κB synergy is still limited. One explanation is that simultaneous occupancy of gene regulatory regions by STAT1 and NF-κB increases recruitment of RNA polymerase II, possibly because of enhanced recruitment of transcriptional co-activators, such as the histone acetyl transferases (HATs) CBP/p300. In addition, increased inflammatory gene expression in IFN-γ primed cells is mediated, at least in part, by extended activation of NF-κB. Recently a new mechanism by which IFN-activated STATs and NF-κB synergize after infection with the intracellular bacterial pathogen *Listeria monocytogenes* to induce *Nos2* expression has been described (Farlik et al. 2010). NF-κB primed *Nos2* for induction by recruiting a complex containing basal transcription factor TFIIH and the CDK7 kinase that is required for activation of RNA polymerase II (pol II) by phosphorylation on residue serine 5. However, NF-κB did not recruit pol II to the *Nos2* promoter. This task, instead, was accomplished by ISGF3, whereupon the recruited pol II could be efficiently phosphorylated and activated by the NF-κB-recruited CDK7. Thus, for *Nos2* induction, NF-κB and STATs cooperate by controlling distinct steps in transcription initiation complex assembly. It is likely that similar mechanisms are utilized for synergistic activation of other genes by NF-κB and STATs.

Inhibition of TLR-Induced Feedback Mechanisms

Excessive activation of inflammatory receptors such as TLRs can lead to toxicity and even lethality. Thus, TLR signaling is subject to negative regulation and feedback inhibition at multiple levels (Liew et al. 2005). Indeed, TLR-induced feedback inhibition is required for the appropriate regulation of the extent of inflammation, and likely plays a role in the subsequent transition to resolution of inflammation and tissue homeostasis. Mechanisms of negative regulation of TLR responses can be categorized into at least three types based on their mode of action: (1) Soluble anti-inflammatory factors. (2) Intracellular inhibitors of signal transduction. (3) Transcriptional repressors and chromatin-modifying enzymes that can inhibit TLR responses in a gene-specific manner.

IL-10 is a potent TLR-induced cytokine that mediates a feedback inhibitory loop that limits inflammatory cytokine production. IL-10 potently deactivates dendritic cells and macrophages by suppressing production of inflammatory cytokines such as TNF and IL-1 (Moore et al. 2001). The suppressive effects of IL-10 on myeloid cells are dependent upon STAT3 (Takeda et al. 1999), which appears to suppress anti-inflammatory gene expression indirectly via induction of transcriptional repressors (Murray 2007). TLR induction of IL-10 is in part mediated by MAPKs and downstream transcriptional factors such as AP1 and CREB, and negatively regulated by glycogen synthase kinase 3 (GSK3) (Hu et al. 2006; Martin et al. 2005). There are a large number of inducible negative regulators of TLR signaling that mediate feedback inhibition, including single immunoglobulin IL-1R-related molecule (SIGIRR), A20, MAP kinase phosphatase 1 (MKP1), IRF4, and SOCS1 (Li and Qin 2005). Signaling inhibitors often act on proximal components of TLR signaling pathways and inhibit expression of multiple downstream genes. In addition, TLRs induce expression of transcription repressors, such as ATF3, that feed back and suppress expression of specific subsets of TLR-inducible genes (Gilchrist et al. 2006).

One mechanism by which IFN-γ-STAT1 signaling potentiates TLR responses is by ablating the afore-described feedback inhibition loops. Strikingly, IFN-γ suppresses TLR-mediated induction of IL-10 protein and mRNA expression, downstream STAT3 activation, and induction of STAT3-dependent genes including SOCS3 and ABIN-3 (Hu et al. 2006; Wang et al. 2010). By inhibiting the IL-10-STAT3 axis, IFN-γ interrupts a TLR-induced feedback inhibition loop and results in increased production of the inflammatory cytokines TNF and IL-6. The mechanism of inhibition involves attenuation of proximal TLR-induced signaling by IFN-γ, which includes attenuation of TLR-induced activation of the PI3K-Akt pathway that had been previously shown to have a rapidly-induced feedback inhibitory function (Fukao and Koyasu 2003; Fukao et al. 2002). The PI3K-Akt pathway is linked to IL-10 production via GSK3 that regulates the function of transcription factors of the AP-1/CREB families that, in turn, regulate IL-10 production. Thus, GSK3 emerges as a potential therapeutic target for anti-inflammatory therapy. IL-27 can also strongly suppress TLR-induced IL-10 expression and prime for enhanced production of proinflammatory cytokine in human macrophages in a STAT1-dependent manner (Kalliolias and Ivashkiv 2008). Both IFN-γ and IL-27 appear to prime macrophages via STAT1, suggesting a central role for STAT in augmenting macrophage activation.

Another feedback mechanism induced by TLRs is activation of canonical Notch target genes, including the transcriptional repressors Hes1 and Hey1 (Hu et al. 2008). Hes1 and Hey1 feed back to attenuate TLR-mediated induction of IL-6 and IL-12 production. Interestingly, IFN-γ disrupts this feedback loop by interrupting proximal Notch signaling to abolish induction of Hes1 and Hey1 (Fig. 1). Release from these repressors can contribute to enhanced expression of inflammatory cytokines such as IL-6 and IL-12. Future work may reveal additional aspects of the feedback inhibition program that are inhibited by Jak-STAT signaling.

Reversal of Endotoxin Tolerance

Some of the most potent homeostatic mechanisms function in "endotoxin tolerance", where pre-exposure of cells to TLR ligands abrogates induction of inflammatory genes on subsequent rechallenge with TLR ligands. Mechanisms of endotoxin tolerance include alterations of TLR signaling and TLR-induced inhibition of inflammatory gene expression that appears to occur via epigenetic regulation of inflammatory gene loci by modification of histones and chromatin that can silence gene expression (Foster et al. 2007; Ramirez-Carrozzi et al. 2006; Saccani et al. 2002). Silencing by such epigenetic modifications during endotoxin tolerance, which is likely mediated by transcriptional repressors (Foster et al. 2007), plays a key role in specifically restraining potentially toxic inflammatory cytokine expression, while allowing beneficial expression of host defense genes. Jak-STAT signaling by IFN-γ-STAT1 reverses endotoxin tolerance. IFN-γ treatment does not affect the decrease in TLR signaling that occurs in tolerized macrophages. Instead, IFN-γ reverses the epigenetic closing of chromatin at inflammatory gene loci, thereby

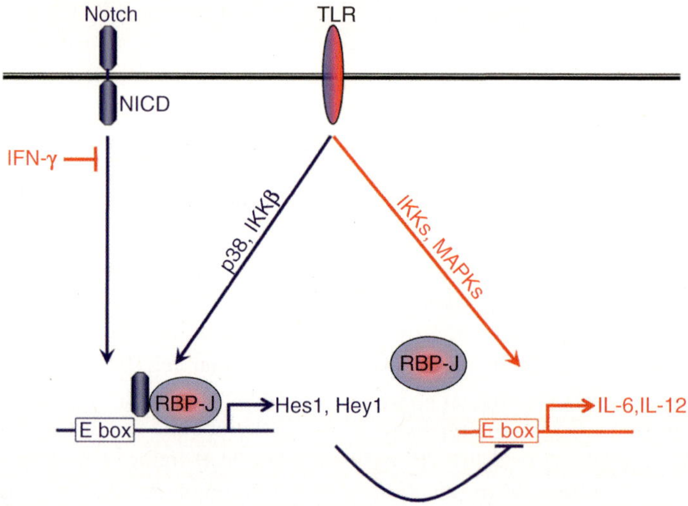

Fig. 1 IFN-γ suppresses Notch-RBP-J-mediated feedback inhibition of TLR-induced inflammatory cytokine production. In macrophages TLRs cooperate with the Notch pathway to activate transcription factor RBP-J that induces expression of canonical Notch target genes including the transcriptional repressors Hes1 and Hey1. Hes1 and Hey1 feed back to restrain inflammatory cytokine expression, in part mediated by interactions with their target E box sequences in the IL-6 and possibly IL-12 promoters. IFN-γ interrupts this feedback loop by suppressing proximal Notch signaling, thereby releasing inflammatory cytokine genes from repression by Hes1 and Hey1

allowing expression of inflammatory genes even in response to attenuated signals in tolerized cells (Chen and Ivashkiv 2010) (Fig. 2).

Suppression of Inflammatory Responses by Jak-STAT Signaling

IL-10-STAT3 signaling strongly suppresses inflammatory responses, establishing STAT3 as a potent anti-inflammatory STAT. The anti-inflammatory roles and mechanisms of STAT3 have been extensively reviewed (Murray 2007) and are not further considered here.

IFNs and IL-27

Cytokines that activate STAT1, such as IFNα/β, IFN-γ and IL-27 can also play a suppressive role in inflammation, including attenuation of associated tissue destruction. Both IFN-γ and IL-27 suppress signaling by IL-1 by mechanisms that involve STAT1-dependent downregulation of IL-1 type I receptor (IL-1RI) expression (Hu et al. 2005; Kalliolias et al. 2010). IFN-γ is protective in animal models of autoimmune arthritis (Guedez et al. 2001; Manoury-Schwartz et al. 1997; Vermeire et al. 1997) and multiple sclerosis (Krakowski and Owens 1996; Vermeire et al. 1997;

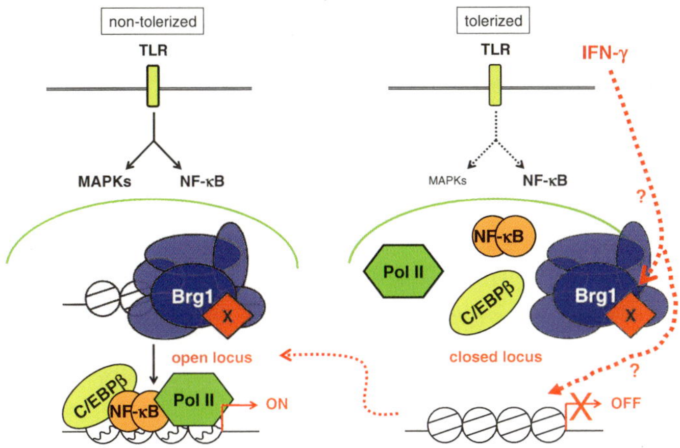

Fig. 2 IFN-γ reverses the block in inflammatory cytokine gene expression that is established during endotoxin tolerance. In naïve macrophages TLR stimulation leads to Brg1-mediated remodeling of chromatin at secondary response inflammatory genes such as *IL6* that results in the 'opening' of chromatin to make it permissive for transcription. This TLR-induced remodeling of chromatin does not occur in macrophages that have been tolerized by previous exposure to a TLR ligand. IFN-γ reverses this block in chromatin remodeling, possibly by modifying components of the Brg1-containing remodeling complex, or by inducing expression of proteins that bind to cytokine gene loci to facilitate chromatin remodeling

Willenborg et al. 1996). Inhibition of IL-1 responses likely contributes to the suppressive effects of IFN-γ and STAT1 on IL-1-dependent diseases, such as arthritis and inflammatory bone resorption (de Hooge et al. 2004; Guedez et al. 2001; Kim et al. 2003; Manoury-Schwartz et al. 1997; Takayanagi et al. 2000, 2002; Vermeire et al. 1997).

Jak-STAT Signaling Crosstalk in Inflammatory Diseases

Cytokines play a key pathogenic role in many inflammatory and autoimmune disorders, including RA and SLE. For example, the resounding success of TNF-α blockade therapy has established a key role for TNF-α in the pathogenesis of RA and other inflammatory conditions such as inflammatory bowel disease, psoriasis, and ankylosing spondylitis (Feldmann et al. 1996; Feldmann and Maini 2001). There is increasing interest in the pathogenic roles of cytokines that activate the Jak-STAT pathway in inflammatory disease pathogenesis, particularly IL-6 and type I IFNs (Ivashkiv 2003; Riese et al. 2010). IL-6 blockade is already an established effective therapy for RA, and Jak inhibitors have shown efficacy in preclinical models (Ghoreschi et al. 2010) and clinical trials (Garber 2011). A potential pathogenic role for type I IFNs in SLE has been proposed (Crow 2003a, b). Mechanisms by which type I IFNs contribute to SLE pathogenesis include promoting DC maturation, up-regulating innate immune, Th1, and antibody responses, suppressing apoptosis, and inducing the expression of other cytokines and chemokines (Ivashkiv 2003). Interestingly, elevated type I IFN expression and function have been observed in RA blood cells and arthritic synovial tissue (Thurlings et al. 2011 and refs. therein), and future work will clarify the pathogenic versus protective role of type I IFNs in this disease.

We have proposed that priming of myeloid cells (DCs and macrophages) and high STAT1 expression may contribute to inflammatory disease pathogenesis (Hu et al. 2002; Ivashkiv 2003; Kono et al. 2003; Taniguchi and Takaoka 2001; van der Pouw Kraan et al. 2003). IFNs are expressed during inflammation at levels that, in addition to direct effects on target cells, modulate immune cell responses by the mechanisms described above. For example low level IFN action could enhance the inflammatory responses to TNF, IL-6, and endogenous activators of TLRs. In fact, many animal models of autoimmune/inflammatory diseases suggest that priming does contribute to disease pathogenesis and progression (Ivashkiv 2003; Klinman 2003; Kono et al. 2003; Santiago-Raber et al. 2003; Taniguchi and Takaoka 2001; Theofilopoulos et al. 2001). In SLE, systemic IFN levels have been shown to enhance antigen presenting cell (APC) function of monocytes, similar to partially activated DCs (Bengtsson et al. 2000; Blanco et al. 2001). Furthermore, leukocytes isolated from SLE patients have been found to express increased levels of STAT1, a key marker in IFN priming (Baechler et al. 2004; Bennett et al. 2003; Ivashkiv 2003; Kuroiwa et al. 2003). Elevated STAT1 expression, measured both at the mRNA and protein levels, and implying priming in vivo, has been observed in other inflammatory processes, such as T-cell mediated hepatitis, RA, and dermatomyositis (Hong et al. 2002; Hu et al. 2002; Kuroiwa et al. 2003; van der Pouw Kraan et al. 2003). In SLE patients who have not undergone treatment, isolated monocytes display an enhanced response to TLR activators, such as LPS and IL-1β (Scuderi et al. 2003; Yuan et al. 2011) and it is thought that such a hyper-responsiveness of TLRs in vivo can be secondary to the effects of IFN priming (Doughty et al. 2001; Durbin et al. 2003; Krutzik et al. 2003; Mohty et al. 2003; Nansen and Randrup Thomsen 2001; Paterson et al. 2003). Moreover, as a consequence of this hyper-responsiveness, patients with rheumatic diseases have been observed to mount exaggerated immune responses to infections (Klinman 2003). Lastly, despite low levels of IFN-γ expression in the synovium of RA patients, RA synovial cells nevertheless strongly express many IFN-γ-inducible genes (van der Pouw Kraan et al. 2003), which can potentially be explained by the priming of synovial cells, increased expression of STAT1, and their hyper-responsiveness to cytokines that activate STAT1. Thus, a substantial body of data supports the notion that priming of immune cells for enhanced Jak-STAT signaling occurs in inflammatory diseases and contributes to disease pathogenesis, and emerging data suggest that Jak inhibitors represent a promising new therapeutic approach to inflammatory diseases.

The rate of progression of inflammatory disease, and its eventual severity and morbidity, is determined by the balance between pro- and anti-inflammatory cytokines that influence whether the inflammatory response progresses or resolves. Immunosuppressive cytokines, such as IL-10, have the potential to suppress inflammation in RA and SLE. However, the anti-inflammatory functions of IL-10 are compromised in RA and SLE (Bussolati et al. 2000; Gergely et al. 2002; Hart et al. 1995; Ji et al. 2003; Mongan et al. 1997). IL-10 signaling and downstream gene expression are suppressed in synovial macrophages isolated from patients with RA, and IL-10 responses are diminished in monocytes derived from SLE patients (Hart et al. 1995; Ji et al. 2003; Mongan et al. 1997; Yuan et al. 2011). Inhibition of IL-10 signaling is most effective in cells exposed to IFNs and correlates with diminished STAT3 activation and function, suggesting the potential involvement of some of the mechanisms that modulate Jak-STAT signaling discussed above. In the context of autoimmune/inflammatory disorders, a loss in IL-10 activity may have adverse consequences for the patient, as this would imply an inability to control and reduce the inflammatory response. Decreased IL-10 activity would lead to increased expression of pro-inflammatory cytokines, such as TNF-α, IL-1 and IL-6, thereby allowing the inflammatory response to proceed unabated.

Concluding Remarks

Jak-STAT signaling by many cytokines plays a key role in the regulation of inflammatory responses and the pathogenesis of inflammatory diseases. The functional outcomes of Jak-STAT signaling in complex inflammatory settings are determined by bidirectional crosstalk with other signaling pathways. Such crosstalk will determine the severity and time course of inflammatory responses, and also contributes to the pathogenesis of inflammatory diseases. Therapeutic targeting of the Jak-STAT pathway holds promise for treating various inflammatory conditions, but development of such therapies needs to take into account the activating versus homeostatic roles of Jak-STAT signaling, and the effects on crosstalk with heterologous signaling pathways.

Acknowledgements This work was supported by grants from the National Institutes of Health, USA.

References

Antoniv TT, Ivashkiv LB (2006) Dysregulation of interleukin-10-dependent gene expression in rheumatoid arthritis synovial macrophages. Arthritis Rheum 54:2711–2721

Baechler EC, Batliwalla FM, Karypis G, Gaffney PM, Ortmann WA, Espe KJ, Shark KB, Grande WJ, Hughes KM, Kapur V et al (2003) Interferon-inducible gene expression signature in peripheral blood cells of patients with severe lupus. Proc Natl Acad Sci USA 100: 2610–2615

Baechler EC, Gregersen PK, Behrens TW (2004) The emerging role of interferon in human systemic lupus erythematosus. Curr Opin Immunol 16:801–807

Bengtsson AA, Sturfelt G, Truedsson L, Blomberg J, Alm G, Vallin H, Ronnblom L (2000) Activation of type I interferon system in systemic lupus erythematosus correlates with disease activity but not with antiretroviral antibodies. Lupus 9:664–671

Bennett L, Palucka AK, Arce E, Cantrell V, Borvak J, Banchereau J, Pascual V (2003) Interferon and granulopoiesis signatures in systemic lupus erythematosus blood. J Exp Med 197: 711–723

Bhattacharya S, Qian J, Tzimas C, Baker DP, Koumenis C, Diehl JA, Fuchs SY (2011) Role of p38 protein kinase in the ligand-independent ubiquitination and down-regulation of the IFNAR1 chain of type I interferon receptor. J Biol Chem 286:22069–22076

Bianchi M, Meng C, Ivashkiv LB (2000) Inhibition of IL-2-induced Jak-STAT signaling by glucocorticoids. Proc Natl Acad Sci USA 97:9573–9578

Blanco P, Palucka AK, Gill M, Pascual V, Banchereau J (2001) Induction of dendritic cell differentiation by IFN-alpha in systemic lupus erythematosus. Science 294:1540–1543

Bode JG, Schweigart J, Kehrmann J, Ehlting C, Schaper F, Heinrich PC, Haussinger D (2003) TNF-alpha induces tyrosine phosphorylation and recruitment of the Src homology protein-tyrosine phosphatase 2 to the gp130 signal-transducing subunit of the IL-6 receptor complex. J Immunol 171:257–266

Bussolati B, Rollino C, Mariano F, Quarello F, Camussi G (2000) IL-10 stimulates production of platelet-activating factor by monocytes of patients with active systemic lupus erythematosus (SLE). Clin Exp Immunol 122:471–476

Chen J, Ivashkiv LB (2010) IFN-gamma abrogates endotoxin tolerance by facilitating Toll-like receptor-induced chromatin remodeling. Proc Natl Acad Sci USA 107:19438–19443

Crow MK (2003a) Interferon-alpha: a new target for therapy in systemic lupus erythematosus? Arthritis Rheum 48:2396–2401

Crow MK (2003b) Type I interferon and autoimmune disease. Autoimmunity 36:445–446

Darnell JE Jr, Kerr IM, Stark GR (1994) Jak-STAT pathways and transcriptional activation in response to IFNs and other extracellular signaling proteins. Science 264:1415–1421

de Hooge AS, van de Loo FA, Koenders MI, Bennink MB, Arntz OJ, Kolbe T, van den Berg WB (2004) Local activation of STAT-1 and STAT-3 in the inflamed synovium during zymosan-induced arthritis: exacerbation of joint inflammation in STAT-1 gene-knockout mice. Arthritis Rheum 50:2014–2023

Doughty L, Nguyen K, Durbin J, Biron C (2001) A role for IFN-alpha beta in virus infection-induced sensitization to endotoxin. J Immunol 166:2658–2664

Du Z, Shen Y, Yang W, Mecklenbrauker I, Neel BG, Ivashkiv LB (2005) Inhibition of IFN-alpha signaling by a PKC- and protein tyrosine phosphatase SHP-2-dependent pathway. Proc Natl Acad Sci USA 102:10267–10272

Durbin J, Doughty L, Nguyen K, Caligiuri M, Van Deusen J, Biron C (2003) The role of STAT1 in viral sensitization to LPS. J Endotoxin Res 9:313–316

Ehrt S, Schnappinger D, Bekiranov S, Drenkow J, Shi S, Gingeras TR, Gaasterland T, Schoolnik G, Nathan C (2001) Reprogramming of the macrophage transcriptome in response to interferon-gamma and *Mycobacterium tuberculosis*: signaling roles of nitric oxide synthase-2 and phagocyte oxidase. J Exp Med 194:1123–1140

Farlik M, Reutterer B, Schindler C, Greten F, Vogl C, Muller M, Decker T (2010) Nonconventional initiation complex assembly by STAT and NF-kappaB transcription factors regulates nitric oxide synthase expression. Immunity 33:25–34

Feldmann M, Maini RN (2001) Anti-TNF alpha therapy of rheumatoid arthritis: what have we learned? Annu Rev Immunol 19:163–196

Feldmann M, Brennan FM, Maini RN (1996) Role of cytokines in rheumatoid arthritis. Annu Rev Immunol 14:397–440

Firestein GS, Zvaifler NJ (2002) How important are T cells in chronic rheumatoid synovitis? II. T cell-independent mechanisms from beginning to end. Arthritis Rheum 46:298–308

Flammer JR, Dobrovolna J, Kennedy MA, Chinenov Y, Glass CK, Ivashkiv LB, Rogatsky I (2010) The type I interferon signaling pathway is a target for glucocorticoid inhibition. Mol Cell Biol 30:4564–4574

Foster SL, Hargreaves DC, Medzhitov R (2007) Gene-specific control of inflammation by TLR-induced chromatin modifications. Nature 447:972–978

Fukao T, Koyasu S (2003) PI3K and negative regulation of TLR signaling. Trends Immunol 24:358–363

Fukao T, Tanabe M, Terauchi Y, Ota T, Matsuda S, Asano T, Kadowaki T, Takeuchi T, Koyasu S (2002) PI3K-mediated negative feedback regulation of IL-12 production in DCs. Nat Immunol 3:875–881

Garber K (2011) Pfizer's JAK inhibitor sails through phase 3 in rheumatoid arthritis. Nat Biotechnol 29:467–468

Gergely P Jr, Niland B, Gonchoroff N, Pullmann R Jr, Phillips PE, Perl A (2002) Persistent mitochondrial hyperpolarization, increased reactive oxygen intermediate production, and cytoplasmic alkalinization characterize altered IL-10 signaling in patients with systemic lupus erythematosus. J Immunol 169:1092–1101

Ghoreschi K, Jesson MI, Li X, Lee JL, Ghosh S, Alsup JW, Warner JD, Tanaka M, Steward-Tharp SM, Gadina M et al (2010) Modulation of innate and adaptive immune responses by tofacitinib (CP-690,550). J Immunol 186:4234–4243

Gilchrist M, Thorsson V, Li B, Rust AG, Korb M, Kennedy K, Hai T, Bolouri H, Aderem A (2006) Systems biology approaches identify ATF3 as a negative regulator of Toll-like receptor 4. Nature 441:173–178

Guedez YB, Whittington KB, Clayton JL, Joosten LA, van de Loo FA, van den Berg WB, Rosloniec EF (2001) Genetic ablation of interferon-gamma up-regulates interleukin-1beta expression and enables the elicitation of collagen-induced arthritis in a nonsusceptible mouse strain. Arthritis Rheum 44:2413–2424

Hart PH, Ahern MJ, Smith MD, Finlay-Jones JJ (1995) Comparison of the suppressive effects of interleukin-10 and interleukin-4 on synovial fluid macrophages and blood monocytes from patients with inflammatory arthritis. Immunology 84:536–542

Hong F, Jaruga B, Kim WH, Radaeva S, El-Assal ON, Tian Z, Nguyen VA, Gao B (2002) Opposing roles of STAT1 and STAT3 in T cell mediated hepatitis: regulation by SOCS. J Clin Invest 110:1503–1513

Hu X, Herrero C, Li WP, Antoniv TT, Falck-Pedersen E, Koch AE, Woods JM, Haines GK, Ivashkiv LB (2002) Sensitization of IFN-gamma Jak-STAT signaling during macrophage activation. Nat Immunol 3:859–866

Hu X, Ho HH, Lou O, Hidaka C, Ivashkiv LB (2005) Homeostatic role of interferons conferred by inhibition of IL-1-mediated inflammation and tissue destruction. J Immunol 175:131–138

Hu X, Paik PK, Chen J, Yarilina A, Kockeritz L, Lu TT, Woodgett JR, Ivashkiv LB (2006) IFN-gamma suppresses IL-10 production and synergizes with TLR2 by regulating GSK3 and CREB/AP-1 proteins. Immunity 24:563–574

Hu X, Chung AY, Wu I, Foldi J, Chen J, Ji JD, Tateya T, Kang YJ, Han J, Gessler M et al (2008) Integrated regulation of Toll-like receptor responses by Notch and interferon-gamma pathways. Immunity 29:691–703

Ivashkiv LB (1996) Cytokine expression and cell activation in inflammatory arthritis. Adv Immunol 63:337–376

Ivashkiv LB (2003) Type I interferon modulation of cellular responses to cytokines and infectious pathogens: potential role in SLE pathogenesis. Autoimmunity 36:473–479

Ivashkiv LB, Hu X (2003) The JAK/STAT pathway in rheumatoid arthritis: pathogenic or protective? Arthritis Rheum 48:2092–2096

Ivashkiv LB, Hu X (2004) Signaling by STATs. Arthritis Res Ther 6:159–168

Ji JD, Tassiulas I, Park-Min KH, Aydin A, Mecklenbrauker I, Tarakhovsky A, Pricop L, Salmon JE, Ivashkiv LB (2003) Inhibition of interleukin 10 signaling after Fc receptor ligation and during rheumatoid arthritis. J Exp Med 197:1573–1583

Kalliolias GD, Ivashkiv LB (2008) IL-27 activates human monocytes via STAT1 and suppresses IL-10 production but the inflammatory functions of IL-27 are abrogated by TLRs and p38. J Immunol 180:6325–6333

Kalliolias GD, Gordon RA, Ivashkiv LB (2010) Suppression of TNF-alpha and IL-1 signaling identifies a mechanism of homeostatic regulation of macrophages by IL-27. J Immunol 185:7047–7056

Kim S, Koga T, Isobe M, Kern BE, Yokochi T, Chin YE, Karsenty G, Taniguchi T, Takayanagi H (2003) Stat1 functions as a cytoplasmic attenuator of Runx2 in the transcriptional program of osteoblast differentiation. Genes Dev 17:1979–1991

Klinman D (2003) Does activation of the innate immune system contribute to the development of rheumatoid arthritis? Arthritis Rheum 48:590–593

Kono DH, Baccala R, Theofilopoulos AN (2003) Inhibition of lupus by genetic alteration of the interferon-alpha/beta receptor. Autoimmunity 36:503–510

Krakowski M, Owens T (1996) Interferon-gamma confers resistance to experimental allergic encephalomyelitis. Eur J Immunol 26:1641–1646

Krutzik SR, Ochoa MT, Sieling PA, Uematsu S, Ng YW, Legaspi A, Liu PT, Cole ST, Godowski PJ, Maeda Y et al (2003) Activation and regulation of Toll-like receptors 2 and 1 in human leprosy. Nat Med 9:525–532

Kuroiwa T, Schlimgen R, Illei GG, Boumpas DT (2003) Monocyte response to Th1 stimulation and effector function toward human mesangial cells are not impaired in patients with lupus nephritis. Clin Immunol 106:65–72

Lee IH, Li WP, Hisert KB, Ivashkiv LB (1999) Inhibition of interleukin 2 signaling and signal transducer and activator of transcription (STAT)5 activation during T cell receptor- mediated feedback inhibition of T cell expansion. J Exp Med 190:1263–1274

Li X, Qin J (2005) Modulation of Toll-interleukin 1 receptor mediated signaling. J Mol Med 83:258–266

Liew FY, Xu D, Brint EK, O'Neill LA (2005) Negative regulation of toll-like receptor-mediated immune responses. Nat Rev Immunol 5:446–458

Manoury-Schwartz B, Chiocchia G, Bessis N, Abehsira-Amar O, Batteux F, Muller S, Huang S, Boissier MC, Fournier C (1997) High susceptibility to collagen-induced arthritis in mice lacking IFN-gamma receptors. J Immunol 158:5501–5506

Martin M, Rehani K, Jope RS, Michalek SM (2005) Toll-like receptor-mediated cytokine production is differentially regulated by glycogen synthase kinase 3. Nat Immunol 6:777–784

Mohty M, Vialle-Castellano A, Nunes JA, Isnardon D, Olive D, Gaugler B (2003) IFN-alpha skews monocyte differentiation into Toll-like receptor 7-expressing dendritic cells with potent functional activities. J Immunol 171:3385–3393

Mongan AE, Ramdahin S, Warrington RJ (1997) Interleukin-10 response abnormalities in systemic lupus erythematosus. Scand J Immunol 46:406–412

Moore KW, de Waal Malefyt R, Coffman RL, O'Garra A (2001) Interleukin-10 and the interleukin-10 receptor. Annu Rev Immunol 19:683–765

Murray PJ (2007) The JAK-STAT signaling pathway: input and output integration. J Immunol 178:2623–2629

Nansen A, Randrup Thomsen A (2001) Viral infection causes rapid sensitization to lipopolysaccharide: central role of IFN-alpha beta. J Immunol 166:982–988

Nguyen KB, Watford WT, Salomon R, Hofmann SR, Pien GC, Morinobu A, Gadina M, O'Shea JJ, Biron CA (2002) Critical role for STAT4 activation by type 1 interferons in the interferon-gamma response to viral infection. Science 297:2063–2066

Paterson HM, Murphy TJ, Purcell EJ, Shelley O, Kriynovich SJ, Lien E, Mannick JA, Lederer JA (2003) Injury primes the innate immune system for enhanced Toll-like receptor reactivity. J Immunol 171:1473–1483

Radtke S, Wuller S, Yang XP, Lippok BE, Mutze B, Mais C, de Leur HS, Bode JG, Gaestel M, Heinrich PC et al (2010) Cross-regulation of cytokine signalling: pro-inflammatory cytokines restrict IL-6 signalling through receptor internalisation and degradation. J Cell Sci 123: 947–959

Ramirez-Carrozzi VR, Nazarian AA, Li CC, Gore SL, Sridharan R, Imbalzano AN, Smale ST (2006) Selective and antagonistic functions of SWI/SNF and Mi-2beta nucleosome remodeling complexes during an inflammatory response. Genes Dev 20:282–296

Riese RJ, Krishnaswami S, Kremer J (2010) Inhibition of JAK kinases in patients with rheumatoid arthritis: scientific rationale and clinical outcomes. Best Pract Res Clin Rheumatol 24:513–526

Rogatsky I, Ivashkiv LB (2006) Glucocorticoid modulation of cytokine signaling. Tissue Antigens 68:1–12

Saccani S, Pantano S, Natoli G (2002) p38-Dependent marking of inflammatory genes for increased NF-kappa B recruitment. Nat Immunol 3:69–75

Santiago-Raber ML, Baccala R, Haraldsson KM, Choubey D, Stewart TA, Kono DH, Theofilopoulos AN (2003) Type-I interferon receptor deficiency reduces lupus-like disease in NZB mice. J Exp Med 197:777–788

Scuderi F, Convertino R, Molino N, Provenzano C, Marino M, Zoli A, Bartoccioni E (2003) Effect of pro-inflammatory/anti-inflammatory agents on cytokine secretion by peripheral blood mononuclear cells in rheumatoid arthritis and systemic lupus erythematosus. Autoimmunity 36:71–77

Shuai K, Liu B (2003) Regulation of JAK-STAT signalling in the immune system. Nat Rev Immunol 3:900–911

Takayanagi H, Ogasawara K, Hida S, Chiba T, Murata S, Sato K, Takaoka A, Yokochi T, Oda H, Tanaka K et al (2000) T-cell-mediated regulation of osteoclastogenesis by signalling cross-talk between RANKL and IFN-gamma. Nature 408:600–605

Takayanagi H, Kim S, Matsuo K, Suzuki H, Suzuki T, Sato K, Yokochi T, Oda H, Nakamura K, Ida N et al (2002) RANKL maintains bone homeostasis through c-Fos-dependent induction of interferon-beta. Nature 416:744–749

Takeda K, Clausen BE, Kaisho T, Tsujimura T, Terada N, Forster I, Akira S (1999) Enhanced Th1 activity and development of chronic enterocolitis in mice devoid of Stat3 in macrophages and neutrophils. Immunity 10:39–49

Taniguchi T, Takaoka A (2001) A weak signal for strong responses: interferon-alpha/beta revisited. Nat Rev Mol Cell Biol 2:378–386

Tassiulas I, Hu X, Ho H, Kashyap Y, Paik P, Hu Y, Lowell CA, Ivashkiv LB (2004) Amplification of IFN alpha induced STAT1 activation and inflammatory function by Syk and ITAM-containing adaptors. Nat Immunol 5:1181–1189

Theofilopoulos AN, Koundouris S, Kono DH, Lawson BR (2001) The role of IFN-gamma in systemic lupus erythematosus: a challenge to the Th1/Th2 paradigm in autoimmunity. Arthritis Res 3:136–141

Thurlings RM, Boumans M, Tekstra J, van Roon JA, Vos K, van Westing DM, van Baarsen LG, Bos C, Kirou KA, Gerlag DM et al (2011) Relationship between the type I interferon signature and the response to rituximab in rheumatoid arthritis patients. Arthritis Rheum 62:3607–3614

van der Pouw Kraan TC, van Gaalen FA, Kasperkovitz PV, Verbeet NL, Smeets TJ, Kraan MC, Fero M, Tak PP, Huizinga TW, Pieterman E et al (2003) Rheumatoid arthritis is a heterogeneous disease: evidence for differences in the activation of the STAT-1 pathway between rheumatoid tissues. Arthritis Rheum 48:2132–2145

Varinou L, Ramsauer K, Karaghiosoff M, Kolbe T, Pfeffer K, Muller M, Decker T (2003) Phosphorylation of the Stat1 transactivation domain is required for full-fledged IFN-gamma-dependent innate immunity. Immunity 19:793–802

Vermeire K, Heremans H, Vandeputte M, Huang S, Billiau A, Matthys P (1997) Accelerated collagen-induced arthritis in IFN-gamma receptor-deficient mice. J Immunol 158:5507–5513

Wang L, Tassiulas I, Park-Min KH, Reid AC, Gil-Henn H, Schlessinger J, Baron R, Zhang JJ, Ivashkiv LB (2008) 'Tuning' of type I interferon-induced Jak-STAT1 signaling by calcium-dependent kinases in macrophages. Nat Immunol 9:186–193

Wang L, Gordon RA, Huynh L, Su X, Park Min KH, Han J, Arthur JS, Kalliolias GD, Ivashkiv LB (2010) Indirect inhibition of Toll-like receptor and type I interferon responses by ITAM-coupled receptors and integrins. Immunity 32:518–530

Wen Z, Zhong Z, Darnell JE Jr (1995) Maximal activation of transcription by Stat1 and Stat3 requires both tyrosine and serine phosphorylation. Cell 82:241–250

Willenborg DO, Fordham S, Bernard CC, Cowden WB, Ramshaw IA (1996) IFN-gamma plays a critical down-regulatory role in the induction and effector phase of myelin oligodendrocyte glycoprotein-induced autoimmune encephalomyelitis. J Immunol 157:3223–3227

Yarilina A, Park-Min KH, Antoniv T, Hu X, Ivashkiv LB (2008) TNF activates an IRF1-dependent autocrine loop leading to sustained expression of chemokines and STAT1-dependent type I interferon-response genes. Nat Immunol 9:378–387

Yuan W, DiMartino SJ, Redecha PB, Ivashkiv LB, Salmon JE (2011) Systemic lupus erythematosus monocytes are less responsive to interleukin-10 in the presence of immune complexes. Arthritis Rheum 63:212–218

Zhu J, Huang H, Guo L, Stonehouse T, Watson CJ, Hu-Li J, Paul WE (2000) Transient inhibition of interleukin 4 signaling by T cell receptor ligation. J Exp Med 192:1125–1134

Activation and Inhibition of JAK-STAT Signal Transduction by RNA Viruses

Curt M. Horvath

Abstract
The investigation of interferon (IFN)-stimulated signal transduction leading to antiviral gene expression revealed the first members of the JAK and STAT protein families. The importance of IFN signaling in the innate cellular response to virus infection is highlighted by the evolution of numerous virus-encoded IFN evasion strategies that can prevent IFN production or antagonize downstream responses. Some RNA viruses in the family *Paramyxoviridae* have evolved the ability to target STAT proteins directly to eliminate antiviral signaling, preventing IFN-stimulated gene expression and innate antiviral responses. The virus-encoded STAT inhibitors are highly homologous to each other, and target STATs through protein interactions. However, in-depth investigations of the biochemical and cellular mechanisms have revealed that individual paramyxovirus genera have evolved distinct mechanisms to mediate STAT destruction.

Interferon Signal Transduction and Antiviral Innate Immune Responses

Immune responses to virus infections are initiated by the recognition of specific pathogen associated molecular patterns (PAMPs), such as viral nucleic acids, by cellular pathogen recognition receptor (PRR) proteins (Akira et al. 2006). For many virus infections, activation of PRR signaling leads to expression of type I interferon (IFN), in humans a single IFNβ and multiple IFNα genes, which in turn engages the IFN receptor-JAK-STAT pathway to induce an IFN-stimulated gene expression program leading to a broadly effective cellular antiviral state.

C.M. Horvath (✉)
Department of Molecular Biosciences, Pancoe Research Pavilion, Northwestern University, 2200 Campus Drive, Evanston, IL 60208, USA
e-mail: horvath@northwestern.edu

Th. Decker and M. Müller (eds.), *Jak-Stat Signaling: From Basics to Disease*, DOI 10.1007/978-3-7091-0891-8_20,

Accumulation of PAMPs during RNA virus infections is a well characterized inducer of the IFN-mediated antiviral response. For RNA viruses that deliver their nucleocapsid to the cytoplasm of host cells, double-stranded and single stranded replication intermediates or defective genomes can potently induce the IFN response by binding to the proteins encoded by the retinoic acid-inducible gene I (RIG-I) and melanoma differentiation-associated gene 5 (MDA5) (Takeuchi and Akira 2008). These proteins, collectively referred to as RIG-I like receptors (RLRs) are characterized by the presence of tandem caspase activation and recruitment domain (CARD) motifs at their N-termini (Hiscott et al. 2006) coupled to a DECH-box RNA helicase domain and a C-terminal regulatory domain (RD) (Kang et al. 2002; Yoneyama et al. 2004, 2005). A third related protein, LGP2, has significant sequence similarity with RIG-I and MDA5 in the helicase region but lacks the N-terminal CARD domains (Fig. 1). The RLRs recognize non-self RNAs, such as double-stranded RNA (dsRNA), 5′-triphosphorylated RNA, or structured RNAs and transmit a signal through the mitochondria-associated adaptor called IFNβ promoter stimulator protein 1 (IPS-1)/mitochondrial antiviral signaling protein (MAVS)/virus-induced signaling adaptor (VISA)/CARD adaptor inducing IFNβ (Cardif), independently identified by several groups (Kawai et al. 2005; Meylan et al. 2005; Seth et al. 2005; Xu et al. 2005), reviewed by Hiscott et al. (2006),

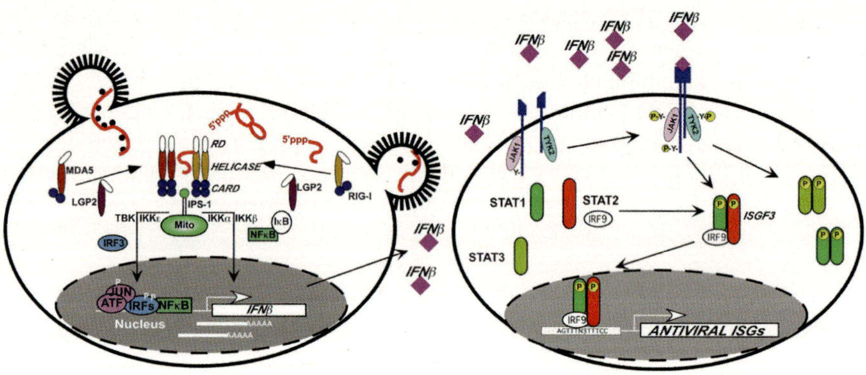

Fig. 1 Activation of the IFN signaling cascade in response to paramyxovirus infection. (*Left*) Viral envelope fusion with the host plasma membrane delivers the viral nucleocapsid to the cytoplasm. RLR proteins RIG-I, MDA5, and LGP2 interact with the foreign RNA via their helicase domain and regulatory domain (RD). RIG-I and MDA5 associate with the mitochondria resident IPS-1 via their caspase activation and recruitment domain (CARD), inducing activation of serine kinases in the Inhibitor of NFκB (IκB) kinase family, TBK1, IKKε, IKKα, and IKKβ. These kinases activate transcription factors IRF3 and NFκB to assemble at the IFNβ enhanceosome, resulting in transcriptional activation. The IFNβ mRNA is translated and secreted from the cell where it can bind to its receptor. (*Right*) When type 1 IFNs bind their receptor, JAK protein tyrosine kinases Tyk2 and JAK1 are activated to produce docking sites for the latent STAT1 and STAT2, which then are phosphorylated on their activating tyrosine residues to induce dimerization and assembly with IFN regulatory factor IRF9 to produce the IFN stimulated gene (ISG) transcription factor 3 (ISGF3). ISGF3 translocates to the nucleus where it binds to the promoter of ISGs, inducing their transcription. The combined effects of the ISG products is production of the an antiviral state and resistance to virus infection

Johnson and Gale (2006). IPS-1 acts as a scaffold that facilitates a serine kinase based signaling cascade leading to the activation of immediate responding transcription factors including NF-κB and IFN regulatory factor-3 (IRF3). These factors translocate to the nucleus and contribute to transcriptional activation of the IFNβ gene (Au et al. 1995; Fitzgerald et al. 2003; Sato et al. 1998; Servant et al. 2001, 2002; Weaver et al. 1998).

The newly synthesized and secreted type I IFNs can bind to the IFN-α/β receptor on the same cell (autocrine signaling) as well as adjacent cells (paracrine signaling). IFN receptor engagement results in activation of STAT1 and STAT2 by tyrosine phosphorylation, which in combination with a DNA binding subunit, IRF9, form the heterotrimeric complex known as the interferon stimulated gene factor 3 (ISGF3) (Fu et al. 1990; Kessler et al. 1990). ISGF3 translocates to the nucleus and binds to the conserved IFN stimulated response element (ISRE) sequences on IFNα/β stimulated gene (ISG) promoters inducing their transcription (Levy et al. 1988; Reich et al. 1987). The accumulated effects of the induced interferon stimulated gene products generate an antiviral state in the IFN-stimulated cell that provides a broadly effective barrier protecting the cell against virus infections.

Virus Evasion of Interferon Mediated Antiviral Responses

The importance of the STAT-mediated IFN signaling systems in mediating antiviral defense is highlighted by the fact that many viruses have evolved mechanisms to evade activation of this innate immune response (Gale and Sen 2009; Goodbourn et al. 2000; Grandvaux et al. 2002; Horvath 2004a, b; Katze et al. 2002; Levy and Garcia-Sastre 2001; Taylor et al. 1999). Each step of the interferon response, from the first recognition of a virus in the primary infected cell to the ability to mount an effective adaptive immune response, is known to be targeted by viral immune suppression. For many viruses, the initial steps of virus detection and IFN induction are targeted by inhibitory mechanisms including dsRNA sequestration or signaling interference to antagonize IRF3 and NF-κB pathways. In addition, virus-encoded IFN receptors or receptor antagonists can block cytokine signaling. Individual viruses can also block specific antiviral effectors or ISG products to preserve key cellular machinery needed for their replication. Viral disruption of IFN production and action is tightly linked with virulence. Understanding the molecular mechanisms by which viruses trigger and evade IFN actions are therefore of paramount importance for defining strategies aimed at controlling virus infection. For RNA viruses in the paramyxovirus family, evasion of IFN signaling by direct interference with the IFN-inducible STAT proteins has been characterized in detail (Gotoh et al. 2002; Horvath 2004a, b). The molecular mechanisms of STAT antagonism and IFN signaling evasion are known to be very diverse among individual viruses in this family (Table 1), and the key features as well as new insights of paramyxovirus STAT targeting are described in this chapter.

Table 1 STAT targeting by paramyxovirus V proteins

Genus	Virus	P gene Product	STAT Target	Notes
Rubulavirus	PIV5	V	STAT1	VDC ubiquitin ligase requires STAT2 to target STAT1
	HPIV2	V	STAT2	VDC ubiquitin ligase requires STAT1 to target STAT2; can also target STAT1
	Mumps	V	STAT1	VDC^{STAT1} ubiquitin ligase requires STAT2 to target STAT1
				VDC^{STAT3} ubiquitin ligase does not require STAT2
Morbillivirus	Measles	V	STAT2>>STAT1	CTD sufficient for binding STAT2; Y110 required to bind STAT1
Henipavirus	Nipah	P, V, W	STAT1>>STAT2	N-terminal STAT binding sites are identical in all proteins
				STAT2 association greatly enhanced by STAT1 binding
	Hendra	P, V, W	STAT1>>STAT2	N-terminal STAT binding sites are identical in all proteins

Paramyxoviruses-STAT Destroyers

The non-segmented negative strand RNA virus family, *Paramyxoviridae,* is divided into two subfamilies, *Paramyxovirinae* and *Pneumovirinae*. Differences in the nucleocapsid and genomic organization as well as virus life cycles distinguish paramyxoviruses from pneumoviruses (Lamb and Parks 2007). The *Paramyxovirinae* are further divided into five genera based on nucleotide sequence similarity, antigenic cross reactivity, and neuraminidase activity of the attachment proteins (Lamb and Parks 2007). The *Respirovirus* genus includes Sendai virus, human parainfluenza virus type I (HPIV1), and human parainfluenza virus type 3 (HPIV3). The *Avulavirus* genus includes bird paramyxoviruses like Newcastle disease virus (NDV). Mumps virus, parainfluenza virus 5 (PIV5, formerly know as Simian virus 5 (SV5)), and human parainfluenza virus type 2 (HPIV2) are all members of the *Rubulavirus* genus. The *Morbillivirus* genus includes measles virus and canine distemper virus (CDV). The *Henipavirus* genus is comprised of Hendra virus and Nipah virus, two paramyxoviruses that elicited an outbreak of fatal encephalitis spread between farm animals and humans in Australia and Malaysia and continue to emerge in southeast Asia (Chua et al. 2000; Selvey et al. 1995). All of these viruses share common structural, biochemical, and genetic elements including a single-stranded antisense, RNA genome that encodes a small number of mRNAs encoding proteins including surface glycoproteins involved in attachment and fusion, subunits of an RNA-dependent RNA polymerase, and host modifying factors (Lamb and Parks 2006). One interesting feature of the *Paramyxovirinae* is the ability of the gene that encodes the phosphoprotein, P, to code for more than one protein (Thomas et al. 1988). An extreme example is Sendai virus, which directs the expression of at least seven proteins from this locus, including P, V, W, C', C, Y1 and Y2. Other paramyxoviruses express fewer, but more than one, proteins from their P/V/C gene, by accessing overlapping open reading frames through alternate translation initiation site selection on a single mRNA, or by generating alternate mRNAs by co-transcriptional insertion of non-templated nucleotides at a precise site that allows access to alternate reading frames (Lamb and Parks 2006; Thomas et al. 1988). This mRNA generating phenomenon produces proteins that share a common amino terminus but differ in their carboxyl terminus downstream of the "editing site". Most commonly, this process produces two proteins called P and V, but in some viruses the third reading frame can produce a protein often referred to as W (Harcourt et al. 2000).

The V protein C-terminal domain (CTD) is highly conserved among the paramyxoviruses. The CTD of the V protein is cysteine rich and binds two zinc (Zn) molecules per V protein (Liston and Briedis 1994; Paterson et al. 1995). A diverse range of host evasion activities, including IFN signaling inhibition (Didcock et al. 1999), prevention of apoptosis (He et al. 2002; Wansley and Parks 2002), cell cycle alterations (Lin and Lamb 2000), inhibition of double-stranded RNA signaling (He et al. 2002; Poole et al. 2002), and prevention of IFN biosynthesis (He et al. 2002; Poole et al. 2002; Wansley and Parks 2002) have been specifically ascribed to the paramyxovirus V proteins. The other proteins derived from this locus, including C and W have also been implicated in evading IFN responses and antiviral signaling. In most cases, a fundamental activity associated with the host evasion proteins is direct interference with STAT protein function by mechanisms that rely on protein interactions, but it is recognized that individual genera within the family exhibit remarkably distinct mechanisms of STAT inhibition.

Rubulaviruses

The *Rubulavirus* genus includes parainfluenza virus 5 (PIV5, formerly known as SV5), mumps virus, and the type 2 human parainfluenza virus (HPIV2). The PIV5 V protein was the first to be recognized as an IFN signaling inhibitor, and sole expression of this protein in human cells results in rapid and specific loss of STAT1 by a proteasome mediated degradation (Didcock et al. 1999). In fact, all of the *Rubulavirus* V proteins have been shown to efficiently target STAT proteins for proteasome mediated destruction with remarkable targeting specificity. The PIV5 V protein targets STAT1, the HPIV2 V protein targets STAT2 (although in some cases can also target STAT1), and the mumps virus V protein targets both STAT1 and STAT3 (Didcock et al. 1999; Kubota et al. 2001; Parisien et al. 2001, 2002a, b; Ulane et al. 2003; Yokosawa et al. 2002; Young et al. 2000). Intact cellular IFN signal transduction is not required for STAT targeting, somatic cell genetic analysis has demonstrated that both PIV5 and mumps virus V proteins require the cellular expression of STAT2 in order to target STAT1, while HPIV2 requires STAT1 to target STAT2. This requirement restricts the host range of PIV5, as the divergent murine STAT2 cannot support targeting of STAT1 (Parisien et al. 2002a). The requirement for human STAT2 in STAT1 destruction provided the first example of a STAT protein that can function as a host range determinant for any virus. This observation was further validated by the creation of a human STAT2 transgenic mouse model system in which human STAT2 was ubiquitously expressed in an immune-competent intact mouse. Expression of human STAT2 enabled targeting of the murine STAT1 protein interference with IFN signaling responses in infected cells, increased virus replication, and enhanced cytokine and ISG induction profiles in the infected animals (Kraus et al. 2008). It has recently been demonstrated that differences in mouse STAT2 are also important for IFN signaling antagonism by the Dengue virus. Reminiscent of PIV5, the Dengue virus protein NS5 is able to bind and degrade human STAT2, but not mouse STAT2. Although mechanistically distinct, it seems likely that this difference can be exploited to create an animal model for the study of this deadly pathogen (Ashour et al. 2009, 2010).

Biochemical studies of *Rubulavirus* V proteins indicate that they achieve STAT degradation by coordinating the assembly of a multi-component E3 ubiquitin ligase complex that includes a number of cellular proteins including DDB1, the cullin family member Cul 4A, and both STAT1 and STAT2 (Andrejeva et al. 2002; Lin et al. 1998; Ulane and Horvath 2002). This V-mediated targeting complex is referred to as VDC, an acronym for *V*-dependent *d*egradation *c*omplex, which fortuitously also denotes its central components, V, DDB1, and Cul4A. (Precious et al. 2005; Ulane et al. 2005). Structural studies strongly support the molecular and biochemical conclusions regarding the Ub ligase complex assembled by Rubulavirus V proteins. Crystallographic visualization of PIV5 V protein in complex with DDB1 revealed molecular details about the V protein and its association with DDB1 (Li et al. 2006) (Fig. 2). Together, the data indicate that V proteins form separate interactions with their STAT substrates and the cellular ubiquitin ligase components. Furthermore, it has been observed that the V proteins can form macromolecular spherical particles visible in the electron microscope suggesting that the V protein particle acts as an enzymatic scaffold for combining the hijacked E3 ubiquitin ligase activities with the V-associated STAT substrate, resulting in highly efficient targeting (Ulane et al. 2005). The obligatory role of STAT2 is to act as a substrate adaptor, bridging STAT1 and the DDB1 complex (Precious et al. 2005; Ulane et al. 2005). Irrespective of the structural considerations, the *Rubulavirus* V protein acts as an efficient catalyst for the degradation of STAT1 to prevent the establishment of the IFN induced antiviral state in the cells (Carlos et al. 2005; Precious et al. 2007).

This PIV5 V protein structure revealed that a unique zinc finger is formed that involves the first histidine of the V CTD, and seven invariant cysteine residues. The structure also verified the importance of both the N and C termini in the interaction with DDB1, consistent the observation that STAT1 can be degraded by only the full length PIV5 V protein and that the P protein has no ability to induce degradation of STAT1 (Didcock et al. 1999). The data are also consistent with the findings that the C-terminal cysteine cluster is important for DDB1 binding (Andrejeva et al. 2002; Lin et al. 1998), though additional contacts are also essential for full VDC activity.

Another well-studied Rubulavirus V protein is from mumps virus. The mumps virus V protein has a unique ability to target both STAT1 and STAT3 for proteasomal

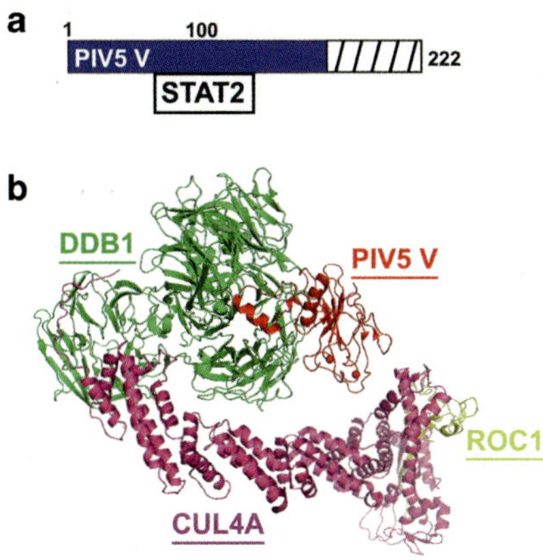

Fig. 2 The PIV5 V protein assembles a STAT-targeting ubiquitin ligase. (**a**) Scale diagram of the PIV5 V protein, illustrating position 100, implicated in STAT2 association, and the conserved zinc-binding CTD (*hatched*). (**b**) Structural representation of the core VDC components, DDB1 (*green*), CUL4A (*pink*), and ROC1 (*yellow*) in association with PIV5 V protein (*red*). Zinc atoms are represented as *red spheres*

destruction (Ulane et al. 2003). This results in a more dramatic suppression of cytokine signaling due to loss of STAT3 in the infected cells (Ulane et al. 2003). Mumps virus is capable of forming separate VDC complexes for targeting STAT1 or STAT3. Like PIV5, mumps V protein is incapable of targeting STAT1 in cells deficient in STAT2, but retains the ability to mediate STAT3 destruction, indicating that STAT2 is not a co-factor for STAT3 degradation. Furthermore, STAT immunoprecipitation of affinity purified mumps virus VDC complexes demonstrated co-fractionation of STAT1 and STAT2, but not STAT3, strongly indicating the assembly of independent mumps VDC complexes capable of targeting STAT1 (in a STAT2-dependent reaction) or STAT3 (in a STAT2-independent reaction).

Although STAT1 targeting and IFN evasion is widely observed among diverse viruses, the ability to degrade STAT3 is rarely observed, and among Paramyxoviruses is unique to the mumps virus. STAT1 and STAT2 are well established antiviral mediators, and are frequently found to be the targets of viral host evasion strategies. In contrast, a role for STAT3 as a mediator of antiviral immune responses is not fully appreciated. It has been observed that STAT3 can be activated in response to IFNα/β stimulation (Caldenhoven et al. 1999; Velichko et al. 2002) and STAT3 has been identified as a target not only for mumps virus but also for measles virus V proteins (Palosaari et al. 2003; Ulane et al. 2003). Like STAT1, activation of STAT3 leads to dimerization and the activated STAT3 dimer binds to a GAS-like response element, but the specific target genes and phenotypic outcomes are distinct from those of IFNγ-activated STAT1. STAT3 has also been identified as a target for virus host evasion, a strong implication as being important in cellular antiviral responses (Palosaari, 2003; Ulane et al. 2003).

Based on the structural and genetic features of the *Rubulaviruses*, a new paramyxovirus, mapuera virus, has been recently isolated from bats. But unlike the

Rubulaviruses, this virus evades the IFN response by sequestering STAT proteins and preventing their nuclear translocation rather than degrading STATs (Hagmaier et al. 2007). Another *Rubulavirus*, HPIV4, has been shown to be completely unable to evade the IFN response (Nishio et al. 2005). Sequence comparison reveals different amino acid composition in the N and C terminus of the virus V protein. However, as cell type and species specific factors are already known to strongly influence the outcome of V-mediated STAT destruction; further work is needed to better understand the nature of these viruses, and their abilities to interface with the host IFN response and STAT proteins.

Henipaviruses

Hendra virus and Nipah virus are the two prototype species in the emerging *Henipavirus* genus that was recently identified and causes lethal diseases in humans. In the *Henipaviruses*, the P protein is derived from the co-linear transcript, and addition of one or two non-templated guanine nucleotides produces mRNAs encoding two additional proteins, V and W with different carboxyl termini (Harcourt et al. 2000). In addition, the C protein is produced from an alternate translational site within the P gene. Immunofluorescence analysis of Nipah virus infected cells with specific antiserum has demonstrated that the P protein localizes throughout the cell, the C protein accumulates in the cytoplasm in a punctate pattern, and that the V protein accumulates exclusively in cytoplasm while the W protein is predominantly in the nucleus. This steady state picture of protein accumulation does not reveal the dynamics of nucleocytoplasmic movement, as it is established that the Henipavirus V protein shuttle between nucleus and cytoplasm with net nuclear export due to the presence of a potent export signal (Lo et al. 2009; Rodriguez et al. 2004; Rodriguez and Horvath 2004).

Henipavirus P, V and W proteins have all been implicated in blocking the host IFN response by targeting STAT proteins. Nipah virus P, V and W proteins all form high affinity interactions with STAT1 through their common N terminal domain (Rodriguez et al. 2004; Shaw et al. 2004), and this activity maps to residues 100–160, which is also present and functional in Hendra virus (Fig. 3). Additional residues, as well as the presence of STAT1 are required for these V proteins to associate with STAT2 (Rodriguez et al. 2004). The *Henipavirus* V proteins inhibit IFN responses by sequestering STAT1 and STAT2 in high molecular weight cytoplasmic complexes, thereby preventing their IFN-mediated tyrosine phosphorylation (Rodriguez et al. 2002, 2003). The short CTD of the Nipah virus W protein possesses a functional nuclear localization signal (NLS) that interacts with host karyopherin alpha 3 and 4 (Shaw et al. 2005), giving rise to nuclear accumulation. It has also been shown that W protein but not V protein can inhibit TLR3 mediated signaling due to loss of activated IRF3, but the mechanism of TLR3 signaling inhibition by W protein remains uncharacterized (Shaw et al. 2005). Another cellular protein, Polo-like kinase 1 (PLK1), has been identified as a binding partner for *Henipavirus* P, V and W proteins. The PLK1 binding site overlaps with the

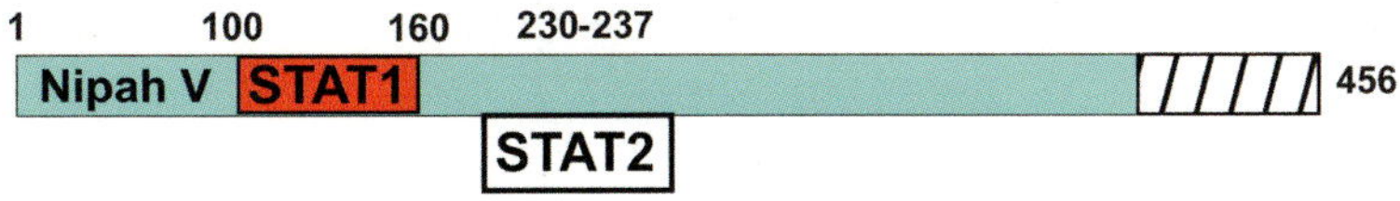

Fig. 3 The Nipah virus V protein binds and inhibits STAT1 and STAT2. Scale diagram of the Nipah virus V protein, illustrating the STAT1 binding region between residues 100 and 160, the residues implicated in STAT2 association between 230 and 237, and the conserved zinc-binding CTD (*hatched*). STAT-binding sequences are identical in V, W, and P proteins

STAT1 binding site between residues 100–160 (Ludlow et al. 2008; Parisien et al. 2002b). The interaction of PLK1 with V and W are independent of the STAT interactions, suggesting V:STAT and V:PLK exist as separate complexes. The study of PLK1 association further refined the known STAT1 binding site specificity to included Serine 130 and Serine 131 which are both essential for binding to STAT1 (and therefore STAT2) and normal IFN signaling interference (Ludlow et al. 2008). A therapeutically relevant outcome of this study was the demonstration that Nipah virus P proteins that no longer interact with STATs or PLK1 retain the ability to function in genome replication assays (Ludlow et al. 2008). While the precise biological significance of PLK1 association with *Henipavirus* V proteins remains to be determined, this kinase was recently demonstrated to be important for regulation of cell death and cytokine expression during PIV5 infection (Sun et al. 2009), suggesting a general role in virus replication. As defective STAT1 interference is an obvious strategy for attenuation of *Henipaviruses* this result suggests that reverse genetics approaches will be successful in creation of vaccine strains. Indeed, it was earlier found that a Nipah virus V protein with a single amino acid substitution fails to engage STAT1 (Hagmaier et al. 2007), and recombinant viruses harboring this mutation are attenuated in replication (Ciancanelli et al. 2009).

Morbilliviruses

Measles virus is the prototype of the *Morbillivirus* genus. The P/V/C locus of measles virus, like that of other paramyxoviruses is associated with host immune evasion, and can disengage IFN-JAK-STAT signaling. The evasion activities are ascribed to the V protein, but specific cases of P and C protein-mediated host evasion have also been revealed (Devaux et al. 2007; Fontana et al. 2008).

The mechanism of measles V protein mediated inhibition of IFN signaling is distinct from other paramyxoviruses. The measles virus V protein does not degrade STATs, but effectively prevents IFN-induced STAT1 and STAT2 nuclear import (Palosaari et al. 2003) by an unknown mechanism. Measles virus V protein can efficiently antagonize IFNα/β signaling but the IFNγ signaling inhibition has not always been confirmed (Caignard et al. 2007; Devaux et al. 2007; Fontana et al. 2008; Palosaari et al. 2003; Takeuchi et al. 2003). Recently, distinct means for disengaging the IFNγ and IFNα/β signal transduction machinery through separate associations with STAT1 and STAT2 have been revealed (Fig. 4). Results indicate that

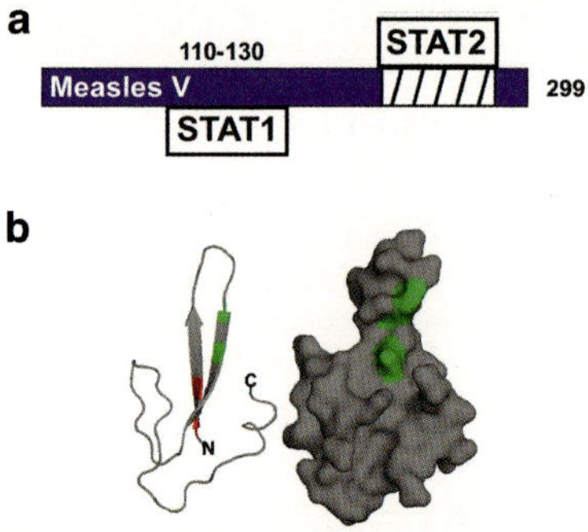

Fig. 4 The Measles virus V protein contacts STAT1 and STAT2 through distinct regions. (**a**) Scale diagram of the measles virus V protein, illustrating residues 110–130, implicated in STAT1 association. These residues are identical and functional in the measles P protein. The V protein zinc-binding CTD (*hatched*) is sufficient for STAT2 interaction and IFN signaling interference. (**b**) Structural model of the measles virus CTD, illustrating the positions of residues implicated in STAT2 associations, defining a contact surface

STAT2 is the primary target for measles V protein-mediated IFNα/β signaling evasion. Measles virus V protein engages STAT2 in the absence of STAT1 and in the absence of the V protein's STAT1 binding site (Ramachandran et al. 2008). Unlike other paramyxovirus V proteins, measles V binds STAT2 specifically through the highly conserved CTD, and this domain is both necessary and sufficient for interaction with STAT2 and IFNα/β signaling evasion. Results demonstrate that both the zinc finger structure and specific amino acids present on the fingers are required for mediating interactions with STAT2, and specific measles virus amino acids in the CTD were implicated in STAT2 association (Ramachandran et al. 2008). Similar findings verify the importance of the CTD for STAT2 interaction, and also implicating conserved tryptophan residues W240 and W250 as important for mediating STAT2 interactions (Caignard et al. 2009). The direct and fundamental association with STAT2 clearly represents a target for design of a small molecule that could disengage the measles virus IFN signaling evasion.

A distinct region of the measles virus V protein associates with STAT1. The STAT1 binding site on measles V has been mapped to residues 110–130 in the region shared between P and V proteins (Caignard et al. 2009; Ramachandran et al. 2008). This region is noted for high sequence conservation amongst *Morbilliviruses* (Devaux et al. 2007). This region originates with tyrosine 110 that has been shown to be important in the P protein mediated block of STAT1 tyrosine phosphorylation and in mediating IFN evasion (Devaux et al. 2007; Caignard et al. 2007; Combredet et al. 2003; Fontana et al. 2008; Ohno et al. 2004; Patterson et al. 2000; Ramachandran et al. 2008).

Though it can interact with STAT1, it has been observed that measles virus V protein more effectively achieves IFNα/β signal interference than IFNγ interference (Ramachandran et al. 2008; Takeuchi et al. 2003). Both P and V have the intrinsic capacity to engage STAT1 via residues 110–130, consistent with the finding that measles virus lacking V protein expression retains the ability to antagonize STAT1 signaling via the P protein in a Y110-dependent reaction (Devaux et al. 2007). It is likely that the differential effects of measles virus on IFNγ signaling may become more apparent during its normal life cycle in whole animals, and may play a role in the immune response to natural measles virus infections.

Concluding Remarks

The investigation of paramyxovirus STAT evasion strategies demonstrates the ability of viruses to exploit cellular processes and protein interactions to create virtual STAT deficient hosts that are more susceptible to virus replication and hidden from the immune system. The growing catalog of paramyxovirus V protein-mediated STAT inhibition strategies provides numerous insights into the vulnerabilities of STAT proteins that can be used to inform the design of therapeutic strategies for virus infections and beyond. Disruption of the interactions between V proteins and STATs by chemical compounds will lead to the creation of virus-specific antiviral drugs, and mutagenesis of the STAT interaction sites in recombinant viruses will produce attenuated strains for use in vaccination. Further, the insights gained through the study of virus-mediated STAT targeting identifies vulnerabilities of STAT proteins could be transferred to other cytokine systems to create new means to disrupt the action of hyper-activated STAT signaling pathways that are causes or contributors to human diseases including cancer and inflammation.

References

Akira S, Uematsu S, Takeuchi O (2006) Pathogen recognition and innate immunity. Cell 124:783–801

Andrejeva J, Poole E, Young DF, Goodbourn S, Randall RE (2002) The p127 subunit (DDB1) of the UV-DNA damage repair binding protein is essential for the targeted degradation of STAT1 by the V protein of the paramyxovirus simian virus 5. J Virol 76:11379–11386

Ashour J, Laurent-Rolle M, Shi PY, Garcia-Sastre A (2009) NS5 of dengue virus mediates STAT2 binding and degradation. J Virol 83:5408–5418

Ashour J, Morrison J, Laurent-Rolle M, Belicha-Villanueva A, Plumlee CR, Bernal-Rubio D, Williams KL, Harris E, Fernandez-Sesma A, Schindler C et al (2010) Mouse STAT2 restricts early dengue virus replication. Cell Host Microbe 8:410–421

Au WC, Moore PA, Lowther W, Juang YT, Pitha PM (1995) Identification of a member of the interferon regulatory factor family that binds to the interferon-stimulated response element and activates expression of interferon-induced genes. Proc Natl Acad Sci USA 92:11657–11661

Caignard G, Guerbois M, Labernardiere JL, Jacob Y, Jones LM, Wild F, Tangy F, Vidalain PO (2007) Measles virus V protein blocks Jak1-mediated phosphorylation of STAT1 to escape IFN-alpha/beta signaling. Virology 368:351–362

Caignard G, Bourai M, Jacob Y, Tangy F, Vidalain PO (2009) Inhibition of IFN-alpha/beta signaling by two discrete peptides within measles virus V protein that specifically bind STAT1 and STAT2. Virology 383:112–120

Caldenhoven E, Buitenhuis M, van Dijk TB, Raaijmakers JA, Lammers JW, Koenderman L, de Groot RP (1999) Lineage-specific activation of STAT3 by interferon-gamma in human neutrophils. J Leukoc Biol 65:391–396

Carlos TS, Fearns R, Randall RE (2005) Interferon-induced alterations in the pattern of parainfluenza virus 5 transcription and protein synthesis and the induction of virus inclusion bodies. J Virol 79:14112–14121

Chua KB, Bellini WJ, Rota PA, Harcourt BH, Tamin A, Lam SK, Ksiazek TG, Rollin PE, Zaki SR, Shieh W et al (2000) Nipah virus: a recently emergent deadly paramyxovirus. Science 288:1432–1435

Ciancanelli MJ, Volchkova VA, Shaw ML, Volchkov VE, Basler CF (2009) Nipah virus sequesters inactive STAT1 in the nucleus via a P gene-encoded mechanism. J Virol 83:7828–7841

Combredet C, Labrousse V, Mollet L, Lorin C, Delebecque F, Hurtrel B, McClure H, Feinberg MB, Brahic M, Tangy F (2003) A molecularly cloned Schwarz strain of measles virus vaccine induces strong immune responses in macaques and transgenic mice. J Virol 77:11546–11554

Devaux P, von Messling V, Songsungthong W, Springfeld C, Cattaneo R (2007) Tyrosine 110 in the measles virus phosphoprotein is required to block STAT1 phosphorylation. Virology 360:72–83

Didcock L, Young DF, Goodbourn S, Randall RE (1999) The V protein of simian virus 5 inhibits interferon signalling by targeting STAT1 for proteasome-mediated degradation. J Virol 73:9928–9933

Fitzgerald KA, McWhirter SM, Faia KL, Rowe DC, Latz E, Golenbock DT, Coyle AJ, Liao SM, Maniatis T (2003) IKKepsilon and TBK1 are essential components of the IRF3 signaling pathway. Nat Immunol 4:491–496

Fontana JM, Bankamp B, Bellini WJ, Rota PA (2008) Regulation of interferon signaling by the C and V proteins from attenuated and wild-type strains of measles virus. Virology 374:71–81

Fu X-Y, Kessler DS, Veals SA, Levy DA, Darnell JE Jr (1990) ISGF3, the transcriptional activator induced by interferon-alpha consists of multiple interacting polypeptide chains. Proc Natl Acad Sci USA 87:8555–8559

Gale M Jr, Sen GC (2009) Viral evasion of the interferon system. J Interferon Cytokine Res 29:475–476

Goodbourn S, Didcock L, Randall RE (2000) Interferons: cell signalling, immune modulation, antiviral response and virus countermeasures. J Gen Virol 81(10):2341–2364

Gotoh B, Komatsu T, Takeuchi K, Yokoo J (2002) Paramyxovirus strategies for evading the interferon response. Rev Med Virol 12:337–357

Grandvaux N, ten Oever BR, Servant MJ, Hiscott J (2002) The interferon antiviral response: from viral invasion to evasion. Curr Opin Infect Dis 15:259–267

Hagmaier K, Stock N, Precious B, Childs K, Wang LF, Goodbourn S, Randall RE (2007) Mapuera virus, a rubulavirus that inhibits interferon signalling in a wide variety of mammalian cells without degrading STATs. J Gen Virol 88:956–966

Harcourt BH, Tamin A, Ksiazek TG, Rollin PE, Anderson LJ, Bellini WJ, Rota PA (2000) Molecular characterization of Nipah virus, a newly emergent paramyxovirus. Virology 271:334–349

He B, Paterson RG, Stock N, Durbin JE, Durbin RK, Goodbourn S, Randall RE, Lamb RA (2002) Recovery of paramyxovirus simian virus 5 with a V protein lacking the conserved cysteine-rich domain: the multifunctional V protein blocks both interferon-beta induction and interferon signaling. Virology 303:15–32

Hiscott J, Lin R, Nakhaei P, Paz S (2006) MasterCARD: a priceless link to innate immunity. Trends Mol Med 12:53–56

Horvath CM (2004a) Silencing STATs: lessons from paramyxovirus interferon evasion. Cytokine Growth Factor Rev 15:117–127

Horvath CM (2004b) Weapons of STAT destruction. Interferon evasion by paramyxovirus V protein. Eur J Biochem 271:4621–4628

Johnson CL, Gale M Jr (2006) CARD games between virus and host get a new player. Trends Immunol 27:1–4

Kang DC, Gopalkrishnan RV, Wu Q, Jankowsky E, Pyle AM, Fisher PB (2002) mda-5: an interferon-inducible putative RNA helicase with double-stranded RNA-dependent ATPase activity and melanoma growth-suppressive properties. Proc Natl Acad Sci USA 99:637–642

Katze MG, He Y, Gale M Jr (2002) Viruses and interferon: a fight for supremacy. Nat Rev Immunol 2:675–687

Kawai T, Takahashi K, Sato S, Coban C, Kumar H, Kato H, Ishii KJ, Takeuchi O, Akira S (2005) IPS-1, an adaptor triggering RIG-I- and Mda5-mediated type I interferon induction. Nat Immunol 6:981–988

Kessler DS, Veals SA, Fu X-Y, Levy DE (1990) Interferon-α regulates nuclear translocation and DNA-binding affinity of ISGF-3, a multimeric transcriptional activator. Genes Dev 4:1753–1765

Kraus TA, Garza L, Horvath CM (2008) Enabled interferon signaling evasion in an immune-competent transgenic mouse model of parainfluenza virus 5 infection. Virology 371:196–205

Kubota T, Yokosawa N, Yokota S, Fujii N (2001) C terminal CYS-RICH region of mumps virus structural V protein correlates with block of interferon alpha and gamma signal transduction pathway through decrease of STAT 1-alpha. Biochem Biophys Res Commun 283:255–259

Lamb RA, Parks GD (2006) Paramyxoviridae: the viruses and their replication. In: Knipe DM, Howley PM (eds) Fields virology. Lippincott Williams and Wilkins, Philadelphia, pp 1449–1496

Lamb RA, Parks GD (2007) Paramyxoviridae: the viruses and their replication, 5th edn. Wolters Kluwer/Lippincott Williams and Wilkins, Philadelphia

Levy DE, Garcia-Sastre A (2001) The virus battles: IFN induction of the antiviral state and mechanisms of viral evasion. Cytokine Growth Factor Rev 12:143–156

Levy DE, Kessler DS, Pine R, Reich N, Darnell JE Jr (1988) Interferon-induced nuclear factors that bind a shared promoter element correlate with positive and negative control. Genes Dev 2:383–393

Li T, Chen X, Garbutt KC, Zhou P, Zheng N (2006) Structure of DDB1 in complex with a paramyxovirus V protein: viral hijack of a propeller cluster in ubiquitin ligase. Cell 124:105–117

Lin GY, Lamb RA (2000) The paramyxovirus simian virus 5 V protein slows progression of the cell cycle. J Virol 74:9152–9166

Lin GY, Paterson RG, Richardson CD, Lamb RA (1998) The V protein of the paramyxovirus SV5 interacts with damage-specific DNA binding protein. Virology 249:189–200

Liston P, Briedis DJ (1994) Measles virus V protein binds zinc. Virology 198:399–404

Lo MK, Harcourt BH, Mungall BA, Tamin A, Peeples ME, Bellini WJ, Rota PA (2009) Determination of the henipavirus phosphoprotein gene mRNA editing frequencies and detection of the C, V and W proteins of Nipah virus in virus-infected cells. J Gen Virol 90:398–404

Ludlow LE, Lo MK, Rodriguez JJ, Rota PA, Horvath CM (2008) Henipavirus V protein association with polo-like kinase reveals functional overlap with STAT1 binding and interferon evasion. J Virol 82:6259–6271

Meylan E, Curran J, Hofmann K, Moradpour D, Binder M, Bartenschlager R, Tschopp J (2005) Cardif is an adaptor protein in the RIG-I antiviral pathway and is targeted by hepatitis C virus. Nature 437:1167–1172

Nishio M, Tsurudome M, Ito M, Garcin D, Kolakofsky D, Ito Y (2005) Identification of paramyxovirus V protein residues essential for STAT protein degradation and promotion of virus replication. J Virol 79:8591–8601

Ohno S, Ono N, Takeda M, Takeuchi K, Yanagi Y (2004) Dissection of measles virus V protein in relation to its ability to block alpha/beta interferon signal transduction. J Gen Virol 85:2991–2999

Palosaari H, Parisien JP, Rodriguez JJ, Ulane CM, Horvath CM (2003) STAT protein interference and suppression of cytokine signal transduction by measles virus V protein. J Virol 77:7635–7644

Parisien J-P, Lau JF, Rodriguez JJ, Sullivan BM, Moscona A, Parks GD, Lamb RA, Horvath CM (2001) The V protein of human parainfluenza virus 2 antagonizes type I interferon responses by destabilizing signal transducer and activator of transcription 2. Virology 283:230–239

Parisien JP, Lau JF, Horvath CM (2002a) STAT2 acts as a host range determinant for species-specific paramyxovirus interferon antagonism and simian virus 5 replication. J Virol 76:6435–6441

Parisien JP, Lau JF, Rodriguez JJ, Ulane CM, Horvath CM (2002b) Selective STAT protein degradation induced by paramyxoviruses requires both STAT1 and STAT2 but is independent of alpha/beta interferon signal transduction. J Virol 76:4190–4198

Paterson RG, Leser GP, Shaughnessy MA, Lamb RA (1995) The paramyxovirus SV5 V protein binds two atoms of zinc and is a structural component of virions. Virology 208:121–131

Patterson JB, Thomas D, Lewicki H, Billeter MA, Oldstone MB (2000) V and C proteins of measles virus function as virulence factors in vivo. Virology 267:80–89

Poole E, He B, Lamb RA, Randall RE, Goodbourn S (2002) The V proteins of simian virus 5 and other paramyxoviruses inhibit induction of interferon-beta. Virology 303:33–46

Precious B, Childs K, Fitzpatrick-Swallow V, Goodbourn S, Randall RE (2005) Simian virus 5 V protein acts as an adaptor, linking DDB1 to STAT2, to facilitate the ubiquitination of STAT1. J Virol 79:13434–13441

Precious BL, Carlos TS, Goodbourn S, Randall RE (2007) Catalytic turnover of STAT1 allows PIV5 to dismantle the interferon-induced anti-viral state of cells. Virology 368:114–121

Ramachandran A, Parisien JP, Horvath CM (2008) STAT2 is a primary target for measles virus V protein-mediated alpha/beta interferon signaling inhibition. J Virol 82:8330–8338

Reich N, Evans B, Levy DE, Fahey D, Knight E Jr, Darnell JE Jr (1987) Interferon-induced transcription of a gene encoding a 15 kDa protein depends on an upstream enhancer element. Proc Natl Acad Sci USA 84:6394–6398

Rodriguez JJ, Horvath CM (2004) Host evasion by emerging paramyxoviruses: Hendra virus and Nipah virus V proteins inhibit interferon signaling. Viral Immunol 17:210–219

Rodriguez JJ, Parisien JP, Horvath CM (2002) Nipah virus V protein evades alpha and gamma interferons by preventing STAT1 and STAT2 activation and nuclear accumulation. J Virol 76:11476–11483

Rodriguez JJ, Wang LF, Horvath CM (2003) Hendra virus V protein inhibits interferon signaling by preventing STAT1 and STAT2 nuclear accumulation. J Virol 77:11842–11845

Rodriguez JJ, Cruz CD, Horvath CM (2004) Identification of the nuclear export signal and STAT-binding domains of the Nipah virus V protein reveals mechanisms underlying interferon evasion. J Virol 78:5358–5367

Sato M, Tanaka N, Hata N, Oda E, Taniguchi T (1998) Involvement of the IRF family transcription factor IRF-3 in virus-induced activation of the IFN-beta gene. FEBS Lett 425:112–116

Selvey LA, Wells RM, McCormack JG, Ansford AJ, Murray K, Rogers RJ, Lavercombe PS, Selleck P, Sheridan JW (1995) Infection of humans and horses by a newly described morbillivirus. Med J Aust 162:642–645

Servant MJ, ten Oever B, LePage C, Conti L, Gessani S, Julkunen I, Lin R, Hiscott J (2001) Identification of distinct signaling pathways leading to the phosphorylation of interferon regulatory factor 3. J Biol Chem 276:355–363

Servant MJ, Grandvaux N, Hiscott J (2002) Multiple signaling pathways leading to the activation of interferon regulatory factor 3. Biochem Pharmacol 64:985–992

Seth RB, Sun L, Ea CK, Chen ZJ (2005) Identification and characterization of MAVS, a mitochondrial antiviral signaling protein that activates NF-kappaB and IRF 3. Cell 122:669–682

Shaw ML, Garcia-Sastre A, Palese P, Basler CF (2004) Nipah virus V and W proteins have a common STAT1-binding domain yet inhibit STAT1 activation from the cytoplasmic and nuclear compartments, respectively. J Virol 78:5633–5641

Shaw ML, Cardenas WB, Zamarin D, Palese P, Basler CF (2005) Nuclear localization of the Nipah virus W protein allows for inhibition of both virus- and toll-like receptor 3-triggered signaling pathways. J Virol 79:6078–6088

Sun D, Luthra P, Li Z, He B (2009) PLK1 down-regulates parainfluenza virus 5 gene expression. PLoS Pathog 5:e1000525

Takeuchi O, Akira S (2008) MDA5/RIG-I and virus recognition. Curr Opin Immunol 20:17–22

Takeuchi K, Kadota SI, Takeda M, Miyajima N, Nagata K (2003) Measles virus V protein blocks interferon (IFN)-alpha/beta but not IFN-gamma signaling by inhibiting STAT1 and STAT2 phosphorylation. FEBS Lett 545:177–182

Taylor DR, Shi ST, Romano PR, Barber GN, Lai MM (1999) Inhibition of the interferon-inducible protein kinase PKR by HCV E2 protein (see comments). Science 285:107–110

Thomas SM, Lamb RA, Paterson RG (1988) Two mRNAs that differ by two nontemplated nucleotides encode the amino coterminal proteins P and V of the paramyxovirus SV5. Cell 54:891–902

Ulane CM, Horvath CM (2002) Paramyxoviruses SV5 and HPIV2 assemble STAT protein ubiquitin ligase complexes from cellular components. Virology 304:160–166

Ulane CM, Rodriguez JJ, Parisien JP, Horvath CM (2003) STAT3 ubiquitylation and degradation by mumps virus suppress cytokine and oncogene signaling. J Virol 77:6385–6393

Ulane CM, Kentsis A, Cruz CD, Parisien JP, Schneider KL, Horvath CM (2005) Composition and assembly of STAT-targeting ubiquitin ligase complexes: paramyxovirus V protein carboxyl terminus is an oligomerization domain. J Virol 79:10180–10189

Velichko S, Wagner TC, Turkson J, Jove R, Croze E (2002) STAT3 activation by type I interferons is dependent on specific tyrosines located in the cytoplasmic domain of interferon receptor chain 2c. Activation of multiple STATS proceeds through the redundant usage of two tyrosine residues. J Biol Chem 277:35635–35641

Wansley EK, Parks GD (2002) Naturally occurring substitutions in the P/V gene convert the noncytopathic paramyxovirus simian virus 5 into a virus that induces alpha/beta interferon synthesis and cell death. J Virol 76:10109–10121

Weaver BK, Kumar KP, Reich NC (1998) Interferon regulatory factor 3 and CREB-binding protein/p300 are subunits of double-stranded RNA-activated transcription factor DRAF1. Mol Cell Biol 18:1359 1368

Xu LG, Wang YY, Han KJ, Li LY, Zhai Z, Shu HB (2005) VISA is an adapter protein required for virus-triggered IFN-beta signaling. Mol Cell 19:727–740

Yokosawa N, Yokota S, Kubota T, Fujii N (2002) C-terminal region of STAT-1alpha is not necessary for its ubiquitination and degradation caused by mumps virus V protein. J Virol 76:12683–12690

Yoneyama M, Kikuchi M, Natsukawa T, Shinobu N, Imaizumi T, Miyagishi M, Taira K, Akira S, Fujita T (2004) The RNA helicase RIG-I has an essential function in double-stranded RNA-induced innate antiviral responses. Nat Immunol 5:730–737

Yoneyama M, Kikuchi M, Matsumoto K, Imaizumi T, Miyagishi M, Taira K, Foy E, Loo YM, Gale M Jr, Akira S et al (2005) Shared and unique functions of the DExD/H box helicases RIG-I, MDA5, and LGP2 in antiviral innate immunity. J Immunol 175:2851–2858

Young DF, Didcock L, Goodbourn S, Randall RE (2000) Paramyxoviridae use distinct virus-specific mechanisms to circumvent the interferon response. Virology 269:383–390

Multifaceted Antiviral Actions of Interferon-stimulated Gene Products

Birgit Strobl, Nicole R. Leitner, and Mathias Müller

Abstract

Interferons (IFNs) are extremely powerful cytokines for the host defence against viral infections. Binding of IFNs to their receptors activates the JAK/STAT signalling pathway with the Janus kinases JAK1, 2 and TYK2 and the signal transducer and activators of transcription (STAT) 1 and STAT2. Depending on the cellular setting, additional STATs (STAT3-6) and additional signalling pathways are activated. The actions of IFNs on infected cells and the surrounding tissue are mediated by the induction of several hundred IFN-stimulated genes (ISGs). Since the cloning of the first ISGs, considerable progress has been made in describing antiviral effector proteins and their many modes of action. Effector proteins individually target distinct steps in the viral life cycle, including blocking virus entry, inhibition of viral transcription and translation, modification of viral nucleic acids and proteins and, interference with virus assembly and budding. Novel pathways of viral inhibition are constantly being elucidated and, additionally, unanticipated functions of known antiviral effector proteins are discovered. Herein, we outline IFN induced antiviral pathways and review recent developments in this fascinating area of research.

B. Strobl (✉) • N.R. Leitner
Institute of Animal Breeding and Genetics, University of Veterinary Medicine Vienna, Veterinärplatz 1, 1210 Vienna, Austria
e-mail: birgit.strobl@vetmeduni.ac.at

M. Müller
Institute of Animal Breeding and Genetics, University of Veterinary Medicine Vienna, Veterinärplatz 1, 1210 Vienna, Austria

Biomodels Austria, University of Veterinary Medicine Vienna, Vienna, Austria

Th. Decker and M. Müller (eds.), *Jak-Stat Signaling: From Basics to Disease*,
DOI 10.1007/978-3-7091-0891-8_21, © Springer-Verlag Wien 2012

Abbreviations[1]

ADAR	Adenosine deaminase acting on RNA
APOBEC	Apolipoprotein B mRNA-editing catalytic polypeptide
BST-2	Bone marrow stromal antigen 2
ds	Double-stranded
eIF	Eukaryotic translation initiation factor
ER	Endoplasmatic reticulum
GAS	IFNγ activated sequence
GBP	Guanylate binding protein
IFN	Interferon
IFNAR	IFNα/β receptor
IL	Interleukin
ISG	IFN-stimulated gene
IFIT	IFN-induced protein with tetratricopeptide repeats
IFITM	IFN-induced transmembrane protein
ISGF3	IFN-stimulated gene factor 3
IRF	IFN regulatory factor
ISRE	IFN-stimulated response element
JAK	Janus kinase
PML	Promyelocytic leukaemia
2-5A_n	2′-5′ oligoadenylate
OAS	2-5A_n synthetase
MDA5	Melanoma differentiation-associated protein 5
PAMP	Pathogen-associated molecular pattern
PKR	dsRNA-dependent protein kinase
PRR	Pattern-recognition receptor
RIG-I	Retinoic acid-inducible protein I
RNase L	Latent ribonuclease
ss	Single-stranded
STAT	Signal transducer and activator of transcription
TRIM	Tripartite motif
TYK2	Tyrosine kinase 2

Introduction

Antiviral activity defines the bioactivity of interferons (IFNs). IFNs induce hundreds of IFN-stimulated genes (ISGs), whereby many are regulated by all IFNs, and others are induced more selectively (de Veer et al. 2001; Der et al. 1998). Even for several

[1] Viruses mentioned in this article, see Table 2

long-known ISG-encoded proteins with direct or indirect antiviral activity, the mechanisms of action are often still poorly understood. Studies on antiviral activities of specific proteins became more complicated than anticipated, largely because of their high virus-specificity and their involvement in multiple pathways, including normal cellular physiology. Moreover, viral counteracting mechanisms can be very efficient and mechanistic discoveries often depended on mutant viruses. As another level of complexity, some of the antiviral effector proteins also have proviral effects.

Recognition of virus infection is a central requirement for the initiation of an efficient host defence. Molecules and signalling cascades involved have been mainly characterized during the last decade and new players are still being identified. A number of germline-encoded pattern-recognition receptors (PRRs) sense the presence of so-called pathogen-associated molecular patterns (PAMPs). PRRs involved in host defence against viruses are localized in several cellular compartments and recognize distinct molecular structures, most prominently viral nucleic acids. Activation of distinct PAMP-induced signalling cascades converge in most cases at the induction of pro-inflammatory cytokines and type I IFNs (Kawai and Akira 2009; McCartney and Colonna 2009).

IFNs are grouped into three classes, called type I, type II and the more recently identified type III IFNs. Type I IFNs are the key cytokines for innate antiviral immunity, as they are rapidly induced upon virus recognition and act on presumably all cell types. Type I IFNs have been discovered more than 50 years ago and comprise a large group of cytokines. Among these, several distinct IFNα subtypes (13 in humans) and IFNβ are induced directly in response to viral infections. Type II IFN consists of only one member, IFNγ, that is mainly produced by activated T cells and natural killer (NK) cells (Pestka et al. 2004). Type III IFNs are structurally more related to the interleukin- (IL-) 10 cytokine family, but have been classified as IFNs based on their similar biological activity as IFNα/β. In humans, this family has three members, IFNλ1–IFNλ3, originally designated as interleukin IL-29, IL-28A and IL-28B, respectively. IFNλs are also directly induced by viral infections by similar, but probably not identical, mechanisms. In contrast to IFNα/β, IFNλs act only on specific cell types due to cell type-restricted expression of their specific receptor chain. The latter appears to be mainly expressed on epithelial cells and thus IFNλs have a more restricted/specific role in the antiviral defence (Commins et al. 2008).

Signalling of type I IFNs is initiated by ligand binding to a common heterodimeric receptor complex consisting of the IFNAR1 and IFNAR2 chains. Ligand binding results in the activation of signal transducers and activators of transcription (STATs) by the receptor-associated Janus kinases (JAKs) TYK2 and JAK1. Mainly STAT1–STAT2 heterodimers, and to a lesser extent STAT1 homodimers, are activated in response to IFNα/β. Dependent on the cell type, other STAT family members can be activated, but their impact on cellular responses is often unclear. STAT1–STAT2 heterodimers, in association with IFN regulatory factor (IRF) 9, form the IFN-stimulated gene factor 3 (ISGF3) and induce transcription of genes containing IFN-stimulated response elements (ISREs) in their regulatory regions. Type III IFNs utilize the IL-10R2 and the IFNλR1 (IL-28Rα) receptor chains and activate TYK2, JAK1 and mainly ISGF3. Accordingly, responses appear very similar to those induced by type I IFNs, although some differences have been reported. IFNγ binds to IFNGR1 and IFNGR2 and activates JAK1, JAK2 and mainly STAT1 homodimers. STAT1 homodimers bind to IFNγ activated sequences (GAS) in promoter regions of responsive genes. IFNγ can also activate other STAT complexes but again, the contribution of these complexes to the overall response requires further characterization. In general, IFNα/β is mainly associated with antiviral and IFNγ with antibacterial activity, but IFNγ certainly also exhibits potent antiviral activity and IFNα/β impacts on the antimicrobial defence. It is important to note that in addition to the induction of an antiviral state in responsive cells, IFNs also exert important immunomodulatory activities, mainly by shaping adaptive immune responses (Borden et al. 2007; Schindler et al. 2007).

Several specific antiviral pathways have been reviewed recently (see below), so the emphasis herein will be on providing an overview about well-known IFN-induced antiviral pathways with a strong focus on recent developments. We present the specific antiviral effector proteins ordered according to their main effects on virus replication (Fig. 1). It has to be mentioned, however, that there are often multiple modes of action

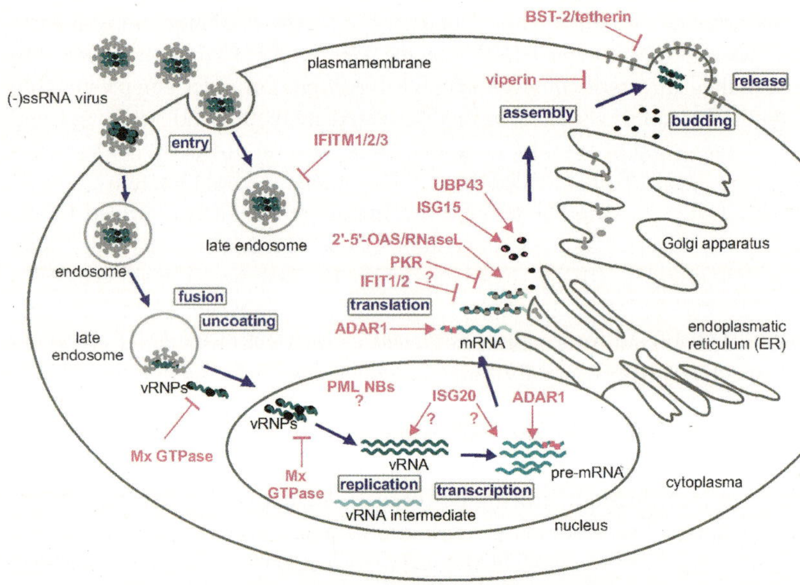

Fig. 1 Schematic overview over antiviral effector mechanisms of IFN-stimulated gene products. Viruses use substantially different replication strategies and we therefore use the life cycle of an enveloped (−)ssRNA virus as an example (depicted in *blue letters*). The virus binds to its cellular receptor at the plasma membrane, enters the cell *via* pH-dependent endocytosis and fusion, subsequently uncoated viral nucleoproteins are transported to the nucleus, where virus replication and transcription occurs. Viral proteins are translated by the cellular machinery, core proteins assemble with the viral genome, envelope proteins are transported via the ER/Golgi apparatus to the plasma membrane where virus assembly and budding occurs. ISG-encoded proteins targeting different steps during viral replication are depicted in *pink letters*. ISG products can either inhibit certain steps (⊥), modify/degrade viral products (↓), or act *via* unknown/putative mechanisms (?). Only the major or best characterized pathways are shown, see text for details and additional functions of IFN-induced proteins and pathways

which are discussed in the respective subsections. Besides, other IFN-induced proteins not mentioned in this article might well have important functions in the cellular resistance against virus infections.

Inhibition of Virus Entry or Uncoating

Interferon-Induced Transmembrane (IFITM) Proteins

The IFITM proteins are relatively small (14–17 kDa) transmembrane proteins with cell type- and tissue-dependent subcellular localization. IFITM proteins have been found in proximity to the endoplasmatic reticulum (ER), the Golgi apparatus, small vesicles, exosomes, or associated with the plasma membrane (Siegrist et al. 2011). They have been originally identified based on their IFN inducibility, although they are also expressed at low levels in untreated cells (Friedman et al. 1984). The antiviral potency of human IFITM1 (9–27, Ifi17, fragilis protein 2) has been first demonstrated by overexpression of *IFITM1*, which led to a marked decrease of vesicular stomatitis virus (VSV) but not influenza A virus (FLUAV) replication (Alber and Staeheli 1996). First mechanistic insights came from a quite recent study (Brass et al. 2009). Using viral pseudoparticles containing the murine leukaemia virus (MLV) genome and unique envelope proteins from different viruses, IFITM actions could be linked to viral surface proteins and to early steps in viral replication. Envelope proteins from several FLUAV strains, three different flaviviruses (i.e. West Nile virus (WNV), yellow fever virus (YFV), Omsk hemorrhagic fever virus (OMSK)), and VSV could confer IFITM1-, IFITM2- and IFITM3-sensitivity to pseudovirus particles, whereas no effect was observed with envelope proteins from three different arenaviruses (i.e. Machupo virus (MACV), Lassa virus (LASV), lymphocytic choriomeningitis virus (LCMV)), or MLV. Consistent with the pseudovirus particles, overexpression of *IFITM1*, *IFITM2* (fragilis protein 3, 1-8D) or *IFITM3* (1-8U, Ifi15, fragilis protein) in human cell lines results in strongly reduced replication of FLUAV. Moreover, siRNA-mediated knockdown of *IFITM3* enhances FLUAV replication and decreases IFNα or IFNγ-mediated anti-FLUAV activity. The effect of *IFITM3* on WNV and Dengue virus (DV) replication was confirmed with overexpression and siRNA-mediated knockdown experiments. No effect was observed on hepatitis C virus (HCV) replication, supporting the notion that IFITM actions are virus-specific (Brass et al. 2009). Further evidence for the inhibition of flavivirus and of rhabdovirus entry and/or uncoating by IFITM1-3 has been provided using stably transfected human cell lines (Jiang et al. 2010; Weidner et al. 2010). Very recently, the IFITM-mediated inhibition *via* viral entry proteins was extended to filoviruses (i.e. Marburg virus (MARV) and Ebola virus (EBOV)), severe acute respiratory syndrome coronavirus (SARS-CoV) (Huang et al. 2011), and human immunodeficiency virus (HIV)-1 (Lu et al. 2011). Consistent with studies in human cells, increased FLUAV replication was observed in embryonic fibroblasts derived from mice lacking the entire *Ifitm* locus (Table 1) (Brass et al. 2009).

Table 1 Gene-targeted mice of ISGs that encode proteins implicated in the IFN-induced antiviral defense: basic phenotype and virus susceptibility (in alphabetical order)

Gene-targeted mice[a, b]	Phenotype	Virus sensitivity in vivo[c, d]	References
Adar1 (p110/p150)$^{-/-}$	Embryonic lethal	–	Hartner et al. (2004, 2009), Wang et al. (2004), XuFeng et al. (2009)
Adar1 (p150)$^{-/-}$	Embryonic lethal	–	Ward et al. (2011)
Apobec3$^{-/-}$	Viable, fertile	Increased levels of integrated MMTV DNA in LN DCs after MMTV infection (*s.c.*) (+/– LPS pre-treatment); increased MMTV DNA levels in LNs after MMTV infection (*i.f.*)	Low et al. (2009), Okeoma et al. (2007, 2009a, b)
		Increased infection levels in bone marrow, spleen and thymus and earlier onset of leukaemias after neonatal infection (*i.p.*) with M-MuLV, no gross difference in M-MuLV-induced tumour types	
Apobec3$^{-/-}$*, HBV-Tg*	Viable, fertile	Normal levels of core-associated HBV DNA in serum; normal dsRNA-induced inhibition of HBV DNA production	Turelli et al. (2008)
Ifit1$^{-/-}$	Viable, fertile	Decreased survival after challenge with 2′-O-methyltransferase mutant WNV (WNV-E218A) (*i.c.*)	Daffis et al. (2010), Pichlmair et al. (2011)
		Decreased survival after infection (*i.n.*) with VSV	
		Normal survival after infection with EMCV (*i.p.*)	
Ifitm1-6$^{-/-}$ *(IfitmDel)*	Viable, fertile	n.r.	Lange et al. (2008)
Ifitm3$^{-/-}$	Viable, fertile	n.r.	Lange et al. (2008)
Igtp$^{-/-}$	Viable, fertile	Normal virus load in liver after MCMV infection (*i.p.*)	Taylor et al. (2000)
Isg15$^{-/-}$	Viable, fertile	Increased lethality and increased virus load after FLUA/B virus (*i.n.*), HSV-1 (*i.cor.*) and γ MHV68 (i.p.) infection	Kim et al. (2008a), Knobeloch et al. (2005), Lai et al. (2009), Lenschow et al. (2007), Osiak et al. (2005)
		Increased susceptibility to SINV infection (*i.c.*)	
		Normal resistance to VSV infection (*i.v.*)	
		Normal clearance of LCMV (*i.v.*) from spleens; no difference in survival after infection with LCMV (*i.c.*)	
		Normal clearance of injected (*tv.*) replication competent HBV DNA	

PKR$^{-/-}$	Viable, fertile	Higher mortality after CV-B4 infection (*i.p.*)	Barry et al. (2009), Nakayama et al. (2010), Sadler and Williams (2008)
		Decreased survival after VSV infection (*i.n.*) (+/− IFNα/β pre-treatment) and higher viral titers in brain and lung	
		Decreased survival after FLUAV strain A/WSN/33 infection (*i.n.*) and higher viral titer in lungs; decreased survival after delNS1 FLUAV infection (*i.n.*)	
		No difference after SFV infection (*i.p.*)	
		Higher viral titers after LCMV infection (*i.p.*) in lung, liver and spleen	
		Higher mortality after HSV-2 infection (*i.vag.*)	
PKR$^{-/-}$, *RNase L*$^{-/-}$ *DKO*	Viable, fertile	Higher mortality after CV-B4 infection (*i.p.*)	Nakayama et al. (2010), Silverman (2007)
		Higher mortality after WNV infection (*i.f.*) with higher viral load in the brain; no difference after WNV infection (*i.c.*)	
		Higher mortality after EMCV infection (*i.p.*); higher mortality after HSV-2 infection (*i.vag.*)	
Pml$^{-/-}$	Viable, fertile; impaired terminal myeloid cell maturation	Increased footpad swelling after LCMV infection (*i.f.*); increased virus load in spleen after LCMV infection (*i.v.*); increased lethality after LCMV infection (*i.cer.*)	Bonilla et al. (2002), Wang et al. (1998)
		Increased neutralizing Ab production after VSV infection (*i.v.*)	
RNase L$^{-/-}$	Viable, fertile	Increased mortality after EMCV infection (*i.p.*) (+/− IFNα pre-treatment)	Silverman (2007)
		Higher mortality after CV-B4 infection (*i.p.*)	
		Increased mortality after HSV-1 infection (*i.cor.*)	
		Increased mortality after WNV infection (*i.f.*) with similar viral load in the brain, no difference after WNV infection (*i.c.*)	
		Increased survival after HSV-2 infection (*i.vag.*); HSV-2 replication is similar in the vaginal epithelium, but lower in the nervous system	

(continued)

Table 1 (continued)

Gene-targeted mice[a, b]	Phenotype	Virus sensitivity in vivo[c, d]	References
Ube1l$^{-/-}$	Viable, fertile	Decreased survival and increased viral load in lungs after FLUBV virus infection *(i.n.)*	Giannakopoulos et al. (2009), Kim et al. (2006), Lai et al. (2009)
		Decreased survival after SINV infection *(s.c., i.cer.)*	
		No difference in survival after LCMV infection *(i.cer.)*	
Ubp43$^{-/-}$	Brain injury, neurological disorders, reduced lifespan	Increased survival and reduced viral load in brain, liver and serum after infection with LCMV *(i.cer.)*	Kim et al. (2008a), Knobeloch et al. (2005), Ritchie et al. (2002, 2004)
		No difference in survival after infection with LCMV *(i.c.)*	
		Increased survival after infection with VSV *(i.cer.)*	
		Increased clearance of injected *(tv.)* replication competent HBV DNA	
Ubp43$^{-/-}$, *Isg15*$^{-/-}$ *DKO*	Brain injury, neurological disorders, reduced lifespan	No difference in survival after infection with LCMV (*i.c.*)	Knobeloch et al. (2005)

[a]Note that most laboratory mouse strains are Mx1-negative, fail to upregulate *Gbp1* in response to IFNs and express a truncated *Oas1b* (see text for details)
[b]*DKO* double knockout
[c]*n.r.* not reported
[d]*i.c.* intracranial, *i.cer.* intracerebral, *i.cor.* intracorneal, *i.f.* intra-footpad, *i.n.* intranasal, *i.p.* intraperitoneal, *i.v.* intravenous, *i.vag.* intravaginal, *s.c.* subcutanous, *tv.* tail vein, *LN* lymph node, *LPS* lipopolysaccharide

Table 2 Viruses mentioned in this article: virus abbreviations, virus, virus family and genome organization

Abbreviation	Virus	Virus family	Genome organization[a]
ASFV	African swine fever virus	*Asfarviridae*	dsDNA
ASLV	Avian sarcoma leukosis virus	*Retroviridae*	ssRNA-RT
BRV	Bovine rotavirus	*Reoviridae*	dsRNA
BVDV	Bovine virus-diarrhoe virus	*Flaviviridae*	(+)ssRNA
CDV	Canine distemper virus	*Paramyxoviridae*	(−)ssRNA
CV	Coxsackie virus	*Picornaviridae*	(+)ssRNA
DV	Dengue virus	*Flaviviridae*	(+)ssRNA
EBOV	Ebola virus	*Filoviridae*	(−)ssRNA
EBV	Epstein-Barr virus	*Herpesviridae*	dsDNA
EMCV	Encephalomyocarditis virus	*Picornaviridae*	(+)ssRNA
FLUAV	Influenza A virus	*Orthomyxoviridae*	(−)ssRNA
FLUBV	Influenza B virus	*Orthomyxoviridae*	(−)ssRNA
FMLV	Friend-murine leukaemia virus	*Retroviridae*	ssRNA-RT
γ MHV68	Murine γ-herpesvirus 68	*Herpesviridae*	dsDNA
HAV	Hepatitis A virus	*Picornaviridae*	(+)ssRNA
HBV	Hepatitis B virus	*Hepadnaviridae*	dsDNA-RT
HCV	Hepatitis C virus	*Flaviviridae*	(+)ssRNA
HCMV	Human cytomegalovirus	*Herpesviridae*	dsDNA
HDV	Hepatitis delta virus	*Unassigned*	(−)ssRNA
HFV	Human foamy virus	*Retroviridae*	ssRNA-RT
HIV	Human immunodeficiency virus	*Retroviridae*	ssRNA-RT
HSV	Herpes simplex virus	*Herpesviridae*	dsDNA
KSHV, HHV8	Kaposi's sarcoma-associated herpesvirus/human herpesvirus 8	*Herpesviridae*	dsDNA
LACV	La Crosse virus	*Bunyaviridae*	(−)ssRNA
LASV	Lassa virus	*Arenaviridae*	(−)ssRNA
LCMV	Lymphocytic choriomeningitis virus	*Arenaviridae*	(−)ssRNA
MACV	Machupo virus	*Arenaviridae*	(−)ssRNA
MARV	Marburg virus	*Filoviridae*	(−)ssRNA
MCMV	Murine cytomegalovirus	*Herpesviridae*	dsDNA
MeV	Measles virus	*Paramyxoviridae*	(−)ssRNA
MHV	Mouse hepatitis virus	*Coronaviridae*	(+)ssRNA
MLV	Murine leukaemia virus	*Retroviridae*	ssRNA-RT
MMTV	Mouse mammary tumor virus	*Retroviridae*	ssRNA-RT
M-MuLV	Moloney murine leukemia virus	*Retroviridae*	ssRNA-RT
MPV	Mouse polyoma virus	*Polyomaviridae*	dsDNA
NDV	Newcastle disease virus	*Paramyxoviridae*	(−)ssRNA
OMSK	Omsk hemorrhagic fever virus	*Flaviviridae*	(+)ssRNA
PV	Poliovirus	*Picornaviridae*	(+)ssRNA

(continued)

Table 2 (continued)

Abbreviation	Virus	Virus family	Genome organization[a]
RSV	Respiratory syncytial virus	*Paramyxoviridae*	(−)ssRNA
RV	Rabies virus	*Rhabdoviridae*	(−)ssRNA
RVFV	Rift valley fever virus	*Bunyaviridae*	(−)ssRNA
SARS-CoV	Severe acute respiratory syndrome coronavirus	*Coronaviridae*	(+)ssRNA
SFV	Semliki Forest virus	*Togaviridae*	(+)ssRNA
SeV	Sendai virus	*Paramyxoviridae*	(−)ssRNA
SINV	Sindbis virus	*Togaviridae*	(+)ssRNA
TMEV	Theiler's murine encephalomyelitis virus	*Picornaviridae*	(+)ssRNA
THOV	Thogoto virus	*Orthomyxoviridae*	(−)ssRNA
VSV	Vesicular stomatitis virus	*Rhabdoviridae*	(−)ssRNA
VACV	Vaccinia virus	*Poxviridae*	dsDNA
WNV	West Nile virus	*Flaviviridae*	(+)ssRNA
YFV	Yellow fever virus	*Flaviviridae*	(+)ssRNA

[a]*ssRNA* single-stranded RNA, *dsDNA* double-stranded DNA, (+) positive-stranded, (−) negative-stranded, *ssRNA-RT* includes a reverse-transcription step in the life cycle

The precise mechanism of IFITM1-, 2- and 3-mediated antiviral activity is still unclear, but IFITM proteins might block virus-receptor interactions, prevent endocytosis or acidification, block fusion or induce signalling to other effector molecules. Recent evidence suggests that the inhibition occurs at late stages of the endocytic pathway. IFITM proteins do not decrease the primary attachment moieties for FLUAV (Brass et al. 2009) and do not affect SARS-CoV receptor expression (Huang et al. 2011). FLUAV still localizes to lysosomal compartments upon ectopic expression of *IFITM1, 2* or *3*. Moreover, the inhibitory effect of IFITMs on SARS-CoV could be circumvented by addition of trypsin, which promotes fusion at or near the plasma membrane (Huang et al. 2011). There is also evidence for different specificities of IFITM family members. Among the murine IFITMs, IFITM3 inhibits FLUAV entry more efficiently than SARS-CoV, MARV, or EBOV, whereas IFITM5 and IFITM6 are more efficient in inhibiting filoviruses (Huang et al. 2011). Interestingly, IFITM3 is posttranslationally modified by S-palmitoylation. Mutation of the palmitoylation site does not alter protein stability or trafficking, but prevents membrane clustering and abrogates its antiviral effect on FLUAV replication (Yount et al. 2010). Although there is accumulating in vitro evidence for an important role of IFITM proteins in the early antiviral defence against a broad range of enveloped viruses, the contribution of IFITM proteins to the antiviral defence in vivo remains to be determined.

Block of Viral Trafficking

Interferon-Inducible GTPases

IFNs induce the expression of the p47, p65, and Mx family of GTPases, and the very large GTPases (VLIG). They all share intrinsic GTPase activity and the capacity for self-assembly. Among these four families, only the Mx proteins have well-described antiviral activities (MacMicking 2004).

The Mx proteins are among the most potent antiviral effector proteins. The Mx family comprises Mx1 and Mx2 in mice, and MxA and MxB in humans. Mx proteins are 70–80 kDa in size, belong to the dynamin protein family and differ from the other IFN-inducible GTPases by their stronger induction by type I and type III IFNs as compared to IFNγ. Mx1 localizes to the nucleus, whereas MxA and Mx2 are located near the smooth ER. MxB is found in the intranuclear and/or cytoplasmic face of nuclear pores and has no detectable antiviral activity. *Mx1* was identified as an important IFN-induced antiviral factor in early studies with mice exhibiting a genetically determined resistance against FLUAV infection (Haller and Kochs 2011; Haller et al. 2007). Importantly, many laboratory inbred mouse strains, including C57BL/6J, BALB/cJ and 129/J, carry non-functional *Mx1* alleles due to large deletions or nonsense mutations (Staeheli et al. 1988). Studies employing *Mx1*- and *MxA*-transgenic mice confirmed the importance of Mx proteins for the antiviral defence in vivo. For example, constitutive expression of *MxA* in *Ifnar1*$^{-/-}$ mice confers full resistance to Thogoto virus (THOV), La Crosse virus (LACV) and Semliki Forest virus (SFV) (Hefti et al. 1999). Mx proteins inhibit replication of a broad range of RNA viruses and some DNA viruses. The mechanism of Mx action has been extensively studied, but is still not entirely clear. Dependent on their localization Mx proteins can recognize and trap essential viral structures, the main targets appear to be viral nucleocapsids. Cytoplasmic MxA blocks the movement of FLUAV and THOV nucleocapsids into the nucleus, while nuclear Mx1 inhibits FLUAV replication prior to the onset of transcription. Mx1 and MxA associate with the viral nucleocapsids and block viral transcription. MxA-mediated changes in trafficking of viral components have also been shown for LACV and African swine fever virus (ASFV). Trapping of viral structures enables Mx proteins to inhibit viral replication already at early steps of infection and provides a very efficient antiviral strategy. It is unclear how exactly Mx proteins interact with viral constituents, but recent structural insights into the basis of Mx oligomerization might help to delineate the mechanism and molecular requirements. Besides, FLUAV strains appear to differ in their sensitivity to MxA actions and this appears to depend on their nucleoproteins (Haller and Kochs 2011; Haller et al. 2007). With respect to human populations, genetic polymorphisms in the *MxA* gene correlate with increased sensitivity to HCV, Hepatitis B virus (HBV) and measles virus (MeV) (Cao et al. 2009; Hijikata et al. 2000; Suzuki et al. 2004; Torisu et al. 2004).

The p65 GTPase family, also known as GBP family, consists of 11 members in mice (GBP1-11) and 7 in humans (GBP1-7). All murine GBPs and at least human

GBP1-5 are induced by IFNγ and to a lesser extent, by IFNα/β (MacMicking 2004; Vestal and Jeyaratnam 2011). Note that similar numbered GBPs are not necessarily the most related ones. Many inbred mouse strains have a dysfunctional allele for *Gbp1* that cannot be induced by either type I or type II IFN (Staeheli et al. 1984). The antiviral functions of GBPs are still undefined, but human GBP1 can form oligomers like Mx proteins. Overexpression of human *GBP1* and its putative murine homolog *Gbp2* inhibits VSV and encephalomyocarditis virus (EMCV) replication. Intriguingly, murine GBP2 GTP-binding activity is thereby essential for the inhibition of EMCV but not VSV. Furthermore, overexpression of human *GBP1* can inhibit replication of an HCV replicon (Vestal and Jeyaratnam 2011).

The p47 GTPase (IRG) family emerged as a crucial pathogen resistance system in mice that is absent in humans (Bekpen et al. 2005). Most characterized members are strongly induced by IFNγ, localize along the phagocytic and secretory pathways and are crucially involved in the control of bacterial and protozoan infections. Antiviral activity has so far only been suggested by in vitro overexpression studies for *Tgtp (Irgb6)* and *Igtp (Irgm3)* against VSV and Coxsackie virus (CV), respectively (Bekpen et al. 2005; Howard 2008). So far, only normal resistance against murine cytomegalovirus (MCMV) has been reported for *Igtp*$^{-/-}$ mice (Taylor et al. 2000) (Table 1).

The VLIG family members are around 280 kDa in size and are the largest GTPases described so far (MacMicking 2004). They are the last IFN-induced GTPases identified and appear to have emerged solely in vertebrates (Li et al. 2009a). To date their functional significance in IFN responses is unclear.

Editing of Viral Nucleic Acids

Adenosine Deaminases Acting on RNA (ADARs)

ADARs catalyze the deamination of adenosin (A) to produce inosine (I) in RNAs with double-stranded (ds) character. A-to-I editing leads to an A- to guanosine (G) nucleotide exchange, since I is decoded as G during translation and RNA-dependent RNA replication. Among the three mammalian ADARs (ADAR1-3) described to date, only ADAR1 is IFN-inducible through an ISRE element in one of the alternative *ADAR1* gene promoters. In mice and humans, alternative splicing leads to the expression of the constitutively expressed p110 and the IFN-inducible p150 isoforms. ADAR1 p150 shuttles between the nucleus and the cytoplasm, ADAR1 p110 is predominantly and ADAR2 and ADAR3 are exclusively found in the nucleus. ADAR3 lacks catalytic activity, shows tissue-restricted expression and has been implicated in negative regulation of ADAR1 and ADAR2. RNA editing by ADAR1 can occur at multiple positions or at highly specific sites. Multiple A-to-G substitutions attributable to ADAR activity have been first described for MeV, followed by a large number of other viruses that mostly contain a negative-stranded RNA or an ambisense genome organization. More site-selective A-to-G exchanges have been reported for example for hepatitis delta virus (HDV), human herpesvirus 8 (HHV8) and Epstein-Barr (EBV) virus (Samuel 2011).

ADARs exert antiviral effects against viruses belonging to several families and using completely different replication strategies. A-to-I editing of viral RNAs can affect virus-host interactions at multiple levels and, importantly, can also have proviral effects. Nucleotide exchanges can lead to amino acid exchanges and altered protein products with different biological activities. This has for example been described for MeV, where extensive hypermutation of the matrix protein is associated with persistent infection of the central nervous system. A-to-I editing can also lead to frame-shift mutations or premature stop-codons and, consequently, to aberrant or reduced viral protein expression, as described for MeV, respiratory syncytial virus (RSV) and LCMV (Samuel 2011). A-to-I conversion can also induce structural changes in RNAs, as RNA duplex structures are less stable when A:uridine (U) base pairs are exchanged by I:U base pairs (Bass 2002; Serra et al. 2004). Reduced duplex-stability likely results in altered dsRNA-mediated activities. In support of this notion, shRNA-mediated downregulation of *ADAR1* leads to enhanced activation of dsRNA-dependent protein kinase (PKR) and IRF3 (Toth et al. 2009). Consistently, synthetic I:U-containing RNAs suppress dsRNA-mediated activation of IRF3 and ISG expression in HeLa cells (Vitali and Scadden 2010). A-to-I substitutions can also alter microRNA processing or silencing capacity, or target RNAs for degradation. These effects have been shown for cellular RNAs and are yet to be demonstrated for viral RNAs. It is also likely that RNA editing can lead to viral genome mutations in the case of single-stranded (ss) RNA viruses that use RNA-dependent RNA replication. Furthermore, RNA editing might also indirectly influence virus replication as it could affect cellular transcripts of proteins involved in the antiviral defence (Samuel 2011).

Adar1(p150 or p110/p150)$^{-/-}$ and *Adar2*$^{-/-}$ mice show severe phenotypes (Table 1). *Adar1*-deficiency leads to embryonic lethality (Hartner et al. 2004, 2009; Wang et al. 2004; Ward et al. 2011; XuFeng et al. 2009), whereas *Adar2*$^{-/-}$ mice are prone to seizures and die young (Higuchi et al. 2000). Thus, ADAR1 and ADAR2 are also crucially involved in cellular processes unrelated to host defence mechanisms. Nevertheless, studies with embryonic fibroblasts derived from these mice clearly established the selective functions of ADAR1 and ADAR2 in editing viral RNAs and in antiviral pathways. *Adar1(p150)*$^{-/-}$ cells ectopically expressing the receptor for MeV show dramatically increased MeV-induced cytopathic effects and markedly increased virus replication (Ward et al. 2011). This is consistent with the reported increased MeV-induced cytotoxicity in HeLa cells after shRNA-mediated *ADAR1* knockdown (Toth et al. 2009). Similarly, other members of the *Paramyxoviridae* (i.e. Newcastle disease virus (NDV), Sendai virus (SeV), canine distemper virus (CDV)) and FLUAV induce less pronounced cytopathic effects in *Adar1(p150)*$^{-/-}$ fibroblasts than in the respective wildtype cells. No effect of *Adar1(p150)* deficiency was found for LCMV and VSV replication (Ward et al. 2011).

Proviral effects of ADAR1 have been shown for VSV, HDV, HIV-1, Karposis's sarcoma-associated herpesvirus (KSHV) and mouse polyoma virus (MPV), although the mechanisms seem to differ (Samuel 2011). For VSV this has been attributed to the interaction of ADAR1 with PKR, inhibition of PKR activity and consequent impairment of eukaryotic translation initiation factor 2α (eIF2α) phosphorylation (Nie et al. 2007). The proviral effect of ADAR1 for HDV is well established. HDV requires site-specific A-to-I editing to switch from the short delta antigen protein, which is essential during early steps of replication, to the longer form, which is crucial for packaging of the viral genome and HDV particle formation (Samuel 2011). Inhibition of HDV occurs upon increased HDV RNA editing conditions, such as overexpression of *ADAR1(p110)* or *ADAR2*, or by IFN-mediated increased *ADAR1 (p150)* expression (Hartwig et al. 2004; Jayan and Casey 2002). Proviral mechanisms in the case of HIV-1, KSHV and MPV are less well characterized, but they might involve site-specific editing of viral mRNA and the switch from early to late transcripts, respectively, for the latter two viruses (Samuel 2011).

Apolipoprotein B mRNA-Editing Catalytic Polypeptide (APOBEC) Family

The APOBEC proteins are tissue-specific cytidine (C) deaminases that exhibit RNA editing and/or DNA mutator activity. In humans, the APOBEC family comprises 11 members that have cellular functions and inhibit the mobility of endogenous retroelements. Apart from that, the APOBEC3 subfamily exerts broad-spectrum anti-retroviral activity. However, retroviruses have evolved very efficient countermeasures and are therefore often insensitive to endogenous APOBEC3 actions unless they lack the respective antagonist (Goila-Gaur and Strebel 2008).

Most human APOBEC3 family members are expressed constitutively and expression is further enhanced by IFNα/β and IFNγ (Goila-Gaur and Strebel 2008; Koning et al. 2009; Refsland et al. 2010; Stenglein et al. 2010; Trapp et al. 2009; Wang et al. 2009). APOBEC3 proteins are packaged into retroviral particles and cause extensive C-to-U mutations in the minus-strand of the viral DNA during reverse transcription. C-to-U editing can lead to mutations in viral structural and non-structural proteins causing replication defects at multiple levels. Degradation of uracilated viral cDNAs by cellular DNA glycosylases is believed to contribute to the APOBEC3-mediated antiviral activity (Goila-Gaur and Strebel 2008). Intriguingly, a very recent report implicates APOBEC3A catalytic activity and the cellular uracil DNA glycosylase UNG2 in the clearance of transfected plasmid DNAs, suggesting that APOBEC3s may act as a restriction factor for a broader range of foreign DNAs (Stenglein et al. 2010). In addition, APOBEC3s exert deaminase-independent antiviral functions and these include interference with tRNA-primed initiation of reverse transcription and reverse transcriptase-mediated DNA elongation (Goila-Gaur and Strebel 2008; Narvaiza et al. 2009). Apart from retroviruses, APOBEC3s can also interfere with the HBV life cycle, however, APOBEC3B, APOBEC3F and APOBEC3G are not required for the anti-HBV actions of IFNγ in human cell lines (Goila-Gaur and Strebel 2008; Proto et al. 2008). In contrast, several reports suggest a contribution of APOBEC3s to the IFNα/β-induced antiviral activity against HIV-1 (Cheney and McKnight 2010; Goila-Gaur and Strebel 2008; Trapp et al. 2009).

Unlike humans, who have seven *APOBEC3* genes, mice only have one (*Apobec3*) and this is induced by type I and type II IFN (Okeoma et al. 2009a; Turelli et al. 2008). Murine APOBEC3 also induces hypermutations in retroviral cDNAs, but restriction of its activity to specific viruses appears different from the human APOBEC3G (Browne and Littman 2008; Rulli et al. 2008). Murine *Apobec3* has been implicated in the in vivo control of Friend-murine leukaemia virus (FMLV) (Santiago et al. 2011; Takeda et al. 2008) and *Apobec3*$^{-/-}$ mice show increased sensitivity to mouse mammary tumour virus (MMTV) (Okeoma et al. 2007, 2009b). *Apobec3*$^{-/-}$ mice infected with Moloney murine leukaemia virus (M-MuLV) show higher virus titers and develop earlier leukaemia in comparison to control animals (Low et al. 2009). Importantly, pre-treatment of murine cells with IFNα results in enhanced *Apobec3* expression and resistance against MMTV. No effect of IFNα on MMTV replication was observed in *Apobec3*$^{-/-}$ cells, demonstrating that *Apobec3* crucially contributes to the anti-MMTV action of IFNs (Okeoma et al. 2009b). Consistent with data from the human system, dsRNA pretreatment inhibits HBV DNA production in HBV-transgenic and *Apobec3*-deficient mice to similar levels as in the control HBV-transgenic mice (Turelli et al. 2008) (Table 1).

Viral RNA Degradation and Translational Inhibition

2′-5′ Oligoadenylate Synthetases (OAS) and Latent Ribonuclease (RNase L)

The OAS/RNase L pathway belongs to the best characterized antiviral pathways to date. It results in the degradation of viral and cellular RNAs and blocks replication of a number of RNA and DNA viruses (Chakrabarti et al. 2011; Kristiansen et al. 2011).

In humans the OAS family consists of four genes (*OAS1, 2, 3* and *L*). As a result of gene duplication, mice have eight different *Oas1* (*Oas1a-h*), one *Oas2*, one *Oas3*, and two *OasL* (*OasL1* and *OasL2*) genes. *OAS* genes are expressed at low levels in resident cells and are induced by type I IFN and upon virus infections. OAS1 proteins contain one unit of the OAS domain, OAS2 and OAS3 contain two and three copies of the catalytic units, respectively. Among the murine OAS1 proteins, only OAS1a and OAS1g are catalytically active. OASL proteins contain one OAS unit which is, with the exception of murine OASL2, without catalytic activity (Kristiansen et al. 2011). Mice and humans harbour one *RNase L* gene, which is widely expressed in most, if not all mammalian tissues and further upregulated by type I IFN exposure in murine cells, but only barely in human cells. RNase L is a latent endoribonuclease that consists of a regulatory ankyrin repeat domain (ARD), a protein kinase (PK)-like domain and the RNase domain (Chakrabarti et al. 2011).

Activation of OAS family members occurs through recognition and binding of viral dsRNA followed by a conformational change. Therefore, OASs are not only antiviral proteins but also considered as PRRs. Activated OAS polymerizes ATP into 2′-5′ oligoadenylates (2-5A$_n$). Binding of these oligomers to the ankyrin domain of monomeric RNase L leads to dimerization and activation of RNase L. Subsequently, the endoribonuclease degrades viral and cellular RNAs with ssRNA loops and thus prevents viral protein synthesis (Fig. 2). Each of the active human OAS family

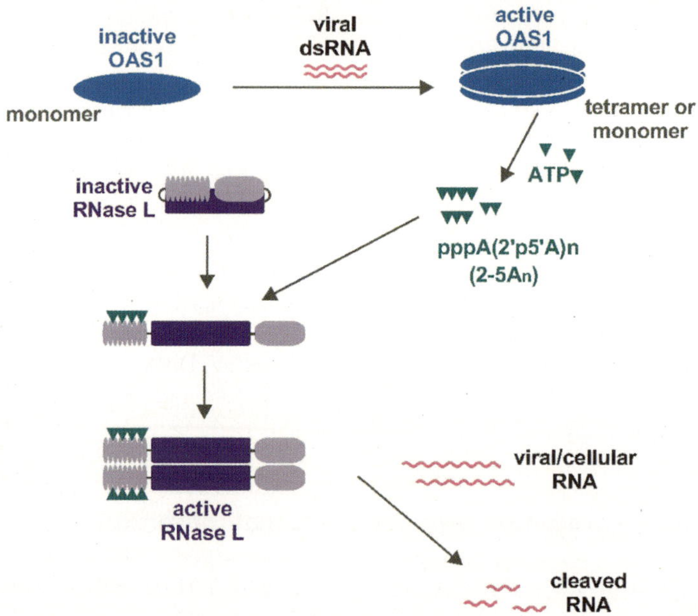

Fig. 2 The OAS1/RNase L pathway. Latent OAS1 is activated by viral dsRNA and, subsequently, oligomerizes ATP into 2′-5′ oligoadenylates (2-5A_n). Binding of 2-5A_n to the ankyrin repeat domain of RNase L leads to activation and dimerization of RNase L through their kinase-like domains. Activated RNase L dimers in turn cleave cellular and/or viral RNAs

members appears to have some unique biological features. For example, catalytically active OAS1 is a monomer or tetramer, OAS2 a dimer and OAS3 functions as a monomer. OAS3 synthesizes dimeric 2-5A_n, whereas OAS1 and OAS2 synthesize trimeric and tetrameric oligomers. As dimeric 2-5A_n do not efficiently activate RNase L, OAS3 might have different functions. Several overexpression studies demonstrated antiviral capacity for human OAS family members. OAS1 and OAS2 inhibit replication of EMCV, but not VSV, OAS3 is effective against alphaviruses, i.e. SFV and Sinbis virus (SINV) (Chakrabarti et al. 2011; Kristiansen et al. 2011; Sadler and Williams 2008). Little is known about the characteristics of dsRNA recognition and activation of OAS family members. Human OAS1 appears to also bind ssRNA, but activation only occurs upon dsRNA binding, with a preference for longer dsRNAs (Kristiansen et al. 2011). There also might be differential specificities in dsRNA recognition among OAS family members, as recently suggested for murine OAS1a and OAS1b (Elbahesh et al. 2011).

RNase L, once activated, can degrade viral and cellular (including ribosomal) RNAs. RNase L cleavage products are small, mostly structured ssRNAs with 5′-hydroxyl and 3′-monophosphate at their termini. Interestingly, these cleavage products can act as PAMPs and induce IFNβ production *via* recognition by retinoic acid-inducible protein I (RIG-I) and/or melanoma differentiation-associated protein 5 (MDA5). The 3′-phosphate of RNase L cleavage products appears to be required at least for the activation of RIG-I (Luthra et al. 2011; Malathi et al. 2007, 2010). Knockout mice for *RNase L* are susceptible to EMCV, CV-B4, herpes simplex virus (HSV)-1 and WNV (Table 1), although the RNase L-specific effects seem to be strongly dependent on the cell type and the virus strain. Several reports showed that RNase L can also negatively influence host survival upon virus infection and promote viral replication as shown for HSV-2, SINV, SFV and some reovirus strains (Silverman 2007). On the other hand, RNase L appears to be protective against virus-induced demyelination. Infection of *RNase* $L^{-/-}$ mice with the neurotropic mouse hepatitis virus (MHV)-JHM strain results in higher susceptibility, but does neither affect virus control in the CNS nor IFNα/β expression (Ireland et al. 2009).

Several studies suggest RNase L-independent antiviral activities of OAS family members. The enzymatically inactive human OAS1L and murine OAS1b can inhibit EMCV and WNV replication, respectively, when expressed in cell culture. Furthermore, OAS1b is required for resistance of mice against flavivirus infection. Dependent on the mouse strain, the OAS1b protein exists in two forms, a full-length OAS1b and a truncated form as a result of a point mutation in the *OAS1b* gene generating a premature stop codon. Only the full-length protein leads to resistance against WNV infection. Notably, most laboratory inbred mouse strains express the truncated version of *OAS1b* (Kristiansen et al. 2011). Recently, lack of enzymatic activity of OAS1b was confirmed and, additionally, full-length OAS1b was shown to inhibit synthetic dsRNA-induced 2-5A_n production both in vivo and in vitro (Elbahesh et al. 2011).

Of potential clinical relevance, exogenously applied OAS1 can induce an antiviral state. Exogenous recombinant porcine OAS1 protects HepG2 cells from the cytopathic effects of EMCV and VSV in a dose-dependent manner and inhibits virus replication in Vero cells. These effects are again independent of both, enzymatic activity of OAS1 and the presence of RNase L. Moreover, injection of OAS1 into mice results in tenfold reduced viral titers in organs upon subsequent EMCV infections (Kristiansen et al. 2010). It will be of interest to determine if and to what extend assumable autocrine/paracrine actions of OAS1 contribute to the in vivo antiviral defence.

Double Stranded RNA-Dependent Protein Kinase (PKR)

Similar to Mx and OAS/RNase L, the antiviral activities of PKR are long-known and have been extensively characterized. PKR is encoded by the *Eif2ak2* gene and is one out of four members of a serine-threonine kinase family that regulates protein synthesis upon diverse stress signals mainly through phosphorylation of eIF2α. PKR consists of two N-terminal RNA binding motifs (RBMs) and a C-terminal catalytic kinase domain (KD). PKR is constitutively expressed in all differentiated cells at low levels and is upregulated upon type I and III IFN through activation of an ISRE binding site in the promoter. In uninfected cells, PKR exists as an inactive monomer by autoinhibition of the kinase domain. DsRNA and other ligands like heparin, ceramide and the PKR-associated factor PACT induce the release of the

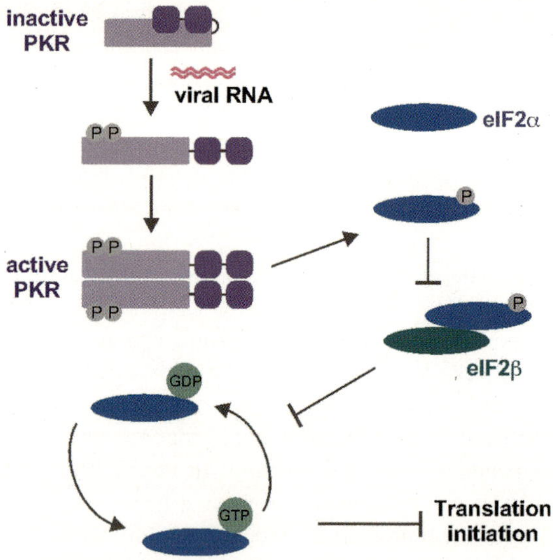

Fig. 3 Scheme of PKR activation and action. In resident cells PKR exists as inactive monomer through intramolecular autoinhibition. Sensing of viral RNA activates PKR by autophosphorylation. Dimerized PKR molecules phosphorylate eIF2α on serine 51, which inhibits the nucleotide exchange factor eIF2β. This disables the exchange of inactive eIF2α-GDP with eIF2α-GTP and leads to a block of translation initiation

inactive molecule and lead to autophosphorylation and dimerization of PKR. Activation of PKR can also occur through binding of ssRNA containing duplex regions and a 5′-triphosphate. Interestingly, endogenous transcripts also efficiently activate PKR, as known for the IFNγ mRNA. Once PKR is activated, it phosphorylates the serine residue 51 of eIF2α leading to inhibition of the guanine nucleotide exchange factor eIF2β. This prevents exchange of GDP to GTP, thus blocking translation initiation (Fig. 3). Intriguingly, PKR regulates its own activity through inhibition of its mRNA translation initiation in the presence of high PKR protein levels (Pindel and Sadler 2011; Sadler 2010).

Two independent knockout mouse models and a transgenic mouse expressing a dominant-negative, kinase-defective PKR gave further insight into PKR functions in antiviral responses (Barry et al. 2009; Nakayama et al. 2010; Sadler and Williams 2008). PKR-deficient mice are susceptible to VSV, FLUA mutant virus, and LCMV (Table 1). Protection against several viruses like HCV, EMCV, WNV, HIV-1, HDV, SINV and HSV-1 have been shown to be dependent on PKR in several in vitro systems (Nakayama et al. 2010; Sadler and Williams 2008).

Although phosphorylation of eIF2α and consequent translational regulation was considered as the major function of PKR, several reports have suggested involvement of PKR in different signalling networks. For example, PKR has been shown to regulate the transcription factors IRF1, STAT1, STAT3, nuclear factor NFκB, c-Jun, cyclic AMP-dependent transcription factor (ATF) 3 and ATF4 (Pindel and Sadler 2011). Two recent reports suggest that PKR is an important regulator of IFNα/β synthesis/secretion. This has been shown in response to infection with a subset of RNA viruses, like EMCV, SFV and Theiler's murine encephalomyelitis virus (TMEV). Mechanistically, absence of PKR results in strongly reduced levels of polyadenylated IFNβ mRNAs. Thus, PKR maintains the integrity of IFNβ mRNA thereby promoting its translation. PKR action in this case seems to be independent of eIF2α phosphorylation. Consistently, serum levels of IFNβ are reduced in EMCV infected *Pkr*$^{-/-}$ mice (Schulz et al. 2010). These data were supported by a later study with in vitro bovine rotavirus (BRV) infections. Virus infection leads to a strong RIG-I-, MDA5- and IRF3-dependent IFNβ mRNA induction. IFNβ mRNA levels are normal in fibroblasts derived from *Pkr*$^{-/-}$ mice, but IFNβ protein secretion is dramatically reduced (Sen et al. 2011). It is unclear, how PKR prevents mRNA de-adenylation and whether other mRNAs are similarly regulated by PKR-dependent stabilizing mechanism.

Interferon-Stimulated Gene Product of 20 kDa (ISG20)

ISG20 (HEM45) was identified independently based on its induction by type I and type II IFNs and by estrogen (Gongora et al. 1997; Mattei et al. 1997; Pentecost 1998). ISG20 is a nuclear 3′-5′ exoribonuclease associated with nuclear bodies that acts on ssRNA and, to a lesser extent, on ssDNA (Degols et al. 2007).

Several overexpression studies of *ISG20* in human cell lines demonstrated inhibition of RNA virus replication. Antiviral activity appears virus-specific and ectopic expression of *ISG20* inhibits VSV, EMCV, FLUAV, DV-2, DV- and WNV-like particles, HCV, Bovine virus-diarrhoea virus (BVDV), hepatitis A virus (HAV) and YFV, but not SARS-CoV replication. For HCV, DV- and WNV-like particles antiviral activity of ISG20 is dependent on its enzymatic activity and catalytically inactive ISG20 reduces IFN-mediated antiviral activity against VSV, but surprisingly not against EMCV or FLUAV. In murine embryonic fibroblast cell lines, overexpression of *Isg20* inhibits SINV replication and siRNA-mediated knockdown of *Isg20* results in enhanced virus replication. Increased survival rates are found in neonatal mice after subcutaneous inoculation with SINV RNA encoding *Isg20* as compared to the respective controls (Zhou et al. 2011). Besides, an HIV-1-derived virus expressing *ISG20* shows strongly delayed replication in a human T-lymphoblastoid cell line and peripheral blood mononuclear cells (Espert et al. 2005).

With respect to DNA viruses, inhibitory effects of *ISG20* on HBV protein synthesis have been suggested in transfected HepG2 cells (Hao and Yang 2008), whereas *ISG20* overexpression in HeLa cells was reported to not affect adenovirus replication (Espert et al. 2003).

It is still unclear how exactly ISG20 inhibits virus replication. The studies outlined above show a requirement for its exoribonuclease activity, thus it seems likely that ISG20 directly degrades viral RNA and/or DNA, however, evidence remains to be provided. ISG20 does not degrade transfected replication-incompetent HCV RNA, suggesting that either viral RNA associated with replication complexes or replication intermediates are targets for ISG20-mediated degradation. It is notable in that context, that ISG20 specifically degrades ssRNA but not RNA containing stem-loop structures at the 3′-end (Nguyen et al. 2001). Alternatively, ISG20 might act indirectly by e.g. targeting other (cellular) factors that are required for viral replication. The contribution of endogenous

ISG20 to IFN-induced antiviral activity and its role in the in vivo defence against specific viruses needs further delineation.

Interferon-Induced Proteins with Tetratricopeptide Repeats (IFITs)

Members of the IFIT gene family were among the first IFN-inducible genes cloned (Chebath et al. 1983; Levy et al. 1986; Wathelet et al. 1986). Four members have been described in humans and three in mice. All IFIT proteins contain multiple tetratricopeptide repeats known to mediate protein-protein interactions. IFIT1 (ISG56) and IFIT2 (ISG54) have well established inhibitory effects on protein synthesis by targeting the translation initiation complex. Although this globally affects protein synthesis, IFIT1 might more specifically block translation of viral RNAs. For example, HCV IRES-driven reporter expression shows increased sensitivity to IFIT1-mediated translational inhibition in comparison to 5′-cap-driven reporter expression in human cell lines (Fensterl and Sen 2011).

A few reports suggested inhibitory functions of IFITs on SeV, WNV, LCMV, VSV, EMCV and HCV (Schmeisser et al. 2010; Wacher et al. 2007; Wang et al. 2003; Zhang et al. 2007), although these were not further characterized. Only very recently, extensive analyses including gene-targeted mice and mutant viruses established that the antiviral functions of IFITs depend on the 5′-structures of viral RNAs (Daffis et al. 2010; Pichlmair et al. 2011). In higher eukaryotes, 2′-O-methylation of cellular mRNA 5′-caps occurs, in addition to the well-known essential methylation at the N-7 position of the cap guanosine residue, at the ribose-2′-O-position of one or two adjoining nucleotides (Langberg and Moss 1981). Several RNA and DNA viruses also have 2′-O-methylated 5′-capped mRNAs (Fechter and Brownlee 2005; Wei and Moss 1975) and mutant WNV, MHV and vaccinia (VACV) viruses lacking their respective 2′-O-methyltransferase activity are sensitive to IFIT1 and/or IFIT2 actions (Table 1) (Daffis et al. 2010; Zust et al. 2011). A mechanistic explanation for the 5′-nucleic acid structure-specific IFIT functions came from the finding that IFITs can form multi-protein complexes that bind to "non-self" 5′-triphosphorylated RNA (PPP-RNA) in human cell lines (Pichlmair et al. 2011). IFIT1 and IFIT5 (ISG58), the latter existing in humans but not in mice, directly bind to PPP-RNA, whereas IFIT2 and IFIT3 (ISG60) can associate with PPP-RNA through interaction with IFIT1. Consistently, siRNA mediated downregulation of *IFIT1* inhibits replication of VSV and Rift Valley fever virus (RVFV), both viruses known to generate PPP-mRNAs, and *Ifit1*$^{-/-}$ mice and fibroblasts show increased sensitivity to VSV infection (Table 1). Proofing the specificity of IFIT1 action, absence of IFIT1 does not affect replication of EMCV, a virus that does not generate PPP-mRNA. Interestingly, sequestration of viral mRNA rather than direct inhibition of translational initiation appeared as the main antiviral effector function in this study, although the fate of the bound viral RNA remained undefined (Pichlmair et al. 2011).

Apart from translational inhibition and viral RNA sequestration, two further and unrelated antiviral effector functions have been ascribed to human IFIT1. Firstly,

IFIT1 can directly inhibit human papillomavirus (HPV) DNA replication by binding to and blocking HPV E1 protein helicase activity (Saikia et al. 2010; Terenzi et al. 2008) and, secondly, IFIT1 reportedly inhibits SeV-induced IFNβ production through interaction with stimulator of IFN genes (STING) (Li et al. 2009b). However, virus-induced IFNβ production was similar in cells derived from *Ifit1*$^{-/-}$ and wildtype mice (Pichlmair et al. 2011).

In summary, IFIT proteins emerged as both virus-sensing and antiviral effector proteins. Their function within "versatile" multi-protein complexes (Pichlmair et al. 2011) might explain some of the inconsistencies in antiviral activities observed between overexpression and siRNA-mediated or genetic knockdown of specific IFIT family members. Stoichiometry of IFITs, association with other RNA-binding or accessory proteins and viral countermeasures likely determine their mode of action and virus specificity. Several IFIT-like genes have been found both in mice and humans (Fensterl and Sen 2011) and their putative protein products might add to the complexity. Future work will be required in order to define the function of the distinct IFIT family members/complexes, the exact nature and fate of their target RNAs and, potential species specifities.

Posttranslational Modification of Viral and Cellular Proteins

Interferon-Stimulated Gene Product of 15 kDa (ISG15)

ISG15 is among the most prominently induced genes during virus infection and in response to type I IFNs. ISG15 is an ubiquitin-like protein that is conjugated to target proteins. The so-called ISGylation of proteins has many common features with protein-ubiquitination. Both involve a series of stepwise enzymatic reactions that result in covalent protein modifications (Fig. 4). ISG15 is synthesised as a precursor protein that is processed to expose a C-terminal LRLRGG motif. The GG motif is adenylated in the presence of ATP and then conjugated sequentially to cystein residues of three enzymes (E1–E3) and, finally, ISG15 is transferred to lysine residues within its target substrates. The E1 enzyme (ISG15-activating enzyme, UBE1L) is specific for the ISGylation pathway, whereas there is an overlap with enzymes involved in ubiquitination for the E2 (ISG15/ubiquitin-conjugating enzyme) and E3 (ISG15/ubiquitin ligase) enzymes. Both ubiquitination and ISGylation are reversible processes and several de-ubiquitinating and de-ISGylating enzymes can cleave off the respective protein modification. Similar to ISG15, many enzymes involved in the ISGylation pathway are induced by IFNs, e.g. the E1 enzyme UBE1L, the E2 enzyme UBCH8, the E3 enzymes HERC5 and tripartite motif protein 25 (TRIM25), and the de-ISGylating enzyme UBP43 (USP18) (Harty et al. 2009; Zhang and Zhang 2011).

Unlike ubiquitination, ISGylation does not target proteins for degradation but rather resembles mono-ubiqitination and mainly affects protein function. ISGylation can also increase protein stability by protecting proteins from degradation. ISGylation has a very broad specificity and over 150 putative cellular ISG15 target

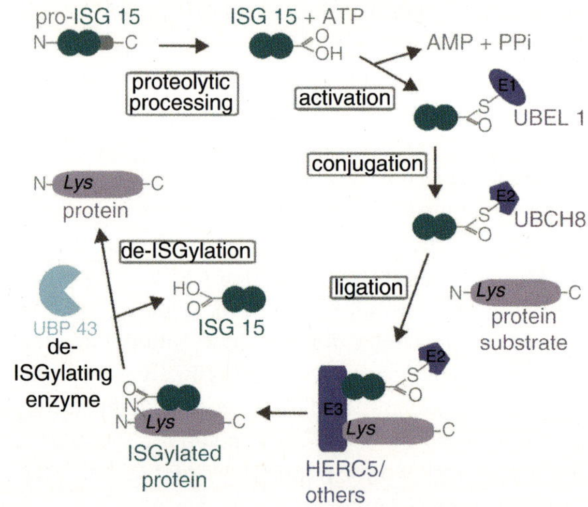

Fig. 4 Schematic diagram of the protein ISGylation system. ISG15 is proteolytically processed by an ISG15-specific protease to expose a C-terminal LRLRGG motif. In the presence of ATP, UBE1L catalyzes adenylation and forms a thioester bond with the C-terminal end of ISG15. ISG15 is then transferred to UBCH8, which is also covalently linked via a thioester bond. Catalyzed by HERC5 or other E3 enzymes, the C-terminus of ISG15 is then linked *via* an isopeptide bond to a lysine residue within the target protein. The de-ISGylation enzyme UBP43 can remove the ISG from the substrate. E1, ISG15-activating enzyme; E2, ISG15-conjugating enzyme; E3, ISG15 ligase

proteins have been identified so far. ISG15 has a broad antiviral activity against both DNA and RNA viruses. *Isg15*$^{-/-}$ mice are highly susceptible to FLUAV and influenza B virus (FLUBV), SINV, HSV, and murine γ-herpesvirus 68 (γMHV68) infections (Harty et al. 2009). Consistent with the involvement of ISGylation, *Ube1l*$^{-/-}$ mice show increased susceptibility to FLUBV and SINV infection. In contrast, ISG15 and UBE1L are redundant for the in vivo defence against LCMV (see also Table 1). Very recently, the FLUAV NS1A protein was identified as ISG15 target protein in two independent studies (Tang et al. 2010; Zhao et al. 2010). FLUAV NS1A protein was shown to directly interact with the major human E3 ISG15 ligase HERC5. Overexpression of *HERC5* in human cell lines enhances the anti-FLUAV activity of IFNβ and, consistently, downregulation of *ISG15*, *UBE1L*, *UBCH8* or *HERC5* by siRNA increases viral protein synthesis and replication (Hsiang et al. 2009; Tang et al. 2010; Zhao et al. 2010). Multiple lysine (K) residues within the NS1A can be modified by ISGylation (Tang et al. 2010; Zhao et al. 2010). In vitro ISGylated NS1A is unable to bind to PKR and to dsRNA and fails to dimerize *via* its RNA-binding domains (Zhao et al. 2010). ISGylated truncated NS1A protein, lacking its second C-terminal nuclear-localization signal, shows impaired interaction with importin-α (Tang et al. 2010). Although the major ISGylated lysine residues identified within the NS1A protein differ among the two studies, FLUAV expressing NS1A mutant for specific lysine residue are more virulent in vivo (Tang et al. 2010) or show enhanced replication in IFNβ treated cell lines, respectively (Zhao et al. 2010).

ISG15 can also inhibit budding of retroviruses, EBOV and VSV. The mechanisms are unclear, but inhibition of ubiquitination might at least be partially involved. ISG15 blocks ubiquitination of HIV-1 Gag protein and the EBOV VP40 matrix protein, both of which have been linked to budding ability (Harty et al. 2009). Besides, ISG15 was reported to inhibit the association of the HIV-1 and the avian sarcoma leukosis virus (ASLV) budding complex with cellular proteins involved in endosome sorting (Pincetic et al. 2010).

Proviral effects of ISG15 have been described for HCV. Several studies using overexpression and/or siRNA-mediated downregulation of *ISG15* or *UEB1L* revealed a positive effect of the ISG15 system for HCV RNA production in HCV replicon cell lines (Broering et al. 2010; Chen et al. 2010b; Chua et al. 2009). In line with this, high ISG15 levels correlate with high HCV load and low responsiveness of patients to IFNα therapy (Broering et al. 2010; Chen et al. 2010a). However, negative effects of ISGylation on HCV replication have also been reported and were attributed to ISGylation of the HCV NS5A protein and decreased NS5A protein stability (Kim and Yoo 2010).

Among the cellular ISG15 target proteins identified to date are many proteins involved in IFN signal transduction (e.g. JAK1, STAT1), virus recognition and downstream signalling (e.g. RIG-I, MDA5, and IRF3), and in the antiviral effector pathways (e.g. Mx, PKR and RNase L) (Malakhov et al. 2003; Zhao et al. 2005). ISGylation of these target proteins can have diverse effects, whereby it generally results in increased antiviral activity. For example, ISGylation has been reported to prevent NDV-induced degradation of IRF3 resulting in enhanced IFNβ production in human fibrosarcoma cells (Lu et al. 2006). This notion was confirmed by a later study showing direct interaction of IRF3 with HERC5 in HEK293 cells (Shi et al. 2010). Ectopic expression of HERC5 potentiates IRF3 transcriptional activity and siRNA-mediated downregulation of HERC5 or ISG15 reduces the expression of IRF3 responsive genes upon SeV infection. Notably, a HERC5 mutant protein that lacks its ligase activity does not affect IRF3 target gene-activation. Downregulation of *HERC5* by siRNA results in increased replication of SeV, VSV and NDV. SeV-induced IRF3 poly-ubiquitination and proteasomal degradation is reduced upon ectopic expression of the ISGylation system. IRF3 was shown to be ISGylated predominantly at three lysine residues and, as expected, the triple lysine mutant does not show a change in ubiquitination upon HERC5 downregulation and SeV infection. Furthermore, the triple lysine IRF3 mutant cannot fully rescue IFNβ reporter gene activation in response to SeV infection in $Irf3^{-/-}$ mouse embryo fibroblasts and displays accelerated degradation in comparison to wildtype IRF3 (Shi et al. 2010).

ISGylation of protein phosphatase 1B (PPM1, PP2Cb) reduces its activity and results in enhanced IκBα degradation and increased NFκB signalling (Takeuchi et al. 2006). ISGylated eIF4E family member 2 (eIF4E2, 4EHP) has increased affinity to 5′-capped RNAs compared to the non-ISGylated form (Okumura et al. 2007). ISGylation of cellular proteins has been recently also shown to be responsible for the inhibition of early steps of the FLUAV replication cycle (Hsiang et al. 2009), although proteins involved remained undefined. Negative regulation by ISGylation has been shown for RIG-I. ISGylation of RIG-I leads to reduced basal and virus-induced IFN production and cells derived from *Ube1l*-deficient mice show significantly lower levels of RIG-I mRNA and protein. Thus, a negative feedback mechanism acting on RIG-I was postulated which might be required to balance cellular innate immune responses (Kim et al. 2008b).

Ubiquitin Carboxyl-terminal Hydrolase 43 (UBP43, USP18)

As mentioned above, UBP43 is an IFN-induced de-ISGylating enzyme. *Ubp43*$^{-/-}$ mice show increased resistance against LCMV and VSV and display strongly increased clearance of injected replication competent HBV DNA (Table 1). However, *Ubp43*$^{-/-}$ mice develop brain injury, accompanied by hydrocephalus and early death and thus in vivo virus challenges have their limitations (Knobeloch et al. 2005; Ritchie et al. 2002, 2004). Fibroblasts derived from *Ubp43*$^{-/-}$ mice exhibit enhanced type I IFN-mediated protection from cytopathic effects caused by VSV and SINV infection (Ritchie et al. 2004). *Ubp43*$^{-/-}$ cells show dramatically increased levels of ISGylated proteins after IFN treatment which is also associated with increased JAK/STAT signalling and hyper-responsiveness to type I IFN (Malakhova et al. 2003). Curiously, the phenotype of *Ubp43*$^{-/-}$ cells is not rescued in fibroblasts derived from *Ubp43*$^{-/-}$/*Isg15*$^{-/-}$ or *Ubp43*$^{-/-}$/*Ube1l*$^{-/-}$ double knockout mice (Kim et al. 2006; Knobeloch et al. 2005), suggesting that the IFN hyper-responsiveness and the virus-resistant phenotype of *Ubp43*$^{-/-}$ cells is not associated with de-ISGylating activity. A possible explanation is provided by studies demonstrating that UBP43 also negatively regulates type I IFN signalling (Malakhova et al. 2003, 2006). UBP43 was shown to directly interact with IFNAR2 and, consequently, inhibit JAK/STAT signalling. This effect is independent of UBP43 isopeptidase activity as complementation of *Ubp43*$^{-/-}$ cells with an enzymatically inactive mutant UBP43 inhibits IFN-induced STAT1 phosphorylation to similar levels as observed in wildtype cells (Malakhova et al. 2006). In line with this, STAT1 phosphorylation is barely affected despite the downregulation of total ISGylation levels in fibroblasts derived from *Ube1l*$^{-/-}$ mice (Kim et al. 2006).

Inhibition of Virus Assembly, Budding and Release

Viperin (RSDA2, CIG5)

Viperin was cloned as human cytomegalovirus (HCMV)-induced gene in human fibroblasts (Zhu et al. 1997) and as IFNγ-activated gene in human macrophages (Chin and Cresswell 2001). The former study defined viperin as an ER-associated protein with anti-HCMV activity in human fibroblasts. Antiviral activity of viperin was subsequently shown for HCV (Helbig et al. 2005; Jiang et al.

2008), HIV-1 (Rivieccio et al. 2006), FLUAV (Wang et al. 2007), SINV (Zhang et al. 2007), WNV and DV (Jiang et al. 2010). Concerning the mechanism of action, detailed studies are only available for FLUAV infections (Wang et al. 2007). Viperin strongly impairs FLUAV release and, consequently, reduces virus replication in stably transfected HeLa cells. The authors demonstrate that inhibition of virus release occurs *via* disruption of plasma membrane lipid rafts. Viperin expression results in higher fluidity of membranes and, correspondingly, in an increased mobility of the viral haemagglutinin protein. No effect of viperin was observed in the course of VSV infection, a virus whose replication is believed to be lipid raft-independent. Additionally, farnesyl diphosphate synthase (FPPS) was identified as viperin-interacting protein and overexpression of FPPS reversed the effects of viperin on FLUAV replication. Consistently, siRNA-mediated knockdown of FPPS reduces virus release. Thus viperin exerts antiviral activity by sequestering FPPS to the ER and inhibiting its enzymatic activity (Wang et al. 2007). FPPS has important roles in isoprenoid biosynthesis by catalyzing the formation of farnesyl diphosphate, the precursor of sterols, dolichols, carotenoids, and ubiquinones (Szkopinska and Plochocka 2005). It remains to be determined which of the FPPS-regulated pathways is essentially involved in facilitating viral release.

Viperin might inhibit other viruses by different mechanisms, however, data are still scarce. Retrovirally expressed viperin inhibits late, but not early, HCMV protein accumulation and strongly reduces HCMV replication in human fibroblasts (Chin and Cresswell 2001). Viperin relocates from the ER to the Golgi and to vacuoles, the sites of viral glycoprotein maturation and viral assembly, respectively. It is unclear if relocation of viperin is beneficial or detrimental for viral replication. In the case of HCV and WNV, ER-association of viperin is important, but not absolutely required, for its antiviral activity (Jiang et al. 2008, 2010). Recently, viperin has been structurally characterized in more detail and its proposed S-adenosyl-L-methionine (SAM) enzyme activity catalyzing the formation of 5′-deoxyadenosyl radicals was confirmed (Shaveta et al. 2010). This is of particular importance, as enzymatic activity is required for the effect against HCV (Jiang et al. 2008), WNV and DV (Jiang et al. 2010).

Bone Marrow Stromal Antigen 2 (BST-2, Tetherin, mPDCA-1, CD137)

BST-2 is a small type II transmembrane protein that associates with lipid rafts at the cell surface or with internal membranes (Andrew and Strebel 2011). The *BST-2* promoter region contains GAS and ISRE consensus sites and is thus likely induced by type I and type II IFN (Ohtomo et al. 1999). BST-2 antiviral function was first suggested for HIV-1. HIV-1 requires its Vpu gene product for virus release in a cell type-specificmanner. BST-2 has been identified as a virus restriction factor by its ability to induce a Vpu-restricted phenotype in cells that otherwise show Vpu-independent HIV-1 release (Andrew and Strebel 2011). BST-2 was also found to inhibit virus release of all retroviruses tested, but also of arenavirus (i.e. LASV), filovirus (i.e. EBOV, MARV) and rhabdovirus (i.e. VSV) family members (Jouvenet et al. 2009; Sakuma et al. 2009; Weidner et al. 2010). The exact mechanism of how BST-2 inhibits virus release is still unclear, but BST-2 tethers virions to the producer cell. BST-2 might contribute to IFN-mediated antiviral activity against a wider range of enveloped viruses that bud from the host cell plasma membrane, however, this remains to be shown. On the other hand, BST-2 might also be proviral in the sense that it enables a shift of the mode of viral transmission, i.e. cell-to-cell versus cell-free spread (Andrew and Strebel 2011).

Additional Pathways of Viral Inhibition

Tripartite Motif (TRIM) Proteins

The TRIM family has been originally defined as proteins that contain a so-called tripartite motif consisting of a RING domain, one or two B-boxes, and a coiled-coil domain. The RING domain of many TRIM proteins has an E3 ubiquitin ligase activity, whereas the other two domains may be involved in mediating protein-protein interactions. TRIM proteins form high-molecular-mass complexes and localize to specific subcellular compartments either in the nucleus or the cytoplasm. TRIM proteins are involved in diverse cellular processes like cell growth, apoptosis and innate immunity (Ozato et al. 2008). The TRIM family consists of 72 genes in humans, whereby 16 *TRIM* genes are induced by type I and/or type II IFN (Carthagena et al. 2009). Among the TRIM proteins, TRIM5, TRIM19 and TRIM22 are strongest upregulated by type I and type II IFN. No homologs of TRIM5 and TRIM22 have been found in mice (Carthagena et al. 2009). Several members have reported antiviral activity that is exerted at multiple levels (Kajaste-Rudnitski et al. 2010), but their contribution to the IFN-mediated antiviral response is largely unknown.

TRIM19, better known as promyelocytic leukaemia (PML) protein, is a constitutive component of PML nuclear bodies (PML NBs), which are small nuclear substructures. PML, Sp100 and small ubiquitin-like modifier (SUMO) are constitutively present in PML NBs and many other proteins are transiently or permanently associated with PML NBs. PML NBs are potentially highly dynamic and the proteins associated can vary between cell types. PML is expressed constitutively, but type I and type II IFNs lead to a strong increase in the size and number of PML NBs. PML exists as many different isoforms that are generated by alternative splicing and grouped into seven classes (PML I-VII). PML isoforms likely have different functions, which still remain to be defined. In addition, Sp100, Daxx and other PML NB proteins are IFN-inducible. PML NBs are involved in the regulation of chromatin structure, transcription and DNA repair, unfolded protein responses and, in the regulation of apoptosis. Since PML NBs are constitutively present, numerous reports described their contribution to the intrinsic antiviral defence against a broad range of DNA and RNA viruses. The exact mechanisms of viral inhibition are still largely unknown, but they involve inhibition of viral mRNA and protein synthesis (Geoffroy and Chelbi-Alix 2011).

Despite the constitutive expression of PML NB constituents, the IFN-mediated induction of PML and Sp100 has in several cases been linked to the antiviral function of PML NBs. For example, the IFN-mediated antiviral effect against an HSV-1 ICP10 mutant virus that lacks the ability to disrupt PML NBs was strongly reduced in *Pml*$^{-/-}$ cells (Chee et al. 2003). Reduced IFN-induced protection against LCMV (Djavani et al. 2001), human foamy virus (HFV) (Regad et al. 2001) and rabies virus (RV) (Blondel et al. 2010) was reported in *Pml*$^{-/-}$ fibroblasts and *Pml*$^{-/-}$ mice are more susceptible to LCMV and VSV infection (Table 1) (Bonilla et al. 2002).

PML and PML NBs can also affect p53 activity. IFN-induced p53 has been shown to be required for the induction of apoptosis and the antiviral defence (Porta et al. 2005; Takaoka et al. 2003). In the case of poliovirus (PV) infections, p53 activation and target gene expression is dependent on the presence of PML. Induction of apoptosis and the resulting inhibition of virus replication is abolished upon siRNA-mediated downregulation of PML and enhanced by PML III overexpression. PV infection induces PML phosphorylation and SUMOylation leading to enhanced recruitment and activation of p53 (Pampin et al. 2006).

In contrast to the direct antiviral activity of TRIM19, TRIM21 and TRIM25 have emerged as crucial components of IFNα/β-inducing pathways (see below). Using ectopic expression and siRNA knockdown experiments, it has just recently been shown that TRIM56 inhibits BVDV replication. The mechanism is unclear, but the antiviral activity was dependent on its ubiquitin ligase activity and not effective against either VSV or HCV (Wang et al. 2011).

Virus Recognition and Amplification of Type I IFN Production

Several proteins involved in virus recognition and IFNα/β production are long-known IFN-inducible proteins whose functions have only recently been identified. Although these do not directly impact on virus replication, their proper function is essential for sensing virus infections and initiating antiviral responses, i.e. by directly or indirectly inducing genes with antiviral effector function and/or by inducing autocrine/paracrine IFNα/β signalling (Kawai and Akira 2009; McCartney and Colonna 2009). Proteins involved in virus recognition, including RNA sensors like RIG-I (DDX58) and MDA5 (IFIH1), as well as proteins implicated in the recognition of cytoplasmic DNA, e.g. DNA-dependent activator of IFN-regulatory factors (DAI, ZBP1, DLM-1) and IFNγ-inducible protein 16 (IFI16, p204) are induced by IFNs (Cui et al. 2004; Fu et al. 1999; Imaizumi et al. 2004; Kang et al. 2002; Trapani et al. 1992). The IFN-inducible protein absent in melanoma 2 (AIM2, IFI210) has also been identified as a cytoplasmic DNA-sensor, but associates with the inflammasome and does not lead to IFNα/β induction upon activation. IFI202 belongs to the same HIN200 family of IFN-inducible proteins as IFI16 and AIM2, but is a negative regulator of AIM2 at least in murine cells. Notably, there are also indications that MCMV and HCMV require IFI16 for replication (Gariglio et al. 2011).

Type I IFN also mediates its own amplification *via* the induction of IRF7. Similar to IRF3, IRF7 gets activated by several PRRs and transcriptionally induces IFNα/β. Although IRF3 and IRF7 recognize similar consensus sequences in promoter regions, IRF3 mainly regulates IFNα4 and IFNβ, whereby IRF7 is required for the induction of other IFNα subtypes. Different from the constitutively expressed IRF3, IRF7 is normally not expressed (except for specialized cell types) unless it is induced by IFNs. Hence, IRF7 mediates a switch from low-level to high-level of IFNα/β production and broadens the range of IFNα subtype expression in response to virus infections (Levy et al. 2002).

TRIM21 was found to interact with IRF3 and to increase IRF3 stability and downstream target gene expression upon SeV infection (Yang et al. 2009). TRIM21 can also negatively impact on IFNα production. Together with FAS-associated death domain (FADD) protein, TRIM21 directly ubiquitinates IRF7 and reduces its phosphorylation, thereby limiting IFNα/β production in response to virus infection (Young et al. 2010). TRIM25 has emerged as central component of the RIG-I pathway. TRIM25 directly interacts and ubiquitinates RIG-I which is required for its binding to the mitochondrial antiviral-signaling protein (MAVS, IPS1, VISA, Cardiff) and the induction of IFNα/β (Gack et al. 2007).

Concluding Remarks and Future Directions

The complexity of host-virus interactions remains a challenging field for future research on innate immune responses. While specificity and molecular mechanisms of viral recognition are important issues, further characterization of antiviral functions of specific ISGs seems equally essential. Many studies have so far only been performed in vitro and/or in overexpression studies and it will be of particular interest how these ISGs impact on antiviral responses under physiological conditions and in vivo. This might not be trivial, as there is most likely a high redundancy of pathways and ISGs can affect viral life cycles at multiple levels. Moreover, several ISGs also exert important, but not fully characterized, cellular functions. Genome-wide association studies in human populations provide another important ongoing future direction. Genetic polymorphisms in any of the ISG loci might reveal association with virus susceptibility or other potential novel immune regulatory functions. Besides, a lot of information has been gained by studies of virus mutants. Viruses have evolved very efficient mechanisms to counteract the host immune system and further analysis from this perspective will help to decipher the multiple levels of virus-host interactions. The more detailed understanding of the exact interplay between viral and cellular factors might help to better control virus infections and to more specifically direct antiviral responses.

Acknowledgements We apologize for not citing work of authors who made important contributions to the field of interferons and antiviral effector proteins. Due to space limitations we were unable to include as many primary publications as we would have liked, instead we refer to a number of excellent reviews. We are grateful to Rita Stiefvater and Caroline Lassnig for critically reading the manuscript. MM and BS are supported by the Austrian Science fund (FWF SFB-28) and GEN-AU III ("Inflammobiota") and MM and NL by GEN-AU II/III ("Austrian Network for Functional Genomics of the Mouse"). MM is further supported by the Austrian Federal Ministry for Science and Research (GZ.200.191/1-II/1a/2008 to Biomodels Austria).

References

Alber D, Staeheli P (1996) Partial inhibition of vesicular stomatitis virus by the interferon-induced human 9–27 protein. J Interferon Cytokine Res 16:375–380

Andrew A, Strebel K (2011) The interferon-inducible host factor bone marrow stromal antigen 2/tetherin restricts virion release, but is it actually a viral restriction factor? J Interferon Cytokine Res 31:137–144

Barry G, Breakwell L, Fragkoudis R, Attarzadeh-Yazdi G, Rodriguez-Andres J, Kohl A, Fazakerley JK (2009) PKR acts early in infection to suppress Semliki Forest virus production and strongly enhances the type I interferon response. J Gen Virol 90:1382–1391

Bass BL (2002) RNA editing by adenosine deaminases that act on RNA. Annu Rev Biochem 71:817–846

Bekpen C, Hunn JP, Rohde C, Parvanova I, Guethlein L, Dunn DM, Glowalla E, Leptin M, Howard JC (2005) The interferon-inducible p47 (IRG) GTPases in vertebrates: loss of the cell autonomous resistance mechanism in the human lineage. Genome Biol 6:R92

Blondel D, Kheddache S, Lahaye X, Dianoux L, Chelbi-Alix MK (2010) Resistance to rabies virus infection conferred by the PMLIV isoform. J Virol 84:10719–10726

Bonilla WV, Pinschewer DD, Klenerman P, Rousson V, Gaboli M, Pandolfi PP, Zinkernagel RM, Salvato MS, Hengartner H (2002) Effects of promyelocytic leukemia protein on virus-host balance. J Virol 76:3810–3818

Borden EC, Sen GC, Uze G, Silverman RH, Ransohoff RM, Foster GR, Stark GR (2007) Interferons at age 50: past, current and future impact on biomedicine. Nat Rev Drug Discov 6:975–990

Brass AL, Huang IC, Benita Y, John SP, Krishnan MN, Feeley EM, Ryan BJ, Weyer JL, van der Weyden L, Fikrig E, Adams DJ, Xavier RJ, Farzan M, Elledge SJ (2009) The IFITM proteins mediate cellular resistance to influenza A H1N1 virus, West Nile virus, and dengue virus. Cell 139:1243–1254

Broering R, Zhang X, Kottilil S, Trippler M, Jiang M, Lu M, Gerken G, Schlaak JF (2010) The interferon stimulated gene 15 functions as a proviral factor for the hepatitis C virus and as a regulator of the IFN response. Gut 59:1111–1119

Browne EP, Littman DR (2008) Species-specific restriction of apobec3-mediated hypermutation. J Virol 82:1305–1313

Cao B, Liu X, Hou F, Li W, Han Z, Zhang Q, Dai Y, Xu C, Qi H (2009) The haplotype of the MxA gene promoter is associated with hepatitis B virus infection in a Chinese population. Liver Int 29:1383–1388

Carthagena L, Bergamaschi A, Luna JM, David A, Uchil PD, Margottin-Goguet F, Mothes W, Hazan U, Transy C, Pancino G, Nisole S (2009) Human TRIM gene expression in response to interferons. PLoS One 4:e4894

Chakrabarti A, Jha BK, Silverman RH (2011) New insights into the role of RNase L in innate immunity. J Interferon Cytokine Res 31:49–57

Chebath J, Merlin G, Metz R, Benech P, Revel M (1983) Interferon-induced 56,000 Mr protein and its mRNA in human cells: molecular cloning and partial sequence of the cDNA. Nucleic Acids Res 11:1213–1226

Chee AV, Lopez P, Pandolfi PP, Roizman B (2003) Promyelocytic leukemia protein mediates interferon-based anti-herpes simplex virus 1 effects. J Virol 77:7101–7105

Chen L, Borozan I, Sun J, Guindi M, Fischer S, Feld J, Anand N, Heathcote J, Edwards AM, McGilvray ID (2010a) Cell-type specific gene expression signature in liver underlies response to interferon therapy in chronic hepatitis C infection. Gastroenterology 138:1123–1133, e1-3

Chen L, Sun J, Meng L, Heathcote J, Edwards AM, McGilvray ID (2010b) ISG15, a ubiquitin-like interferon-stimulated gene, promotes hepatitis C virus production in vitro: implications for chronic infection and response to treatment. J Gen Virol 91:382–388

Cheney KM, McKnight A (2010) Interferon-alpha mediates restriction of human immunodeficiency virus type-1 replication in primary human macrophages at an early stage of replication. PLoS One 5:e13521

Chin KC, Cresswell P (2001) Viperin (cig5), an IFN-inducible antiviral protein directly induced by human cytomegalovirus. Proc Natl Acad Sci USA 98:15125–15130

Chua PK, McCown MF, Rajyaguru S, Kular S, Varma R, Symons J, Chiu SS, Cammack N, Najera I (2009) Modulation of alpha interferon anti-hepatitis C virus activity by ISG15. J Gen Virol 90:2929–2939

Commins S, Steinke JW, Borish L (2008) The extended IL-10 superfamily: IL-10, IL-19, IL-20, IL-22, IL-24, IL-26, IL-28, and IL-29. J Allergy Clin Immunol 121:1108–1111

Cui XF, Imaizumi T, Yoshida H, Borden EC, Satoh K (2004) Retinoic acid-inducible gene-I is induced by interferon-gamma and regulates the expression of interferon-gamma stimulated gene 15 in MCF-7 cells. Biochem Cell Biol 82:401–405

Daffis S, Szretter KJ, Schriewer J, Li J, Youn S, Errett J, Lin TY, Schneller S, Zust R, Dong H, Thiel V, Sen GC, Fensterl V, Klimstra WB, Pierson TC, Buller RM, Gale M Jr, Shi PY, Diamond MS (2010) 2′-O methylation of the viral mRNA cap evades host restriction by IFIT family members. Nature 468:452–456

de Veer MJ, Holko M, Frevel M, Walker E, Der S, Paranjape JM, Silverman RH, Williams BR (2001) Functional classification of interferon-stimulated genes identified using microarrays. J Leukoc Biol 69:912–920

Degols G, Eldin P, Mechti N (2007) ISG20, an actor of the innate immune response. Biochimie 89:831–835

Der SD, Zhou A, Williams BR, Silverman RH (1998) Identification of genes differentially regulated by interferon alpha, beta, or gamma using oligonucleotide arrays. Proc Natl Acad Sci USA 95:15623–15628

Djavani M, Rodas J, Lukashevich IS, Horejsh D, Pandolfi PP, Borden KL, Salvato MS (2001) Role of the promyelocytic leukemia protein PML in the interferon sensitivity of lymphocytic choriomeningitis virus. J Virol 75:6204–6208

Elbahesh H, Jha BK, Silverman RH, Scherbik SV, Brinton MA (2011) The Flvr-encoded murine oligoadenylate synthetase 1b (Oas1b) suppresses 2-5A synthesis in intact cells. Virology 409:262–270

Espert L, Degols G, Gongora C, Blondel D, Williams BR, Silverman RH, Mechti N (2003) ISG20, a new interferon-induced RNase specific for single-stranded RNA, defines an alternative antiviral pathway against RNA genomic viruses. J Biol Chem 278:16151–16158

Espert L, Degols G, Lin YL, Vincent T, Benkirane M, Mechti N (2005) Interferon-induced exonuclease ISG20 exhibits an antiviral activity against human immunodeficiency virus type 1. J Gen Virol 86:2221–2229

Fechter P, Brownlee GG (2005) Recognition of mRNA cap structures by viral and cellular proteins. J Gen Virol 86:1239–1249

Fensterl V, Sen GC (2011) The ISG56/IFIT1 gene family. J Interferon Cytokine Res 31:71–78

Friedman RL, Manly SP, McMahon M, Kerr IM, Stark GR (1984) Transcriptional and posttranscriptional regulation of interferon-induced gene expression in human cells. Cell 38:745–755

Fu Y, Comella N, Tognazzi K, Brown LF, Dvorak HF, Kocher O (1999) Cloning of DLM-1, a novel gene that is up-regulated in activated macrophages, using RNA differential display. Gene 240:157–163

Gack MU, Shin YC, Joo CH, Urano T, Liang C, Sun L, Takeuchi O, Akira S, Chen Z, Inoue S, Jung JU (2007) TRIM25 RING-finger E3 ubiquitin ligase is essential for RIG-I-mediated antiviral activity. Nature 446:916–920

Gariglio M, Mondini M, De Andrea M, Landolfo S (2011) The multifaceted interferon-inducible p200 family proteins: from cell biology to human pathology. J Interferon Cytokine Res 31:159–172
Geoffroy MC, Chelbi-Alix MK (2011) Role of promyelocytic leukemia protein in host antiviral defense. J Interferon Cytokine Res 31:145–158
Giannakopoulos NV, Arutyunova E, Lai C, Lenschow DJ, Haas AL, Virgin HW (2009) ISG15 Arg151 and the ISG15-conjugating enzyme UbE1L are important for innate immune control of Sindbis virus. J Virol 83:1602–1610
Goila-Gaur R, Strebel K (2008) HIV-1 Vif, APOBEC, and intrinsic immunity. Retrovirology 5:51
Gongora C, David G, Pintard L, Tissot C, Hua TD, Dejean A, Mechti N (1997) Molecular cloning of a new interferon-induced PML nuclear body-associated protein. J Biol Chem 272:19457–19463
Haller O, Kochs G (2011) Human MxA protein: an interferon-induced dynamin-like GTPase with broad antiviral activity. J Interferon Cytokine Res 31:79–87
Haller O, Stertz S, Kochs G (2007) The Mx GTPase family of interferon-induced antiviral proteins. Microbes Infect 9:1636–1643
Hao Y, Yang D (2008) Cloning, eukaryotic expression of human ISG20 and preliminary study on the effect of its anti-HBV. J Huazhong Univ Sci Technol Med Sci 28:11–13
Hartner JC, Schmittwolf C, Kispert A, Muller AM, Higuchi M, Seeburg PH (2004) Liver disintegration in the mouse embryo caused by deficiency in the RNA-editing enzyme ADAR1. J Biol Chem 279:4894–4902
Hartner JC, Walkley CR, Lu J, Orkin SH (2009) ADAR1 is essential for the maintenance of hematopoiesis and suppression of interferon signaling. Nat Immunol 10:109–115
Hartwig D, Schoeneich L, Greeve J, Schutte C, Dorn I, Kirchner H, Hennig H (2004) Interferon-alpha stimulation of liver cells enhances hepatitis delta virus RNA editing in early infection. J Hepatol 41:667–672
Harty RN, Pitha PM, Okumura A (2009) Antiviral activity of innate immune protein ISG15. J Innate Immun 1:397–404
Hefti HP, Frese M, Landis H, Di Paolo C, Aguzzi A, Haller O, Pavlovic J (1999) Human MxA protein protects mice lacking a functional alpha/beta interferon system against La crosse virus and other lethal viral infections. J Virol 73:6984–6991
Helbig KJ, Lau DT, Semendric L, Harley HA, Beard MR (2005) Analysis of ISG expression in chronic hepatitis C identifies viperin as a potential antiviral effector. Hepatology 42:702–710
Higuchi M, Maas S, Single FN, Hartner J, Rozov A, Burnashev N, Feldmeyer D, Sprengel R, Seeburg PH (2000) Point mutation in an AMPA receptor gene rescues lethality in mice deficient in the RNA-editing enzyme ADAR2. Nature 406:78–81
Hijikata M, Ohta Y, Mishiro S (2000) Identification of a single nucleotide polymorphism in the MxA gene promoter (G/T at nt-88) correlated with the response of hepatitis C patients to interferon. Intervirology 43:124–127
Howard J (2008) The IRG proteins: a function in search of a mechanism. Immunobiology 213:367–375
Hsiang TY, Zhao C, Krug RM (2009) Interferon-induced ISG15 conjugation inhibits influenza A virus gene expression and replication in human cells. J Virol 83:5971–5977
Huang IC, Bailey CC, Weyer JL, Radoshitzky SR, Becker MM, Chiang JJ, Brass AL, Ahmed AA, Chi X, Dong L, Longobardi LE, Boltz D, Kuhn JH, Elledge SJ, Bavari S, Denison MR, Choe H, Farzan M (2011) Distinct patterns of IFITM-mediated restriction of filoviruses, SARS coronavirus, and influenza A virus. PLoS Pathog 7:e1001258
Imaizumi T, Yagihashi N, Hatakeyama M, Yamashita K, Ishikawa A, Taima K, Yoshida H, Inoue I, Fujita T, Yagihashi S, Satoh K (2004) Expression of retinoic acid-inducible gene-I in vascular smooth muscle cells stimulated with interferon-gamma. Life Sci 75:1171–1180
Ireland DD, Stohlman SA, Hinton DR, Kapil P, Silverman RH, Atkinson RA, Bergmann CC (2009) RNase L mediated protection from virus induced demyelination. PLoS Pathog 5:e1000602

Jayan GC, Casey JL (2002) Increased RNA editing and inhibition of hepatitis delta virus replication by high-level expression of ADAR1 and ADAR2. J Virol 76:3819–3827

Jiang D, Guo H, Xu C, Chang J, Gu B, Wang L, Block TM, Guo JT (2008) Identification of three interferon-inducible cellular enzymes that inhibit the replication of hepatitis C virus. J Virol 82:1665–1678

Jiang D, Weidner JM, Qing M, Pan XB, Guo H, Xu C, Zhang X, Birk A, Chang J, Shi PY, Block TM, Guo JT (2010) Identification of five interferon-induced cellular proteins that inhibit West Nile virus and dengue virus infections. J Virol 84:8332–8341

Jouvenet N, Neil SJ, Zhadina M, Zang T, Kratovac Z, Lee Y, McNatt M, Hatziioannou T, Bieniasz PD (2009) Broad-spectrum inhibition of retroviral and filoviral particle release by tetherin. J Virol 83:1837–1844

Kajaste-Rudnitski A, Pultrone C, Marzetta F, Ghezzi S, Coradin T, Vicenzi E (2010) Restriction factors of retroviral replication: the example of Tripartite Motif (TRIM) protein 5 alpha and 22. Amino Acids 39:1–9

Kang DC, Gopalkrishnan RV, Wu Q, Jankowsky E, Pyle AM, Fisher PB (2002) mda-5: an interferon-inducible putative RNA helicase with double-stranded RNA-dependent ATPase activity and melanoma growth-suppressive properties. Proc Natl Acad Sci USA 99:637–642

Kawai T, Akira S (2009) The roles of TLRs, RLRs and NLRs in pathogen recognition. Int Immunol 21:317–337

Kim MJ, Yoo JY (2010) Inhibition of hepatitis C virus replication by IFN-mediated ISGylation of HCV-NS5A. J Immunol 185:4311–4318

Kim KI, Yan M, Malakhova O, Luo JK, Shen MF, Zou W, de la Torre JC, Zhang DE (2006) Ube1L and protein ISGylation are not essential for alpha/beta interferon signaling. Mol Cell Biol 26:472–479

Kim JH, Luo JK, Zhang DE (2008a) The level of hepatitis B virus replication is not affected by protein ISG15 modification but is reduced by inhibition of UBP43 (USP18) expression. J Immunol 181:6467–6472

Kim MJ, Hwang SY, Imaizumi T, Yoo JY (2008b) Negative feedback regulation of RIG-I-mediated antiviral signaling by interferon-induced ISG15 conjugation. J Virol 82:1474–1483

Knobeloch KP, Utermohlen O, Kisser A, Prinz M, Horak I (2005) Reexamination of the role of ubiquitin-like modifier ISG15 in the phenotype of UBP43-deficient mice. Mol Cell Biol 25:11030–11034

Koning FA, Newman EN, Kim EY, Kunstman KJ, Wolinsky SM, Malim MH (2009) Defining APOBEC3 expression patterns in human tissues and hematopoietic cell subsets. J Virol 83:9474–9485

Kristiansen H, Scherer CA, McVean M, Iadonato SP, Vends S, Thavachelvam K, Steffensen TB, Horan KA, Kuri T, Weber F, Paludan SR, Hartmann R (2010) Extracellular 2′-5′ oligoadenylate synthetase stimulates RNase L-independent antiviral activity: a novel mechanism of virus-induced innate immunity. J Virol 84:11898–11904

Kristiansen H, Gad HH, Eskildsen-Larsen S, Despres P, Hartmann R (2011) The oligoadenylate synthetase family: an ancient protein family with multiple antiviral activities. J Interferon Cytokine Res 31:41–47

Lai C, Struckhoff JJ, Schneider J, Martinez-Sobrido L, Wolff T, Garcia-Sastre A, Zhang DE, Lenschow DJ (2009) Mice lacking the ISG15 E1 enzyme UbE1L demonstrate increased susceptibility to both mouse-adapted and non-mouse-adapted influenza B virus infection. J Virol 83:1147–1151

Langberg SR, Moss B (1981) Post-transcriptional modifications of mRNA. Purification and characterization of cap I and cap II RNA (nucleoside-2′-)-methyltransferases from HeLa cells. J Biol Chem 256:10054–10060

Lange UC, Adams DJ, Lee C, Barton S, Schneider R, Bradley A, Surani MA (2008) Normal germ line establishment in mice carrying a deletion of the Ifitm/Fragilis gene family cluster. Mol Cell Biol 28:4688–4696

Lenschow DJ, Lai C, Frias-Staheli N, Giannakopoulos NV, Lutz A, Wolff T, Osiak A, Levine B, Schmidt RE, Garcia-Sastre A, Leib DA, Pekosz A, Knobeloch KP, Horak I, Virgin HW (2007)

IFN-stimulated gene 15 functions as a critical antiviral molecule against influenza, herpes, and Sindbis viruses. Proc Natl Acad Sci USA 104:1371–1376
Levy D, Larner A, Chaudhuri A, Babiss LE, Darnell JE Jr (1986) Interferon-stimulated transcription: isolation of an inducible gene and identification of its regulatory region. Proc Natl Acad Sci USA 83:8929–8933
Levy DE, Marie I, Smith E, Prakash A (2002) Enhancement and diversification of IFN induction by IRF-7-mediated positive feedback. J Interferon Cytokine Res 22:87–93
Li G, Zhang J, Sun Y, Wang H, Wang Y (2009a) The evolutionarily dynamic IFN-inducible GTPase proteins play conserved immune functions in vertebrates and cephalochordates. Mol Biol Evol 26:1619–1630
Li Y, Li C, Xue P, Zhong B, Mao AP, Ran Y, Chen H, Wang YY, Yang F, Shu HB (2009b) ISG56 is a negative-feedback regulator of virus-triggered signaling and cellular antiviral response. Proc Natl Acad Sci USA 106:7945–7950
Low A, Okeoma CM, Lovsin N, de Las Heras M, Taylor TH, Peterlin BM, Ross SR, Fan H (2009) Enhanced replication and pathogenesis of Moloney murine leukemia virus in mice defective in the murine APOBEC3 gene. Virology 385:455–463
Lu G, Reinert JT, Pitha-Rowe I, Okumura A, Kellum M, Knobeloch KP, Hassel B, Pitha PM (2006) ISG15 enhances the innate antiviral response by inhibition of IRF-3 degradation. Cell Mol Biol (Noisy-le-grand) 52:29–41
Lu J, Pan Q, Rong L, Liu SL, Liang C (2011) The IFITM proteins inhibit HIV-1 infection. J Virol 85:2126–2137
Luthra P, Sun D, Silverman RH, He B (2011) Activation of IFN-{beta} expression by a viral mRNA through RNase L and MDA5. Proc Natl Acad Sci USA 108:2118–2123
MacMicking JD (2004) IFN-inducible GTPases and immunity to intracellular pathogens. Trends Immunol 25:601–609
Malakhov MP, Kim KI, Malakhova OA, Jacobs BS, Borden EC, Zhang DE (2003) High through put immunoblotting. Ubiquitiin-like protein ISG15 modifies key regulators of signal transduction. J Biol Chem 278:16608–16613
Malakhova OA, Yan M, Malakhov MP, Yuan Y, Ritchie KJ, Kim KI, Peterson LF, Shuai K, Zhang DE (2003) Protein ISGylation modulates the JAK-STAT signaling pathway. Genes Dev 17:455–460
Malakhova OA, Kim KI, Luo JK, Zou W, Kumar KG, Fuchs SY, Shuai K, Zhang DE (2006) UBP43 is a novel regulator of interferon signaling independent of its ISG15 isopeptidase activity. EMBO J 25:2358–2367
Malathi K, Dong B, Gale M Jr, Silverman RH (2007) Small self-RNA generated by RNase L amplifies antiviral innate immunity. Nature 448:816–819
Malathi K, Saito T, Crochet N, Barton DJ, Gale M Jr, Silverman RH (2010) RNase L releases a small RNA from HCV RNA that refolds into a potent PAMP. RNA 16:2108–2119
Mattei MG, Tissot C, Gongora C, Mechti N (1997) Assignment of ISG20 encoding a new interferon-induced PML nuclear body-associated protein, to chromosome 15q26 by in situ hybridization. Cytogenet Cell Genet 79:286–287
McCartney SA, Colonna M (2009) Viral sensors: diversity in pathogen recognition. Immunol Rev 227:87–94
Nakayama Y, Plisch EH, Sullivan J, Thomas C, Czuprynski CJ, Williams BR, Suresh M (2010) Role of PKR and type I IFNs in viral control during primary and secondary infection. PLoS Pathog 6:e1000966
Narvaiza I, Linfesty DC, Greener BN, Hakata Y, Pintel DJ, Logue E, Landau NR, Weitzman MD (2009) Deaminase-independent inhibition of parvoviruses by the APOBEC3A cytidine deaminase. PLoS Pathog 5:e1000439
Nguyen LH, Espert L, Mechti N, Wilson DM III (2001) The human interferon- and estrogen-regulated ISG20/HEM45 gene product degrades single-stranded RNA and DNA in vitro. Biochemistry 40:7174–7179
Nie Y, Hammond GL, Yang JH (2007) Double-stranded RNA deaminase ADAR1 increases host susceptibility to virus infection. J Virol 81:917–923

Ohtomo T, Sugamata Y, Ozaki Y, Ono K, Yoshimura Y, Kawai S, Koishihara Y, Ozaki S, Kosaka M, Hirano T, Tsuchiya M (1999) Molecular cloning and characterization of a surface antigen preferentially overexpressed on multiple myeloma cells. Biochem Biophys Res Commun 258:583–591

Okeoma CM, Lovsin N, Peterlin BM, Ross SR (2007) APOBEC3 inhibits mouse mammary tumour virus replication in vivo. Nature 445:927–930

Okeoma CM, Low A, Bailis W, Fan HY, Peterlin BM, Ross SR (2009a) Induction of APOBEC3 in vivo causes increased restriction of retrovirus infection. J Virol 83:3486–3495

Okeoma CM, Petersen J, Ross SR (2009b) Expression of murine APOBEC3 alleles in different mouse strains and their effect on mouse mammary tumor virus infection. J Virol 83:3029–3038

Okumura F, Zou W, Zhang DE (2007) ISG15 modification of the eIF4E cognate 4EHP enhances cap structure-binding activity of 4EHP. Genes Dev 21:255–260

Osiak A, Utermohlen O, Niendorf S, Horak I, Knobeloch KP (2005) ISG15, an interferon-stimulated ubiquitin-like protein, is not essential for STAT1 signaling and responses against vesicular stomatitis and lymphocytic choriomeningitis virus. Mol Cell Biol 25:6338–6345

Ozato K, Shin DM, Chang TH, Morse HC III (2008) TRIM family proteins and their emerging roles in innate immunity. Nat Rev Immunol 8:849–860

Pampin M, Simonin Y, Blondel B, Percherancier Y, Chelbi-Alix MK (2006) Cross talk between PML and p53 during poliovirus infection: implications for antiviral defense. J Virol 80:8582–8592

Pentecost BT (1998) Expression and estrogen regulation of the HEM45 MRNA in human tumor lines and in the rat uterus. J Steroid Biochem Mol Biol 64:25–33

Pestka S, Krause CD, Walter MR (2004) Interferons, interferon-like cytokines, and their receptors. Immunol Rev 202:8–32

Pichlmair A, Lassnig C, Eberle CA, Gorna MW, Baumann CL, Burkard TR, Burckstummer T, Stefanovic A, Krieger S, Bennett KL, Rulicke T, Weber F, Colinge J, Muller M, Superti-Furga G (2011) IFIT1 is an antiviral protein that recognizes 5′-triphosphate RNA. Nat Immunol 12:624–630

Pincetic A, Kuang Z, Seo EJ, Leis J (2010) The interferon-induced gene ISG15 blocks retrovirus release from cells late in the budding process. J Virol 84:4725–4736

Pindel A, Sadler A (2011) The role of protein kinase R in the interferon response. J Interferon Cytokine Res 31:59–70

Porta C, Hadj-Slimane R, Nejmeddine M, Pampin M, Tovey MG, Espert L, Alvarez S, Chelbi-Alix MK (2005) Interferons alpha and gamma induce p53-dependent and p53-independent apoptosis, respectively. Oncogene 24:605–615

Proto S, Taylor JA, Chokshi S, Navaratnam N, Naoumov NV (2008) APOBEC and iNOS are not the main intracellular effectors of IFN-gamma-mediated inactivation of hepatitis B virus replication. Antiviral Res 78:260–267

Refsland EW, Stenglein MD, Shindo K, Albin JS, Brown WL, Harris RS (2010) Quantitative profiling of the full APOBEC3 mRNA repertoire in lymphocytes and tissues: implications for HIV-1 restriction. Nucleic Acids Res 38:4274–4284

Regad T, Saib A, Lallemand-Breitenbach V, Pandolfi PP, de The H, Chelbi-Alix MK (2001) PML mediates the interferon-induced antiviral state against a complex retrovirus via its association with the viral transactivator. EMBO J 20:3495–3505

Ritchie KJ, Malakhov MP, Hetherington CJ, Zhou L, Little MT, Malakhova OA, Sipe JC, Orkin SH, Zhang DE (2002) Dysregulation of protein modification by ISG15 results in brain cell injury. Genes Dev 16:2207–2212

Ritchie KJ, Hahn CS, Kim KI, Yan M, Rosario D, Li L, de la Torre JC, Zhang DE (2004) Role of ISG15 protease UBP43 (USP18) in innate immunity to viral infection. Nat Med 10:1374–1378

Rivieccio MA, Suh HS, Zhao Y, Zhao ML, Chin KC, Lee SC, Brosnan CF (2006) TLR3 ligation activates an antiviral response in human fetal astrocytes: a role for viperin/cig5. J Immunol 177:4735–4741

Rulli SJ Jr, Mirro J, Hill SA, Lloyd P, Gorelick RJ, Coffin JM, Derse D, Rein A (2008) Interactions of murine APOBEC3 and human APOBEC3G with murine leukemia viruses. J Virol 82:6566–6575

Sadler AJ (2010) Orchestration of the activation of protein kinase R by the RNA-binding motif. J Interferon Cytokine Res 30:195–204

Sadler AJ, Williams BR (2008) Interferon-inducible antiviral effectors. Nat Rev Immunol 8:559–568

Saikia P, Fensterl V, Sen GC (2010) The inhibitory action of P56 on select functions of E1 mediates interferon's effect on human papillomavirus DNA replication. J Virol 84:13036–13039

Sakuma T, Noda T, Urata S, Kawaoka Y, Yasuda J (2009) Inhibition of Lassa and Marburg virus production by tetherin. J Virol 83:2382–2385

Samuel CE (2011) Adenosine deaminases acting on RNA (ADARs) are both antiviral and proviral. Virology 411:180–193

Santiago ML, Smith DS, Barrett BS, Montano M, Benitez RL, Pelanda R, Hasenkrug KJ, Greene WC (2011) Persistent friend virus replication and disease in Apobec3-deficient mice expressing functional B-cell-activating factor receptor. J Virol 85:189–199

Schindler C, Levy DE, Decker T (2007) JAK-STAT signaling: from interferons to cytokines. J Biol Chem 282:20059–20063

Schmeisser H, Mejido J, Balinsky CA, Morrow AN, Clark CR, Zhao T, Zoon KC (2010) Identification of alpha interferon-induced genes associated with antiviral activity in Daudi cells and characterization of IFIT3 as a novel antiviral gene. J Virol 84:10671–10680

Schulz O, Pichlmair A, Rehwinkel J, Rogers NC, Scheuner D, Kato H, Takeuchi O, Akira S, Kaufman RJ, Reis e Sousa C (2010) Protein kinase R contributes to immunity against specific viruses by regulating interferon mRNA integrity. Cell Host Microbe 7:354–361

Sen A, Pruijssers AJ, Dermody TS, Garcia-Sastre A, Greenberg HB (2011) The early interferon response to rotavirus is regulated by PKR and depends on MAVS/IPS 1, RIG I, MDA 5, and IRF3. J Virol 85:3717–3732

Serra MJ, Smolter PE, Westhof E (2004) Pronounced instability of tandem IU base pairs in RNA. Nucleic Acids Res 32:1824–1828

Shaveta G, Shi J, Chow VT, Song J (2010) Structural characterization reveals that viperin is a radical S-adenosyl-L-methionine (SAM) enzyme. Biochem Biophys Res Commun 391:1390–1395

Shi HX, Yang K, Liu X, Liu XY, Wei B, Shan YF, Zhu LH, Wang C (2010) Positive regulation of interferon regulatory factor 3 activation by Herc5 via ISG15 modification. Mol Cell Biol 30:2424–2436

Siegrist F, Ebeling M, Certa U (2011) The small interferon-induced transmembrane genes and proteins. J Interferon Cytokine Res 31:183–197

Silverman RH (2007) A scientific journey through the 2-5A/RNase L system. Cytokine Growth Factor Rev 18:381–388

Staeheli P, Prochazka M, Steigmeier PA, Haller O (1984) Genetic control of interferon action: mouse strain distribution and inheritance of an induced protein with guanylate-binding property. Virology 137:135–142

Staeheli P, Grob R, Meier E, Sutcliffe JG, Haller O (1988) Influenza virus-susceptible mice carry Mx genes with a large deletion or a nonsense mutation. Mol Cell Biol 8:4518–4523

Stenglein MD, Burns MB, Li M, Lengyel J, Harris RS (2010) APOBEC3 proteins mediate the clearance of foreign DNA from human cells. Nat Struct Mol Biol 17:222–229

Suzuki F, Arase Y, Suzuki Y, Tsubota A, Akuta N, Hosaka T, Someya T, Kobayashi M, Saitoh S, Ikeda K, Matsuda M, Takagi K, Satoh J, Kumada H (2004) Single nucleotide polymorphism of the MxA gene promoter influences the response to interferon monotherapy in patients with hepatitis C viral infection. J Viral Hepat 11:271–276

Szkopinska A, Plochocka D (2005) Farnesyl diphosphate synthase; regulation of product specificity. Acta Biochim Pol 52:45–55

Takaoka A, Hayakawa S, Yanai H, Stoiber D, Negishi H, Kikuchi H, Sasaki S, Imai K, Shibue T, Honda K, Taniguchi T (2003) Integration of interferon-alpha/beta signalling to p53 responses in tumour suppression and antiviral defence. Nature 424:516–523

Takeda E, Tsuji-Kawahara S, Sakamoto M, Langlois MA, Neuberger MS, Rada C, Miyazawa M (2008) Mouse APOBEC3 restricts friend leukemia virus infection and pathogenesis in vivo. J Virol 82:10998–11008

Takeuchi T, Kobayashi T, Tamura S, Yokosawa H (2006) Negative regulation of protein phosphatase 2Cbeta by ISG15 conjugation. FEBS Lett 580:4521–4526

Tang Y, Zhong G, Zhu L, Liu X, Shan Y, Feng H, Bu Z, Chen H, Wang C (2010) Herc5 attenuates influenza A virus by catalyzing ISGylation of viral NS1 protein. J Immunol 184:5777–5790

Taylor GA, Collazo CM, Yap GS, Nguyen K, Gregorio TA, Taylor LS, Eagleson B, Secrest L, Southon EA, Reid SW, Tessarollo L, Bray M, McVicar DW, Komschlies KL, Young HA, Biron CA, Sher A, Vande Woude GF (2000) Pathogen-specific loss of host resistance in mice lacking the IFN-gamma-inducible gene IGTP. Proc Natl Acad Sci USA 97:751–755

Terenzi F, Saikia P, Sen GC (2008) Interferon-inducible protein, P56, inhibits HPV DNA replication by binding to the viral protein E1. EMBO J 27:3311–3321

Torisu H, Kusuhara K, Kira R, Bassuny WM, Sakai Y, Sanefuji M, Takemoto M, Hara T (2004) Functional MxA promoter polymorphism associated with subacute sclerosing panencephalitis. Neurology 62:457–460

Toth AM, Li Z, Cattaneo R, Samuel CE (2009) RNA-specific adenosine deaminase ADAR1 suppresses measles virus-induced apoptosis and activation of protein kinase PKR. J Biol Chem 284:29350–29356

Trapani JA, Browne KA, Dawson MJ, Ramsay RG, Eddy RL, Show TB, White PC, Dupont B (1992) A novel gene constitutively expressed in human lymphoid cells is inducible with interferon-gamma in myeloid cells. Immunogenetics 36:369–376

Trapp S, Derby NR, Singer R, Shaw A, Williams VG, Turville SG, Bess JW Jr, Lifson JD, Robbiani M (2009) Double-stranded RNA analog poly(I:C) inhibits human immunodeficiency virus amplification in dendritic cells via type I interferon-mediated activation of APOBEC3G. J Virol 83:884–895

Turelli P, Liagre-Quazzola A, Mangeat B, Verp S, Jost S, Trono D (2008) APOBEC3-independent interferon-induced viral clearance in hepatitis B virus transgenic mice. J Virol 82:6585–6590

Vestal DJ, Jeyaratnam JA (2011) The guanylate-binding proteins: emerging insights into the biochemical properties and functions of this family of large interferon-induced guanosine triphosphatase. J Interferon Cytokine Res 31:89–97

Vitali P, Scadden AD (2010) Double-stranded RNAs containing multiple IU pairs are sufficient to suppress interferon induction and apoptosis. Nat Struct Mol Biol 17:1043–1050

Wacher C, Muller M, Hofer MJ, Getts DR, Zabaras R, Ousman SS, Terenzi F, Sen GC, King NJ, Campbell IL (2007) Coordinated regulation and widespread cellular expression of interferon-stimulated genes (ISG) ISG-49, ISG-54, and ISG-56 in the central nervous system after infection with distinct viruses. J Virol 81:860–871

Wang ZG, Delva L, Gaboli M, Rivi R, Giorgio M, Cordon-Cardo C, Grosveld F, Pandolfi PP (1998) Role of PML in cell growth and the retinoic acid pathway. Science 279:1547–1551

Wang C, Pflugheber J, Sumpter R Jr, Sodora DL, Hui D, Sen GC, Gale M Jr (2003) Alpha interferon induces distinct translational control programs to suppress hepatitis C virus RNA replication. J Virol 77:3898–3912

Wang Q, Miyakoda M, Yang W, Khillan J, Stachura DL, Weiss MJ, Nishikura K (2004) Stress-induced apoptosis associated with null mutation of ADAR1 RNA editing deaminase gene. J Biol Chem 279:4952–4961

Wang X, Hinson ER, Cresswell P (2007) The interferon-inducible protein viperin inhibits influenza virus release by perturbing lipid rafts. Cell Host Microbe 2:96–105

Wang YJ, Wang X, Zhang H, Zhou L, Liu S, Kolson DL, Song L, Ye L, Ho WZ (2009) Expression and regulation of antiviral protein APOBEC3G in human neuronal cells. J Neuroimmunol 206:14–21

Wang J, Liu B, Wang N, Lee YM, Liu C, Li K (2011) TRIM56 is a virus- and interferon-inducible E3 ubiquitin ligase that restricts pestivirus infection. J Virol 85:3733–3745

Ward SV, George CX, Welch MJ, Liou LY, Hahm B, Lewicki H, de la Torre JC, Samuel CE, Oldstone MB (2011) RNA editing enzyme adenosine deaminase is a restriction factor for controlling measles virus replication that also is required for embryogenesis. Proc Natl Acad Sci USA 108:331–336

Wathelet M, Moutschen S, Defilippi P, Cravador A, Collet M, Huez G, Content J (1986) Molecular cloning, full-length sequence and preliminary characterization of a 56-kDa protein induced by human interferons. Eur J Biochem 155:11–17

Wei CM, Moss B (1975) Methylated nucleotides block 5′-terminus of vaccinia virus messenger RNA. Proc Natl Acad Sci USA 72:318–322

Weidner JM, Jiang D, Pan XB, Chang J, Block TM, Guo JT (2010) Interferon-induced cell membrane proteins, IFITM3 and tetherin, inhibit vesicular stomatitis virus infection via distinct mechanisms. J Virol 84:12646–12657

XuFeng R, Boyer MJ, Shen H, Li Y, Yu H, Gao Y, Yang Q, Wang Q, Cheng T (2009) ADAR1 is required for hematopoietic progenitor cell survival via RNA editing. Proc Natl Acad Sci USA 106:17763–17768

Yang K, Shi HX, Liu XY, Shan YF, Wei B, Chen S, Wang C (2009) TRIM21 is essential to sustain IFN regulatory factor 3 activation during antiviral response. J Immunol 182:3782–3792

Young JA, Sermwittayawong D, Kim HJ, Nandu S, An NS, Erdjument-Bromage H, Tempst P, Coscoy L, Winoto A (2011) FAS-associated death domain (FADD) and the E3 ubiquitin-protein ligase trim21 interact to negatively regulate virus-induced interferon production. J Biol Chem 286:6521–6531

Yount JS, Moltedo B, Yang YY, Charron G, Moran TM, Lopez CB, Hang HC (2010) Palmitoylome profiling reveals S-palmitoylation-dependent antiviral activity of IFITM3. Nat Chem Biol 6:610–614

Zhang D, Zhang DE (2011) Interferon-stimulated gene 15 and the protein ISGylation system. J Interferon Cytokine Res 31:119–130

Zhang Y, Burke CW, Ryman KD, Klimstra WB (2007) Identification and characterization of interferon-induced proteins that inhibit alphavirus replication. J Virol 81:11246–11255

Zhao C, Denison C, Huibregtse JM, Gygi S, Krug RM (2005) Human ISG15 conjugation targets both IFN-induced and constitutively expressed proteins functioning in diverse cellular pathways. Proc Natl Acad Sci USA 102:10200–10205

Zhao C, Hsiang TY, Kuo RL, Krug RM (2010) ISG15 conjugation system targets the viral NS1 protein in influenza A virus-infected cells. Proc Natl Acad Sci USA 107:2253–2258

Zhou Z, Wang N, Woodson SE, Dong Q, Wang J, Liang Y, Rijnbrand R, Wei L, Nichols JE, Guo JT, Holbrook MR, Lemon SM, Li K (2011) Antiviral activities of ISG20 in positive-strand RNA virus infections. Virology 409:175–188

Zhu H, Cong JP, Shenk T (1997) Use of differential display analysis to assess the effect of human cytomegalovirus infection on the accumulation of cellular RNAs: induction of interferon-responsive RNAs. Proc Natl Acad Sci USA 94:13985–13990

Zust R, Cervantes-Barragan L, Habjan M, Maier R, Neuman BW, Ziebuhr J, Szretter KJ, Baker SC, Barchet W, Diamond MS, Siddell SG, Ludewig B, Thiel V (2011) Ribose 2′-O-methylation provides a molecular signature for the distinction of self and non-self mRNA dependent on the RNA sensor Mda5. Nat Immunol 12:137–143

Interferon Regulation of the Innate Response to Bacteria

Ali Abdul-Sater and Christian Schindler

Abstract

Characterization of potent antiviral response to Interferons (IFNs) led to identification of the JAK-STAT signaling cascade almost two decades ago. More recently, studies have begun to explore how IFNs participate in the innate response to bacterial infections. This includes the activation of classic antibacterial responses, like expression on inducible nitric oxidase (iNOS) and GTPases, as well as the induction of autophagy. Not surprisingly, studies on its anti-bacterial activity of IFNs have provided important new insights into IFN biology and JAK-STAT signaling.

Introduction

The potent antiviral activity of Interferons (IFNs) was first described by Isaacs and Lindenmann well over 50 years ago (Isaacs and Lindenmann 1957). Subsequent purification and cloning revealed that these four-helix bundle cytokines could be divided into two major groups, the type I and type II IFNs (reviewed in Uze et al. 2007). Type I IFNs (IFN-Is), the larger and more pleiotropic group included both fibroblast (a.k.a. – IFN-β) and leukocyte (a.k.a. – IFN-α's) IFNs, whereas type II IFN was represented by a single member, immune IFN (a.k.a. – IFN-γ). Consistent with these structural differences, type I and II IFNs were subsequently found to bind to distinct receptors and activate unique, but partially overlapping programs of

A. Abdul-Sater
Department of Microbiology and Immunology, Columbia University, HHSC 1314, 701 West 168th Street, New York, NY 10032, USA

C. Schindler (✉)
Departments of Microbiology & Immunology and Medicine, Columbia University, HHSC 1504, 701 West 168th Street, New York, NY 10032, USA
e-mail: cws4@columbia.edu

Th. Decker and M. Müller (eds.), *Jak-Stat Signaling: From Basics to Disease*,
DOI 10.1007/978-3-7091-0891-8_22,

signal transduction and gene expression (reviewed in Schindler and Plumlee 2008). A new family of type III IFNs (a.k.a. – IFN-λs or IL-28/29) was identified much more recently and found to bind a distinct epithelially expressed receptor (Kotenko et al. 2003; Sheppard et al. 2003). Although the antiviral activity mediated by each of these IFN classes was critical in their characterization, as well as their receptors and downstream signaling components, more recent studies have begun to explore the ability of IFNs to regulate both the innate response to bacteria and the subsequent transition to adaptive immunity (Douville and Hiscott 2010; Monroe et al. 2010; Trinchieri 2010). Intriguingly, in contrast to their essentially universal antiviral activity, the relationship between IFNs and an effective innate response to a bacterial infection is more complicated. In general, type II IFN has been found to potently activate the antibacterial program of phagocytes, especially towards bacteria that have evolved the capacity to replicate within cells (e.g. *Chlamydiae*, *Fransciella tularensis*, *Mycobacterium tuberculosis*, *Listeria monocytogenes* and *Legionella pneumophila*; (Monroe et al. 2010; Trinchieri 2010). Yet, the antibacterial response directed by type I IFNs is far more complicated, likely owing to the biphasic nature of IFN-I response. Although IFN-Is rapidly induce many important inflammatory genes, prolonged IFN-I stimulation has been associated with anti-inflammatory, anti-proliferative and pro-apoptotic activities, which a number of bacterial pathogens have learned to exploit (Stockinger and Decker 2008; Trinchieri 2010); see also section on *Staphylococcus* and *Streptococcus*). More recently, IFN-Is have also been shown to contribute to the innate response to a number of extracellular bacteria (e.g. Streptococci and *Staphylococcus aureus* [Monroe et al. 2010; Trinchieri 2010]). The role epithelially active IFN-IIIs play in the innate response to bacterial pathogens is a question many groups are now beginning to explore.

IFN-I Mediated Antiviral Activity and the JAK-STAT Pathway

Initial biochemical and molecular analysis of IFN response was almost exclusively focused on its potent antiviral activity (Pestka et al. 2004). However, with the development of pure IFN preparations it became possible to investigate the biology of IFN response. This included the important discovery that IFNs were able to directly stimulate the expression of specific target genes (Friedman et al. 1984; Larner et al. 1984). Characterization of IFN-I and IFN-II mediated gene expression led to the identification of the JAK-STAT signaling pathway (Fig. 1), where JAKs (JAunus Kinases) are IFN receptor associated kinases, and STATs (Signal Transducers and Activators of Transcription) are the transcription factors they activate (Schindler and Plumlee 2008). Underscoring their critical antiviral activity, viruses have evolved sophisticated strategies to subvert all aspects of IFN response (reviewed in Versteeg and Garcia-Sastre 2010). Similarly, genetic studies have underscored the important role IFN-γ plays in the response to intracellular bacterial pathogens (Zhang et al. 2008). Yet, the strategies by which bacteria subvert IFN response remain largely unexplored.

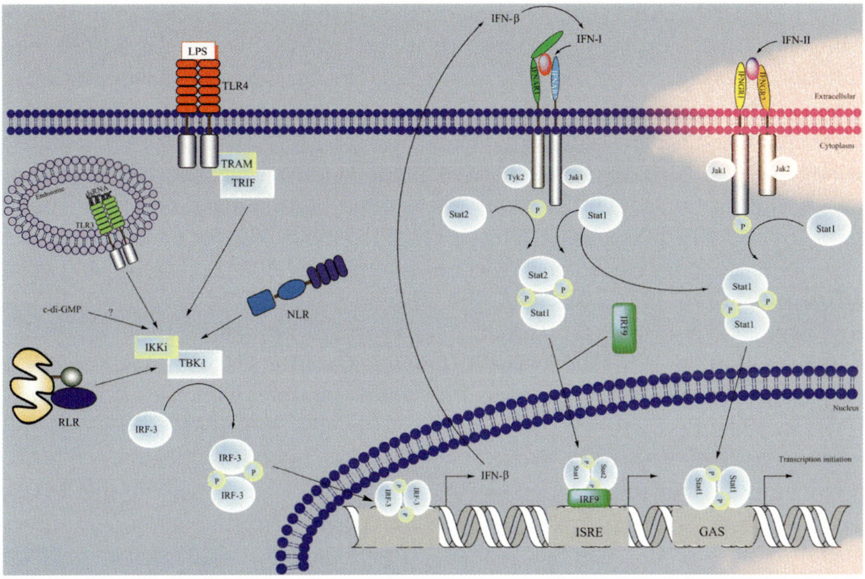

Fig. 1 The IFN-I and IFN-γ signaling paradigms. Upon binding their corresponding PAMPs, both transmembrane spanning (e.g. TLR3 and TLR4) and cytoplasmically located PRRs (e.g. RLRs, NLRs, and unknown receptors) direct the TBK1 (or IKKi/IKKε) dependent activation of IRF3 (or IRF5), which drives the expression of IFN-β. Secreted IFN-β binds to the dimeric IFN-α receptor (IFNAR1 and IFNAR2), directing activation of two receptor associated JAKs (Jak1 and Tyk2). Once activated, the JAKs sequentially phosphorylate receptor tyrosine and then of Stat1 and Stat2, which are recruited to the receptor in an SH2 domain dependent manner. Phosphorylated Stat1 and Stat2 heterodimerize and associate with IRF9, whereupon they translocate to the nucleus, bind to ISREs and drive the expression of target genes. One important target gene, IRF7 becomes activated by TBK1 (or IKKi/IKKε) to drive the expression of IFN-α's. Phosphorylated Stat1 can also from active homodimers, which translocate to the nucleus, directly bind GAS elements and drive the expression of a distinct set of genes. IFN-IIIs signal bind to a distinct receptor (IL-28R and IL-10R2), activating the same JAKs and STATs and target genes as IFN-Is (not shown). In contrast, IFN-γ binds to a unique dimeric receptor (i.e. IFNGR1 and IFNGR2) that promotes the activation of two distinct JAKs (i.e. Jak1 and Jak2). These JAKs phosphorylate a single IFNGR1 tyrosine, which directs the SH2 domain dependent recruitment and activation of Stat1. Phosphorylated Stat1 forms homodimers, as above, which translocate to the nucleus, bind GAS elements and drive target expression. However, the kinetics of Stat1 activation is prolonged in IFN-γ treated cells

IFN-II (i.e. IFN-γ), which is expressed by a limited number of immune cells, mediates its response through two receptor chains that are expressed by most cell types (Fig. 1; reviewed in Schindler and Plumlee 2008). Upon stimulation, receptor associated Janus Kinases (i.e. Jak1 and Jak2) sequentially activate each other and then tyrosine 440 of the α-receptor chain. Phosphotyrosine 440, in the context of 3–4 carboxy terminal amino acids (a.k.a. a tyrosine motif), directs the SH2 dependent recruitment of Stat1 to the receptor complex, where it is phosphorylated, in a JAK dependent manner on tyrosine 701. Phosphorylated Stat1 forms an active homodimer that translocates to the nucleus and binds to a member of the GAS (Gamma Activation

Site; TTTCCNGGAAA) family of enhancers (reviewed in Decker et al. 1997), culminating in the expression of target genes. Stat1/GAS driven genes include transcription factors (e.g. *IRF1 and CIITA*), iNOS, GTPases, chemokines and cell surface immuno-regulatory molecules (Decker et al. 1997; Ehrt et al. 2001; Gil et al. 2001).

In contrast, members of the large family of type I IFNs, which are expressed by most cell types, signal through the two chains of the IFN-α receptor (IFNAR; Fig. 1; reviewed in Schindler et al. 2007; Uze et al. 2007). Intriguingly, the expression of IFN-Is is controlled through an autocrine loop, which is initiated by a divergent set of pattern recognition receptors (PRRs) and culminates in the sequential expression of immediate IFN-Is (i.e. IFN-β & IFN-α_4) and subsequently the delayed IFN-Is (i.e. all other IFN-α's; Fig. 1; [Hiscott 2007; Hiscott et al. 2006; Marie et al. 1998]). Virtually all of the IFN-I inducing PRR/microbial sensor systems direct the TBK1 (TANK Binding Kinase 1) dependent activation/phosphorylation of IRF3 (Interferon Regulatory Factor 3), which along with NF-κB and AP-1 drive IFN-β expression (Seth et al. 2006; Wang et al. 2007). Secreted IFN-β binds to IFNAR, promoting the activation of two receptor associated JAKs (i.e. Jak1 and Tyk2). These tyrosine kinases phosphorylate specific receptor tyrosine motifs, and subsequently Stat1 (on Y^{701}) and Stat2 (on Y^{689}; Improta et al. 1994; Shuai et al. 1993; Zhao et al. 2008). Phosphorylated Stat1 and Stat2 form active heterodimers, which associate with IRF9 to form ISGF3 (Interferon Stimulated Gene Factor 3). ISGF3 translocates to the nucleus, where it directs the expression of genes featuring ISRE (IFN Stimulated Response Element; AGTTTN$_3$TTTCC) enhancers (Kessler et al. 1988). These genes include transcription factors (e.g. IRF1 and IRF7), many PRR/sensor components, some GTPases, as well as a growing number of well-characterized antiviral genes (Liu et al. 2011; Sadler and Williams 2008). IFN-Is also transduce signals through Stat1 homodimers, especially in macrophages, resulting in the expression many GAS-driven genes (see above; [Pine et al. 1994]).

IFN-IIIs (i.e. IFN-λ1, IFN-λ2 and IFN-λ3; a.k.a., IL-29, IL-28a and IL-28b) are largely active on mucosal epithelium, a specificity that has been attributed to tissue restricted receptor expression (Ank et al. 2008; Dumoutier et al. 2004). Even though IFN-IIIs transduce their signals through a distinct receptor, they activate the same intracellular signaling pathways as IFN-Is, culminating in ISGF3 and Stat1 dependent gene expression (Donnelly and Kotenko 2010). Analogous to studies of type I and II IFNs, these signaling pathways have been extensively validated through the analysis of numerous knockout mice (i.e. receptors, JAK, STAT and IRF knockouts; reviewed in Donnelly and Kotenko 2010; Schindler and Plumlee 2008; Uze et al. 2007).

Chlamydia

Chlamydia species, important agents of sexually transmitted and pulmonary diseases, were the first non-viral pathogens found to be inhibited by type I IFN (IFN-I; Sueltenfuss and Pollard 1963). These early studies, employing partially purified IFNs, determined that IFN-Is suppress chlamydial replication and maturation in a variety of human and murine cell types (Hanna et al. 1966; Kazar et al. 1971). These initial findings were subsequently confirmed and extended with purified and recombinant IFNs. In addition, cells infected with *Chlamydia* were found to secrete autocrine/paracrine IFN-Is, culminating in the expression of IFN target genes (Nagarajan et al. 2005). A number of IFN-I target genes, which are also synergistically activated by other inflammatory cytokines (e.g. TNF-α and IL-1), have been shown to suppress *Chlamydia* growth. These includes iNOS (NOS2) and the tryptophan-degrading enzyme (Indoleamine 2,3-dioxygenase; IDO), which deprives *Chlamydia* of an essential metabolite (Carlin et al. 1989; Carlin and Weller 1995; Devitt et al. 1996; Ishihara et al. 2005; Rothfuchs et al. 2001; Shemer-Avni et al. 1989). However, in vivo systemic IFN-Is appear to exert a largely immuno-suppressive activity, leading to an exacerbation of chlamydial infections (Nagarajan et al. 2005; Qiu et al. 2008).

Studies exploring the mechanism by which *Chlamydia* induce IFN-I expression have yielded conflicting results. Some studies have highlighted an important role for TLR3 and TRIF (Derbigny et al. 2007, 2010), whereas others implicate a MyD88 dependent pathway (Nagarajan et al. 2005; Naiki et al. 2005). There is however agreement that the TBK1-IRF3 axis is required for *Chlamydia* dependent IFN-I expression (Derbigny et al. 2007; Nagarajan et al. 2005; Prantner et al. 2010). More recent studies have explored the role inflammasome activation may play in contributing to an effective innate response towards this important human pathogen (Abdul-Sater et al. 2009; Abdul-Sater et al. 2010).

Staphylococcus and *Streptococcus*

Although neutrophils play a critical role in the host response to gram-positive, phagocytic bacteria, like *Staphylococcus aureus* and Group A/B Streptococcus, several studies have highlighted a role for IFNs (reviewed in Monroe et al. 2010; Trinchieri 2010). More limited studies on *Staphylococcus aureus* reveal that both type I and II IFNs are able to render hosts more susceptible to lung infections (Martin et al. 2009; Shahangian et al. 2009; Sun and Metzger 2008). This effect is particularly pronounced in mice recovering from an antecedent influenza pneumonia. Mechanistic studies suggest that virally induced IFN-γ suppresses the expression of MARCO, a phagocytic scavenger receptor, on resident alveolar macrophages, whereas virally induced IFN-Is suppress the expression of CXCL1 and CXCL2, two important neutrophil chemo-attractants (Martin et al. 2009; Shahangian et al. 2009; Sun and Metzger 2008). Other studies have highlighted the ability of IFN-Is to induce the expression of IL-10 and IL-27, two immuno-suppressive cytokines that may also contribute to this phenotype (Chang et al. 2007; Shinohara et al. 2008). However, there is also evidence that *Staphylococcus aureus* directly induces IFN-I expression in host tissues, rendering them more susceptible to bacterial infection, even absent an antecedent influenza infection (Martin et al. 2009). The potent immuno-suppressive activity associated with prolonged IFN-I expression can be attributed to the growing list of IFN dependent negative regulators (Schindler et al. 2007; Schindler and Plumlee 2008).

In stark contrast to *Staphylococcus aureus*, IFN-Is play an important role in promoting an effective innate response to infection by Group A and B Streptococcus (Mancuso et al. 2007), whereas IFN-γ plays a more important role in the subsequent adaptive response (Chaussee et al. 2011). Studies exploring the IFN-I response have largely focused on macrophages, ascribing the ability of these bacteria to induce IFN-Is in a TLR-independent pathway that senses endosomally accumulating foreign DNA (Charrel-Dennis et al. 2008; Mancuso et al. 2007; Parker et al. 2011). Although the specific DNA sensor remains controversial, several studies support a role for STING (STimulator of INterferon Genes; Ishikawa and Barber 2011), an endosomal IFN regulator, as well as TBK1 and IRF3 (Gratz et al. 2011; Ishikawa and Barber 2011). Intriguingly, studies on conventional DCs have attributed IFN-I induction to a TLR-dependent pathway that senses Streptococcal RNA and activates IRF5 (Gratz et al. 2011; Mancuso et al. 2009).

Mycobacterium tuberculosis

Mycobacterium tuberculosis (Mtb), an evolutionarily ancient scourge, is a potent stimulator of the immune system, yet persists in hosts, owing to its abilities to: take up residence macrophages; subvert their activity; and to become dormant (Gutierrez et al. 2005; Harris et al. 2009). Even though IFN-γ secreting Th1 cells activate an effective antibacterial program within macrophages, dormant Mtb persist in many hosts. Consistent with the important role that IFN-γ secreting Th1 cells play in the immune response to Mtb, analysis of patients with a predisposition to mycobacterial infections has identified several distinct inborn errors in IFN-γ response (Zhang et al. 2008). These human pedigrees feature mutations in genes either regulating IFN-γ production (i.e. the IL-12 receptor and NEMO) or IFN-γ response (i.e. the IFN-γ receptor and Stat1). Careful studies have shown that IFN-γ suppresses Mtb growth through both iNOS dependent and independent mechanisms. In sum, these enable macrophages to overcome the blockade in phagosome maturation that is mediated by bacterial products. This includes the IFN-γ dependent induction of Irgm1, a p47 GTPase, LC3 (microtubule-associated protein 1 light chain 3) and an associated autophagocytic response (Harris et al. 2009; Tiwari et al. 2009).

Although induced during an infection, the role type I IFNs play during the innate response to Mtb remains more controversial (Giacomini et al. 2001; Remoli et al. 2002; Weiden et al. 2000). Several studies, exploiting either ectopic IFN-I treatment or IFN-I unresponsive mice suggest that IFN-Is are beneficial to the host during a Mtb infection (Denis 1991; Giosue et al. 1998; Kuchtey et al. 2006). Consistent with this, IFN-Is induce the expression of two genes associated with Mtb suppression, iNOS (a.k.a. NOS2) and Lpr1 (Kuchtey et al. 2006; Pan et al. 2005). In addition, IFN-I signaling appears to be suppressed in Mtb infected cells (Prabhakar et al. 2005). Alternatively however, IFN-I induction has been associated with a hypervirulent Mtb strain (Manca et al. 2001). Moreover, IFN-I pretreatment was found to increase pulmonary Mtb loads and mortality (Manca et al. 2001). Similarly, IFN-I pretreated monocytes were reported to exhibit enhanced *M. bovis* BCG growth (Bouchonnet et al. 2002). More recently, a potent IFN-I stimulating agent (i.e. Poly-ICLC) was found to exacerbate Mtb growth in wild type, but not IFNAR mutant mice through the recruitment of a permissive population of myelocytes (Antonelli et al. 2010). However, IFN-I levels were considerably higher than during a native MTB infection. In another high profile study, active Mtb infection was shown to correlate with an IFN-I transcriptional profile within neutrophils, but this study was not able to distinguish between cause and effect (Berry et al. 2010; Prabhakar et al. 2005). Future studies, rigorously controlling for the biphasic nature response to IFN-Is (i.e. the inflammatory response associated with acute treatment vs. the anti-inflammatory activity associated with prolonged stimulation; see above) may provide important insight into this controversy.

Another active area of investigation has been the mechanism by which Mtb stimulates IFN-I expression. Consistent with studies from other intracellular bacteria, the TLR-MyD88 pathway does not appear to play an important role in IFN-I induction (Shi et al. 2005; Stanley et al. 2007). Likewise, IFN-I expression was found to be dependent on the Mtb ESX secretion system, which is critical for bacterial virulence (Stanley et al. 2007). More recently, genetic studies have provided evidence that Mtb stimulated IFN-β expression is directed by a Nod2-Rip2-Tyk2 dependent activation of IRF5, with an uncharacteristically modest contribution from IRF3 (Pandey et al. 2009). These studies have also identified an N-glycolated mural dipeptide (N-glycolyc MDP) as one critical PAMP initiating this response (Pandey et al. 2009).

Francisella tularensis

Francisella tularensis is an aggressive facultative intracellular bacterium that has been shown to induce both type I and II IFNs (reviewed in Sjostedt 2003). Since as few as ten inhaled organisms can cause a lethal infection, most studies have been carried out on attenuated vaccine strains (*F. tularensis* subspecies *novicida or holarctica).* Analogous to Mtb, IFN-γ is an important component of an effective response towards this virulent pathogen. It appears to suppress intracellular *F. tularensis* growth by stimulating both reactive oxygen species (ROS) and reactive nitrogen species (RNS) production, as well as by IDO dependent depletion of tryptophan stores (Chu et al. 2011; Lindgren et al. 2004; Peng and Monack 2010). In contrast, the role of IFN-Is, whose expression is robustly induced when *F. tularensis* escapes from the phagosome, is more complicated. On a cellular level, autocrine/paracrine IFN-Is promote the pyroptosis of infected macrophages, removing the replicative niche (Henry et al. 2007). However, systemically IFN-Is appear to antagonize the normal accumulation and activity of IL-17A+ γδ T cells and neutrophil (Henry et al. 2010; Navarini et al. 2006). Notably, IFNAR1 knockout mice are less susceptible to infection with *F. tularensis*, suggesting the overall effect of IFN-Is is harmful to the host.

Genetic studies exploring the mechanism by which cytosolic *F. tularensis* stimulates IFN-I (i.e., IFN-β) expression have highlighted an important role for IRF3, but not for RIG-I, MDA5, Nod1 or Nod2 (Henry et al. 2007). A potential role for TLRs remains more controversial. A study with *F. tularensis ssp novicida* infected MyD88/Trif double knockout bone marrow macrophages convincingly excluded an important role for TLRs in IFN-β induction (Henry et al. 2007). Consistent with this, *F. tularensis ssp novicida* stimulated IFN-β expression was dependent on STING (STimulator of INterferon Genes), a component of a partially characterized cytosolic DNA sensor (Ishikawa and Barber 2011; Jones et al. 2010). However, a second study exploiting knockout peritoneal macrophages implicated TLR2 in the ability of live vaccine strain (LVS) *F. tularensis* to induce IFN-β expression (Cole et al. 2007). More intriguing has been recent evidence that the cytosolic DNA sensor AIM2 (Absent in Melanoma 2) and ASC (apoptosis-associated speck-like protein containing a CARD; a.k.a. PYCARD) directs the *F. tularensis* dependent activation of the caspase1 inflammasome, which culminates in IL-1β and IL-18 secretion (Fernandes-Alnemri et al. 2010; Jones et al. 2010). This response is enhanced but not dependent on autocrine IFN-Is. Moreover, AIM2 knockout mice mounted a defective response to *F. tularensis* infection, suggesting the inflammasome plays in important role in the host response to this pathogen (Fernandes-Alnemri et al. 2010; Jones et al. 2010).

Listeria monocytogenes

Listeria monocytogenes is a stubbornly persistent food bourne pathogen that causes severe disease in immunocompromised hosts through its capacity to replicate within macrophages (reviewed in Stockinger and Decker 2008; Trinchieri 2010). Of the virulence factors that enable *L. monocytogenes* to infect phagocytes, Listeriolysin O (LLO), a pore-forming cytolysin that facilitates escape from the phagosome, has been characterized most extensively. Similar to other facultative bacteria, *L. monocytogenes* is sensitive to the robust inflammatory response stimulated by IFN-γ. This has been attributed to the Stat1 dependent expression of specific inflammatory genes (Varinou et al. 2003). Among these target genes, two members of the p47 family of GTPases (i.e. LRG-47 and IGTP), but not iNOS (NOS2), play a particularly important role in suppressing *L. monocytogenes* growth (Stockinger and Decker 2008; Taylor 2007). Intriguingly, NO production may actually impede an effective host response towards this pathogen (Bogdan et al. 2000).

Like many intracellular microbes, *L. monocytogenes* has been shown to induce the expression of IFN-Is. However, in contrast to IFN-γ, this autocrine IFN-I undermines the immune response towards this pathogen. Several mechanisms have been identified by which IFN-Is suppress the immune response to *L. monocytogenes*. First, IFN-Is have been shown to increase the susceptibility of macrophages and T-cells to apoptotic death. Second, IFN-Is appear to antagonize both the production and response to IFN-γ (Rayamajhi et al. 2010). There is also evidence that IFN-Is limit T-cell dependent secretion of IL-17 during infection (Guo et al. 2008; Henry et al. 2010). Finally, IFN-Is subvert the innate response by limiting the specific recruitment, activity and lifespan of critical populations of myeloid cells (Jia et al. 2009; Navarini et al. 2006; Shahangian et al. 2009).

The mechanism by which *L. monocytogenes* induces IFN-I expression remains an active area of investigation. Although genetic studies have excluded significant roles for both the TLR and Nod (i.e. Nod1 and Nod2) families of PRRs, the response is dependent on IRF3, especially in macrophages, which are the most important source of IFN-Is during a *L. monocytogenes* infection (O'Connell et al. 2004; Stockinger et al. 2004). As with other bacterial pathogens, a number of studies have suggested that foreign nucleic acids may initiate the IFN-I response (i.e. induce IFN-β expression; Fig. 1), but there are conflicting results over whether this is through a known or novel sensor system (Pollpeter et al. 2011; Stetson and Medzhitov 2006). Intriguingly, a recent unbiased genetic screen has identified MdrM, a multi drug resistance transporter encoded by *L. monocytogenes*, as contributing to the IFN-I response (Crimmins et al. 2008). Moreover, a search for the critical substrate transported through this channel has identified 3′,5′-cyclic diadenylate (c-diAMP) as a new and effective IFN-I stimulating PAMP (Woodward et al. 2010; A. Abdul-Sater, unpublished observation). The potential relationship between c-diAMP and the structurally analogous c-diGMP (3′,5′-cyclic diguanylate, see below), also produced by *L. monocytogenes*, is an active area of investigation.

Legionella pneumophila

Legionella pneumophila, a facultative intracellular bacterium, is the causative agent of Legionnaires' disease. Prevalent in the environment (*i.e., water aerosols*), this bacterium continues to be an important cause of nosocomial and community-acquired pneumonias, especially in the elderly and immuno-compromised host (Marston et al. 1997; Mykietiuk et al. 2005). Like other facultative intracellular pathogens, *L. pneumophila* avoids being targeted to the lysosome through the formation of a specialized replicative compartment (*reviewed in* Franco et al. 2009; Isberg et al. 2009). Its ability to subvert normal vesicular trafficking within phagocytes is dependent on the Icm/Dot type IV secretion system, which serves to inject > 140 "effector" proteins into the host cell. Like most other intracellular microbes, IFN-γ effectively suppresses *L. pneumophila* growth in wild type, but not Stat1 deficient macrophages (Plumlee et al. 2009). However, the IFN-γ stimulated target genes critical for suppressing *L. pneumophila* growth have not yet been identified.

Intriguingly, the ability of IFN-Is to antagonize *L. pneumophila* growth is not affected by the loss of Stat1, the canonical component of both IFN-I stimulated transcription factors (Fig. 1; Plumlee et al. 2009). Likewise, IFN-Is effectively suppress *L. pneumophila* replication in Stat2 knockout macrophages, but this response is lost in Stat1/Stat2 double knockout macrophages (Plumlee, unpublished observation). These observations suggest that Stat1 and Stat2 are functionally redundant in their ability to suppress *L. pneumophila* growth. While responses mediated solely by Stat1 appear similar to those stimulated by IFN-γ (Fig. 1), the responses directed by Stat2 that are independent of Stat1 (i.e. "Stat2 signaling") are not well characterized (Hahm et al. 2005; Hofer et al. 2010; Perry et al. 2011; Sarkis et al. 2006). Recent additional characterization of the "independent Stat2 pathway" has revealed that it entails the formation of Stat2-Stat2-IRF9 complex, which directs the expression of ISRE-driven genes, albeit with considerably delayed kinetics (A. Abdul-Sater, unpublished observation). It is intriguing to speculate that this pathway evolved as a defense towards pathogens that target Stat1 for degradation (Versteeg and Garcia-Sastre 2010).

Both *L. pneumophila's* sensitivity towards IFN-Is and its ability to stimulate IFN-I production underscores the important role the IFN-I autocrine/paracrine loop plays in protecting infected macrophages (Coers et al. 2007; Plumlee et al. 2009; Schiavoni et al. 2004). As is the case with other intracellular bacterial pathogens, neither the PAMP nor PRR(s) that direct *L. pneumophila* dependent IFN-I secretion have been characterized. Consistent with other bacteria, recent studies have however revealed that TBK1 and IRF3 are critical in IFN-β induction (Plumlee et al. 2009). The dearth of progress in the IFN-I response is contrasted by elegant studies highlighting the role *L. pneumophila* flagellin plays in directing a Naip5–IPAF (NLRC4) dependent activation of the caspase1 inflammasome, culminating in IL-1β secretion and pyroptosis (Lightfield et al. 2008; Molofsky et al. 2006; Vance 2010). Intriguingly, several recent studies have implicated *L. pneumophila*, DNA, RNA and 3′,5′-cyclic diguanylate (c-diGMP) as critical IFN-I stimulating PAMPs (McWhirter et al. 2009; Monroe et al. 2009; Stetson and Medzhitov 2006; Abdul-Sater, unpublished observation). Moreover, "foreign" cytosolic nucleic acids appear to stimulate both IFN-I expression and inflammasome activation, where secreted IFN-Is appear to facilitate inflammasome activation (Fernandes-Alnemri et al. 2010; Goubau et al. 2010; Jones et al. 2010; Sander et al. 2011). It seems likely that characterizing this IFN-I response will provide important new insight into innate immunity.

Conclusion

The evolutionary contest between host and pathogen drove the development of an elaborate host defense system that features an adaptive immune response in addition to the innate response found in more primitive eukaryotes. This entailed the development of leukocyte lineages, as well as cytokines, like Interferons, that regulate the activity of these immune cells. Initially characterized for their antiviral activity, the IFN family is increasingly recognized for its role in the host response to bacterial pathogens. Thus, IFN-γ, largely produced by lymphocytes, potently stimulates the antibacterial program of phagocytes, whereas IFN-Is direct more pleiotropic antibacterial responses, as well as the subsequent transition towards adaptive immunity. Consistent with this role, a growing number of intracellular PPRs that stimulate IFN-I production have been identified. Moreover, PRRs and components of the IFN-I response vary significantly between cell types. Although time will tell, new tools to specifically probe IFN-III response are likely to highlight an important role for these IFNs, especially at mucosal surfaces. In addition, studies on all three classes of IFNs are likely to reveal why this family of ligands is so large, as well as how, when and why exuberant IFN response contributes to chronic inflammatory disease.

Note added in Proof Some as the data cited as unpublished observation has now been published in: Abdul-Sater A, Grajkowski A, Erdjument-Bromage H, Plumlee C, Levi A, Schreiber MT, Lee C, Shuman H, Beaucage SL, Schindler C (2012) The overlapping host responses to bacterial cyclic dinucleotides. Microbes & Infection 14(2):188–197.

References

Abdul-Sater AA, Koo E, Hacker G, Ojcius DM (2009) Inflammasome-dependent caspase-1 activation in cervical epithelial cells stimulates growth of the intracellular pathogen *Chlamydia trachomatis*. J Biol Chem 284:26789–26796

Abdul-Sater AA, Said-Sadier N, Padilla EV, Ojcius DM (2010) Chlamydial infection of monocytes stimulates IL-1 beta secretion through activation of the NLRP3 inflammasome. Microb Infect 12:652–661

Ank N, Iversen MB, Bartholdy C, Staeheli P, Hartmann R, Jensen UB, Dagnaes-Hansen F, Thomsen AR, Chen Z, Haugen H et al (2008) An important role for type III interferon (IFN-lambda/IL-28) in TLR-induced antiviral activity. J Immunol 180:2474–2485

Antonelli LR, Gigliotti Rothfuchs A, Goncalves R, Roffe E, Cheever AW, Bafica A, Salazar AM, Feng CG, Sher A (2010) Intranasal Poly-IC treatment exacerbates tuberculosis in mice through the pulmonary recruitment of a pathogen-permissive monocyte/macrophage population. J Clin Invest 120:1674–1682

Berry MP, Graham CM, McNab FW, Xu Z, Bloch SA, Oni T, Wilkinson KA, Banchereau R, Skinner J, Wilkinson RJ et al (2010) An interferon-inducible neutrophil-driven blood transcriptional signature in human tuberculosis. Nature 466:973–977

Bogdan C, Rollinghoff M, Diefenbach A (2000) The role of nitric oxide in innate immunity. Immunol Rev 173:17–26

Bouchonnet F, Boechat N, Bonay M, Hance AJ (2002) Alpha/beta interferon impairs the ability of human macrophages to control growth of *Mycobacterium bovis* BCG. Infect Immun 70:3020–3025

Carlin JM, Weller JB (1995) Potentiation of interferon-mediated inhibition of *Chlamydia* infection by interleukin-1 in human macrophage cultures. Infect Immun 63:1870–1875

Carlin JM, Borden EC, Byrne GI (1989) Interferon-induced indoleamine 2,3-dioxygenase activity inhibits *Chlamydia psittaci* replication in human macrophages. J Interferon Res 9:329–337

Chang EY, Guo B, Doyle SE, Cheng G (2007) Cutting edge: involvement of the type I IFN production and signaling pathway in lipopolysaccharide-induced IL-10 production. J Immunol 178:6705–6709

Charrel-Dennis M, Latz E, Halmen KA, Trieu-Cuot P, Fitzgerald KA, Kasper DL, Golenbock DT (2008) TLR-independent type I interferon induction in response to an extracellular bacterial pathogen via intracellular recognition of its DNA. Cell Host Microbe 4:543–554

Chaussee MS, Sandbulte HR, Schuneman MJ, Depaula FP, Addengast LA, Schlenker EH, Huber VC (2011) Inactivated and live, attenuated influenza vaccines protect mice against influenza: *Streptococcus pyogenes* super-infections. Vaccine 29:3773–3781

Chu P, Rodriguez AR, Arulanandam BP, Klose KE (2011) Tryptophan prototrophy contributes to *Francisella tularensis* evasion of gamma interferon-mediated host defense. Infect Immun 79:2356–2361

Coers J, Vance RE, Fontana MF, Dietrich WF (2007) Restriction of *Legionella pneumophila* growth in macrophages requires the concerted action of cytokine and Naip5/Ipaf signalling pathways. Cell Microbiol 9:2344–2357

Cole LE, Shirey KA, Barry E, Santiago A, Rallabhandi P, Elkins KL, Puche AC, Michalek SM, Vogel SN (2007) Toll-like receptor 2-mediated signaling requirements for *Francisella tularensis* live vaccine strain infection of murine macrophages. Infect Immun 75:4127–4137

Crimmins GT, Herskovits AA, Rehder K, Sivick KE, Lauer P, Dubensky TW Jr, Portnoy DA (2008) *Listeria monocytogenes* multidrug resistance transporters activate a cytosolic surveillance pathway of innate immunity. Proc Natl Acad Sci USA 105:10191–10196

Decker T, Kovarik P, Meinke A (1997) GAS elements: a few nucleotides with a major impact on cytokine-induced gene expression. J Interferon Cytokine Res 17:121–134

Denis M (1991) Recombinant murine beta interferon enhances resistance of mice to systemic *Mycobacterium avium* infection. Infect Immun 59:1857–1859

Derbigny WA, Hong SC, Kerr MS, Temkit M, Johnson RM (2007) *Chlamydia muridarum* infection elicits a beta interferon response in murine oviduct epithelial cells dependent on interferon regulatory factor 3 and TRIF. Infect Immun 75:1280–1290

Derbigny WA, Johnson RM, Toomey KS, Ofner S, Jayarapu K (2010) The *Chlamydia muridarum*-induced IFN-beta response is TLR3-dependent in murine oviduct epithelial cells. J Immunol 185:6689–6697

Devitt A, Lund PA, Morris AG, Pearce JH (1996) Induction of alpha/beta interferon and dependent nitric oxide synthesis during *Chlamydia trachomatis* infection of McCoy cells in the absence of exogenous cytokine. Infect Immun 64:3951–3956

Donnelly RP, Kotenko SV (2010) Interferon-lambda: a new addition to an old family. J Interferon Cytokine Res 30:555–564

Douville RN, Hiscott J (2010) The interface between the innate interferon response and expression of host retroviral restriction factors. Cytokine 52:108–115

Dumoutier L, Tounsi A, Michiels T, Sommereyns C, Kotenko SV, Renauld JC (2004) Role of the interleukin (IL)-28 receptor tyrosine residues for antiviral and antiproliferative activity of IL-29/interferon-lambda 1: similarities with type I interferon signaling. J Biol Chem 279: 32269–32274

Ehrt S, Schnappinger D, Bekiranov S, Drenkow J, Shi S, Gingeras TR, Gaasterland T, Schoolnik G, Nathan C (2001) Reprogramming of the macrophage transcriptome in response to interferon-gamma and *Mycobacterium tuberculosis*: signaling roles of nitric oxide synthase-2 and phagocyte oxidase. J Exp Med 194:1123–1140

Fernandes-Alnemri T, Yu JW, Juliana C, Solorzano L, Kang S, Wu J, Datta P, McCormick M, Huang L, McDermott E et al (2010) The AIM2 inflammasome is critical for innate immunity to *Francisella tularensis*. Nat Immunol 11:385–393

Franco IS, Shuman HA, Charpentier X (2009) The perplexing functions and surprising origins of *Legionella pneumophila* type IV secretion effectors. Cell Microbiol 11:1435–1443

Friedman RL, Manly SP, McMahon M, Kerr IM, Stark GR (1984) Transcriptional and posttranscriptional regulation of interferon-induced gene expression in human cells. Cell 38:745–755

Giacomini E, Iona E, Ferroni L, Miettinen M, Fattorini L, Orefici G, Julkunen I, Coccia EM (2001) Infection of human macrophages and dendritic cells with *Mycobacterium tuberculosis* induces a differential cytokine gene expression that modulates T cell response. J Immunol 166: 7033–7041

Gil MP, Bohn E, O'Guin AK, Ramana CV, Levine B, Stark GR, Virgin HW, Schreiber RD (2001) Biologic consequences of Stat1-independent IFN signaling. Proc Natl Acad Sci USA 98: 6680–6685

Giosue S, Casarini M, Alemanno L, Galluccio G, Mattia P, Pedicelli G, Rebek L, Bisetti A, Ameglio F (1998) Effects of aerosolized interferon-alpha in patients with pulmonary tuberculosis. Am J Respir Crit Care Med 158:1156–1162

Goubau D, Rehwinkel J, Reis e Sousa C (2010) PYHIN proteins: center stage in DNA sensing. Nat Immunol 11:984–986

Gratz N, Hartweger H, Matt U, Kratochvill F, Janos M, Sigel S, Drobits B, Li X-D, Knapp S, Kovarik P (2011) Type I interferon production induced by *Streptococcus pyogenes*-derived nucleic acids is required for host protection. PLoS Pathog 7(5):e1001345

Guo B, Chang EY, Cheng G (2008) The type I IFN induction pathway constrains Th17-mediated autoimmune inflammation in mice. J Clin Invest 118:1680–1690

Gutierrez MC, Brisse S, Brosch R, Fabre M, Omais B, Marmiesse M, Supply P, Vincent V (2005) Ancient origin and gene mosaicism of the progenitor of *Mycobacterium tuberculosis*. PLoS Pathog 1:e5

Hahm B, Trifilo MJ, Zuniga EI, Oldstone MB (2005) Viruses evade the immune system through type I interferon-mediated STAT2-dependent, but STAT1-independent, signaling. Immunity 22:247–257

Hanna L, Merigan TC, Jawetz E (1966) Inhibition of TRIC agents by virus-induced interferon. Proc Soc Exp Biol Med 122:417–421

Harris J, Hope JC, Lavelle EC (2009) Autophagy and the immune response to TB. Transbound Emerg Dis 56:248–254

Henry T, Brotcke A, Weiss DS, Thompson LJ, Monack DM (2007) Type I interferon signaling is required for activation of the inflammasome during *Francisella* infection. J Exp Med 204: 987–994

Henry T, Kirimanjeswara GS, Ruby T, Jones JW, Peng K, Perret M, Ho L, Sauer JD, Iwakura Y, Metzger DW et al (2010) Type I IFN signaling constrains IL-17A/F secretion by gammadelta T cells during bacterial infections. J Immunol 184:3755–3767

Hiscott J (2007) Triggering the innate antiviral response through IRF-3 activation. J Biol Chem 282:15325–15329

Hiscott J, Lin R, Nakhaei P, Paz S (2006) MasterCARD: a priceless link to innate immunity. Trends Mol Med 12:53–56

Hofer MJ, Li W, Lim SL, Campbell IL (2010) The type I interferon-alpha mediates a more severe neurological disease in the absence of the canonical signaling molecule interferon regulatory factor 9. J Neurosci 30:1149–1157

Improta T, Schindler C, Horvath CM, Kerr IM, Stark GR, Darnell JE Jr (1994) Transcription factor ISGF-3 formation requires phosphorylated Stat91 protein, but Stat113 protein is phosphorylated independently of Stat91 protein. Proc Natl Acad Sci USA 91:4776–4780

Isaacs A, Lindenmann J (1957) Virus interference. I. The interferon. Proc R Soc Lond B Biol Sci 147:258–267

Isberg RR, O'Connor TJ, Heidtman M (2009) The *Legionella pneumophila* replication vacuole: making a cosy niche inside host cells. Nat Rev Microbiol 7:13–24

Ishihara T, Aga M, Hino K, Ushio C, Taniguchi M, Iwaki K, Ikeda M, Kurimoto M (2005) Inhibition of *Chlamydia trachomatis* growth by human interferon-alpha: mechanisms and synergistic effect with interferon-gamma and tumor necrosis factor-alpha. Biomed Res 26: 179–185

Ishikawa H, Barber GN (2011) The STING pathway and regulation of innate immune signaling in response to DNA pathogens. Cell Mol Life Sci 68:1157–1165

Jia T, Leiner I, Dorothee G, Brandl K, Pamer EG (2009) MyD88 and type I interferon receptor-mediated chemokine induction and monocyte recruitment during *Listeria monocytogenes* infection. J Immunol 183:1271–1278

Jones JW, Kayagaki N, Broz P, Henry T, Newton K, O'Rourke K, Chan S, Dong J, Qu Y, Roose-Girma M et al (2010) Absent in melanoma 2 is required for innate immune recognition of *Francisella tularensis*. Proc Natl Acad Sci USA 107:9771–9776

Kazar J, Gillmore JD, Gordon FB (1971) Effect of interferon and interferon inducers on infections with a nonviral intracellular microorganism, *Chlamydia trachomatis*. Infect Immun 3:825–832

Kessler DS, Levy DE, Darnell JE Jr (1988) Two interferon-induced nuclear factors bind a single promoter element in interferon-stimulated genes. Proc Natl Acad Sci USA 85: 8521–8525

Kotenko SV, Gallagher G, Baurin VV, Lewis-Antes A, Shen M, Shah NK, Langer JA, Sheikh F, Dickensheets H, Donnelly RP (2003) IFN-lambdas mediate antiviral protection through a distinct class II cytokine receptor complex. Nat Immunol 4:69–77

Kuchtey J, Fulton SA, Reba SM, Harding CV, Boom WH (2006) Interferon-alphabeta mediates partial control of early pulmonary *Mycobacterium bovis* bacillus Calmette-Guerin infection. Immunology 118:39–49

Larner AC, Jonak G, Cheng YS, Korant B, Knight E, Darnell JE Jr (1984) Transcriptional induction of two genes in human cells by beta interferon. Proc Natl Acad Sci USA 81:6733–6737

Lightfield KL, Persson J, Brubaker SW, Witte CE, von Moltke J, Dunipace EA, Henry T, Sun YH, Cado D, Dietrich WF et al (2008) Critical function for Naip5 in inflammasome activation by a conserved carboxy-terminal domain of flagellin. Nat Immunol 9:1171–1178

Lindgren H, Stenmark S, Chen W, Tarnvik A, Sjostedt A (2004) Distinct roles of reactive nitrogen and oxygen species to control infection with the facultative intracellular bacterium *Francisella tularensis*. Infect Immun 72:7172–7182

Liu SY, Sanchez DJ, Cheng G (2011) New developments in the induction and antiviral effectors of type I interferon. Curr Opin Immunol 23:57–64

Manca C, Tsenova L, Bergtold A, Freeman S, Tovey M, Musser JM, Barry CE III, Freedman VH, Kaplan G (2001) Virulence of a *Mycobacterium tuberculosis* clinical isolate in mice is determined by failure to induce Th1 type immunity and is associated with induction of IFN-alpha/beta. Proc Natl Acad Sci USA 98:5752–5757

Mancuso G, Midiri A, Biondo C, Beninati C, Zummo S, Galbo R, Tomasello F, Gambuzza M, Macri G, Ruggeri A et al (2007) Type I IFN signaling is crucial for host resistance against different species of pathogenic bacteria. J Immunol 178:3126–3133

Mancuso G, Gambuzza M, Midiri A, Biondo C, Papasergi S, Akira S, Teti G, Beninati C (2009) Bacterial recognition by TLR7 in the lysosomes of conventional dendritic cells. Nat Immunol 10:587–594

Marie I, Durbin JE, Levy DE (1998) Differential viral induction of distinct alpha-interferon genes by positive feedback through interferon regulatory factor-7. EMBO J 17:6660–6669

Marston BJ, Plouffe JF, File TM Jr, Hackman BA, Salstrom SJ, Lipman HB, Kolczak MS, Breiman RF (1997) Incidence of community-acquired pneumonia requiring hospitalization. Results of a population-based active surveillance study in Ohio. The Community-Based Pneumonia Incidence Study Group. Arch Intern Med 157:1709–1718

Martin FJ, Gomez MI, Wetzel DM, Memmi G, O'Seaghdha M, Soong G, Schindler C, Prince A (2009) *Staphylococcus aureus* activates type I IFN signaling in mice and humans through the Xr repeated sequences of protein A. J Clin Invest 119:1931–1939

McWhirter SM, Barbalat R, Monroe KM, Fontana MF, Hyodo M, Joncker NT, Ishii KJ, Akira S, Colonna M, Chen ZJ et al (2009) A host type I interferon response is induced by cytosolic sensing of the bacterial second messenger cyclic-di-GMP. J Exp Med 206:1899–1911

Molofsky AB, Byrne BG, Whitfield NN, Madigan CA, Fuse ET, Tateda K, Swanson MS (2006) Cytosolic recognition of flagellin by mouse macrophages restricts *Legionella pneumophila* infection. J Exp Med 203:1093–1104

Monroe KM, McWhirter SM, Vance RE (2009) Identification of host cytosolic sensors and bacterial factors regulating the type I interferon response to *Legionella pneumophila*. PLoS Pathog 5:e1000665

Monroe KM, McWhirter SM, Vance RE (2010) Induction of type I interferons by bacteria. Cell Microbiol 12:881–890

Mykietiuk A, Carratala J, Fernandez-Sabe N, Dorca J, Verdaguer R, Manresa F, Gudiol F (2005) Clinical outcomes for hospitalized patients with *Legionella pneumonia* in the antigenuria era: the influence of levofloxacin therapy. Clin Infect Dis 40:794–799

Nagarajan UM, Ojcius DM, Stahl L, Rank RG, Darville T (2005) *Chlamydia trachomatis* induces expression of IFN-gamma-inducible protein 10 and IFN-beta independent of TLR2 and TLR4, but largely dependent on MyD88. J Immunol 175:450–460

Naiki Y, Michelsen KS, Schroder NW, Alsabeh R, Slepenkin A, Zhang W, Chen S, Wei B, Bulut Y, Wong MH et al (2005) MyD88 is pivotal for the early inflammatory response and subsequent bacterial clearance and survival in a mouse model of Chlamydia pneumoniae pneumonia. J Biol Chem 280:29242–29249

Navarini AA, Recher M, Lang KS, Georgiev P, Meury S, Bergthaler A, Flatz L, Bille J, Landmann R, Odermatt B et al (2006) Increased susceptibility to bacterial superinfection as a consequence of innate antiviral responses. Proc Natl Acad Sci USA 103:15535–15539

O'Connell RM, Saha SK, Vaidya SA, Bruhn KW, Miranda GA, Zarnegar B, Perry AK, Nguyen BO, Lane TF, Taniguchi T et al (2004) Type I interferon production enhances susceptibility to *Listeria monocytogenes* infection. J Exp Med 200:437–445

Pan H, Yan BS, Rojas M, Shebzukhov YV, Zhou H, Kobzik L, Higgins DE, Daly MJ, Bloom BR, Kramnik I (2005) Ipr1 gene mediates innate immunity to tuberculosis. Nature 434:767–772

Pandey AK, Yang Y, Jiang Z, Fortune SM, Coulombe F, Behr MA, Fitzgerald KA, Sassetti CM, Kelliher MA (2009) NOD2, RIP2 and IRF5 play a critical role in the type I interferon response to *Mycobacterium tuberculosis*. PLoS Pathog 5:e1000500

Parker D, Martin FJ, Soong G, Harfenist BS, Aguilar JL, Ratner AJ, Fitzgerald KA, Schindler C, Prince A (2011) *Streptococcus pneumoniae* DNA initiates type I interferon signaling in the respiratory tract. mBio 17;2(3):e00016 11

Peng K, Monack DM (2010) Indoleamine 2, 3-dioxygenase 1 is a lung-specific innate immune defense mechanism that inhibits growth of *Francisella tularensis* tryptophan auxotrophs. Infect Immun 78:2723–2733

Perry ST, Buck MD, Lada SM, Schindler C, Shresta S (2011) STAT2 mediates innate immunity to dengue virus in the absence of STAT1 via the type I interferon receptor. PLoS Pathog 7: e1001297

Pestka S, Krause CD, Walter MR (2004) Interferons, interferon-like cytokines, and their receptors. Immunol Rev 202:8–32

Pine R, Canova A, Schindler C (1994) Tyrosine phosphorylated p91 binds to a single element in the ISGF2/IRF 1 promoter to mediate induction by IFN alpha and IFN gamma, and is likely to autoregulate the p91 gene. EMBO J 13:158–167

Plumlee CR, Lee C, Beg AA, Decker T, Shuman HA, Schindler C (2009) Interferons direct an effective innate response to *Legionella pneumophila* infection. J Biol Chem 284:30058–30066

Pollpeter D, Komuro A, Barber GN, Horvath CM (2011) Impaired cellular responses to cytosolic DNA or infection with *Listeria monocytogenes* and vaccinia virus in the absence of the murine LGP2 protein. PLoS One 6:e18842

Prabhakar S, Qiao Y, Canova A, Tse DB, Pine R (2005) IFN-alpha beta secreted during infection is necessary but not sufficient for negative feedback regulation of IFN-alpha beta signaling by *Mycobacterium tuberculosis*. J Immunol 174:1003–1012

Prantner D, Darville T, Nagarajan UM (2010) Stimulator of IFN gene is critical for induction of IFN-beta during *Chlamydia muridarum* infection. J Immunol 184:2551–2560

Qiu H, Fan Y, Joyee AG, Wang S, Han X, Bai H, Jiao L, Van Rooijen N, Yang X (2008) Type I IFNs enhance susceptibility to *Chlamydia muridarum* lung infection by enhancing apoptosis of local macrophages. J Immunol 181:2092–2102

Rayamajhi M, Humann J, Penheiter K, Andreasen K, Lenz LL (2010) Induction of IFN-alphabeta enables *Listeria monocytogenes* to suppress macrophage activation by IFN-gamma. J Exp Med 207:327–337

Remoli ME, Giacomini E, Lutfalla G, Dondi E, Orefici G, Battistini A, Uze G, Pellegrini S, Coccia EM (2002) Selective expression of type I IFN genes in human dendritic cells infected with *Mycobacterium tuberculosis*. J Immunol 169:366–374

Rothfuchs AG, Gigliotti D, Palmblad K, Andersson U, Wigzell H, Rottenberg ME (2001) IFN-alpha beta-dependent, IFN-gamma secretion by bone marrow-derived macrophages controls an intracellular bacterial infection. J Immunol 167:6453–6461

Sadler AJ, Williams BR (2008) Interferon-inducible antiviral effectors. Nat Rev Immunol 8:559–568

Sander LE, Davis MJ, Boekschoten MV, Amsen D, Dascher CC, Ryffel B, Swanson JA, Muller M, Blander JM (2011) Detection of prokaryotic mRNA signifies microbial viability and promotes immunity. Nature 474:385-389

Sarkis PT, Ying S, Xu R, Yu XF (2006) STAT1-independent cell type-specific regulation of antiviral APOBEC3G by IFN-alpha. J Immunol 177:4530–4540

Schiavoni G, Mauri C, Carlei D, Belardelli F, Pastoris MC, Proietti E (2004) Type I IFN protects permissive macrophages from *Legionella pneumophila* infection through an IFN-gamma-independent pathway. J Immunol 173:1266–1275

Schindler C, Plumlee C (2008) Inteferons pen the JAK-STAT pathway. Semin Cell Dev Biol 19:311–318

Schindler C, Levy DE, Decker T (2007) JAK-STAT signaling: from interferons to cytokines. J Biol Chem 282:20059–20063

Seth RB, Sun L, Chen ZJ (2006) Antiviral innate immunity pathways. Cell Res 16:141–147

Shahangian A, Chow EK, Tian X, Kang JR, Ghaffari A, Liu SY, Belperio JA, Cheng G, Deng JC (2009) Type I IFNs mediate development of postinfluenza bacterial pneumonia in mice. J Clin Invest 119:1910–1920

Shemer-Avni Y, Wallach D, Sarov I (1989) Reversion of the antichlamydial effect of tumor necrosis factor by tryptophan and antibodies to beta interferon. Infect Immun 57:3484–3490

Sheppard P, Kindsvogel W, Xu W, Henderson K, Schlutsmeyer S, Whitmore TE, Kuestner R, Garrigues U, Birks C, Roraback J et al (2003) IL-28, IL-29 and their class II cytokine receptor IL-28R. Nat Immunol 4:63–68

Shi S, Blumenthal A, Hickey CM, Gandotra S, Levy D, Ehrt S (2005) Expression of many immunologically important genes in *Mycobacterium tuberculosis*-infected macrophages is independent of both TLR2 and TLR4 but dependent on IFN-alphabeta receptor and STAT1. J Immunol 175:3318–3328

Shinohara ML, Kim JH, Garcia VA, Cantor H (2008) Engagement of the type I interferon receptor on dendritic cells inhibits T helper 17 cell development: role of intracellular osteopontin. Immunity 29:68–78

Shuai K, Stark GR, Kerr IM, Darnell JE Jr (1993) A single phosphotyrosine residue of Stat91 required for gene activation by interferon-gamma. Science 261:1744–1746

Sjostedt A (2003) Virulence determinants and protective antigens of *Francisella tularensis*. Curr Opin Microbiol 6:66–71

Stanley SA, Johndrow JE, Manzanillo P, Cox JS (2007) The type I IFN response to infection with *Mycobacterium tuberculosis* requires ESX-1-mediated secretion and contributes to pathogenesis. J Immunol 178:3143–3152

Stetson DB, Medzhitov R (2006) Recognition of cytosolic DNA activates an IRF3-dependent innate immune response. Immunity 24:93–103

Stockinger S, Decker T (2008) Novel functions of type I interferons revealed by infection studies with *Listeria monocytogenes*. Immunobiology 213:889–897

Stockinger S, Reutterer B, Schaljo B, Schellack C, Brunner S, Materna T, Yamamoto M, Akira S, Taniguchi T, Murray PJ et al (2004) IFN regulatory factor 3-dependent induction of type I IFNs by intracellular bacteria is mediated by a TLR- and Nod2-independent mechanism. J Immunol 173:7416–7425

Sueltenfuss EA, Pollard M (1963) Cytochemical assay of interferon produced by duck hepatitis virus. Science 139:595–596

Sun K, Metzger DW (2008) Inhibition of pulmonary antibacterial defense by interferon-gamma during recovery from influenza infection. Nat Med 14:558–564

Taylor GA (2007) IRG proteins: key mediators of interferon-regulated host resistance to intracellular pathogens. Cell Microbiol 9:1099–1107

Tiwari S, Choi HP, Matsuzawa T, Pypaert M, MacMicking JD (2009) Targeting of the GTPase Irgm1 to the phagosomal membrane via PtdIns(3,4)P(2) and PtdIns(3,4,5)P(3) promotes immunity to mycobacteria. Nat Immunol 10:907–917

Trinchieri G (2010) Type I interferon: friend or foe? J Exp Med 207:2053–2063

Uze G, Schreiber G, Piehler J, Pellegrini S (2007) The receptor of the type I interferon family. Curr Top Microbiol Immunol 316:71–95

Vance RE (2010) Immunology taught by bacteria. J Clin Immunol 30:507–511

Varinou L, Ramsauer K, Karaghiosoff M, Kolbe T, Pfeffer K, Muller M, Decker T (2003) Phosphorylation of the Stat1 transactivation domain is required for full-fledged IFN-gamma-dependent innate immunity. Immunity 19:793–802

Versteeg GA, Garcia-Sastre A (2010) Viral tricks to grid-lock the type I interferon system. Curr Opin Microbiol 13:508–516

Wang X, Hussain S, Wang EJ, Wang X, Li MO, Garcia-Sastre A, Beg AA (2007) Lack of essential role of NF-kappaB p50, RelA, and cRel subunits in virus-induced type 1 IFN expression. J Immunol 178:6770–6776

Weiden M, Tanaka N, Qiao Y, Zhao BY, Honda Y, Nakata K, Canova A, Levy DE, Rom WN, Pine R (2000) Differentiation of monocytes to macrophages switches the *Mycobacterium tuberculosis* effect on HIV-1 replication from stimulation to inhibition: modulation of interferon response and CCAAT/enhancer binding protein beta expression. J Immunol 165: 2028–2039

Woodward JJ, Iavarone AT, Portnoy DA (2010) c-di-AMP secreted by intracellular *Listeria monocytogenes* activates a host type I interferon response. Science 328:1703–1705

Zhang SY, Boisson-Dupuis S, Chapgier A, Yang K, Bustamante J, Puel A, Picard C, Abel L, Jouanguy E, Casanova JL (2008) Inborn errors of interferon (IFN)-mediated immunity in humans: insights into the respective roles of IFN-alpha/beta, IFN-gamma, and IFN-lambda in host defense. Immunol Rev 226:29–40

Zhao W, Lee C, Piganis R, Plumlee C, de Weerd N, Hertzog PJ, Schindler C (2008) A conserved IFN-alpha receptor tyrosine motif directs the biological response to type I IFNs. J Immunol 180:5483–5489

Index

Th. Decker and M. Müller (eds.), *Jak-Stat Signaling: From Basics to Disease*, DOI 10.1007/978-3-7091-0891-8, © Springer-Verlag Wien 2012

Printed by Printforce, the Netherlands